Principles and Practices of Irrigation Water Management

Principles and Practices of Irrigation Water Management

Editor: Davis Twomey

www.callistoreference.com

Callisto Reference,
118-35 Queens Blvd., Suite 400,
Forest Hills, NY 11375, USA

Visit us on the World Wide Web at:
www.callistoreference.com

ISBN: 978-1-64116-068-1 (Hardback)

Trademark Notice: Registered trademark of products or corporate names are used only for explanation and identification without intent to infringe.

Cataloging-in-Publication Data

Principles and practices of irrigation water management / edited by Davis Twomey.
   p. cm.
Includes bibliographical references and index.
ISBN 978-1-64116-068-1
1. Irrigation water--Management. 2. Irrigation water. 3. Irrigation farming. I. Twomey, Davis.
S618 .P75 2019
631.7--dc21

Table of Contents

Preface

It is often said that books are a boon to mankind. They document every progress and pass on the knowledge from one generation to the other. They play a crucial role in our lives. Thus I was both excited and nervous while editing this book. I was pleased by the thought of being able to make a mark but I was also nervous to do it right because the future of students depends upon it. Hence, I took a few months to research further into the discipline, revise my knowledge and also explore some more aspects. Post this process, I begun with the editing of this book.

Irrigation involves the use and application of water resources to facilitate crop production. The conventional techniques of agriculture required excessive water for irrigation which resulted in wastage. This has necessitated an economic usage of water and implementation of innovative techniques for an economic irrigation water management framework. Research is being conducted to develop practices for sustainable agricultural production and water management. This book is a compilation of chapters that discuss the most vital concepts and emerging trends in the field of irrigation water management. It also explores the principles and practices of irrigation water management with an emphasis on the issue of environmental sustainability. This book will be useful to agronomists, agriculture scientists, ecologists, experts and students as it offers innovative insights into this field.

I thank my publisher with all my heart for considering me worthy of this unparalleled opportunity and for showing unwavering faith in my skills. I would also like to thank the editorial team who worked closely with me at every step and contributed immensely towards the successful completion of this book. Last but not the least, I wish to thank my friends and colleagues for their support.

Editor

System of Rice Intensification (SRI): Packages of Technologies Sustaining the Production and Increased the Rice Yield in Tamil Nadu, India

B.J. Pandian*, T. Sampathkumar and R. Chandrasekaran

Water Technology Centre, Tamil Nadu Agricultural University, Coimbatore, India

Abstract

System of Rice Intensification (SRI) is a holistic agro-ecological crop management technique seeking alternatives to the high-input oriented agriculture and one among the scientific management tool of allocating irrigation water based on soil and climatic condition to achieve maximum crop production per unit of water applied over a unit area in unit time. System of rice intensification was the main focus technology demonstrated by Water Technology Centre (WTC), Tamil Nadu Agricultural University (TNAU) under Irrigated Agriculture Modernization and Water Bodies Restoration and Management (TN-IAMWARM) Project. In general, an increase in rice productivity under SRI over the conventional system of rice cultivation was observed in all the demonstrations. The widespread adoption of SRI showed increasing trend in yield (from 28.3% in 2007-08 to 32.4% in 2010-11). The results of beneficiary wise analysis indicated that more beneficiaries reaped 40-50% yield increase followed by 20-30% yield increase over conventional. The data obtained from large scale demonstrations clearly indicated that the water requirement was less under SRI (885 mm) as compared to conventional (1180 mm). The demonstration of SRI technologies registered higher grain yield and Water Use Efficiency (WUE) of 6,406 kg ha^{-1} and 7.31 kg ha^{-1} mm^{-1}, respectively as compared to conventional (5,284 kg ha^{-1} and 4.51 kg ha^{-1} mm^{-1}). The water productivity in SRI was found to be 1,398 as against 2,274 lit. kg^{-1} in conventional irrigation.

Keywords: SRI; Water productivity; Square planting; Alternate wetting; Drying irrigation

Introduction

Rice is the predominant crop accounting nearly 65% of total irrigated area in Tamil Nadu. The rice cropping system uses water in a wide variety of ways, both beneficial and non-beneficial. Tamil Nadu is one of the water starving states in India where total water resource available is 44361 Million cubic meter (Mm³) as against the demand of 51813 Mm³ leaving a gap of 7452 Mm³. Hence, it is necessary to develop suitable technologies that recognize and adequately address the challenges we face and are going to face in the years to come since achieving food security has been the overriding goal of agricultural policies. The intensified efforts to improve both crop and water productivity and subsequently the farmers' income have resulted in many efficient water management practices in wetland rice.

Among the production constraints, availability of irrigation water is a major one, since rice is a predominant crop in Tamil Nadu consumes 70% of the water available for agriculture. The gap between water supply and water demand for irrigated crops in Tamil Nadu is projected to reach 21,000 Mm³ by 2025 [1]. Irrigation water is used inefficiently in lowland rice cultivation, whilst farms that do not have access to irrigation water experience water scarcity. Whenever the monsoon fails, lowland rice faces water scarcity, leading either to crop failure or poor crop yield. Although rice yields have shown increasing trend over the years, there is a need to economize the water use in rice production. Rice research in India during the last century has resulted in the development of important technologies, adoption of which has helped in keeping the rice production growth rate ahead of the population growth rate. However, the appalling paucity of water threatens the sustainability of the irrigated rice eco-system. Such water shortage in many rice-growing areas is prompting a search for production systems that use less water to produce rice. Although several strategies are being pursued to save water in irrigated rice ecosystems, water losses still remain high since all those systems use prolonged periods of flooding.

Under modern methods of rice cultivation, 3000-5000 l of water was used to produce one kilogram of rice [2]. A significant portion of the total water requirement for rice production is used for land preparation alone. In recent years, wherever water is scarce, deficit irrigation is being recommended if it is economically tenable. Keeping in view, the need to maximize the production and providing minimum sustainable income, water-intensive crops such as sugarcane and wetland rice are being given up or go for alternate crops in the areas of water scarcity in spite of their vital importance in meeting people's needs. Farmers have a need for irrigated rice-based systems with technologies that save water by improving water productivity. The intensified efforts to improve both crop and water productivity and subsequently the farmers' income have resulted in many efficient water management practices in wetland rice. System of Rice Intensification (SRI) is a holistic agro-ecological crop management technique seeking alternatives to the high-input oriented agriculture and one among the scientific management tool of allocating irrigation water based on soil and climatic condition to achieve maximum crop production per unit of water applied over a unit area in unit time by converting conventional agronomic principles synergistically into higher yield production process.

Genesis of SRI in Tamil Nadu

The components of SRI were first tested in Tamil Nadu Agricultural

---

*Corresponding author: BJ Pandian, Water Technology Centre, Tamil Nadu Agricultural University, Coimbatore, India, Email: bjp1402@yahoo.co.in

University (TNAU) in 2000 and later on in 2003 under adoptive research trails in two major river basins of Tamiraparani and Cauvery of Tamil Nadu. In terms of rice productivity, SRI method of rice cultivation has registered higher yield levels than the farmers' conventional method of rice cultivation. The results from the study conducted at Dharmapuri and Krishnagiri districts of Tamil Nadu showed an increase by 21.9% by SRI over conventional method in these districts and net income of the farmers increased from 15.30 to 42.40% by adoption of SRI. The partial budgeting analysis results revealed that SRI adoption would bring net gain to the tune of INR (Indian rupees).13725/- per ha [3]. After implementation and execution of the trials, the components *viz.*, less seed rate 7.5 kg ha^{-1}, raising mat nursery, young seedling (14-15 days old), square and single seedling transplanting per hill in wider spacing (25 x 25 cm), mechanical weeding 4 times (10, 20, 30 and 40 DAT), limited irrigation and nutrient management through Leaf Colour Chart (LCC) were counseled for adoption. Initial adoption and spread remained low in Tamil Nadu from 2004 to 2006. Considering the lower acceptance and existing scope for adoption, TNAU included SRI as one of the water saving technologies in Irrigated Agriculture Modernization and Water Bodies Restoration and Management (TN-IAMWARM) Project funded by World Bank with the objective of increasing the both crop and water productivity in rice crop.

In a predominantly agricultural state like Tamil Nadu, there is an urgent need for intensifying efforts to improve productivity, and sustainable farm income. Long-term growth in agriculture depends in large part on increasing the efficiency and productivity of use of water. With these aspects in mind this project was formulated and implemented in rice crop to increase the use efficiency of applied water and rice production in Tamil Nadu through large scale demonstration of SRI as one of the objectives.

Materials and Methods

Tamil Nadu is the southernmost state of India, surrounded by Andhra Pradesh from the North, Karnataka and Kerala from the west, Indian Ocean from the south and Bay of Bengal from the East. Cape Comorin or Kanyakumari, the southernmost point of India lies in the state of Tamil Nadu. The state of Tamil Nadu roughly extends between the 8° 04' N latitude (Cape Comorin) and the 78° 0' E longitude. Geographically, Tamil Nadu is situated on the eastern side of the Indian Peninsula between the northern latitude of 8.5" and 13.35" and the eastern longitude of 76.15" and 80.20". The climate of Tamil Nadu is tropical in nature with little variation in summer and winter temperatures. Tamil Nadu gets all its rains from the North-east Monsoon between October and December, when the rest of India remains dry. The project area, Tamil Nadu is one of the driest states in India, averaging only 925 millimeters of rainfall in a year.

Tamil Nadu's geographic area can be grouped into 17 river basins (127 Sub Basins) a majority of which are water-stressed. The project was implemented from 2007 in IV phases (I, II, III and IV) including 63 sub basins Figure 1 System of rice intensification was the main focus technology demonstrated by TNAU under TN-IAMWARM Project in Tamil Nadu. During the first year (2007-08), SRI was demonstrated in 1311 ha including 1456 farmers (Phase I in 9 sub basins). Subsequently SRI demonstrations were extended in remaining 54 sub basins. The size of the study area varied from 0.5 to 1.0 ha. The types of soil in the study area were loamy to clay loam. The fertility status of the soil in the study area was medium in available nitrogen, low to medium in available phosphorus and medium to high in available potassium. The farmers were trained to adopt the five major components of SRI in

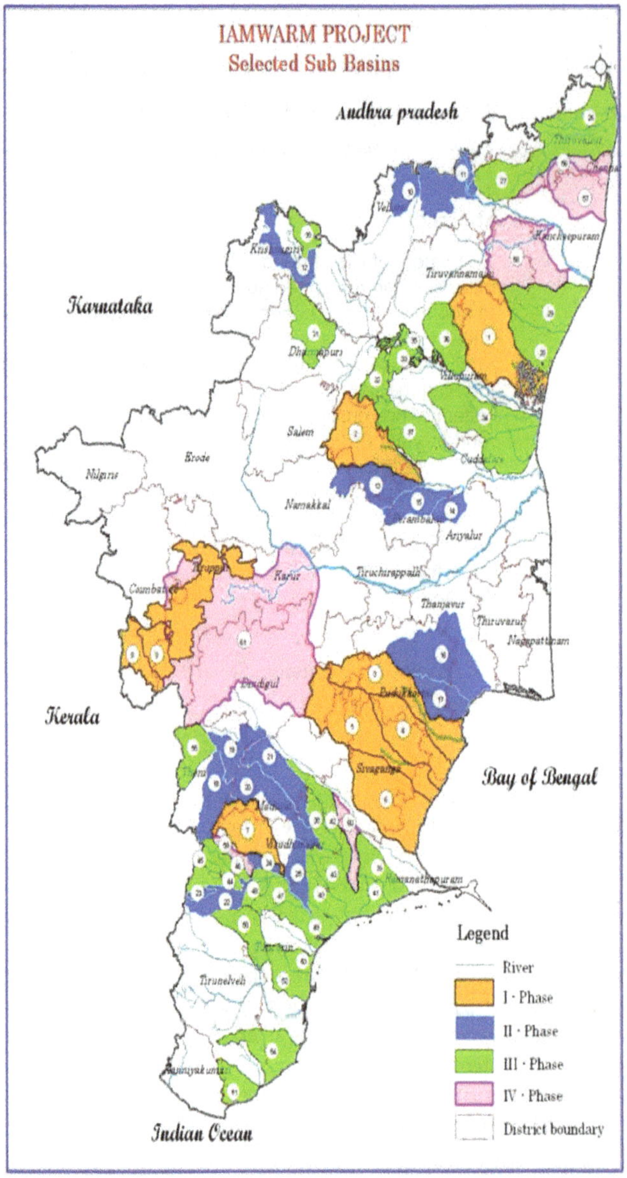

Figure 1: Selected Sub Basins of Tamilnadu Region.

demonstration fields. All the demonstration farmers were provided with inputs viz., seeds, cono weeder, and markers for square planting and LCC for nitrogen management. The major components demonstrated in SRI were less seed rate 7.5 kg ha^{-1}, raising mat nursery, young seedling (14-15 days old), square and single seedling transplanting per hill in wider spacing (25 x 25 cm), mechanical weeding 4 times (10, 20, 30 and 40 DAT), limited irrigation and nutrient management through LCC. Conventional method of cultivation was also demonstrated in the same field for comparison. The conventional system of cultivation includes adoption of 30-35 days old seedlings for planting and adoption of flood irrigation (continuous submergence with 5 cm depth of water) until harvest of the crop. Thus the project has provided an ambient platform for up-scaling the technology to every villages of the project area. The varieties used in the demonstrations were ADT 36, ADT 39, ADT 45, ASD 16, CR-1009, BPT 5204 and CORH-3. To study the effect of SRI on crop productivity water use studies were taken up in selected fields in demonstrations. Studies on water productivity with SRI were undertaken as a part of the project from 2007 to 2010. Studies were conducted in four locations, viz, Varaganadhi sub basin, Karumaniar sub-basin (Tirunelveli district, L1), Sevalaperiar sub-basin (Virudhunagar district, L2), Ongur sub-basin (Chengalpattu district, L3) and Nallavur sub-basin (Villuppuram district, L4) to compare the efficiency of SRI irrigation with the conventional irrigation (CI).

Alternate Wetting and Drying (AWD) irrigation system was the recommended water management practice under SRI. Under AWD irrigation system, water is applied to flood the field for a certain number of days after the disappearance of ponded water. The field was allowed to be dry for a few days between water applications. The number of days under AWD irrigation can vary from 1 day to more than 10 days. From planting to panicle initiation stage, field was irrigated to a depth of 2.5 cm after the previously irrigated water disappears and hairline cracks develop. After panicle initiation, irrigation was given to a depth of 2.5 cm one day after the previously ponded water disappears from the surface. At hairline crack stage, soil will not be completely dry, but yet moist. The measurement of irrigation water in fields was carried out by parshall flume.

Traditional mindset of farmers, lack of awareness, non-cooperation of planting labours, inability to deliver regulated irrigation water and non-availability of critical implements (Markers/Weeder) at appropriate season was the main bottlenecks identified in SRI adoption. Few innovative measures viz., exposure visit, field days, community nursery and field visit of high Command Government Officials were followed to expose SRI on large scale in the state. The SRI demonstrations were carried out in 17981 ha during the period from 2007 to 2011 in 19497 farmers' holdings.

Results and Discussion

The overall performance of SRI introduced at the project area of TN-IAMWARM indicated that an increase in rice productivity in SRI over the conventional system of rice cultivation was observed in all the years (Table 1). It is incredible that it has created such a remarkable consciousness among the rice growers of the State. The widespread adoption of SRI at field level showed an increasing trend in yield (from 28.3% in 2007-08 to 32.4% in 2010-11). Irrespective of years, the yield contributing parameters viz., No. of tillers per hill, No. of productive tillers per hill and grains per panicle were also higher in SRI demonstration than conventional practice (Table 2). Beneficiary wise analysis indicated that more no. of beneficiaries reaped 40-50% yield increase followed by 20-30% yield increase over conventional (Table 3).

Maximum grain yield achieved in SRI was due to higher Leaf Area Index (LAI) and light interception at wider spacing between plants gained from open plant structure. This resulted in higher LAI and greater leaf size leading to vigorous root system [4]. Whereas in conventional method at closer spacing between the rice plants, the number of panicles in unit area increases but with shorter panicles containing lesser grains resulted in lesser yield as shown in (Table 4). Planting of younger seedlings with optimal growing conditions is responsible for accelerated growth rate in SRI plants as these make possible to complete more phyllochrons before entering into their reproductive phase [5,6]. Completion of more phyllochrons at early seedling stage resulted in more number of tillers and effective tillers per hill. Moreover, younger seedlings have improved root characteristics like root length density and root weight after transplanting than do aged seedlings [7]. Rice grown under conventional system creates hypoxic soil condition and its roots degenerate under flooding, losing three-fourth of their roots by the time the plants reach the flowering stage [8]. Unflooded condition (alternate wetting and drying) in SRI, combined with mechanical weeding resulted in better aeration in the soil and greater root growth for better access to nutrients as compared to conventional planting. The SRI plants have deeper and stronger root systems, supported by intermittent irrigation and without physical

Year	Rice yield (kg ha⁻¹)		% increase	Demo. Area (ha)	No. of Demonstrations
	SRI	Conventional			
2007-08	3710	2902	28.3	1311	1456
2008-09	4361	3237	33.3	2581	3029
2009-10	4588	3340	37.3	4000	5245
2010-11	4456	3365	32.4	10089	9767
Average/Total	4454	3220	32.9	17981	19497

Table 1: Performance of SRI demonstration in TN-IAMWARM Project in Tamil Nadu, India.

Year	No. of tillers per hill		No. of productive tillers per hill		No. of grains per panicle	
	SRI	Conventional	SRI	Conventional	SRI	Conventional
2007-08	22.7	10.6	18.6	8.0	135	118
2008-09	23.5	11.2	19.3	8.4	138	127
2009-10	25.8	12.5	21.2	9.4	152	134
2010-11	24.6	11.8	20.2	8.9	145	128
Average	24.2	11.5	19.8	8.6	142	126

Table 2: Effect of SRI on yield parameters (average values) in demonstrations of TN-IAMWARM Project in Tamil Nadu, India.

Year	% increase over conventional method of cultivation						Total
	<10	10-20	20-30	30-40	40-50	> 50	
2007-08	337	311	363	301	144	-	1456
2008-09	-	568	678	1004	387	392	3029
2009-10	71	567	543	331	2790	943	5245
2010-11	105	1158	2662	1918	1503	2421	9767
Total	513	2604	4246	3554	4824	3756	19497

Table 3: Beneficiary wise analyses of SRI farmers in TN –IAMWARM Project, Tamil Nadu, India.

barriers to root growth [9]. A number of previously published reports on SRI have showed enhancement of rice yield [10-13]. This study found SRI practices increasing grain yield at the range of 28.3 to 32.4%, from 4.954 t ha^{-1} to 6.583 t ha^{-1}, while utilizing fewer seeds and less water.

Water productivity of rice in SRI demonstrations

Irrigation studies conducted at Varaghanadhi sub-basin (Villupuram District, Tamil Nadu) during 2007-08 revealed that the irrigation water required for conventional method of rice cultivation was 9,204 m^{3}ha^{-1}, whereas it was 4,306 m^{3}ha^{-1} under SRI, thereby saving irrigation water to the tune of 41% (Table 5). Apart from the economy in water use, the alternate wetting and drying method of irrigation had a positive influence on yield and water productivity. The data clearly indicated that the water requirement was less under SRI (885 mm) as compared to conventional (1180 mm) (Table 6). SRI registered higher grain yield and WUE of 6,406 kg ha^{-1} and 7.31 kg ha^{-1} mm^{-1}, respectively as compared to conventional (5,284 kg ha-1 and 4.51 kg ha^{-1} mm^{-1}). The water productivity in SRI was found to be 1,398 litres kg^{-1} as against 2,274 litres kg^{-1} in conventional irrigation.

In SRI, Alternate Wetting and Drying (AWD) irrigation is practiced and it provides the water-loving rice plant with the primary conditions necessary for transplanting of single young seedling and mechanical weeding technologies leads to high yield. The AWD irrigation is a water-saving technology that lowland rice farmers are practicing to reduce their water use in irrigated fields. Shallow irrigation under AWD system could save water up to 40% without any yield loss. Irrigation intervals vary with soil texture. Fine textured clayey soil with

higher field capacity need irrigation at longer intervals while coarse textured light soils with lower water holding capacity require irrigation frequently [14]. Practice of alternate wetting and drying under SRI had potential to save the water upto 24% and increment in grain yield to the tune of 71% [15,14]. The mechanisms for evoking these changes remain to be studied and determined in satisfactory detail.

Impact of SRI in rice production

Rice production and productivity in Tamil Nadu: Rice area in Tamil Nadu has been hovering around 18-20 lakh hectares for the past few years and except during 2007-08 (Figure 2). Remaining other years witnessed relatively larger positive deviation over the base year. The area under SRI is in the increasing trend in Tamil Nadu. In the past four years the SRI area has almost doubled and it was 4.2 lakhs during 2007-08 and rose to 10.01 lakhs in 2011-12. The share of SRI in total rice area increased from 23.48 (2007-08) to 50.0% (2011-12) during the above period. A favourable policy environment on promotion of SRI in the State could be cited as the prime reason for the increasing trend in area under SRI.

The rice production in Tamil Nadu has got stabilized due to upscaling of SRI on a mission mode approach. In the initial two years of SRI adoption, the percentage increase over the base year average (2002 to 2006) was less than 9% and it could be due to the occurrence of natural calamities like floods, unseasonal rainfall and cyclones. But, the last two years (2010-2011), the percentage increase in rice production has witnessed a conspicuous increase over the base period and the SRI coverage had been increased more than 35 to 50% in the total rice area (Figure 2). After the mass introduction of SRI in Tamil Nadu, the

Year	Rice area (Lakh ha.)	Area under SRI (Lakh ha.)	Productivity (kg ha^{-1})	Total production (lakh MT)
Base year (2002 to 2006)	17.54	-	2667	47.36
2007-08	17.89	4.20	2817	50.40
2008-09	19.32	5.38	2682	51.83
2009-10	18.46	6.49	3070	56.65
2010-11	19.06	8.50	3039	57.02
2011-12	19.05	10.01	3915	74.59

Base year – 5 year average (normal); Source: Season and Crop Report, Directorate of Economics and Statistics, GoTN, Chennai

Table 4: Impact of SRI on Rice area, Production, Productivity in Tamil Nadu.

Parameters	Conventional practice	SRI
Number of irrigations	18	13
Irrigation water (m^3 ha^{-1})	9,204	4,306
Rainfall (mm)	281	281
Total water used (m^3 ha^{-1})	12,014	7,116
Water saving (%)	-	40.8
Grain yield (kg ha^{-1})	5,120	7,528
Water productivity (kg m^{-3})	0.426	1.058

Table 5: Water requirement studies in SRI at Varaganadhi sub-basin during 2007-08.

Parameters	L1		L2		L3		L4		Average	
	SRI	CI	SRI	CI	SRI	CI	SRI	CI	SRI	CI
Water used (mm)	923	1,252	973	1223	827	1,148	818	1,098	885	1,180
Productive tillers hill^{-1}	31	20	48	31	36	23	39	24	38	24
Productive tillers m^{-2}	470	350	612	505	720	580	780	704	646	535
Grain yield (kg ha^{-1})	5,810	4,450	5,982	5,032	7,046	5,965	6,784	5,689	6,406	5,284
Water use efficiency (kg ha^{-1}mm^{-1})	6.28	3.55	6.14	4.11	8.51	5.20	8.29	5.18	7.31	4.51
Water productivity (l kg^{-1})	1,588	2,813	1,626	2,430	1,174	1,924	1,205	1,930	1,398	2,274

L – Locations; CI – Conventional Irrigation

Table 6: Water productivity of rice in SRI in selected sub-basins during 2010-11.

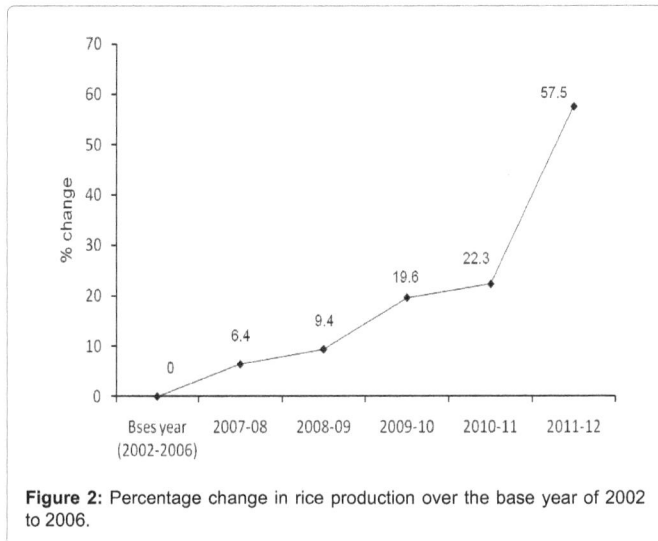

Figure 2: Percentage change in rice production over the base year of 2002 to 2006.

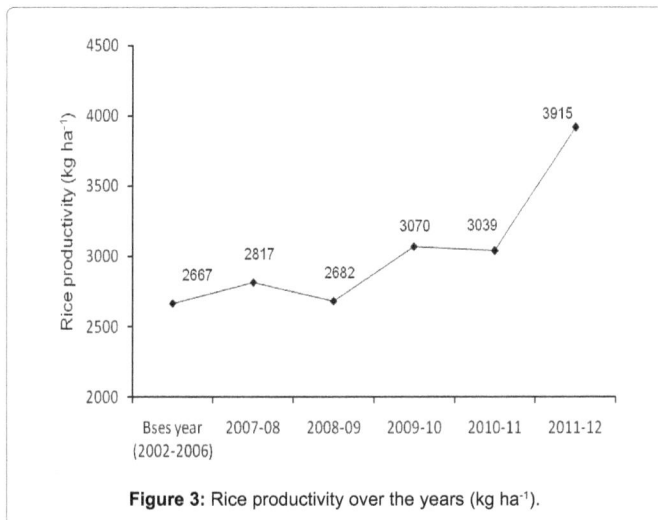

Figure 3: Rice productivity over the years (kg ha⁻¹).

yield levels had a positive trend except for year 2008-09 during that year the State experienced Nisha cyclone which devastated rice crops in 4.71 lakh hectares and there by seriously reducing the production and productivity levels (Figure 3). Even during the year 2007-08, the rice productivity levels were higher than the base year and the role of SRI could never be under estimated in the stress period caused by the heavy rains during December, 2007 and unusual rains during March 2008 in which nearly 1.68 lakh ha under rice crop had been directly affected. Later, the yield levels were consistently higher than 3.00 t ha⁻¹ and it could be due to large scale adoption of SRI. It was also evidenced from the increased total rice production of the state. The favourable monsoon environment and SRI acreage led to increase in rice productivity during 2011-12.

Conclusions

System of rice intensification is not a fixed package of technical specifications, but a system of production with four main components, viz., planting method, weed control, soil fertility management and water (irrigation) management. Several field practices have been developed around these components. Drastic changes in the agronomic practices of SRI initially failed to build confidence amongst the farmers. Farmers

themselves should become more experimental and entrepreneurial as a result of their engagement with SRI. This is not a technology like that of the Green Revolution, where farmers were expected simply to adopt a 'package.' SRI in its core conception involves adaptation rather than adoption, and farmers are expected to become more innovative. Though farmers who involved are committed to continue SRI, immense steps have to be taken to disseminate not only the technology but also the scientific facts behind the technology as a means of promoting SRI. It was realized that rice does not require flooded water or standing water, and it is enough to keep the soil moist from the demonstrations. The farmers using ground water will certainly realize a saving of water, time and electricity under SRI irrigation. If SRI is adopted in an entire command area, saved water encourage to bring more area under cultivation. Though SRI is a promising and successful technology in increasing the yield, it has not yet become a major method of cultivation owing to existing institutional and behavioural factors. Some of the principal challenges viz., resistance to accept SRI, non-co-operation of planting labourer, lack of training and extension facilities, absence of precise water management and non-availability of essential tools. If we address these issues, SRI would be a successful technology to boost the rice yields and the income of farmer.

Acknowledgement

The authors acknowledge the financial support provided by the World Bank, Funded Project TN-IAMWARM, TNAU and Government of Tamil Nadu.

References

1. Palanisami K, Paramasivam P (2000) Water scenario in Tamil Nadu-present and future. Tamil Nadu Agricultural University, Coimbatore: 36-46.

2. ICRISAT-WWF Project (2009) SRI Fact Sheet, International Crops Research Institute for Semi-Arid Tropics, India.

3. Anjugam M, Varatha Raj S, Padmarani S (2008) Cost-Benefit Analysis of SRI technique in Paddy Cultivation, Department of Agricultural Economics, TNAU, Coimbatore.

4. Thakur AK, Rath S, Roychowdhury S, Uphoff N (2010) Comparative Performance of Rice with System of Rice Intensification (SRI) and Conventional Management using Different Plant Spacings. J Agron Crop Sci 196: 146–159.

5. Berkelaar D (2001) SRI the System of Rice Intensification: Less Can Be More. ECHO Development Notes, Durrance Rd North Ft. Myers, Florida, USA.

6. Nemoto K, Morita S, Baba T (1995) Shoot and root development in rice related to the phyllochron. Crop Sci 35: 24-29.

7. Mishra A, Salokhe V (2010) The Effects of Planting Pattern and Water Regime on Root Morphology, Physiology and Grain Yield of Rice. J Agron Crop Sci 196: 368-378.

8. Kar S, Varade SB, Subramanyam TK, Ghildyal BP (1974) Nature and growth pattern of rice root system under submerged and unsaturated conditions 23: 173-179.

9. Stoop AW, Adam A, Kassam A (2009) Comparing rice production systems: A challenge for agronomic research and for dissemination of knowledge-intensive farming practices. Agr Water Manage 96: 1491-1501.

10. Ceesay M, Reid WS, Fernandes ECM, Uphoff NT (2006) The effects of repeated soil wetting and drying on lowland rice yield with system of rice intensification (SRI) methods. International Journal of Agricultural Sustainability 4: 5-14.

11. Kabir H, Uphoff N (2007) Results of disseminating the System of Rice Intensification (SRI) with farmer field school methods in northern Myanmar. Exp Agr 43:463-476.

12. Namara R, Bossio D, Weligamage P, Herath I (2008) The practice and effects of the System of Rice Intensification (SRI) in Sri Lanka. Quarterly Journal of International Agriculture 47: 5-23.

13. Zhao LM, Wu LH, Wu MY, Li YS (2011) Nutrient uptake and water use efficiency as affected by modified rice cultivation methods with reduced irrigation. Paddy and Water Environment 9: 25–32.

2

Flood Control and Flood Management of Sarbaz and Kajo Rivers in Makoran

Zainudini MZ* and Sardarzaei A

Faculty of Marine Science, Basic Science Department, Chabahar Maritime University, Iran

Abstract

Floods are among the most devastating natural hazards in the world. Iran and specially Makoran has big rivers and flood planning and management is mainly concentrated on riverine floods occurring during seasonal rains. This research investigate on the Lower Sarbaz and Kajo rivers and used discharge analysis of river channel-gauging time series to assess the regional water supplies and agricultural purposes. This analysis reveals that for all flood conditions especially on the lower Sarbaz river, have systematically risen for monthly maximum discharge volumes over the period of record. The Sarbaz and Kajo rivers have served as important routes for transportation and commerce since the formation. These two rivers need to be re-assessed especially in light of the predicted large increase in rainfall and water in the rivers over the next 25 years or so as predicted by the global climate change models. Iran has big rivers and flood planning and management is mainly concentrated on riverine floods occurring during monsoon. However, flash floods in hilly and mountainous areas are also common with demonstrated damage potential. Flood events in arid areas can be extremely damaging with increasing development, particularly in Garmbeet and Bahowkalat area (Down-stream of Pishin Dam). Flood protection and drainage design are considerable importance. Existing flood risk models are inadequate, and predicted changes in the climate show that there may be much more water in the system in the near future. New models are being built to test how the river-floodplain systems will respond to large increases in the discharge in the future.

Keywords: Agriculture; Catchment; Discharge; Flood; Hydrology; Kajo; Makoran; Pishin dam; Rainfall; Rate of flow; Sarbaz

Introduction

Stream and rivers provide water for drinking, watering animals, irrigation and for human habitation for over thousands of years. They are dynamic environments and their banks are prone to erosion, and periodically flood over their banks. This is part of a rivers normal cycle that relied upon the generations of farmers for replenishing and fertilizing their fields. However, the use of river floodplains has changed dramatically in the past few decades. Thousands of miles of earthen levees, flood walls, and river control structures have been built along both rivers. Dikes and levees have been built around the two rivers in attempts to keep the flood waters out of towns. However, these tend to make the flooding problem worse because it confines the river to a narrow channel, and the water rise more quickly and cannot seep in the ground of the floodplain. In the light of flood control modifications to dam or reservoir as outlined in alternatives are intended to support limited flood control operation at reservoir in both rivers in Makoran. Furthermore, it could be anticipated that this flood control capacity would remain until a flood event occurs in that particular corresponding area. During a flood event, out flows from the dam could be reduced in order to prevent flow of water at reservoirs, the flood water was uncontrollable because the Pishin dam does not have any levees or by pass canals or dykes to protect the dam from cracking or even from complete destruction . There could be several flood control strategies being analysed but commonly to modify operations reservoirs in Makoran which may require new outlets, increased storage and new discharge guidelines [1]. The objective of this study is to provide engineers, sponsors and other interested for water management some discussion of the potential environmental impacts associated with these proposed modifications and to provide some related recommendations and participations. The need to develop a re-operation plan based on the dam modification alternatives presented would be an integral part to the overall success towards flood control [2,3]. Water discharge during flood 1997 and 2005 Water resources authority released excess flood water without any consideration to downstream agricultural farms

and villages which damaged many lives and farms land, in fact it was a deliberate mismanagement by the authority and also there is lack of any flood by pass or diversion canal to that particular area on the other hand there were also occurs drought because entire water released from Pishin dam. Then in this case minimum water should be release from the reservoir and maximum water can be stored. As the reservoir rises and reaches each intake, the corresponding outflows adjust on a continuum with one outlet submerged then it could be possibility in the case of flooding all outlet perhaps would submerged , where after the reservoir fills, so discharge is passed both through the sluiceways and over the spillway (Figure 1). Increasingly speaking it varies each year, monthly outflow averages common range but also depending on the rainy month. In this regard the maximum storage pool elevation would be require the use of a five meters high rubber weir added to the spillway crest. So five meters high spillway structure would be inflated or dropped into place only during events that could require use of the additional flood control storage [4,5].

Materials and Methods

The study was conducted at Pishin and Zirdan Dams in Makoran. The area is located near the borders of Iran and Pakistan. Thus location up on which this study concentrates is bounded by the coastline of southern Iran and Western of Pakistan, approximately, by the line of latitude 25 degree to the South and the line of longitude 60 degree to the

---

*Corresponding author: Zainudini MA, Faculty of Marine science, Basic Science Department, Chabahar Maritime University, Iran
E-mail: mazainudini@yahoo.com

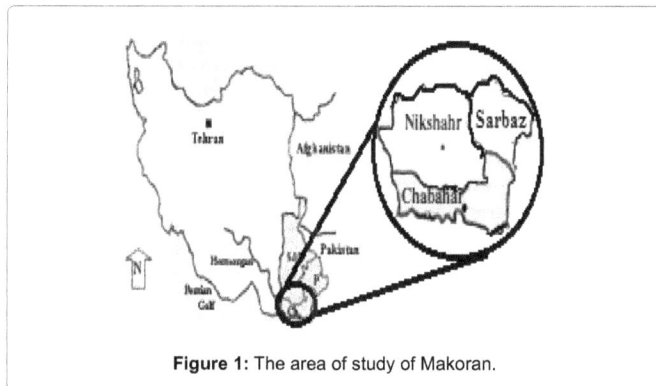

Figure 1: The area of study of Makoran.

west. The area consists of an inland chain of steeply sloping bare rock (mountains) which drain onto a coastal alluvial plain. The analysis is based on a multi-sites analysis approach, since the two rivers locations are not considered sufficiently similar to be pooled together. The flow data was obtained from 12 months from January to December has been tested by computer excel program the peak discharge was ranging from 3012 m³/sec the same river on the month of April suddenly rate of flow dropped to 6 m³/sec even though for other months during summer rate of flow for Sarbaz and Kajo rivers are completely drought or nil discharge according to the dry months when there is no rainfall at all in Makoran region.

Reservoir Maintenance, Operations and Flood Control in Makoran

The reservoir operations would be tied up primarily to flood control, a requirement could be necessary in place to ensure the reservoir elevation near the living communities in the areas is at or below to corresponding prior to the onset of the flood control therefore mostly in rainy season of year public property and lives could be protected. Thus, during the summer to fall drawdown period, flows from the river or reservoir would be passed through the outlet structures such that the reservoir lowers to the desired elevation of dam. Furthermore, whenever draw-down is complete, inflow will be passed through the outlet works to maintain reservoir elevation therefore it flows nearly to the Sarbaz, Rask and Chabahar carriage way [6].

The reservoir would remain relatively constant throughout the entire period of year and improvement of flood control and reduction of peak flow could be manageable. Where flow operations for reservoirs during non-flood events will be similar to the operation that is in place of major period non-rainy season. But, except for flood events, out flows should continue to follow historic outflows. Increasingly, wide flow variations would occur through the years. In general, daily discharge trends show flow increasing in all reservoirs in the province from a low flow in the summer but to a mean or average monthly flow in January and February and also sometimes it could be exceeding. Thus, this pattern would vary widely by year although maximum flows can be much higher than the average mean, so during flood season in Makoran, could be high water releases, which are common. These events tend to be relatively short in duration lasting around 3 to 5 days. However, these relationships are much more complex for a multi-purpose reservoir since they involve the seasonal distribution of stream flow and the reconciliation between it and seasonal and other varying demands for the several purposes perhaps for which the Pishin reservoir accordingly. The bank-full flows in the upper reaches below the dam occur at high discharges rate as in Figure 2.

Natural and Physical Environment of Makoran

In Makoran, the flood study could be said with location of Sarbaz and Kajo rivers the upstream parts of the rivers flow through heavily mountainous and hilly regions while the remainder of rivers flows through broad, flat valleys. The upland, villages landscapes are dominated by a mix of forests, which forests comprised of Nannorrhops (Mazari palm trees) and tamarixes etc. Also the upstream half of the rivers below Pishin and Zirdan flood control dams are dominated by coarse gravels, with areas of fines and larger sands also present to a lesser degrees. The downstream half of both rivers below the dams are predominately small to medium sized sand or cobbles. Therefore, interspersed with areas of fine sediment and larger sands and boulders, the coarse gravels are also present in those particular areas. Furthermore, the historic water quality problems stemmed from drainage and rainfall runoff, which also caused low dissolved oxygen, elevated water temperatures and high turbidity. Low summer flows have also caused some problems in the river but these are primarily a natural function arising from the small drainage size. Although, there are many reaches existed, so the majority of reaches are located of the side of the rivers. However urban development along these rivers in Makoran is important feature, providing crucial life history for flood control and also for such as rearing and holding habitat for fish, and over wintering habitat for migrating waterfowl. The existing fill and spill for flood control dams operations result in seasonal fluctuations surface elevations [7].

Impacts of Natural Environment From Flood Water in Makoran

The flood control could be appropriate to have and build high weir and dam may be useful during floods for approaching the 100 years event. Therefore, the control structure would engage for flood protection only and cause the reservoir to rise above the historic full pool elevation. Although in the light of flood it causes the flood waters to inundate a variety of upland and wetland habitats including, steep brush and timber slopes, shallow slopes and vegetated lowlands as well. It would be depending on the magnitude of the flood, it could also increase additional reaches and pools. However, floodwater may increase sediment deposition in the newly in-undated vegetation may affect growth rates [8]. Thus, it is anticipated that the timber and underbrush between different elevations will remain intact although and increase in woody debris recruitment would occur as a result of the higher pool of floodwater. The increased size of the reservoir during large flood events would be most noticeable at the head of the flood control reservoir. The extent of lost habitat from this additional

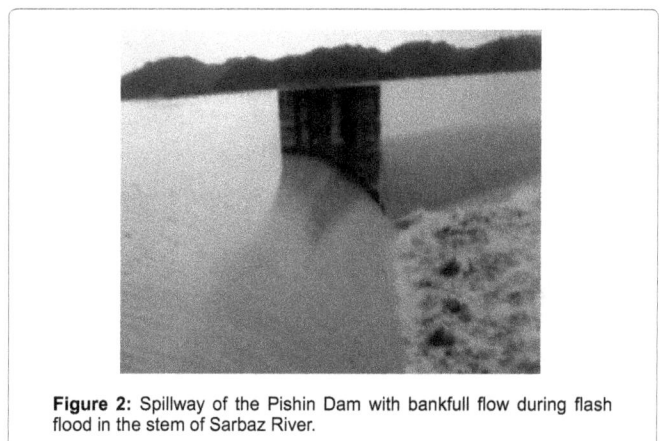

Figure 2: Spillway of the Pishin Dam with bankfull flow during flash flood in the stem of Sarbaz River.

main stem inundation is not likely to be significant given the timing of the events, so the existing steelhead may be infrequent nature of the flooding event. However, flood control operations under the proposed project would produce more frequent fluctuations of the winter reservoir. This proposed methods calls for the flood reservoir to be evacuated through the new outlet gates after each flood event to reclaim flood storage. Therefore, it could be appropriate and would strive to keep the winter pool at management desire elevation, and reduce the availability of shoreline vegetation to riparian animals and aquatic resources. In the light of flood control it could be good enough keeping the reservoir away from the shoreline during the lean winter months will make prey contributions of the tributaries and main-stem much more important. Also modifications to the dams under the flood control event include a small structural changes as well as the addition of a weir structure and additional low level outlet gates so impact from structural modifications would be limited to concrete work in and around the dam. Then after the weir would be employed only during the largest events to provide the extra reservoir storage space when is needed.

Flood Control and Reservoirs for Storage Purposes in Makoran

Floodwater storage reservoir could be stored such an excess water from periods of high flow for use during periods of drought. In addition to conserving water for later use, so the storage of floodwater would also reduce flood damage below the reservoir. Furthermore, regards of floodwater control so storage reservoirs are used to control floods, to conserve water, and to regulate stream flow. Reservoirs could be of two types: Single-purpose or multi-purpose. It would be dependent on the location and structural problems, the problems for a single-purpose reservoir leads to simple relationships among the available water supply, the water demand and the volume of reservoir storage to be provided. However, these relationships are much more complex for a multi-purpose reservoir since they involve the seasonal distribution of stream flow and the reconciliation between it and seasonal and other varying demands for the several purposes are intended (Figures 3 and 4) [8]. These streams water, which would carry little or no water during periods of the year in Makoran often, becomes a raging torrent after heavy rains and a hazard to all activities along its bank. By testing river flow to get better and actual rainfall runoff characteristics by obtaining appropriate monthly maximum flow to view and improve monthly peak discharge of Sarbaz and Kajo rivers in four different sites gauged by hydrometric recorded and tested by excel for wet and dry periods for flood purposes, though it shown the result of Makoran water flow in rivers analysed. Thus Sarbaz and Kajo rivers during storm water with monthly maximum discharge as indicated in (Figures 5 and 6).

Figure 3: River Kajoo Overtopped levee at District Ghasrghand during flash flood 2013.

Figure 4: River Sarbaz Peak Discharge Overtopped levee At Jakigovar, During Storm 2013 Up Stream OfPishin Dam.

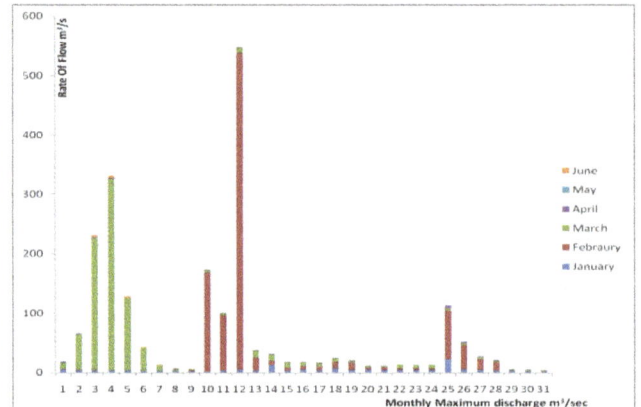

Figure 5: Duration 6 months from January to June (test 1).

In the light of discharge of water there could be drawn hydrograph, so an inflow hydrograph is simple a graphical plot of the river flow with time on the horizontal scale and discharge on the vertical scale. Although a typical hydrograph resulting from an isolated storm, which could be consisted of a rising limb, a crest and a recession curve. Also some ground water may enter the stream as base flow.

Flood Control and Reservoir Development in Makoran

The Reservoirs reduces people's dependence on the natural availability of water in rivers and streams flow. In recent years, it has become a common perception that reservoirs must serve for the protection of the environment during the life time of this generation as well as the future generations. Thus, how do we construct and manage flood reservoirs to meet these requirements? This is the crucial question of sustainable flood control reservoir development and management. The International commission on water resources systems also set up international guidelines to cope flood problem to reduce hazard for people life. However, management of annual peak flows to restore development of dam reservoirs in the Makoran regions. The contributions provide decision makers and flood control management with a collection of scientific contributions to be taken into account before setting up any project for flood control so dealing with water resources uses and management [8]. Strategy and different methodology for flood water shed management, also it is important for area under consideration soil erosion control and land water or underground water management use, water shed and river processes and their modelling. Furthermore, in the light of flood occurring

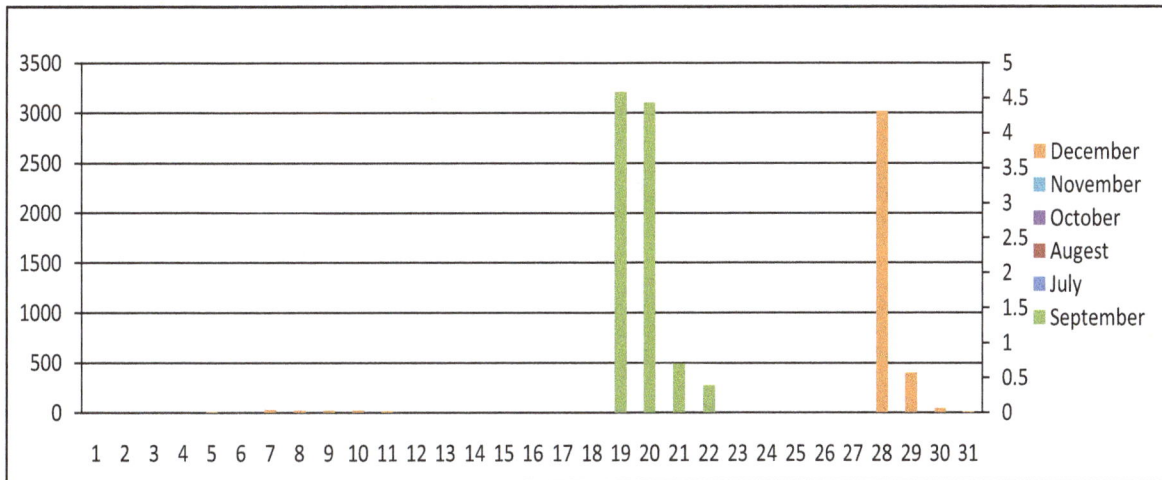

Figure 6: Monthly Maximum Discharge m³/s (test 1) From July to December.

and environmental problems associated with land water use and monitoring, planning and rivers with flood water occur could be organized and properly managed. The activities towards flood control area, which under study is extensive with priority on functional techniques and management methods for the sustainable development of rural areas. It could be greatly benefits and concerning towards for the Baloch people life and also for their agricultural purposes of region towards agricultural production for food. The income of farmers in demonstration sites, which are using for agricultural and horticultural purposes, could also be increased sharply for aim of food production [9].

Flood and Drought Management for Makoran Regional Steering

The floods and droughts have been common events throughout the world. Many countries experience such catastrophic events each year. Beyond all doubt these natural hazards affect human living conditions and sustainable socio-economic development. Therefore, the flood and drought both are very important matter to be concern and could be under significant consideration for better solution towards safe place for human being and environment in the region. These issues of steering for water resources and flood management became very important aspects towards handling safe water resources management and organization of the corresponding areas. Hydro- meteorological hazards, towards monthly or daily rainfall data, as far as including drought event, are also main research activities for Makoran regions. To reduce the catastrophic destruction brought about water resources management, floods and droughts concentrated global action was necessary. Thus, floods and droughts for the region are two extreme events in the domain of hydrology and water resources. However, people live along rivers, large river such as Sarbaz and Kajo usually attract high-density populations and economic growth. This study show incorrect use of water resources and flood water plains because there would be due to wrong and un-sufficient water management towards flood plains by diverting water ways so there is no bypass canal, and land erosion activities lead to changes in river configuration and reduction in water storage facilities and flood ways [10]. Rainfall is very scarce in the Makoran regions, shortage of available water resources leads to insufficient water supply. There would be many factors contribute to water shortage, such as inadequate water distribution systems, insufficient water supply facilities in the regions.

Drought intensifies water shortage and for some reason drought prediction is less accurate than flood forecasting. Therefore, Makoran regions have experienced serious flood or drought over the past few decades and periods of drought may last for several years and may re-appear after several years of more normal rainfall. However, right now the attention of the International Scientific Community has been drawn to the problem of severe drought or flood conditions with the related water resources management problems. It is also necessary for Makoran to reduce drought and flood disaster. Furthermore, water resources management and development is a key step to achieving this goal. In some part of Makoran areas where drought is frequent and it is essential to establish long and short term water resources management planning and to improve long term hydro-meteorological forecasting and so mitigate flood and drought disaster.

Classification and to Asses Planning of Rainfall – Runoff Processes

The hydrological processes in the above sections could be represented in rainfall runoff reviews in order to account for the transformation of rainfall into stream flow and other water balance components. This could be in a form of a monthly based rate of flow from Rivers in all Balochistan such as Sarbaz and Kajo rivers [11]. To a great extent the purpose of solving runoff problem and the level of complexity that may be required. There are mainly two purposes; (1) to test maximum discharge which represented rate of flow of water or flood that is held about a process or system and (2) to predict the character of a system under a given set of conditions. Most of the river Sarbaz rate of flow of water level tends to concentrate on the latter purpose. The least that can be expected is a rainfall runoff analysing procedure that relates the storm flow at the gauging station to the amount of rainfall that produced it [9].

By testing river flow to get better and actual rainfall runoff characteristics by obtaining appropriate monthly maximum flow to view and improve monthly peak discharge of river Sarbaz and Kajo in four different sites gauged by hydrometric recorded and tested by excel for wet and dry periods and so for flood purposes, though it shown the result of Makoran water flow in rivers analysed. Focusing the river Sarbaz for the purposes of water flow and gauged at two gauging station called Shirgwaz and Homeiri gauging stations though the catchment for

Pishin station is 6850 km[2] and the catchment area for the Bahawkalat hydrometric station is 3910 km[2] however, the river network is a complex inter relationship of a historically. A conceptual approach that allowed some degree of perception of the hydrological processes to be expressed in mathematical form. The establishment and development of distributed monthly maximum flow analysis that account for the spatial variability of hydrological processes is appropriate to achieve river discharge in Makoran, thus the different monthly maximum discharge m^3/sec are illustrated in Figures 7-11 which indicated various rate of flow for Sarbaz River during wet months [12].

Thus, the significant of the figures illustrated for different wet months shown the rate of flow for the region under study and a good translation lagging behind river Sarbaz during entire periods of the year 2013. In the light of the storms and flood monthly maximum the calibration of river Sarbaz basically illustrated during month of December 2013 has been tested by computer the peak discharge was 3012 m^3/sec, for the February 2013 was tested 532 m^3/sec, for the

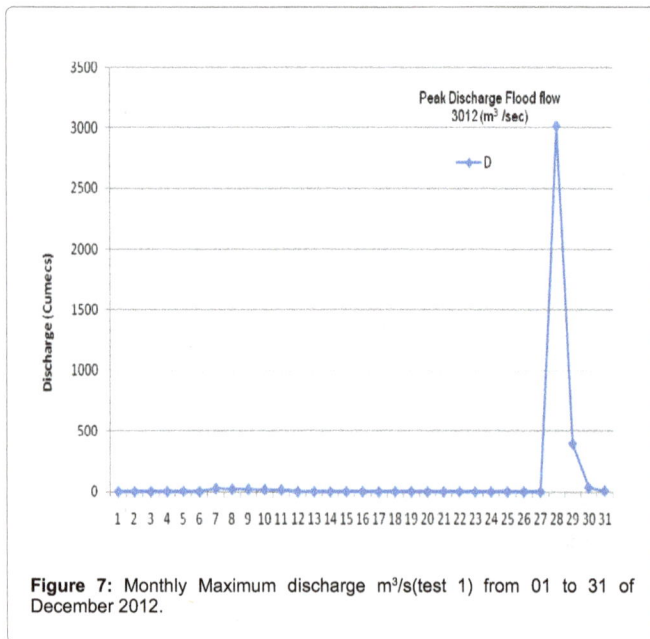

Figure 9: Monthly Maximum discharge m^3/s(test 3) from 01 to 31 of February 2013.

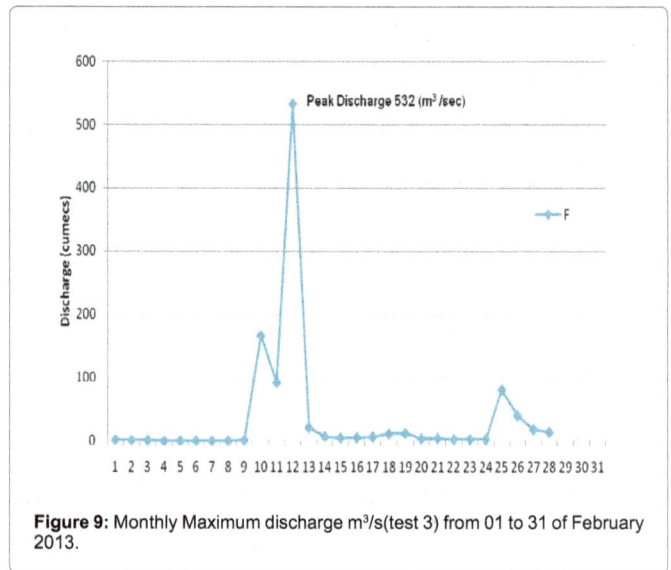

Figure 7: Monthly Maximum discharge m^3/s(test 1) from 01 to 31 of December 2012.

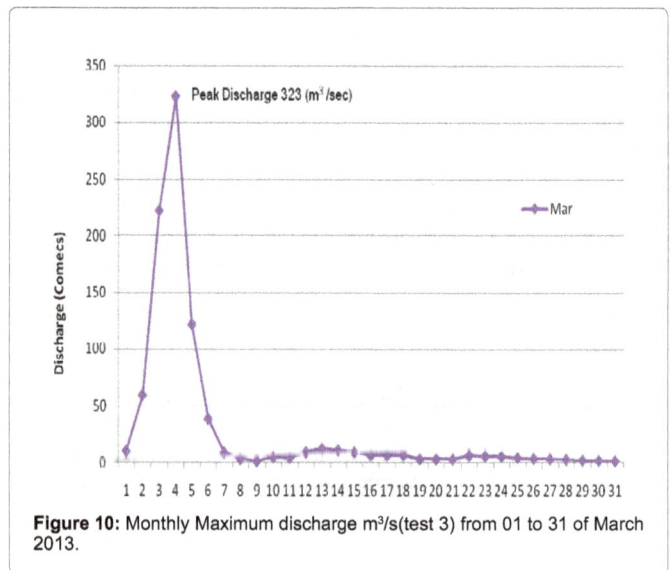

Figure 10: Monthly Maximum discharge m^3/s(test 3) from 01 to 31 of March 2013.

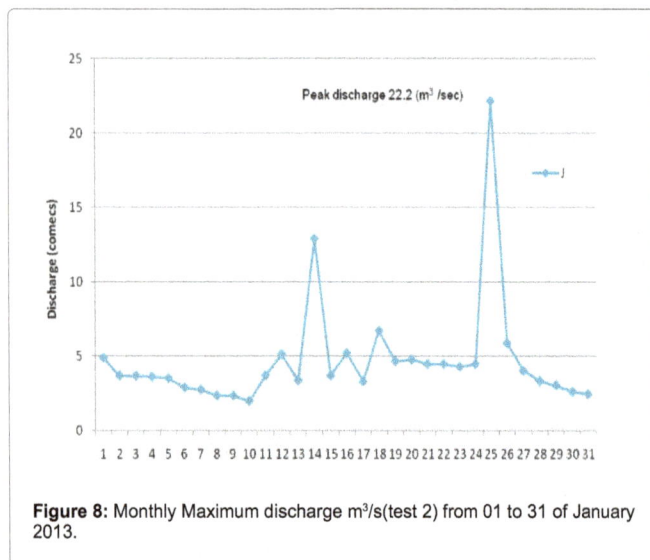

Figure 8: Monthly Maximum discharge m^3/s(test 2) from 01 to 31 of January 2013.

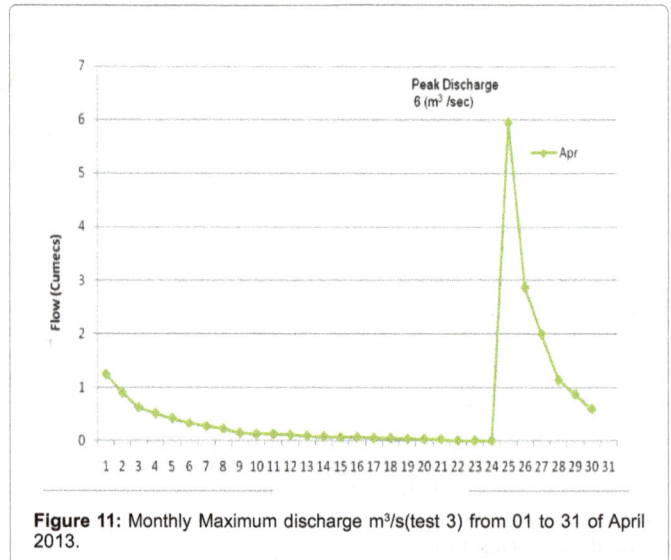

Figure 11: Monthly Maximum discharge m^3/s(test 3) from 01 to 31 of April 2013.

March peak discharge has been tested by excel was 323 m³/sec where as for the same river on the month of April suddenly rate of flow dropped to 6 m³/sec even though for other months during summer rate of flow for Sarbaz river is completely drought or nil discharge accordingly to the dry months when there is no rainfall in Makoran region, this is the link between rainfall and storm characteristics and its effect on monthly maximum discharge have been dealt with in the past also the storms characteristics mainly considered were the storm pattern, might be speed and direction of rainstorm moving in the downstream direction produces a higher peak flow than storms moving upstream which can be concluded that storms moving at the same speed as the stream velocity have more impact on peak discharge than rapidly moving storms.

Conclusion

However, measures to avoid flooding or to alleviate the damage it causes within a river catchments are best to taken into context of the interrelationship between the water courses, land use and developments within that catchments. Whenever the options could be involve both river and coastal defences, for example estuarine barrages, should be important to bring together both coastal and sites inland catchments, as well then process to observe the outcome towards the result of flood. Although where such works have significant effects on the environment these should be taken into account. Therefore, it is often beneficial to operating authorities and relevant interest groups to undertake a voluntary, environmental assessment, In this research the corresponding area shows climate change or global warming may be affecting flooding and flood frequency could be change. So measures to avoid flooding or to alleviate the damage it causes with in a river catchments are best taken in the context as well by the recent analysis towards flooding may be occurring sooner [1]. Unfortunately local authority costs are not based largely towards flood control and coastal defence activity due to less management of central government of Iran. Thus financial support for flood and flood defence works involves significant sums of expenditure it is therefore, essential that to put good maintenance towards flood protection. The research and analysing data shows the requirement that the range of options must be considered for new or improved defence measures encourages and more rigorous approach to cost effectiveness [13]. Thus it could be concluded that if progress is to be made with flood work periodically towards benefits of environment and to save people life. In response to heightened public concern for the environment and the effect of global climate change then the water authorities could be the best place to make information available to the general public on a range of local issues affecting flood and flood defence [14].

Acknowledgement

The authors are extremely grateful to the following companies for their assistance with the collection of data: Water and Sanitation Organisation, Water Resources Organisation Chabahar and Agricultural Organisation. The authors would also like to express their great appreciation to Professor Mohammad Safar Mirjat and Dr. Abdolsamad Chandio for their valuable and constructive suggestions during the planning and development of this research work.

References

1. Blackl, Veatch (2002) Independent Consultant Safety and Inspection Report. FERC Project No 4441. Completed in accordance with FERC Rules and Regulations, Part 12, Subpart D.

2. Valipour M (2014) Future of the area equipped for irrigation. Archives of Agronomy and Soil Science. 60: 1641-1660

3. Valipour M (2014) Land use policy and agricultural water management of the previous half of century in Africa. Applied Water Science.

4. Valipour M (2013) Evolution of Irrigation-Equipped Areas as Share of Cultivated Areas. Irrigation and Drainage Systems Engineering 2: e114.

5. Valipour M (2013) Necessity of Irrigated and Rainfed Agriculture in the World. Irrigation and Drainage Systems Engineering.

6. Pacific International Engineering (PIE) (2001) Fish, Riparian and Wildlife Habitat Study. General Revaluation Report and Environmental Impact Statement. Pacific International Engineering, Washington State Department of Ecology, Tacoma, WA.

7. Fernandez CA (1998) Corps of Engineers Centralia Flood Damage Reduction project Skookumchuck dam flood operations and management.

8. Ivan Johnson, Carlos A, Fernande J (1998) Hydrology in The Humid Tropical Environment.

9. Beven KJ (1992) Changing ideas in hydrology-the case of physically-based analysis. Journal of Hydrology 105: 157-172.

10. Kusky TM (2008) Floods: Hazards of Surface and Groundwater Systems, The Hazardous Earth Set, Facts on Files, New York.

11. Bodurtha T (1989) Centralia Flood Damage Reduction Study, Planning Aid Report.

12. Marriott MJ (2002) "Probability of Flooding", The Mathematical Gazette 86: 506-509.

13. Beven KJ, Calver A, Morris EM (1989) The Institute of Hydrology Distributed Analysis, Institute of Hydrology report 98, Institute of Hydrology, Oxan, U.K.

14. Valipour M (2012) Number of Required Observation Data for Rainfall Forecasting According to the Climate Conditions. American Journal of Scientific Research 74: 79-86.

3

Green Concept in Storm Water Management

James C.Y. Guo*

Professor and Director, Dept. of Civil Engineering, University of Colorado Denver, USA

Abstract

Storm runoff is considered one of important water resources for urban areas. Over a geologictime, streams and lakes are periodically refreshed with flood water and continually shaped with the flood flows. Urban development always results in increases of runoff peak rates, volumes, and frequency of higher flows. As a result, flood mitigation has become a major task in urban developments. Before 1970's, storm water drainage systems in an urban area were designed to remove flood water from streets as quickly as possible. From 1970 to 1980, the US EPA conducted a nationwide stormwater data collection and reached a conclusion that stream stability is more related to frequent, small storm events rather than the extreme, large events. Since man-made stormwater systems were designed to pass extreme events, the large inlets and outlets release frequent events without any detention effect. As a result, urban pollutant and sediment solids are transported and deposited in the receiving water bodies. Under a US Congress mandate starting in 1980's, a nationwide stormwater best-management-practices (BMPs) program was developed and implanted in major metropolitan areas. The tasks in BMPs include: (1) retrofitting the existing drainage facilities to achieve a full-spectrum control on peak flows, and (2) applications of Low-Impact -Development (LID) designs to reduce runoff volumes under the post-development condition. With the latest observations in climate change, the uncertainty in the design floodhas imposed unprecedented challenges in flood mitigation designs. The flexibility in freeboards and easements need to be refined in order to accommodate the changes in extreme rainfall events. This paper presents a summary of the Green approach in stormwater management and LID designs as the engineering measures to preserve the watershed regime.

Keywords: Stormwater; BMP; LID; Detention; Urban drainage; Green concept

Introduction

Before 1970's, designs of urban drainage systems were mainly aimed at efficient removal of stormwater from streets for traffic safety [1]. Urban drainage systems were essentially sized to pass extreme events. Under the concept of "bigger is better", urban areas were equipped with street gutters, inlets, culverts, and storm drains. From 1970 to 1980, the US EPA conducted a nation-wide investigation on urban storm water [2]. As reported, urbanization process results in tremendous increases in storm runoff rates, volumes, and frequencies of high flows. Also it is confirmed that man-made drainage systems are efficiently transport urban pollutants into receiving water bodies. These findings trigger the 1972 Federal Clean Water Act. Under a Congress mandate, all metro areas in the US must improve the urban drainage systems to protect urban water environment. This paper documents the evolvement of the green concept from Best Management Practices (BMPs) to Low-Impact-Development (LID) in storm water management and flood mitigation.

Basic problems in urban storm water

Since 1970's, storm water detention was introduced to mitigate urban flooding problems [3,4]. As suggested, a sewer trunk line shall drain into a detention basin before the storm runoff is released into the downstream water body [5]. The storage effect in a detention basin reduces the peak flow and also delays the time to peak. Both conveyance and storage systems are utilized to drain and to store excess storm water. This practice implies that both runoff volume and flow rate shall be taken into consideration when sizing an urban drainage network [6,4].

As illustrated in Figure 1, an urban lot is composed of impervious and pervious surfaces. Under a rainfall event, overland flows are produced from roofs, pavements, and driveways. As a shallow and wide sheet flow, overland flows sweep streets, and carry urban pollutants and debris. After a concentrated flow is formed, the peak flow is accumulated along the waterway. Wherever the peak flow exceeds the capacity of the drainage system, flooding problems occur. A shown in Figure 2,

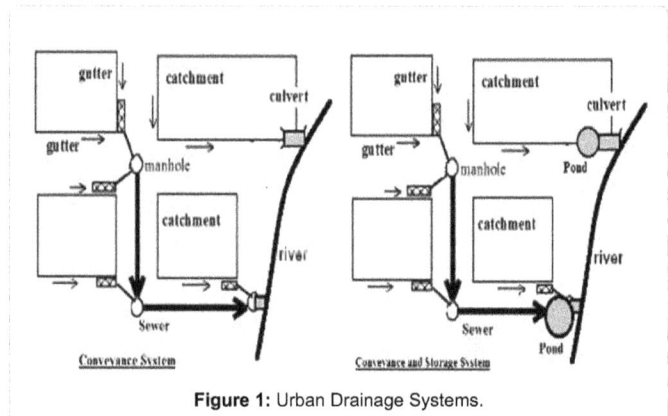

Figure 1: Urban Drainage Systems.

the *V-problem* is referred to as the storm water quality issues that are directly related to the shallow water depths in overland flows, while the *Q-problem* is referred to as the flooding issues that are caused with the concentrated flow [7]. Under the mandate of the 1972 Federal Clean Water Act, the *V-problem* is associated with water quality enhancement, while the Q-problem is related to flood mitigation. Under the green concept for storm water management, there are two distinct approaches developed to cope with these two problems:

(1) How to reduce the increased on-site runoff volume from the

*Corresponding author: James C.Y. Guo, Professor and Director, Dept. of Civil Engineering, University of Colorado Denver, USA
Email: James.Guo@ucdenver.edu

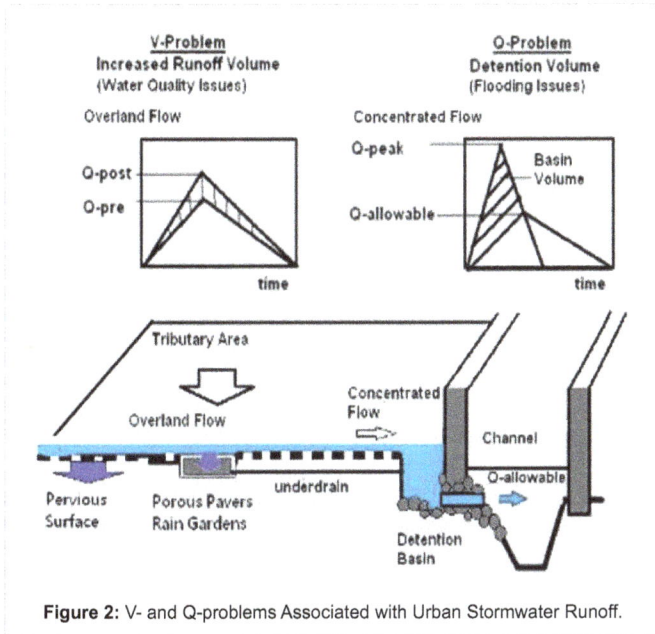

Figure 2: V- and Q-problems Associated with Urban Stormwater Runoff.

post-development to the predevelopment condition using LID devices such as porous pavers, rain gardens etc., [8] and

(2) How to regulate the flow releases from the post-development peak flows to the allowable flow rates using detention and retention facilities at strategic locations [9,10].

The key factor in urban hydrology is watershed imperviousness. Urban development always leads to more pavements, roofs, driveways, and parking lots. All these changes in land use increase the area-imperviousness percent. Figure 3 presents the impact of increased watershed imperviousness on the increases of runoff volumes and peak flows. Using the case of imperviousness of 5% as the basis, the peak flow will be increased 3.25 times and the runoff volume will be increased 1.5 times after the watershed is developed to an area-imperviousness of 90% [11,12]. Figure 3 implies that an effective urban drainage system should be designed to dispose the local increased runoff volume through the on-site infiltration practices, and then to convey the excess runoff flows to the strategic locations where storm water storage practices can be implemented to reduce the post-development flows to its pre-development condition. In the last decade, there are various methods developed to mitigate stormwater *V-and Q-problems*. In general, the *V-problem* is alleviated with on-site infiltration -based devices, while the Q-problem is managed with extended storm water detention process [13].

Green approach for urban drainage planning

The 1972 *US Federal Clean Water Act* has significantly expanded storm water management in the United States from flood mitigation into both storm water quality and quantity controls. As recommended, an urban drainage plan shall observe the following steps [14]:

(1) Minimize the Directly Connected Impervious areas (MDCIA),

(2) Dispose on-site runoff volume using LID devices,

(3) Convey concentrated runoff using a cascading flow system,

(4) Store runoff flows at strategic locations,

(5) Control flow release at the pre-development rate and frequency,

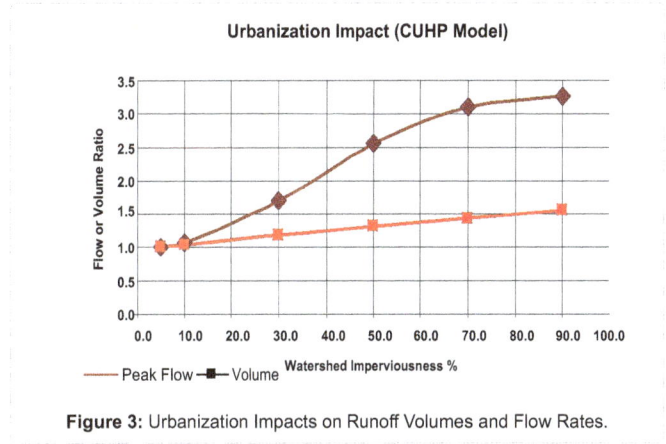

Figure 3: Urbanization Impacts on Runoff Volumes and Flow Rates.

(6) Apply erosion and sediment controls at all construction sites.

Applications of the above are discussed in details in the following sections.

Land use under Mdcia practice: The watershed's response to a rainfall event is very sensitive to how the storm drains are networked together. Conventionally, roof areas are connected together through roof gutters that collect storm runoff from roofs and then drain onto the driveways. All driveways are linked through storm drains to pass storm water directly to the adjacent streets. This drainage pattern is termed *Distributed System*. A distributed flow system is efficient to remove storm water, but it tends to result in higher peak and faster runoff flows. As illustrated in Figure 4, a distributed flow system uses two independent flow paths to drain storm water from the pervious and impervious areas separately, while a cascading flow system in Figure 5 is laid to spread storm water from the upper impervious area onto the lower pervious area.

Under the concept of MDCIA, a LID device or grass swale is placed between two adjacent impervious areas to slow down runoff flows for the purpose of filtering and infiltration benefits. As a rule of thumb, the impervious area is 2 to 3 times the receiving pervious area. For instance, a case of 3 units of impervious area draining onto one unit pervious area will result in an area-impervious percentage of 75% [7].

In practice, the land uses within the project site hardly result in a complete interception of the cascading flow. As recommended in Figure 6 EPA SWMM, the catchment is divided into the upper impervious and the lower pervious areas. Mathematically, the intercepted runoff volume generated from the upper impervious area is directly added to the lower pervious area as:

$$V_R = PA \qquad\qquad 1$$
$$V_P = m[r(P - D_{vi})I_a A + (P - D_{vp} - F)(1 - I_a)A] \qquad 2$$

Where V_R=rainfall volume in [L^3], P=precipitation depth in [L per watershed], A= watershed area in [L^2], V_p=runoff volume from pervious area in [L^3], D_{vi} = depression loss on impervious area in [L], I_a = impervious area ratio, D_{vp}=depression loss on pervious area in [L], F=infiltration amount in [L], m = 1 if V_p>0 or 0 if $V_p \leq 0$, and r = flow interception ratio of V_m. When r=1, Eq (2) represents a complete flow interception, while r=0, Eq (2) reproduces the flow condition in a distributed flow system. For 0<r <1, the residual runoff volume is directly released to the street as:

$$V_m = (1 - r)(P - D_{vi})I_a A \qquad\qquad 3$$

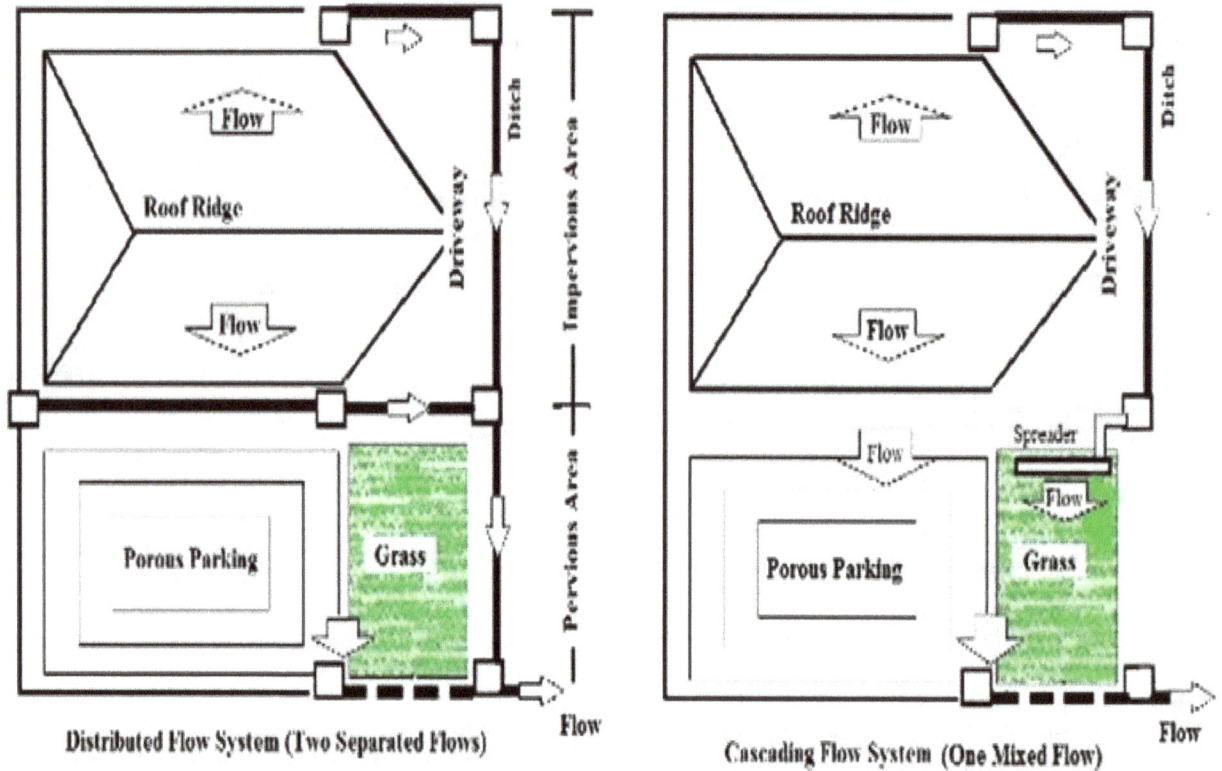

Figure 4: Distributed and Cascading Drainage Systems.

| Porous Pavement | Infiltrating Bed |

Figure 5: Examples of MDCIA Practice.

V_m = runoff volume from impervious area in [L^3]. The resultant runoff coefficient is calculated as:

$$V_F = V_P + V_m \qquad\qquad 4$$

$$C = \frac{V_F}{V_R} = (1-r)(1-\frac{D_{vi}}{P})I_a + m[r(1-\frac{D_{vi}}{P})I_a + (1-\frac{D_{vp}}{P}-\frac{F}{P})(1-I_a)] \qquad 5$$

Where C=runoff coefficient. For a distributed flow system, r=0 or Eq (5) is reduced to

$$C = \frac{V_F}{V_R} = (1-\frac{D_{vi}}{P})I_a + m(1-\frac{D_{vp}}{P}-\frac{F}{P})(1-I_a) \qquad 6$$

Like the rational method, the runoff coefficients in Eq (5) and (6) are linear with respect to watershed imperviousness. As a sum of two separated flows, Eq (6) is always dominated by the impervious areas or V_m. Numerically, runoff coefficients in Eq (6) is always greater than zero as long as P>D_{vi}.

Eq 6 has been tested and accepted in UDSWCM 2001 update. Considering Denver's hydrologic parameters: D_{vi} =0.1 inch, D_{vp}=0.4 inch, F=0.88 inch, P_1= 2.6 inch for the 100-yr event, 1.35 inch for the 5-yr event, and 0.95 inch for the 2-yr event, the 100-, 5- and 2-yr runoff coefficients for a cascading flow system are produced and presented in Fig 6. For comparison, the effect of cascading flow was tested for 3 cases, including r=0 (no flow interception), r=0.5 (50% of flow intercepted), and r=1.0 (100% flow interception) . Under an impervious percent of 45%, a case of complete flow interception results in a reduction of runoff coefficient from 0.4 to 0.2.

On-Site Stormwater disposal using lid design: Porous pavements and rain gardens (RG) were first tested in the State of Maryland in 1993. Over the years, they have spread out as the most popular infiltrating practice in the USA for storm runoff on-site treatment devices [15-17]. As illustrated in Figure 7, both rain gardens and pavers are structured as a two-layered basin. The surface basin in a RG is designed to intercept the water quality capture volume (WQCV) with a maximum water depth from 12 to 15 inches (30.5 to 38 cm).

A RG is often covered with selected grass and plants. During an intense event, the surface basin will be filled up to its maximum capacity, and then the excess storm water overflows into the downstream manhole. The subsurface filtering layers underneath a RG consist of an upper sand-mix layer of 18 inches (45.7 cm), a lower gravel layer of 8 inches (20.3 cm), and a sub-drain system that is formed with 4 -inch (10.2 cm) perforated pipes networked together to drain infiltrating water into an adjacent manhole [18].

Water Quality Capture Volume (WQCV) in a RG is the storage volume reserved for the water quality treatment. The infiltration pool is constructed with a filtering and infiltrating bed to dispose WQCV into the local groundwater table for water recharge. As reported [19],WQCV was empirically derived from the break-even point on the distribution of runoff-depth population. A WQCV is found to be equivalent to the rainfall amount of 3- to 6-month event. Furthermore, the one-parameter exponential distribution was adopted to describe the frequency distribution of rainfall event depths [20]. The exponential distribution is described as:

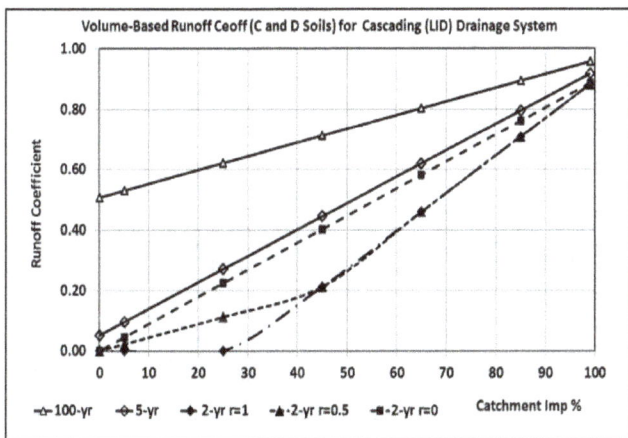

Figure 6: Runoff Coefficients for Cascading Flow System.

$$f(D) = \frac{1}{D_m} e^{\frac{-D}{D_m}} \qquad\qquad 7$$

in which $f(D)$ = frequency of rainfall event-depth, D, and D_m = average rainfall event-depth. The WQCV can then be related to its design rainfall depth, D, as:

$$V_o = C(D - D_i) \qquad\qquad 8$$

in which V_o = WQCV in mm per watershed, C= runoff coefficient, D = design rainfall depth, and $8D_i$ = incipient runoff depth recommended to be 2.5 mm. Aided with Eq 7, Eq 7 can be integrated into

$$C_v = P_D(0 \leq V \leq V_o) = P_D(0 \leq d \leq D) = 1 - ke^{\frac{-V_o}{CD_m}} \text{ and } k = e^{\frac{-D_i}{D_m}} \qquad 9$$

in which C_v = runoff volume capture rate between zero and unity, V_o= WQCV selected for design, $P_D(0V V_o)$ = probability to have an event that produces a runoff depth less than V_o. The value of k is defined by the incipient runoff depth and the average event rainfall depth. The value of k varies in a narrow range between 0.80 and 0.90. Eq 9 represents the synthetic runoff capture curves normalized by local average rainfall event-depth, runoff coefficient, and runoff incipient depth. Figure 8 presents a set of generalized runoff capture curves produced using Eq 9 with runoff coefficients of 0.4, 0.6, 0.8, 0.9, and 1.0. It is noticed that the curvature of runoff capture curve increases when the runoff coefficient decreases. The runoff capture curve becomes almost a linear response between rainfall depth and runoff amount when C=1.0. This tendency reflects the fact that the higher the imperviousness in a watershed, the less the surface depression and detention. As a result, the response of a watershed to rainfall is quick and direct. As recommended [19],a WQCV basin will intercept up to 80% of runoff flow population, and bypasses the top 20% larger events.

The uncertainty in a RG's operation is directly related to its infiltration rate through the filtering layers. Considering clogging effects, the design infiltration rate is defaulted to be 1.0 inch/hr (2.5 cm/hr) [21]. In fact, a newly constructed RG may have an infiltration rate as high as 10.0 to 15 inch per hour (25.4 to 38.1 cm/hr) [18]. Over the years in service, the infiltration rate in the RG is gradually reduced due to clogging effects. In practice, a RG can be an independent unit, or nested in the bottom of an extended detention basin [22].

Conveyance system for multiple design events: Urban stormwater drainage systems are designed or renewed to have three layers of cascading flows. They are Micro, Minor and Major Flow Systems as shown in Figure 9 and Figure 10. Storm runoff generated from

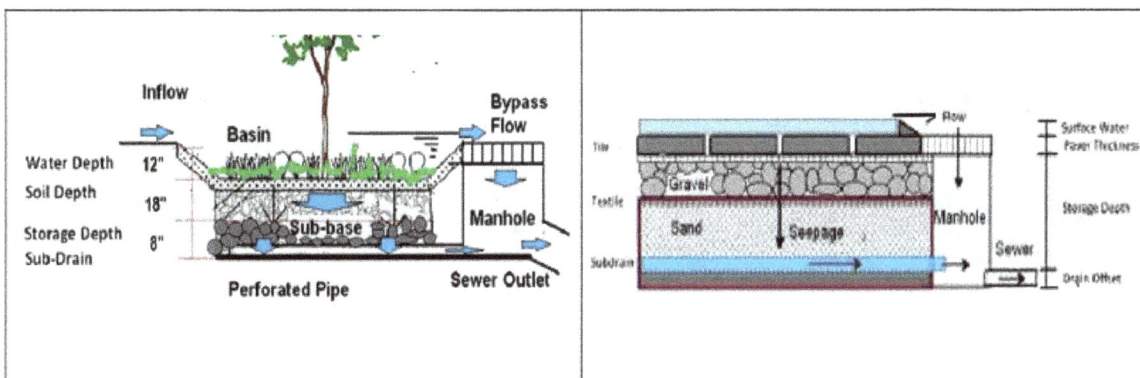

Figure 7: Layout Of Rain Garden And Porous Paver.

impervious surfaces shall be drained onto a Micro Flow System for water filtering and infiltration. A micro flow system consists of porous pavers, grass swales, bio-retention basins that are designed to treat the

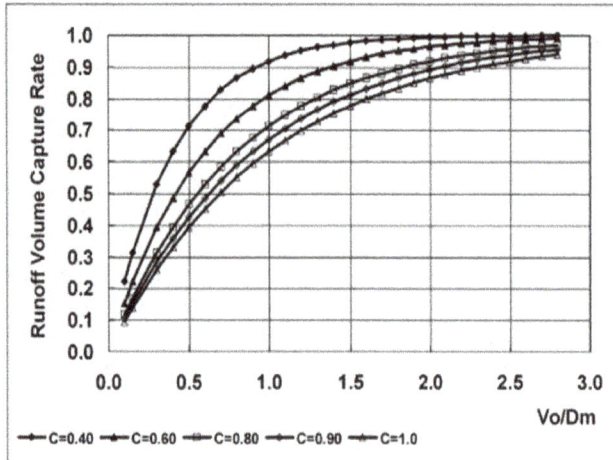

Figure 8: Water Quality Capture Volume.

WQCV [23]. Overflows from a micro facility will be drained into the that consists of street inlets and storm drains. After the underground storm sewers become full, the excess storm water will be carried on the streets which are considered the Major Flow System. A micro drainage system is also termed LID facility. In practice, a LID facility is composed of a surface storage basin and subsurface filtering layers. Most porous pavers are conveyance-based with a thin water depth on the surface, while a bio-basin is a storage -based facility with 12- to 18-inch (30- to 45-cm) depth of water in the surface basin. A storage-based LID is also called bio-retention, rain garden, or landscaping detention basin.

Detention system for flow release control: As illustrated in Figure 11, the storm water detention volumes for the 10- and 100 -yr events are determined using the post-development hydrographs and allowable release rates [24,25]. In practice, the allowable release rate is directly related or equal to the pre-development peak flow. For convenience, the after-detention hydrograph is approximated using a linear rising limb to the allowable flow. The required detention volume is the difference between the post-development and after-detention hydrographs.

Similarly, a detention system in Figure 12 is designed to have 3 layers of storage volumes when shaping the basin's geometry. The bottom layer provides the required WQCV for micro events. The mid and upper layers are designed to control flow releases for the minor (or 10-yr) and

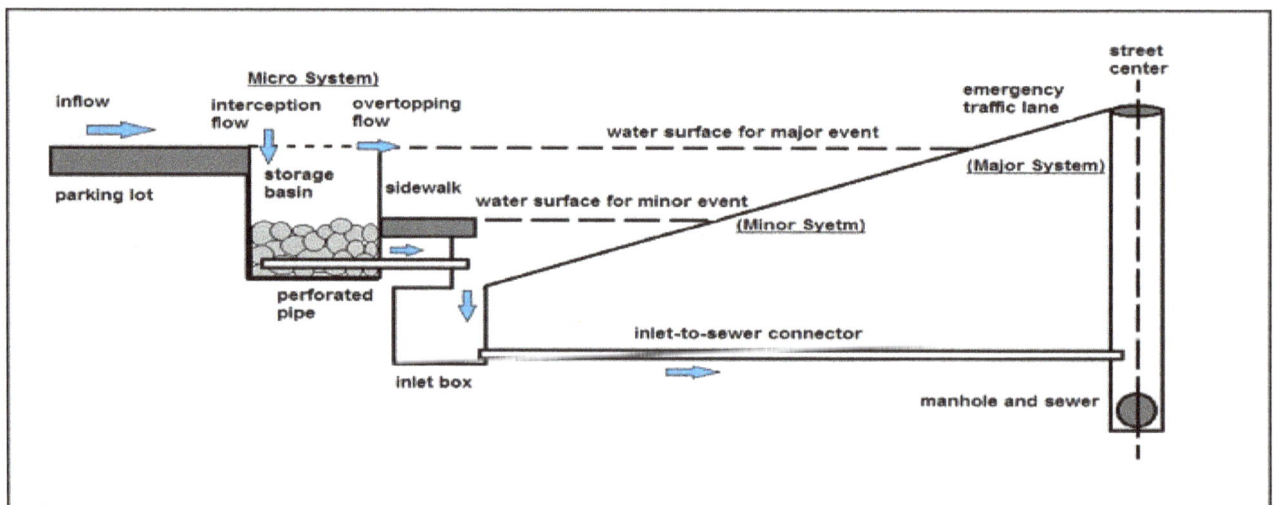

Figure 9: Micro-Minor-Major Conveyance Systems In Street.

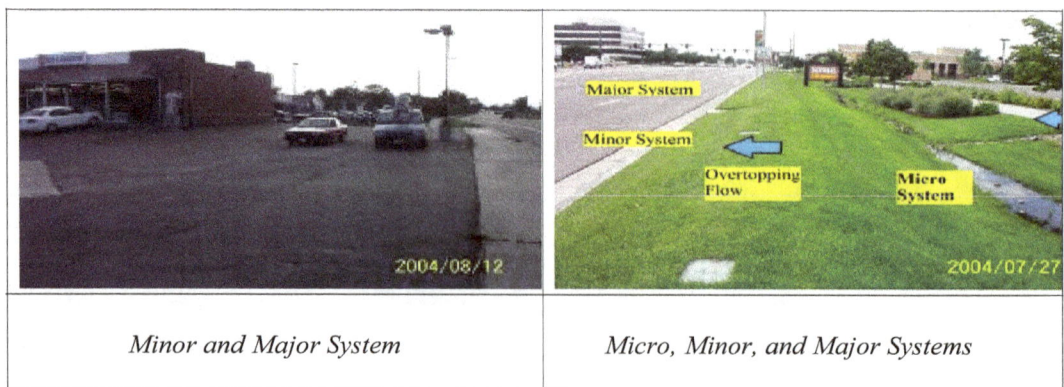

| Minor and Major System | Micro, Minor, and Major Systems |

Figure 10: Comparison of Conventional and Renewed Urban Drainage System.

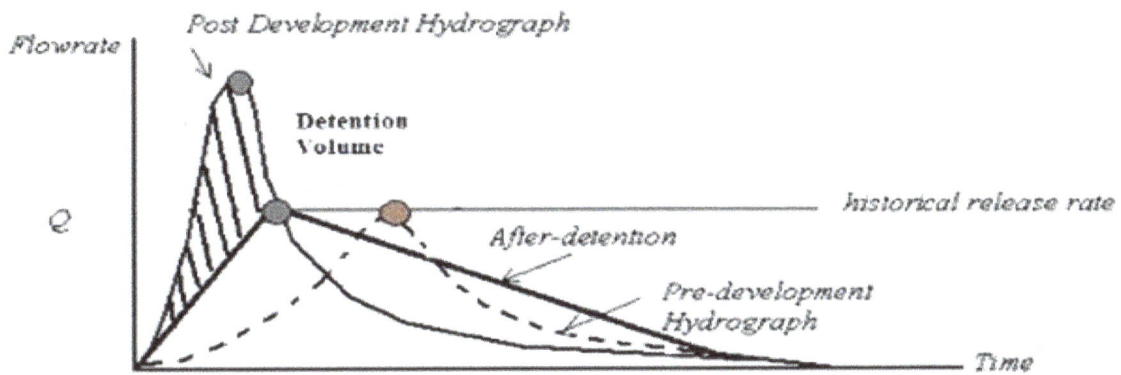

Figure 11: Stormwater Detention Volume.

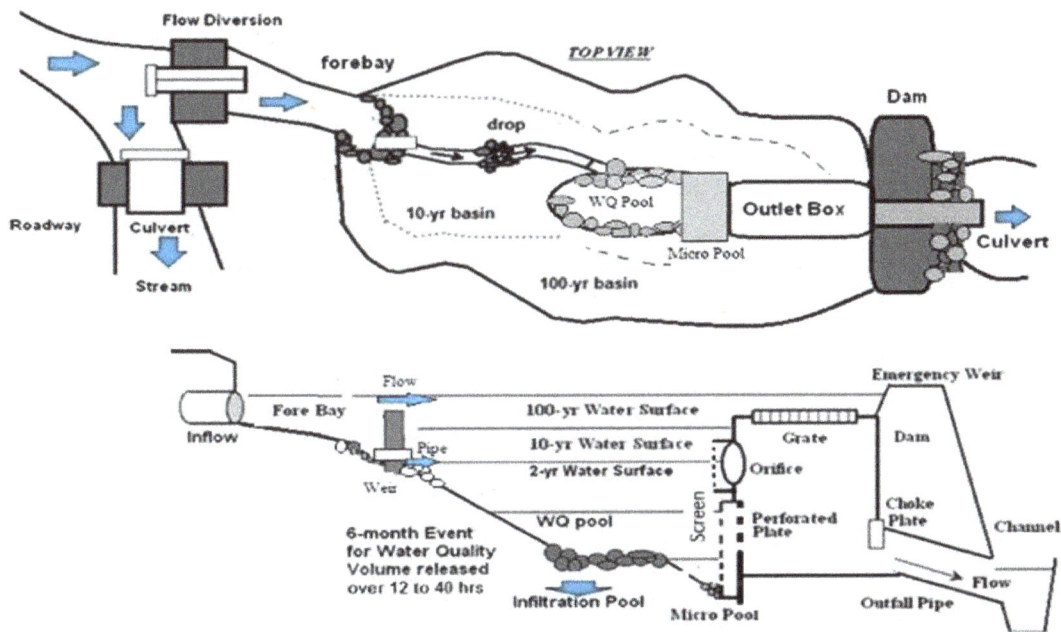

Figure 12: Multiple Layered Detention System.

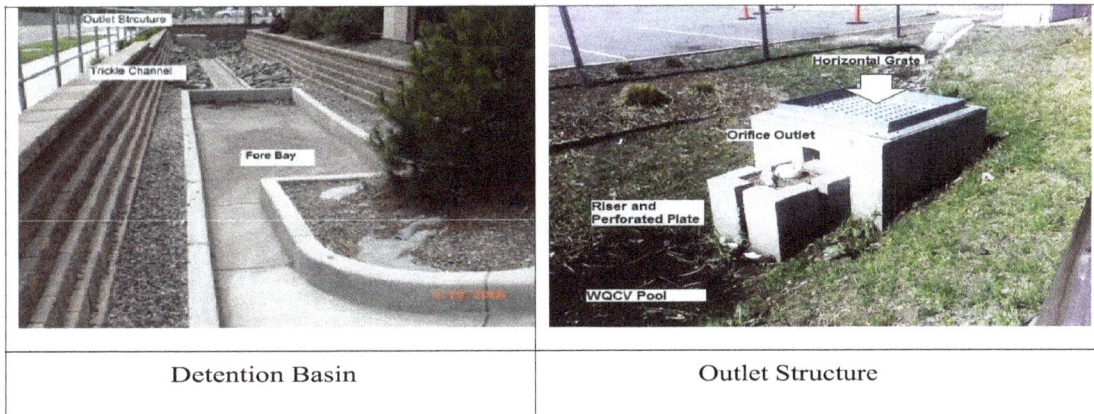

| Detention Basin | Outlet Structure |

Figure 13: Example of Urban Detention Basin and Outlet Structure.

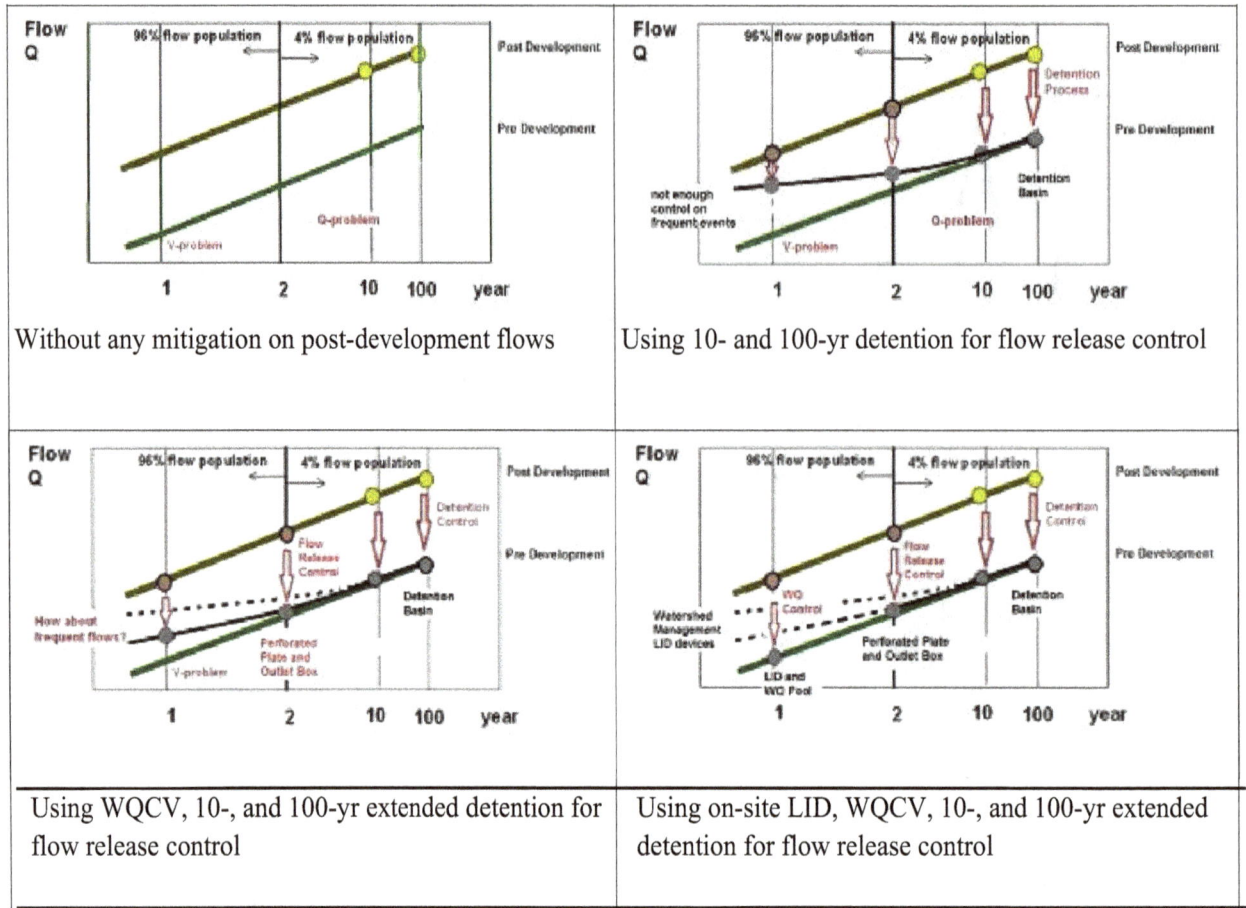

Without any mitigation on post-development flows	Using 10- and 100-yr detention for flow release control
Using WQCV, 10-, and 100-yr extended detention for flow release control	Using on-site LID, WQCV, 10-, and 100-yr extended detention for flow release control

Figure 14: Impact of Stormwater Mitigation on Watershed Regime Preservation.

major (or 100-yr) events. These two layers add more storage volumes to control the flow releases from the 10- to 100-yr events. A fore bay at the entrance is designed to have a low flow pipe and an overtopping weir. All low flows will bed rained through the pipe opening, while high flows overtop the weir. The settlement process at the fore bay will trap solids>1 mm in diameter. From the fore bay to the WQCV pool is a lined trickle channel. The WQCV pool is sized for the purpose of water quality enhancement, and placed immediately upstream of the outlet structure. The outlet structure in a detention basin is also designed as a 3 -layer outlet system, including a micro-flow outlet using a perforated plate or a vertical riser, a minor-flow outlet using a vertical orifice, and a major-flow outlet using a horizontal grate on top of the structure. A micro pool in Figure 13 is always preferable because it serves as a siphon in case that the orifice and riser are clogged [26,27].

Evaluation of green storm water management: The Green concept in storm water management is to apply a micro-scale on-site design strategy with a goal of maintaining or replicating the predevelopment hydrologic regime [28]. The natural hydrologic functions of storage, infiltration, and ground water recharge, as well as the volume and frequency of runoff flows are maintained using integrated and cascading flow systems. In practice, the qualitative goal for a Green storm water strategy is translated into various functional landscapes that act as on-site or regional storm water facilities for storm water flow, volume, frequency, and WQ controls. Although many hydrologic methods have been developed for event-based analyses [29],the ultimate goal of a LID

design is in fact to warrant the preservation of the hydrologic regime [30]. For Instance, the long-term runoff statistics may be employed as the basis to quantify the impact of the development on the watershed hydrologic regime [31]. A standard detention volume is defined by the storm water storage volume required to preserve the mean and standard deviation of runoff volume population under the predevelopment condition. Consequently, a detention basin is considered oversized if the after-detention runoff volume population has a lower mean flow, while a undersized detention basin produces a mean flow higher than that under the pre-development condition.

As illustrated in Figure 14, the upper left case is the 1950 to 1970 conventional approach that has neither flood mitigation nor WQ control. The upper right case is the 1970 to 1980 detention approach that was developed for extreme-event controls only. The lower left case is the 1980 to 2000 extended detentions approached that provides a full -spectrum flow control (EPA 2007). The last one shows the complete mitigation using on-site LID's for watershed improvements and regional extended detention basins for peak flow reduction. The flow-frequency relationship represents the watershed's response in flow rates to local rain storms. Figure14 is a recommended measure to quantify the preservation of watershed regime [31]. Of course, watershed regime is characterized more than flow rates. The latest development in the US EPA's studies, flow-duration curves are also recommended as one of the basic approaches to quantify the impact of development. A flow duration curve presents the distribution of both flow rate and flow frequency [32]. Flow-duration curves have to be produced from the long-term continuous storm water simulations for both pre- and post-development conditions. This new approach will set a higher standard for storm water simulation to become more a full-spectrum flow release control rather than the extreme events only.

Conclusion

In the last 3 decades, the Green concept for storm water management has been evolved from BMP's into LID. Many Green innovative ideas for urban renewal are still on the rising swing. The ultimate goals of Green storm water management are to protect, maintain and enhance the public health, safety, and general welfare by establishing minimum requirements and procedures to reduce the adverse impacts associated with increased storm water runoff. Many innovative engineering concepts and methods have been developed to apply environmental on-site facilities to the maximum extent practicable (MEP) to reduce stream channel erosion, pollution, siltation, sedimentation, and local flooding, and to use appropriate structural best management practices (BMPs) only when necessary. The Green storm water approach will restore, enhance, and maintain the chemical, physical, and biological integrity of streams, and to minimize damage to public and private property, and reduce the impacts of land development. Apparently, the trend in storm water engineering practice is continually being shifted from an event-based approach to a long- term continuous simulation, and also from flood flow control to storm water quality and quantity controls. Any and all urban drainage facilities must be designed to mimic the pre-development hydrologic condition for all events. A new innovative stormwater approach will be tested, monitored, and then evaluated with its outcomes for both stormwater management and flood mitigation. In the near future, retrofitting the existing drainage facilities and maintaining the new systems will become joint efforts for urban renewal projects. The Green concept will lead to a softer, cooler, cleaner, and more balanced water environment in urban cities.

References

1. Chow VT, Maidment David R, Mays Larry W (1998) Applied Hydrology, McGraw Hill Book Company, New York, USA.

2. EPA Report (1983) Results of the Nationwide Urban Runoff Program. U.S. Environmental Protection Agency, Washington DC, USA.

3. Guo James CY (2004) Hydrology-Based Approach to Storm Water Detention Design Using New Routing Schemes. J Hydrol Eng 9: 333-336.

4. Guo James CY (2007) Stormwater Detention and Retention LID systems. Journal of Urban Water Management.

5. USWDCM (2001) Urban Storm Water Design Criteria Manual, Urban Drainage and Flood Control District, Denver, Colorado, USA.

6. Guo James CY (2003) Design of Infiltrating Basin by Soil Storage and Conveyance Capacities. Int J Water 28: 411-415.

7. Guo James CY, Blackler EG, Earles A, MacKenzie K (2010) Effective Imperviousness as Incentive Index for Stormwater LID Designs. J Environ Eng 136: 1341-1346.

8. Guo James CY, Hughes William (2001) Runoff Storage Volume for Infiltration Basin. J Irrig Drain Eng 127: 170-175.

9. Guo James CY (2009) Retrofitting Detention Basin for LID Design with a Water Quality Control Pool. J Irrig Drain Eng 135: 671-675.

10. Guo James CY, Cheng JYC (2008) Retrofit Storm Water Retention Volume for Low Impact Development (LID). J Irrig Drain Eng 134: 872-876.

11. Blackler G, Guo James CY (2013a) Field Test of Paved Area Reduction Factors using a Storm Water Management Model and Water Quality Test Site. J Irrig Drain Eng.

12. Blackler G, Guo James CY (2013b) Paved Area Reduction Factors under Temporally Varied Rainfall and Infiltration. J Irrig Drain Eng 139: 173-179.

13. Guo James CY, Urbonas B, MacKenzie K (2011) The Case for a Water Quality Capture Volume for Stormwater BMP. J Stormwater.

14. Guo James CY, Urbonas B (2013) Volume-based Runoff Coefficient. J Irrig Drain Eng.

15. Booth DB (1990) Stream-channel Incision Following Drainage Basin Urbanization. Water Resour Bull 26: 407-417.

16. EPA Report (2006) Low Impact Development Methods. U.S Environmental Protection Agency, Washington DC, USA.

17. EPA Report (2007) Reducing Stormwater Costs through Low Impact Development (LID) Strategies and Practices. U.S. Environmental Protection Agency, Washington DC, USA.

18. Guo James CY, Kocman S, Ramaswami A (2009) Design of Two-layered Porous Landscaping LID Basin. J Environ Eng 135: 1268-1274.

19. Guo James CY, Urbonas Ben (2002) Runoff Capture and Delivery Curves for Storm Water Quality Control Designs. J Water Resour Plann Manage 128: 208-215.

20. Guo James CY (2002) Overflow Risk of Storm Water BMP Basin Design. J Hydrol Eng 7.

21. Guo James CY (2012b) Cap-orifice as a Flow Regulator for Rain Garden Design. J Irrig Drain Eng 138: 198-202.

22. Guo James CY, Shih HM, MacKenzie K (2012) Stormwater Quality Control LID Basin with Micro pool. J Irrig Drain Eng 138:

23. Guo James CY (2008) Runoff Volume-Based Imperviousness Developed for Storm Water BMP and LID Designs. J Irrig Drain Eng 134: 193-196.

24. Guo James CY, Clark J (2006) Rational Volumetric Method for Off-line Storm Water Detention Design. Journal of PB Network for Water Engineering and Management 21.

25. Guo James CY (2012a) Off-stream Detention Design for Stormwater Management. J Irrig Drain Eng 138: 371-376.

26. Jones J, Guo JCY, Urbonas B (2006) Safety on Detention and Retention Pond Designs. J Storm Water.

27. Waugh PD, Jones JE, Urbonas BR, MacKenzie KA, Guo JCY (2002) Denver Urban Storm Drainage Criteria Manual. Journal of Urban Drainage 112.

28. LID-Practice (2003) The Practice of Low Impact Development, US Department of Housing and Urban Development, Office of Policy Development and Research, Washington DC, USA.

29. LID-Hydro Analysis (1999) Low--Impact Development Hydrologic Analysis, Prince George's County, Department of Environmental Resources Programs and Planning Division, Maryland, USA.

30. LID-Strategy (1999) Low-Impact Development Design Strategies: An Integrated Design Approach, Prince George's County, Department of Environmental Resources Programs and Planning Division, Maryland, USA.

31. Guo James CY (2010) Preservation of Watershed Regime for Low Impact Development using (LID) Detention. J Hydrol Eng 15: 15-19.

32. EPA Report (2008) Development of Duration-Curve Based Methods for Quantifying Variability and Change in Watershed Hydrology and Water Quality. U.S. Environmental Protection Agency, Washington DC, USA.

4

Statistical Assessment of a Numerical Model Simulating Agro Hydro-chemical Processes in Soil under Drip Fertigated Mandarin Tree

Phogat V[1]*, Skewes MA[1], Cox JW[1,2] and Simunek J[3]

[1]South Australian Research and Development Institute, Australia
[2]The University of Adelaide, Australia
[3]Department of Environmental Sciences, University of California, USA

Abstract

Qualitative assessment of model performance is essential because reliable statistical comparison of observed data with simulated behaviour of a model reflects the performance and consistency of the mathematical tool under defined conditions. In this study we compared the measured temporal and spatial distribution of water content, soil solution salinity (EC_{sw}), and nitrate (NO_3-N) concentration in the soil beneath a drip-fertigated mandarin tree during a complete season with corresponding HYDRUS-2D simulated values using a range of standard statistical techniques, comprising mean error (*ME*), mean absolute error (*MAE*), root mean square error (*RMSE*), paired *t*-test (t_{cal}), coefficient of determination (R^2), Nash and Sutcliffe model efficiency (*E*), index of agreement (*IA*), relative model efficiency (E_{rel}), relative index of agreement (IA_{rel}), modified *E* (E_1) and *IA* (IA_1).

Temporal and spatial values of *ME*, *MAE*, and *RMSE* for water content (-0.04 to 0.05 cm^3.cm^{-3}) and salinity (-0.42-0.93 dSm^{-1}) were within an acceptable range. However, a relatively wider range in *MAE* (1.44-27.65 mg.L^{-1}) and *RMSE* (2.00-39.57 mg.L^{-1}) values were obtained for NO_3-N concentrations measured weekly or at the 25-cm depth (*MAE* = 21.2 and *RMSE* = 30.7 mg.L^{-1}). Temporal and spatial *RMSE* were higher than *MAE*, which suggests a slight bias in *RMSE* due to squared differences between measured and simulated values. Similarly, the paired *t*-test (t_{cal}) showed significant differences for NO_3-N during the mid-season (85-140 DOY) for temporal (weekly) comparison and at several depths for water content (10, 25, 80 and 110 cm), salinity (100 and 150 cm) and NO_3-N concentration (25, 100 and 150 cm).

The R^2 values varied in a narrow range (0.5 to 0.59). Similarly, values for *E* (0.12-0.43), *IA* (0.80-0.84), and E_1 (0.26-0.32) and IA_1 (0.61-0.69) suggest that the model precisely predicted water content, salinity and nitrate concentration over the season, however, E_{rel} (-319.25) and IA_{rel} (-71.3) values were highly negative for nitrate concentration, indicating a mismatch. It was concluded that none of the evaluated measures described and tested the performance of the model for water, salinity and nitrate ideally. Each criterion had its specific advantages and disadvantages, which should be taken into account. Hence, sound model performance evaluation requires the use of a combination of different statistical criteria, which consider both absolute and relative errors. Judicious use of statistical criteria should lead to improvements in the modelling assessment of water, salinity and nitrate dynamics in soil under cropped conditions.

Keywords: HYDRUS-2D; Water content; Salinity; Nitrate; Statistical errors; Model efficiency

Introduction

The advent of high speed computers, which enable enhanced modelling capabilities and rapid development in mathematical software, has transformed the mathematical evaluation of natural processes. Models are extensively used in almost every field of science for problem solving and decision making, and the vadose zone of agricultural soils is not spared from this revolutionary change. Due to the advancement of micro-irrigation systems such as sprinkler and/or surface/subsurface drip irrigation, which has transformed irrigation and fertilizer practices, there is an increasing interest in evaluating and optimizing these high frequency systems for water and fertilizer use efficiency [1-3].

Development of mathematical tools has contributed towards improving irrigation system design and installation, and the monitoring of water and solute movement through the soil from point source applications. Additionally, models can save time and money because of their ability to perform long term simulations evaluating the effects of root zone processes and management activities on water quality, water quantity, and soil quality. Recently, a number of numerical codes have been developed or improved, such as SWAP, FEHM, HYDROBIOGEOCHEM, RZWQM, TOUGH2, APSIM, HydroGeoSphere, and HYDRUS, to cater to the needs of soil physicists/irrigation experts for evaluating irrigation systems and vadose zone

processes for water movement, climate variability, water quality, and solute transport in the soil [4-11]. Especially, the HYDRUS model has been used extensively for evaluating the effects of soil hydraulic properties, soil layering, dripper discharge rates, irrigation frequency, water quality, and timing of nutrient applications on wetting patterns and solute distribution [1-3,12-22]. The most significant aspect of these studies, which determines the utility of these models, is the evaluation of modelling outputs against field observed values. The model should generate data which closely mirror field observations, so that reliable conclusions can be drawn about real world processes.

There are a number of reasons why there is a need to evaluate model performance: (1) to provide a quantitative estimate of the model's ability to reproduce historic and future behaviour of agricultural/environmental systems; (2) to provide a means for evaluating

*Corresponding author: Phogat V, South Australian Research and Development Institute, Australia, E-mail: vinod.phogat@sa.gov.au

improvements to the modelling approach through adjustment of model parameter values, model structural modifications, the inclusion of additional observational information, and representation of important spatial and temporal characteristics of the domain; (3) to compare current modelling efforts with previous studies [23]. Field calibration and validation of the model requires conducting tests based on statistical measures, and is the most important aspect of testing the goodness of fit of values generated by the model. This process of assessing the performance of a model requires evaluation of the closeness of the simulated behaviour of the model to field measurements made within the domain.

Exhaustive evaluations and objective analyses have been carried out for models used in various fields of hydrological and hydraulic modelling [23-30]. Accepting the wide recognition and utility of HYDRUS for modelling water and solute movement under irrigation applications, there is a need to assess the performance of HYDRUS for potential sources of deviation using appropriate and simple indicators. Most field evaluation studies using HYDRUS either present graphical comparisons or subjective assessments [15,31-33], and generally considered only limited error and correlation estimates, i.e., an objective assessment [16,18,20,34] of the performance of the model to evaluate water, salt and nitrate movement in soils. These criteria may place emphasis only on a particular behaviour of the model, and may not be able to assess the overall efficacy of the model on a long term basis. Hence, there is a need to evaluate the performance of HYDRUS more vigorously, utilizing different error analyses, test of significance, regression analyses, and efficiency testing to clearly assess the model's sustained performance. It is important to compare the suitability and relative importance of each of these techniques for evaluating modelling predictions of water and solute transport under high efficiency irrigation systems.

In the present investigation, the performance of HYDRUS-2D in simulating water movement, soil solution salinity, and nitrate movement under a mandarin tree during one season was assessed, using eleven statistical measures: mean error (ME), mean absolute error (MAE), root mean square error ($RMSE$), paired t-test (t_{cal}), coefficient of determination (R^2), model efficiency (E), index of agreement (IA), relative model efficiency (E_{rel}), relative index of agreement (IA_{rel}), modified E (E_1) and modified IA (IA_1)). The aim of this comparison was to identify which subset of the statistical measures is most appropriate for evaluating model performance.

Materials and Methods

The statistical tests were employed on the measured and simulated data generated from the field experiment on mandarin and modelling simulations illustrated in our earlier paper Phogat et al. [3]. However, a brief description of experimental details and modelling technique is presented here.

Experimental detail

Modelling evaluation was performed on field experimental data collected at Dareton Agricultural and Advisory Station (34.10oS and 142.04oE), located in the Coomealla Irrigation Area, New South Wales, Australia for one season during 2006-2007. The field experiment involved surface drip irrigation of mandarin, established in October 2005. The trees were planted at a spacing of 5 m x 2 m between rows and plants, respectively. The trees were managed and fertilized following current commercial practices. The total yearly rainfall during the experimental period was 187 mm, which was significantly less than annual potential evapotranspiration (1400 mm).

Irrigation water was supplied through a surface drip system, with drip lines placed at a distance of 60 cm on both sides of a tree line. The laterals had 1.6 L.h^{-1} pressure compensating drippers spaced at an interval of 40 cm. Irrigation was performed weekly, and the total seasonal irrigation was 432.8 mm. The salinity of the irrigation water (EC_w) was monitored daily, and ranged between 0.09 and 0.19 dS.m^{-1}, well below the EC_w threshold for irrigation of orange, a close relative of mandarin (1.1 dS.m^{-1}). Daily water content measurements were performed using Sentek® EnviroSCAN® capacitance soil water sensors, and soil water was sampled on a weekly basis using SoluSAMPLERs™ [35]. The extracted soil solution was analysed to determine soil solution salinity (EC_{sw}) and nitrate-nitrogen (NO_3^--N) content.

Modelling technique

The HYDRUS-2D software package was used to simulate the transient two-dimensional movement of water and solutes in the soil [11]. Refer to the HYDRUS technical manual for a detailed description of the governing equations describing variably-saturated flow using the Richards' equation, solute transport using the advection–dispersion equation, and root water uptake, as well as various initial and boundary conditions that can be implemented. In this approach, the drip tubing was considered as a line source, because in this twin line drip irrigation system the wetted patterns from adjacent drippers merge to form a continuous wetted strip along both sides of the tree [36,37]. Modelled observation nodes corresponded to the locations where EnviroSCAN probes (at depths of 10, 25, 50, 80, and 110 cm) and SoluSAMPLERs (at depths of 25, 50, 100, and 150 cm) were installed.

Soil hydraulic properties were described using the van Genuchten-Mualem constitutive relationships [38]. The spatial root distribution is defined in HYDRUS-2D according to Vrugt et al. [39]. We considered a simple root distribution model, in which the roots of mandarin trees expanded horizontally into all available space between the tree lines (x_m = 200 cm), were concentrated mainly below the drip emitter (x^* = 60 cm, z^* = 20 cm) where water and nutrients were applied, and extended to a depth of 60 cm (z_m = 60 cm).

Reduction of root water uptake due to water stress was described using the piecewise linear relation developed by Feddes et al. [40]. The following parameters in the Feddes et al. model were used: h_1 = -10, h_2 = -25, h_3 = -200 to -1000, h_4 = -8000 cm, which were taken from Taylor and Ashcroft for orange. Reduction of root water uptake due to salinity stress, $\alpha_2(h_\phi)$, was described by adopting the Maas and Hoffmann salinity threshold and slope function [40,41]. The salinity threshold (EC_T) for orange (closely related to mandarin) corresponds to a value of the electrical conductivity of the saturation extract (EC_e) of 1.7 dS.m^{-1}, and a slope (s) of 16%.

The longitudinal dispersivity (ε_L) was considered to be 20 cm, and the transverse dispersivity (ε_T) was taken as one-tenth of ε_L, optimised in similar studies involving solute transport in soils [12,20]. Since NH_4NO_3 and mono-ammonium phosphate were the fertilizers used in our study, nitrification of NH_4^+-N to NO_3^--N was assumed to be the main N process occurring in the soil. HYDRUS-2D incorporates this process by means of a sequential first-order decay chain.

A time-variable flux boundary condition was applied to a 20 cm long boundary directly below the dripper, centred on 60 cm from the top left corner of the soil domain. During irrigation, the drip line boundary was held at a constant water flux, q. The atmospheric boundary condition was assumed for the remainder of the soil surface during periods of irrigation, and for the entire soil surface during periods between irrigation. A no-flow boundary condition was established at the left and right edges of the soil profile, to account for flow and transport symmetry. A free drainage boundary condition was assumed at the bottom of the soil profile. Initial conditions for water, salinity and nitrate simulations were based on measured data which are described in details in Phogat et al. [3].

HYDRUS-2D requires daily values of potential evaporation (Es) and transpiration (T_p), which were obtained using the dual crop coefficient approach and local meteorological data [42,43].

Statistical indicators

Error estimates: The model's performance was evaluated by comparing measured (M) and HYDRUS-2D simulated (S) values of water content, electrical conductivity of the soil solution (EC_{sw}), and nitrate concentration (NO_3^{-}-N) in the soil, and calculating a range of error estimates, tests of significance, regression analyses, and dimensionless efficiency tests. The error estimates included mean error (ME), mean absolute error (MAE), and root mean square error ($RMSE$), given by:

$$ME = \frac{1}{N}\sum_{i=1}^{N}\left(M_i - S_i\right) \tag{1}$$

$$MAE = \frac{1}{N}\sum_{i=1}^{N}\left|M_i - S_i\right| \tag{2}$$

$$RMSE = \sqrt{\frac{1}{N}\sum_{i=1}^{N}\left(M_i - S_i\right)^2} \tag{3}$$

The test of significance was conducted using the paired t-test (t_{cal}) and given as:

$$t_{cal} = \frac{d}{S_m\sqrt{\frac{1}{n_1}+\frac{1}{n_2}}}; d = \overline{x_1} - \overline{x_2} \text{ and } S_m = \sqrt{\frac{n_1 s_1^2 + n_2 s_2^2}{n_1 + n_2 - 2}} \tag{4}$$

Here, n and s are the number of comparable paired points and standard deviation respectively; subscripts 1 and 2 are indicative respectively of measured and predicted values; S_m is the standard deviation of the mean, and t_{cal} is the calculated paired t-test value.

Efficiency criteria: The coefficient of determination (R^2) was applied for testing the proportion of variance in the measured data explained by the model, and is defined as the square of the coefficient of correlation (r) according to Bravais-Pearson, calculated as:

$$R^2 = \left(\frac{\sum_{i=1}^{N}\left(M_i - \overline{M}\right)\left(S_i - \overline{S}\right)}{\sqrt{\sum_{i=1}^{N}\left(M_i - \overline{M}\right)^2}\sqrt{\sum_{i=1}^{N}\left(S_i - \overline{S}\right)^2}}\right)^2 \tag{5}$$

Values of R^2 can vary between 0 and 1, with higher values indicating less variance, and values greater than 0.5 typically considered acceptable [24].

Efficiency measures for the evaluation of model performance investigated in this study were: model efficiency (E), index of agreement (IA), relative model efficiency (E_{rel}), and relative index of agreement (IA_{rel}).

Model efficiency (E), as proposed by Nash and Sutcliffe [44], is defined as one minus the sum of absolute squared differences between

simulated and measured values, normalized by the variance of measured values during the period under investigation:

$$E = 1 - \frac{\sum_{i=1}^{N}\left(M_i - S_i\right)^2}{\sum_{i=1}^{N}\left(M_i - \overline{M}\right)^2} \tag{6}$$

The range of E lies between $-\infty$ and 1.0 (perfect fit). An efficiency value between 0 and 1 is generally viewed as an acceptable level of performance. Efficiency lower than zero indicates that the mean value of the observed time series would be a better predictor than the model, and denotes unacceptable performance [45].

The index of agreement (IA) was proposed by Willmot (1981), and represents the ratio of the mean square error to the potential error:

$$IA = 1 - \frac{\sum_{i=1}^{N}\left(M_i - S_i\right)^2}{\sum_{i=1}^{N}\left(\left|S_i - \overline{M}\right| + \left|M_i - \overline{M}\right|\right)^2} \tag{7}$$

The value of IA varies between 0 and 1. A value of 1 indicates a perfect agreement between measured and simulated values, and 0 signifies no agreement at all.

Relative efficiency criteria: Various criteria described above (R^2, E, and IA) quantify the difference between observations and predictions in absolute values. As a result, an over- or under-prediction of larger values has, in general, a greater influence than that of smaller values. To counteract this, efficiency measures based on relative deviations can be derived from E and IA as:

$$E_{rel} = 1 - \frac{\sum_{i=1}^{N}\left(\frac{M_i - S_i}{M_i}\right)^2}{\sum_{i=1}^{N}\left(\frac{M_i - \overline{M}}{\overline{M}}\right)^2} \tag{8}$$

$$IA_{rel} = 1 - \frac{\sum_{i=1}^{N}\left(\frac{M_i - S_i}{\overline{M}}\right)^2}{\sum_{i=1}^{N}\left(\frac{\left|S_i - \overline{M}\right| + \left|M_i - \overline{M}\right|}{\overline{M}}\right)^2} \tag{9}$$

where, E_{rel} and IA_{rel} represent the relative efficiency and a relative index of agreement, respectively. These parameters can also range between the values described for E and IA, respectively.

Modified form of E and IA: The modified form of E and IA are extensively used to overcome the problem of squared differences and oversensitivity to extreme values induced by the mean squared error in E and IA as given below:

$$E_j = 1 - \frac{\sum_{i=1}^{N}\left(\left|Mi - Si\right|\right)^j}{\sum_{i=1}^{N}\left(\left|Mi - \overline{M}\right|\right)^j}, j \in N \tag{10}$$

$$IA_j = 1 - \frac{\sum_{i=1}^{N}\left(\left|Mi - Si\right|\right)^j}{\sum_{i=1}^{N}\left(\left|Si - \overline{M}\right|\right) + \left(\left|Mi - \overline{M}\right|\right)^j} \quad , j \in N \qquad (11)$$

where j represents an arbitrary power i.e. a positive integer (N). Especially when j=1, the errors and differences are given their appropriate weighting, not inflated by their squared values. Hence E_1 and IA_1 represent modified form of efficiency and index of agreement. Squaring in statistics (E_2 and IA_2) is useful because squares are easier to manipulate mathematically than are absolute values, but use of squares forces an arbitrarily greater influence on the statistic by way of the larger values [23]. These parameters can also range between the values described for E and IA, respectively.

The error measures (*ME*, *MAE* and *RMSE*) and *t*-test were computed temporally (across all measurement/simulation depths on weekly basis), spatially (across all weekly measurement/simulation for each depth), and across all individual measurements/simulations for the entire dataset. However, regression and efficiency (R^2, E, IA, E_{ref}, IA_{ref}, E_1 and IA_1) measures were only evaluated on the entire dataset.

Results and Discussion

Data sets for error parameter comparison

The total data set of weekly measured (*M*) and HYDRUS-2D simulated (*S*) water content, soil solution salinity (EC_{sw}), and nitrate-nitrogen concentration (NO_3^--N) at different depths and their graphical comparisons are described in Phogat et al. [3]. These data sets provide an ideal basis for comparing the range of error parameters, given that the three data sets (water content, salinity and nitrate concentration) represent a range from good matching between simulated and measured data (water content) to relatively poorly matched (nitrate concentration). This allows comparison of the various error parameters across a range of error values and efficiency testing.

Comparison of error indices

Numerous error estimation methods are in use for comparing simulation results with measured data. These indices are valuable tools because they evaluate the error in the units of the constituent of interest, which helps in the analysis of results and describe the performance and utility of the modelling exercise. Weekly computed temporal error indices (*ME*, *MAE*, and *RMSE*) on measured and simulated water content, soil solution salinity (EC_{sw}) and nitrate-nitrogen content (NO_3^--N) are depicted in box plots in Figure 1 and range of these parameters are shown in Table 1. Spatial values of error indices are shown in Figure 2. Seasonal values of all statistical measures are displayed in Table 2.

Mean error (*ME*) is the signed measure of deviations between measured and simulated values, indicating whether the deviations tend to be positive or negative. The *ME* in temporal data ranged from -0.04 to 0.04 cm³.cm⁻³, -0.42 to 0.54 dS.m⁻¹, and -11.31 to 12.38 mg.L⁻¹ for water contents, EC_{sw}, and NO_3^--N concentrations, respectively in Table 1 and Figure 1. Similarly, spatial *ME*s for water contents, EC_{sw}, and NO_3^--N concentrations ranged from -0.01 to 0.04 cm³.cm⁻³, -0.24 to 0.34 dSm⁻¹, and -21.09 to 4.21 mg.L⁻¹, respectively in Table 1 and Figure 2). Seasonal *ME* values for water content, EC_{sw}, and NO_3^--N for the entire data set were 0.003 cm³.cm⁻³, 0.12 dSm⁻¹, and -6.7 mg.L⁻¹ (Table 2). This comparison revealed that *ME* was smallest for water content and greatest for NO_3^--N, with EC_{sw} in between. Although NO_3^--N errors indicated the widest variability, a low value of *ME* may still conceal simulation inaccuracy due to the offsetting effect of large positive and negative errors, hence the need to also consider *MAE* and *RMSE*.

Comparing *MAE* and *RMSE* in Figure 2 reveals that the magnitude of the inter-quartile range (IQR) was higher in *RMSE* as compared to *MAE*. IQR is the difference between the 25th and 75th percentile, and indicates the magnitude of variation in the mid-range of error values. The magnitude of IQR in *RMSE* was 0.01 cm³.cm⁻³, 0.36 dS.m⁻¹, and 19.05 mg.L⁻¹, as compared to 0.01 cm³.cm⁻³, 0.33 Ds.m⁻¹, and 14.98 mg.L⁻¹ in *MAE*, for water content, EC_{sw}, and NO_3^--N, respectively. Similarly, the

Soil Parameter	Scale	Range	ME	MAE	RMSE
Water content (cm³cm⁻³)	Temporal	Max	0.04	0.04	0.05
		Min	-0.04	0.005	0.01
	Spatial	Max	0.04	0.04	0.05
		Min	-0.01	0.02	0.02
Salinity (ECsw) (dSm⁻¹)	Temporal	Max	0.54	0.76	0.93
		Min	-0.42	0.08	0.09
	Spatial	Max	0.34	0.47	0.56
		Min	-0.24	0.19	0.25
Nitrate (NO3-N) (mg L⁻¹)	Temporal	Max	12.31	27.65	39.57
		Min	-11.38	1.44	2.00
	Spatial	Max	4.21	21.24	30.75
		Min	-21.09	5.06	5.89

Table 1: Maximum and minimum values of *ME*, *MAE*, and *RMSE* obtained from spatial and temporal data sets of water content, EC_{sw} and NO_3^--N

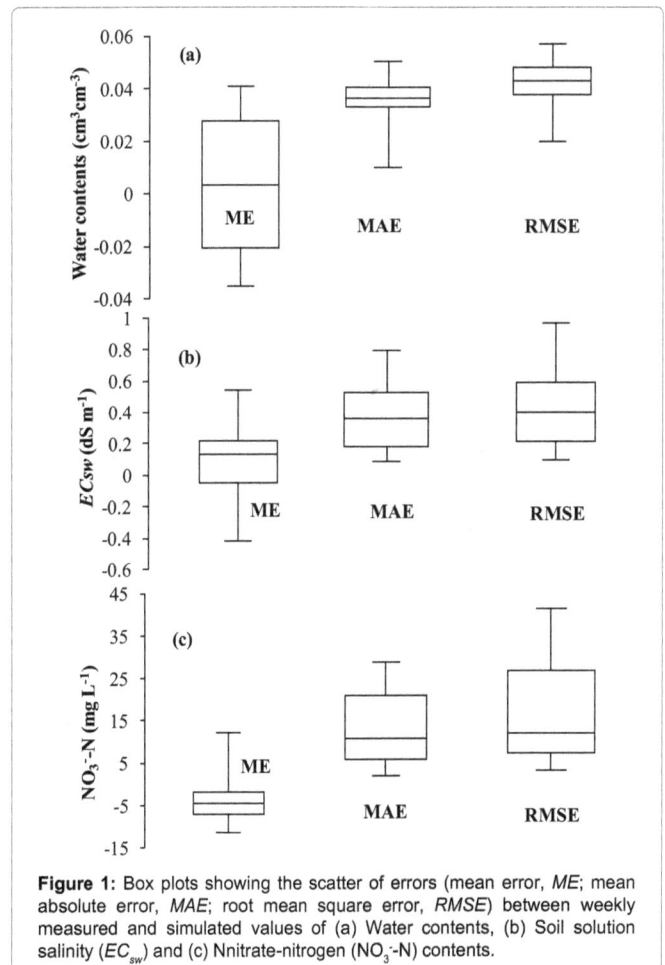

Figure 1: Box plots showing the scatter of errors (mean error, *ME*; mean absolute error, *MAE*; root mean square error, *RMSE*) between weekly measured and simulated values of (a) Water contents, (b) Soil solution salinity (EC_{sw}) and (c) Nnitrate-nitrogen (NO_3^--N) contents.

Statistical Measures	Moisture content (cm³cm⁻³)	Salinity (EC_{sw}) (dSm⁻¹)	Nitrate (NO_3^--N) (mg L⁻¹)
N#	240	188	192
ME	0.003	0.12	-6.70
MAE	0.03	0.34	24.49
RMSE	0.03	0.46	26.76
hSD_m	0.023	0.28	7.35
t_{cal}	0.84	1.79	3.42*
R^2	0.50	0.59	0.56
E	0.43	0.30	0.12
E_{rel}	0.36	0.49	-319.25
E_1 (j=1)	0.26	0.32	0.28
IA	0.84	0.85	0.80
IA_{rel}	0.82	0.80	-71.3
IA_1 (j=1)	0.61	0.69	0.68

#Number of measured and simulated data pairs; *Significant at 5% level

Table 2: Values of *ME, MAE,* and *RMSE,* half of the standard deviation of measured data (hSD_m), and a paired *t*-test (t_{cal}), coefficient of determination (R^2), model efficiency (*E*), index of agreement (*IA*), relative model efficiency (E_{rel}), relative index of agreement (IA_{rel}), modified *E* (E_1), and modified *IA* (IA_1) estimated for the entire (seasonal) dataset

spatial *RMSE* values were also higher than the *MAE* values at all depths shown in Figure 2, but their magnitude was much wider at shallow depths (10-25 cm) where *RMSE* values were 0.01 cm³.cm⁻³, 0.2 dS.m⁻¹, and 9.49 mg.L⁻¹ higher than *MAE* values for water content, EC_{sw}, and NO_3^--N, respectively. However, higher variations in all types of errors at the surface depth (10-25 cm) reflect the assumption in the model of a constant atmospheric boundary flux during daily time steps, which deviates from actual conditions at the surface boundary, particularly the diurnal fluctuation in evaporation, which peaks in day time and decreases during the night [2].

Comparison of *MAE* and *RMSE* further indicated that as the magnitude of variations between measured and predicted values increased, *RMSE* increased disproportionately, as is evident from the NO_3^--N values. Similar trends for NO_3^--N were also obtained in the *MAE* and *RMSE* analysis of the whole data set, where the values of these parameters were 24.49 mg.L⁻¹ and 26.76 mg.L⁻¹, respectively in Table 2. *RMSE* was always larger than *MAE*, and varied with the variability of the error magnitude, because the errors are squared in *RMSE* before they are averaged. *RMSE* varies with the variability within the distribution of error magnitudes and with the square root of the number of errors ($n^{1/2}$), as well as with the average-error magnitude (as *MAE*) [26]. Hence, *RMSE* gives a relatively high weight to large errors, as obtained in the case of NO_3^--N. On the other hand, *MAE* is a linear measure, which means that all individual differences are weighted equally in the average. Hence, *MAE* may be preferred over *RMSE* as a more natural measure of the average error, and for an unambiguous assessment of model predictions. Legates and McCabe expressed similar views as *RMSE* produces inflated values when large outliers are present [23].

The use of these two measures (*RMSE* and *MAE*) suffers from a significant drawback, in that they do not indicate the direction of the error. However, this discrepancy may be ignored where the main focus of the comparison is the magnitude of the error rather than its direction.

There is no universally accepted threshold limit for error magnitude when judging the degree of accuracy of model performance. However, Singh et al. stated that *RMSE* and *MAE* values smaller than half of the standard deviation of the measured data (hSD_m) may be considered

low and appropriate for model evaluation [46]. Therefore, *RMSE, MAE,* and hSD_m values obtained using temporal water content data are compared in Figure 3. It can be seen that these errors (*RMSE* and *MAE*) were higher than hSD_m except on a few occasions during mid season (DOY 36 to 64) and during the terminal period (DOY 194 onward). Similarly, *MAE* and *RMSE* values were higher than hSD_m in all analyses of the complete data set shown in Table 2. Hence, model performance was relatively poor in view of this criterion.

In the context of such definitive measures of model performance, it is important to consider the natural variability inherent in the measured data against which the simulations are judged [47]. Real world variability of the natural environment, such as soil variations, as well as measurement inaccuracies can cause measured data to vary relative to the best simulation. For example, in our study EnviroSCAN® sensors were used to measure the profile water content. The probes were properly calibrated during installation; however, measurements with capacitance probes are highly variable and sensitive to bulk electrical conductivity, temperature, and change in storage estimates [48]. The capacitance sensors used in access tubes may generate consistent errors ≤ 0.05 cm³ cm⁻³, which is similar to the variation observed in our study between EnvironSCAN measured and simulated values.

The error associated with simulated data and outliers in observed data can be further minimised by optimizing the input data by complex weighing techniques and probability distribution based uncertainty analysis techniques like Monte Carlo simulation and Bayesian analysis framework which deals with both random and systemic errors in the simulations [30,49].

Test of significance

The paired *t*-test was used to evaluate the level of significance between measured and simulated data on water content, EC_{sw}, and NO_3^--N content which is shown in Figure 4. It showed non-significant differences (*p* = 0.05) between mean values of temporal measured and simulated water content. However, positive t_{cal} values during the early period showed that the measured values were higher than the corresponding simulated values, and the opposite was true later in the season shown in Figure 4a. Similarly, insignificant differences were observed for soil salinity, except at DOY 36 and 43, where differences between measured and simulated mean EC_{sw} were significant. However, significant differences were observed in NO_3^--N content from DOY 78 to 134, which corresponded to a period from March 2007 to early May 2007.

However, *t*-test showed significant differences in the spatial data set at depths of 10, 25, 80, and 110 cm for water content, 25, 100, and 150 cm for NO_3^--N content and 100 and 150 cm for EC_{sw} (data not shown here). These revelations conform to visual observation of the dataset [3]. Additionally, the t_{cal} values for water content and EC_{sw} for the whole season (Table 1) were non-significant at *p* = 0.05, whereas the t_{cal} value for NO_3^--N showed significant difference, indicating a relatively poor performance of the model for nitrate simulation. However, *t*-test represents variation between the mean values of measured and simulated data, and therefore a single seasonal figure may not reflect the degree of spatial and temporal heterogeneity across the season. Moreover, *t*-test assumes that the measured and simulated values are normally distributed, and that both groups have equal variance. These assumptions may not be perfectly satisfied, and this calls into question the reliability of this statistical measure.

It is also important to understand that hypothesis driven tests, such as paired *t*-test, should not be relied on solely to measure reliability of

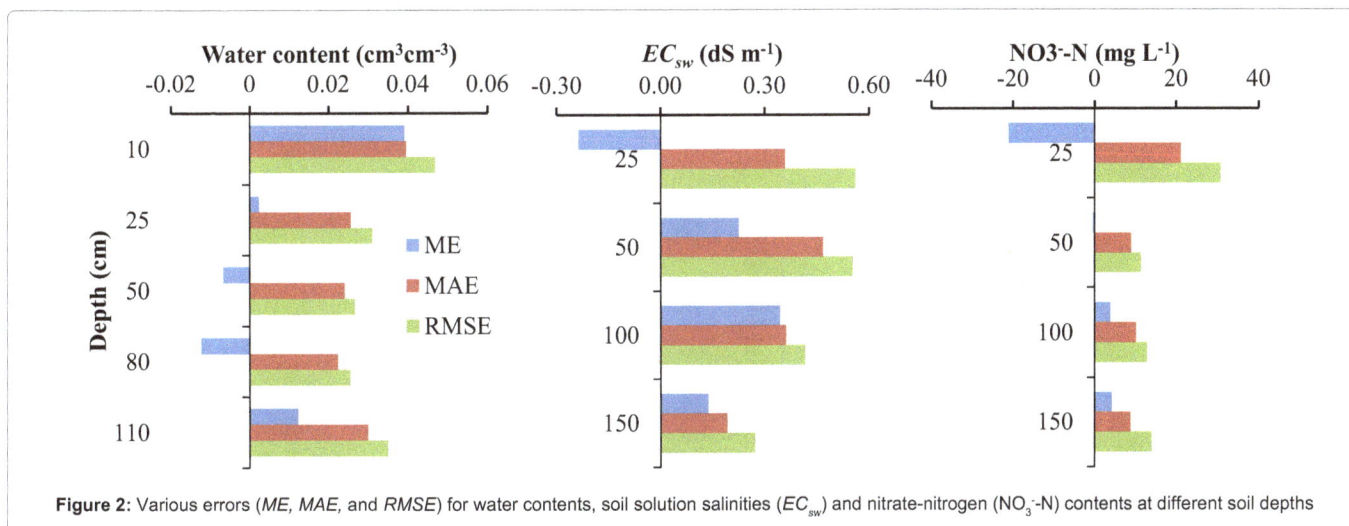

Figure 2: Various errors (*ME*, *MAE*, and *RMSE*) for water contents, soil solution salinities (EC_{sw}) and nitrate-nitrogen (NO_3^--N) contents at different soil depths

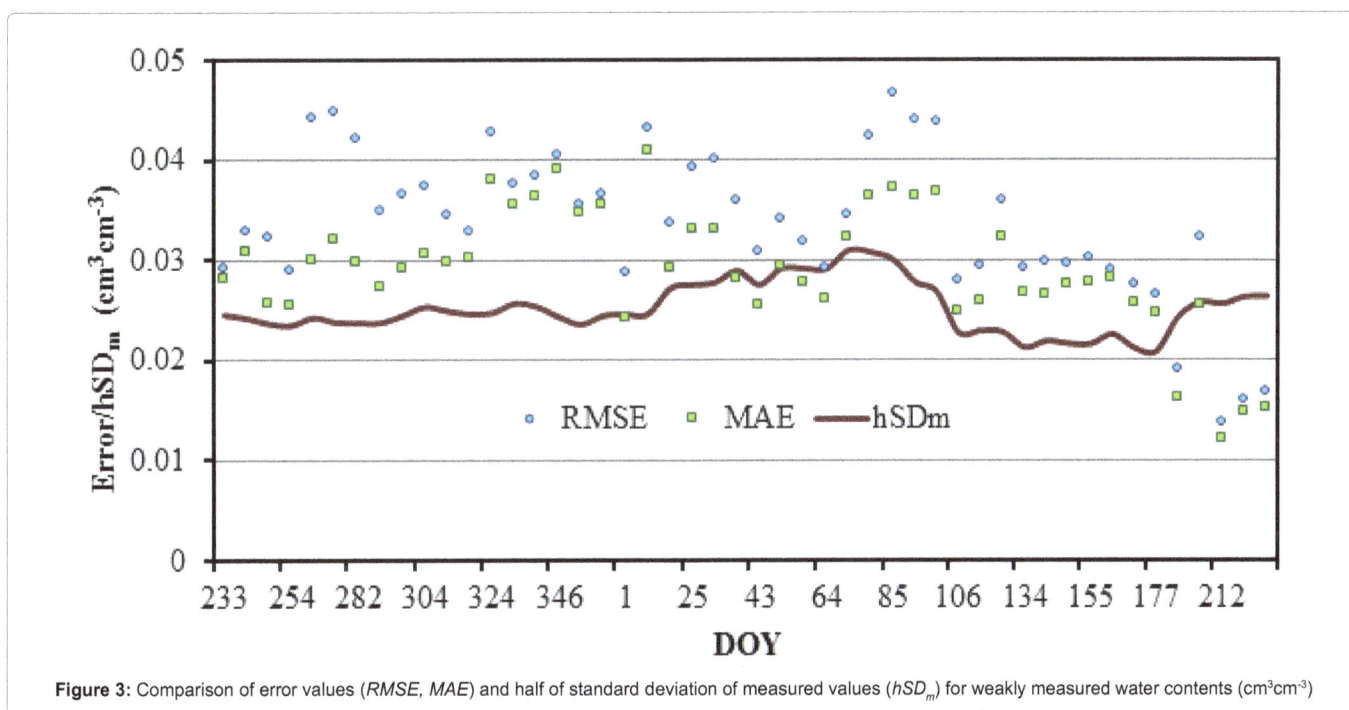

Figure 3: Comparison of error values (*RMSE*, *MAE*) and half of standard deviation of measured values (hSD_m) for weakly measured water contents (cm^3cm^{-3})

simulations, as the degree of random variation determines the detection of a significant difference; significant systematic bias will be less likely to be detected if it is accompanied by large random errors [50,51].

Regression analysis and efficiency testing

Model performance was also assessed using regression analysis [coefficient of determination (R^2)] and various efficiency testing indices [model efficiency (*E*), index of agreement (*IA*), relative model efficiency (E_{rel}), relative index of agreement (IA_{rel}), modified form of *E* (E_1), and modified form of *IA* (IA_1)]. These measures were computed using the complete data set of measured and corresponding simulated values of water content (240 values), soil solution salinity (EC_{sw}; 188 values), and nitrate nitrogen content (NO_3^--N; 192 values) at all spatial and temporal positions (Table 2).

The R^2 value of 0.5 for water content was just at the margin of the satisfactory level [24]. However, its values for EC_{sw} (0.59) and the NO_3^-

-N content (0.56) were within the acceptable limit. Hence, R^2 produced relatively similar results across all simulated processes (water content, EC_{sw}, and NO_3^--N) (Table 2). This is contrary to our results using error tests, where the variability in water content was comparatively smaller than in NO_3^--N content (Table 2). This reveals a serious drawback in considering R^2 values alone for model performance evaluation, in that it only quantifies dispersion among values. Krause et al. reported that a model which systematically over- or under-predicts at all times will still result in R^2 values close to 1.0, even if all predictions are wrong, which undermines the reliability of R^2 values [25]. Similarly, Legates and McCabe suggested that correlation based measures are inappropriate and should not be used to evaluate the goodness-of-fit of model simulations, as these measures are oversensitive to extreme values and are insensitive to additive and proportional differences between model predictions and observations [23]. Hence, consideration of R^2 alone for model performance assessment sometimes leads to a flawed acceptance of modelling results.

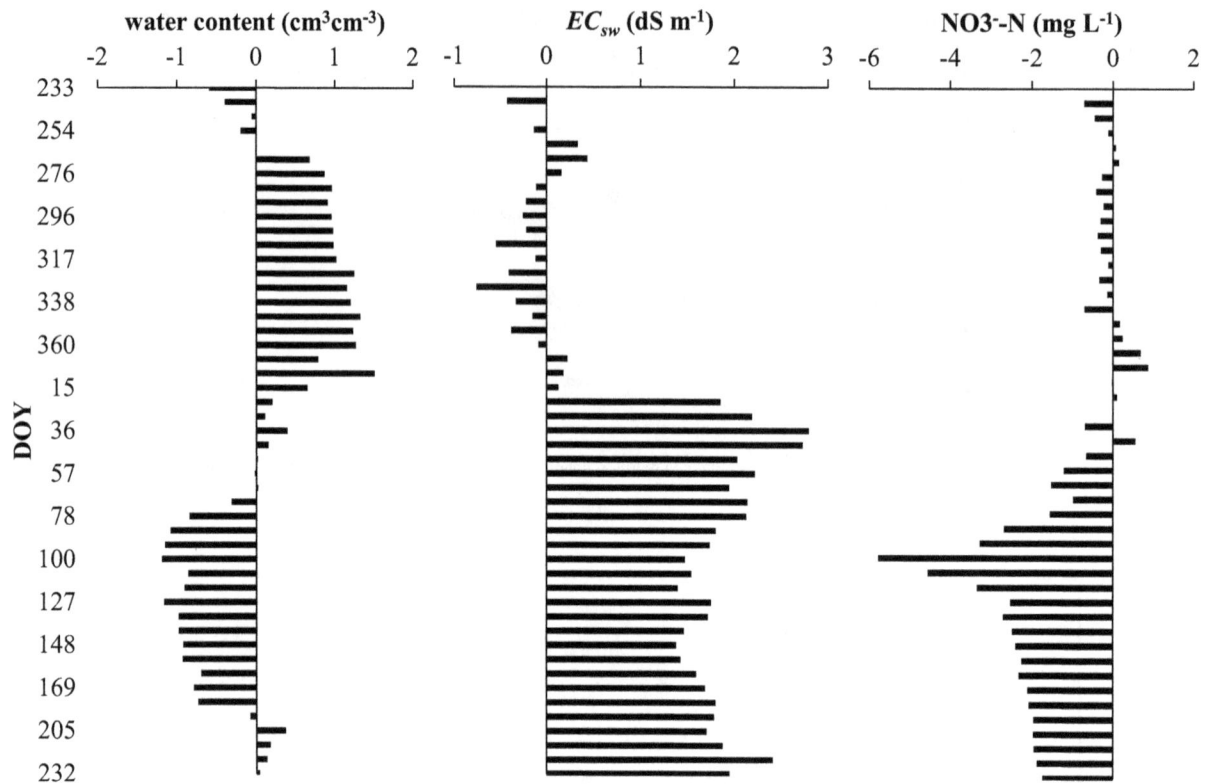

Figure 4: Temporal paired t-test (t_{cal}) values between measured and simulated a) Water content, b) Soil solution salinity (EC_{sw}) and c) Nitrate-nitrogen (NO_3^--N) content

Six efficiency tests were applied to the data set to evaluate their relative performance. The E, E_{rel}, IA_{rel}, E_l and IA_l values for water content were 0.43, 0.84, 0.82, 0.20 and 0.59 respectively (Table 2), suggesting a good match between measured and simulated data, as indicated by previous parameters. On the contrary, IA (0.36) showed relatively poor efficiency of the model for simulating water content distribution. However, values of modified efficiency(E_l) and index of agreement (IA_l) were lower than relative estimates because these statistics utilize absolute values rather than squared differences in their computation which makes them more conservative measures [23]. The E value (0.43) was within the satisfactory limit reported in other studies [24,45]. Similarly, the E, IA, E_{rel}, IA_{rel}, E_l and IA_l values for weekly soil solution salinity (EC_{sw}) data were 0.30, 0.49, 0.85, 0.80, 0.32 and 0.69, respectively, which are well within acceptable limits, and match previous indicators relatively well. Relative efficiency (E_{rel}) for water and relative index of agreement (IA_{rel}) for water and salinity were relatively close to their precursors, E and IA, respectively (Table 2).

Conversely, the nitrate simulation provides a more complicated picture. The Nash and Sutcliffe efficiency ($E = 0.12$) and index of agreement ($IA = 0.80$) values are within acceptable limits, and in fact IA is quite high, in contrast to previous error parameters for nitrate concentration. The modified estimates (E_l and IA_l) fall within acceptable range. However, large negative values of E_{rel} (-319.25) and IA_{rel} (-71.3) reflect the wide divergence between measured and simulated values at certain times during the simulation. However, relative deviations reduce the influence of absolute differences among the measures and simulated values.

It is significant that Nash-Sutcliffe efficiency (E), the most frequently used indicator in hydrologic studies, is much more sensitive

to errors in higher values, as the differences between measured and simulated values are squared. As a result larger values in a data series have a much higher weighting, whereas lower values are neglected [23]. This comparison suggests that E and/or IA may not always be suitable parameters for describing model performance. Additionally, the large negative values of E_{rel} and IA_{rel} for nitrate showed disproportionately high under-prediction, as reported in Krause et al. [25]. Hence, these parameters proved to be sensitive only to large variations in values and not at all to small divergences because, due to the summation of the absolute or squared errors in efficiency testing methods, emphasis is placed on larger errors while smaller errors tend to be neglected. Hence modified efficiency (E_l) and index of agreement (IA_l) could be the more appropriate measures for model's efficiency testing than their precursors (E and IA) and relative statistics (E_{rel} and IA_{rel}). Nevertheless IA_l has advantages due to its bounds between 0.0 and 1.0 [23]. But good modelling efficiency shown by E_l and IA_l statistics is contradictory to the poor modelling simulation for NO_3^--N revealed in error estimates. Additionally, tests of significance and efficiency measures, similar to t-test, evaluate the mean variability in the domain and are unable to capture the modelling divergence at a particular point.

Overall, it can be stated that none of the efficiency parameters which were evaluated in this study adequately described and tested the reliability of model predictions. Each has specific pros and cons, which have to be taken into account during model calibration and validation. Hence, for a sound model performance evaluation, a combination of different statistical efficiency criteria, complemented by the assessment of the absolute or relative error, may need to be included.

Conclusion

Simulation models have been increasingly used in high efficiency drip irrigation systems to evaluate the water and solute dynamics under cropped conditions, and suggest necessary management options to optimise the system efficiency. In this study, eleven statistical measures were used to compare HYDRUS-2D simulated values of water content, soil solution salinity (EC_{sw}), and nitrate-nitrogen (NO_3^--N) dynamics with field measured values obtained under drip irrigated mandarin crop over a season. as the statistical parameters compared were mean error (ME), mean absolute error (MAE), root mean square error ($RMSE$), paired t-test (t_{cal}), coefficient of determination (R^2), model efficiency (E), index of agreement (IA), relative model efficiency (E_{rel}), relative index of agreement (IA_{rel}), modified E (E_l), and modified IA (IA_l). The purpose of applying all of these parameters to the same data sets was to evaluate the relative importance of these parameters in model performance testing.

The error parameters (ME, MAE and $RMSE$) remained within acceptable limits when applied to measured and simulated values of water content and EC_{sw}, whilst a wider range of values of MAE (1.44 to 27.65 mg.$^{L-1}$) and $RMSE$ (2.00 to 39.57 mg.$^{L-1}$) obtained for nitrate (NO_3^--N) indicated poor agreement between simulated and measured values for this data set. Low ME values may conceal simulation inaccuracy due to the offsetting effect of large positive and negative errors. The results revealed that $RMSE$ values were consistently higher than MAE due to squaring of the difference between measured and simulated values. Hence, it was concluded that among the error tests, MAE may be preferred over ME and $RMSE$ for evaluating goodness-of-fit of the simulated values.

Paired t-test values revealed a non-significant difference ($p = 0.05$) between weekly measured and predicted water content and EC_{sw} distributions. However, differences were significant for the NO_3^--N distribution during the mid-season and at several spatial depths. Similarly, regression analysis (R^2) and efficiency testing methods (E, IA, E_{rel}, IA_{rel}, E_l and IA_l) also indicated that the model accurately predicted seasonal changes in water and salinity distributions in the soil. However, negative values of E_{rel} (-319.25) and IA_{rel} (-71.3) for NO_3^--N reflected the relatively poor prediction of NO_3^--N dynamics in the soil. However, relative deviations reduce the influence of absolute differences among the measures and simulated values.

It was concluded that, for reliable model performance evaluation, a combination of different statistical efficiency criteria, along with the assessment of the absolute or relative volume error, must be included. Taken together, these comparisons were able to provide an objective assessment of the closeness of the simulated behaviour to the observed measurements of water, salinity and nitrate distribution in the soil under mandarin. It is expected that such studies would help in improving the performance evaluation and reliability of modelling data on irrigation and fertigation programme of horticultural crops, and contribute to improving system efficiency, and reducing environmentally harmful agro-hydrological practices.

Acknowledgement

Dareton Agricultural Research and Advisory Station farm staff for managing the site and plantings; and financial support of the National Program for Sustainable Irrigation is gratefully acknowledged.

References

1. Hanson BR, Šimůnek J, Hopmans JW (2006) Evaluation of urea–ammonium–nitrate fertigation with drip irrigation using numerical modeling. Agric Water Manage 86: 102-113.

2. Ramos TB, Šimůnek J, Goncalves MC, Martins JC, Prazeres A, et al. (2012) Two-dimensional modeling of water and nitrogen fate from sweet sorghum irrigated with fresh and blended saline waters. Agric Water Manage 111: 87-104.

3. Phogat V, Skewes MA, Cox JW, Sanderson G, Alam J et al. (2014) Seasonal simulation of water, salinity and nitrate dynamics under drip irrigated mandarin (Citrus reticulata) and assessing management options for drainage and nitrate leaching. J Hydrol 513: 504-516.

4. Van Dam JC, Huygen J, Wesseling JG, Feddes RA, Kabat P, et al. (1997) Theory of SWAP version 2.0. Simulation of water low, solute transport and plant growth in the soil-water-atmosphere-plant environment. WAGENINGEN UR.

5. Zyvoloski GA, Robinson BA, Dash ZV, Trease LL (1997) Use's Manual for the FEHM application- A finite-element heat- and mass-transfer code. Geothermal.

6. Yeh GT, Salvage KM, Gwo JP, Zachara JM, Szecsody JE (1998) HydroBioGeoChem: A coupled model of hydrologic transport and mixed biogeochemical kinetic/equilibrium reactions in saturated-unsaturated media. UNT digital library.

7. Ahuja LR, Hanson JD, Rojas KW, Shaffer MJ (1999) The Root zone water quality model. Water Resources Publications LLC.

8. Pruess K, Oldenburg C, Moridis G (1999) TOUGH2 User's Guide, Version 2.0.

9. Keating BA, Carberry PS, Hammer GL, Probert ME, Robertson MJ, et al. (2003) An overview of APSIM, a model designed for farming systems simulation. Eur J Agron 18: 267-288.

10. Therrien R, McLarren RG, Sudicky EA, Panday SM (2010) HydroGeoSphere: A three-dimensional numerical model describing fully-integrated subsurface and surface flow and solute transport.

11. Šimůnek J, van Genuchten MTh, Šejna M (2008) Development and applications of the HYDRUS and STANMOD software packages and related codes. Vadose Zone J 7: 587-600.

12. Cote CM, Bristow KL, Charlesworth PB, Cook FJ, Thorburn PJ (2003) Analysis of soil wetting and solute transport in subsurface trickle irrigation. Irrig Sci 22: 143-156.

13. Lazarovitch N, Šimůnek J, Shani U (2005) System-dependent boundary condition for water flow from subsurface source. Soil Sci Soc Am J 69: 46-50.

14. Assouline S, Moller M, Cohen S, Ben-Hur M, Grava A, et al. (2006) Soil–plant system response to pulsed drip irrigation and salinity: Bell pepper-Case study. Soil Sci Soc Am J 70: 1556-1568.

15. Ajdary K, Singh DK, Singh AK, Khanna M (2007) Modelling of nitrogen leaching from experimental onion field under drip fertigation. Agric Water Manage 89: 15-28.

16. Patel N, Rajput TBS (2008) Dynamics and modeling of soil water under subsurface drip irrigated onion. Agric Water Manage 95: 1335-1349.

17. Šimůnek J, Hopmans JW (2009) Modeling compensated root water and nutrient uptake. Ecol Model 220: 505-521.

18. Doltra J, Munoz P (2010) Simulation of nitrogen leaching from a fertigated crop rotation in a Mediterranean climate using the EU-Rotate-N and Hydrus-2D models. Agric Water Manage 97: 277-285.

19. Li J, Liu Y (2011) Water and nitrate distributions as affected by layered-textural soil and buried dripline depth under subsurface drip fertigation. Irri Sci 29: 469-478.

20. Phogat V, Mahadevan M, Skewes M, Cox JW (2012) Modeling soil water and salt dynamics under pulsed and continuous surface drip irrigation of almond and implications of system design. Irri Sci 30 : 315-333.

21. Phogat V, Skewes MA, Mahadevan M, Cox JW (2013a) Evaluation of soil plant system response to pulsed drip irrigation of an almond tree under sustained stress conditions. Agric Water Manage 118: 1-11.

22. Phogat V, Skewes M, Cox JW, Alam J, Grigson G, et al. (2013b) Evaluation of water movement and nitrate dynamics in a lysimeter planted with an orange tree. Agric Water Manage 127: 74-84.

23. Legates DR, McCabe GJ (1999) Evaluating the use of goodness-of-fit measures in hydrologic and hydroclimatic model validation. Water Resour Res 35: 233-241.

24. Van Liew MW, Arnold JG, Garbrech JD (2003) Hydrologic simulation on agricultural watersheds: Choosing between two models. Trans ASAE 46: 1539-1551.

25. Krause P, Boyle DP, Base F (2005) Comparison of different efficiency criteria for hydrological model assessment. Adv Geosci 5: 89-97.

26. Willmott CJ, Matsuura K (2005) Advantages of the mean absolute error (MAE) over the root mean square error (RMSE) in assessing average model performance. Climate Res 30: 79-82.

27. Schaefli B, Gupta H (2007) Do Nash values have value? Hydrol Processes 21: 2075-2080.

28. Jain SK, Sudheer KP (2008) Fitting of hydrologic models: A close look at the Nash-Sutcliffe index. J Hydrol Eng 13: 981-986.

29. Gupta HV, Kling H, Yilmaz KK, Martinez GF (2009) Decomposition of the mean squared error and NSE performance criteria: Implications for improving hydrological modelling. J Hydrol 377: 80-91.

30. Pechlivanidis IG, Jackson B, McIntyre N, Wheater HS (2011) Catchment scale hydrological modelling: A review of model types, calibration approaches and uncertainty analysis methods in the context of recent developments in technology and applications. Global NEST J 13: 193-214.

31. Hassan G, Persaud N, Reneau RB (2005) Utility of HYDRUS-2D in modeling profile soil moisture and salinity dynamics under saline water irrigation of soybean. Soil Sci 170: 28-37.

32. Goncalves MC, Šimůnek J, Ramos TB, Martins JC, Neves MJ, et al. (2006) Multicomponent solute transport in soil lysimeters irrigated with waters of different quality. Water Resour Res.

33. Crevoisier D, Popova Z, Mailhol JC, Ruelle P (2008) Assessment and simulation of water and nitrogen transfer under furrow irrigation. Agric Water Manage 95: 354-366.

34. Roberts T, Lazarovitch N, Warrick AW, Thompson TL (2009) Modeling salt accumulation with subsurface drip irrigation using HYDRUS-2D. Soil Sci Soc Am J 73: 233-240.

35. Biswas T, Schrale G (2007) Sentek SoluSAMPLER: A tool for managing salt and nutrient movement in the root zone, Instruction Manual Version 2.0. Sentek Pvt Ltd, Australia.

36. Skaggs TH, Trout TJ, Šimůnek J, Shouse PJ (2004) Comparison of Hydrus2D simulations of drip irrigation with experimental observations. J Irrig Drainage Eng 130: 304-310.

37. Falivene S, Goodwin I, Williams D, Boland A (2005) Introduction to Open Hydroponics. NPSI Fact sheet.

38. Van Genuchten MTh (1980) A closed form equation for predicting the hydraulic conductivity of unsaturated soils. Soil Sci Soc Am J 44: 892-898.

39. Vrugt JA, Hopmans JW, Šimůnek J (2001) Calibration of a two dimensional root water uptake model. Soil Sci Soc Am J 65: 1027-1037.

40. Feddes RAP, Kowalik J, Zaradny H (1978) Simulation of field water use and crop yield.

41. Taylor SA, Ashcroft GM (1972) Physical Edaphology. Freeman and Co, USA.

42. Allen RG, Pereira LS, Raes D, Smith M (1998) Crop evapotranspiration guidelines for computing crop water requirements. FAO Irrigation and Drainage Paper, Italy.

43. Allen RG, Pereira LS (2009) Estimating crop coefficients from fraction of ground cover and height. Irrig Sci 28: 17-34.

44. Nash JE, Sutcliffe JV (1970) River flow forecasting through conceptual models part I-A discussion of principles. J Hydrol 10: 282-290.

45. Moriasi DN, Arnold JG, Van Liew MW, Bingner RL, Harmel RD et al. (2007) Model evaluation guidelines for systematic quantification of accuracy in watershed simulations. Trans ASABE 50: 885-900.

46. Singh J, Knapp HV, Demissie M (2004) Hydrologic modelling of the Iroquois river watershed using HSPF and SWAT.

47. Harmel RD, Cooper RJ, Slade RM, Haney RL, Arnold JG (2006) Cumulative uncertainty in measured stream flow and water quality data for small watersheds. Trans ASAE 49: 689-701.

48. Evett SR, Schwartz RC, Casanova JJ, Heng LK (2012) Soil water sensing for water balance, ET and WUE. Agric Water Manage 104: 1-9.

49. Finsterle S, Najita J (1998) Robust estimation of hydrogeologic model parameters. Water Resour Res 34: 2939-2947.

50. Altman DG (1991) Practical statistics for medical research. Chapman and Hall, London.

51. Bland JM, Altman DG (1995) Comparing two methods of clinical measurement: a personal history. Internat J Epidemiol 24: 7-14.

5

Impacts of Climate Change on Soybean Irrigation Water Requirements in Northwest Region of Rio Grande do Sul, Brazil

Tirzah Moreira de Melo*, José Antônio Saldanha Louzada and Olavo Correa Pedrollo

Institute of Hydraulic Researches - Federal University of Rio Grande do Sul (IPH/UFRGS), Bento Gonçalves Av, 9500, P.O. Box 15029, Porto Alegre City, State of Rio Grande do Sul, Brazil

Abstract

Higher temperatures and a larger variability in precipitation will cause, in general, higher irrigation water requirements. The most important non-irrigated crops for the economy of the state of Rio Grande do Sul, Brazil, are corn and soybeans and the mesoregion which most contributes to the annual harvests of these crops is the Northwest region. This article aims to assess whether the impacts of climate change on agriculture in this region will be positive or negative and in what intensity they may occur. Hence, data from future climate projections generated by different climate models, as well as soil sampling for characterizing physical and hydraulic soil properties were considered. The one-dimensional SWAP model was used to estimate the irrigation water requirements. The results of the hypothesis tests performed for all simulations supports the premise that the irrigation water requirements in the near future (2025s) are not statistically different from the baseline period (1960-1990). On the other hand, water irrigation requirements in 2055s and 2085s reject this hypothesis.

Keywords: Future water demands; SWAP model; Climate change impacts; Uncertainty analysis

Introduction

There is no more doubt that the climate is changing and the Earth is warming, which is mainly attributed to human activities. The current concentration of CO_2 in the atmosphere (387 ppm) is higher than it was during the past 800,000 years before the Industrial Revolution, when it ranged from 170 to 280 ppm [1]. This resulted in an increase in average global temperature of $0.74 \pm 0.18°C$ over the last 100 years [2].

Such climate changes are mainly caused by the increase in gas emissions that contribute to the greenhouse effect as a result from human activities, such as burning fossil fuels and deforestation, as well as natural events such as volcanic eruptions [3]. When climate changes occur, all aspects of agriculture should be reviewed, including water demand, irrigation systems, land use, as well as seasonal characteristics of cropping systems [4].

Higher temperatures and a larger temporal variability in precipitation may cause higher water demands for irrigation. This is likely to occur even if the total precipitation during the growing period remains the same [5], and it is mainly due to a higher rate of evapotranspiration.

A recent study [6] indicates that the temperature in Brazil increased by approximately 0.75°C till the end of the past century for all seasons, but more markedly in the period from June to August. It is believed that climate changes subject developing countries, such as Brazil, more vulnerable and that the greatest impacts are likely to be experienced by their ecosystems and agriculture.

In general, with respect to precipitation, it is possible that the rainy season will become wetter, while dry seasons will become even drier. Locations where there is an increase in water deficit for agriculture during the dry months will require much larger quantities of water, intensifying conflicts over its use [7].

To better understand the severity of the consequences, a detailed and careful analysis about the region, the farming practices, the crops and the land uses is of utmost importance. The analysis of historical series of meteorological data allows us to detect local or regional climate patterns and trends over time and space.

During the last decades, the Intergovernmental Panel on Climate Change (IPCC) has summarized the most important researches carried out by the use of several climate models developed around the world. Climate projections are provided by general (GCMs) or regional (RCM) circulation models as a result of future scenarios of climate forcings caused by the emission of greenhouse gases and aerosols. Such models include systems of partial differential equations based on physical laws of fluid motion and chemistry [8], ranging from simple approaches of local energy balance to three-dimensional GCM models (horizontal cell resolution of between 250 and 600 km wide, and 10 to 20 vertical layers in the atmosphere), which attempt to model all complexities of Earth's climate system [9].

Simulation results of different models point to the fact that the effects of climate change on agriculture are also a function of climate scenarios, time slices (for which the uncertainties increase the further we deviate from the baseline period), current local weather and crop management systems and practices [10].

Many authors simulated reductions in crop duration of various crops along this century due to an increase in temperature [4,11,12]. While it had been shown the change rates in rice demands for irrigation fell below 3% in South Korea [13], an increase of 26-32% of annual demands for irrigation in California due to a warmer and drier climate was also reported [14], especially at the end of the century. As noted, these results are a function of increasing temperatures and different precipitation regimes predicted at each location.

---

*Corresponding author: Tirzah Moreira de Melo, Institute of Hydraulic Researches - Federal University of Rio Grande do Sul (IPH/UFRGS), Bento Gonçalves Av., 9500, P.O. Box 15029, Porto Alegre City, State of Rio Grande do Sul, Brazil
E-mail: tirzahmelo@hotmail.com

The study presented by [6] considered six GCMs projections for Brazil and the scenarios A2 and B2. They have shown an increase in temperature of up to 4°C in 2085s, while larger differences are likely to be experienced from June to October, i.e., winters will become warmer in comparison with the anomalies expected for the summers.

Specifically in the state of Rio Grande do Sul, the accumulated losses of soybean due to long drought periods were registered during summer months, because the crop development coincides with the period of frequent droughts, i.e., from November to March.

The most important non-irrigated crops of the state of Rio Grande do Sul are corn and soybeans and the mesoregion which most contributes to the annual harvests of these crops is the Northwest region. The corn and soybean planted area are 610,442 and 2,747,600 ha, respectively [15]. A serious aggravating factor in this region that justifies conducting this research is the fact that the cropping system is essentially non-irrigated.

Although knowledge about the causes of global warming and its unquestionable consequences are well known, it is extremely complex to define the uncertainties on future projections. Firstly, it is crucial that the model used to simulate the crop development can accurately estimate historical changes in water demands, before simulating the impacts of climate change [16]. Secondly, there is no consensus on how closely the climate models are to represent the uncertainties associated with the generation of meteorological data for future scenarios.

Uncertainties inherent in such models can be attributed to different discretizations, parameterizations and carbon cycle models [8,17] also add that the initial conditions and limitations of the models, as well as the forcings that define different scenarios, are potential uncertainties, but known.

Considering the abovementioned aspects of climate change and some of its most significant consequences on agriculture, this article aims to assess whether the impacts of climate change on agriculture predicted for the Northwest region of Rio Grande do Sul will be positive or negative and in what intensity they may occur. Hence, data from future climate projections generated by different climate models (A1B scenario), as well as soil sampling conducted in the area for spatial characterization of the physical and hydraulic soil properties were considered. This evaluation was carried out by using the SWAP model to determine future irrigation requirements. Since this is a relevant variable, an uncertainty analysis was conduct in other to investigate how different the results are and how these differences arc likely to contribute to decision-making in the future.

Materials and Methods

Study area

The study area is located in Pejuçara city, Northwest region of the state of Rio Grande do Sul and comprises the Donato basin, with an area of 1.10 km² (Figure 1). The mean of maximum temperatures in

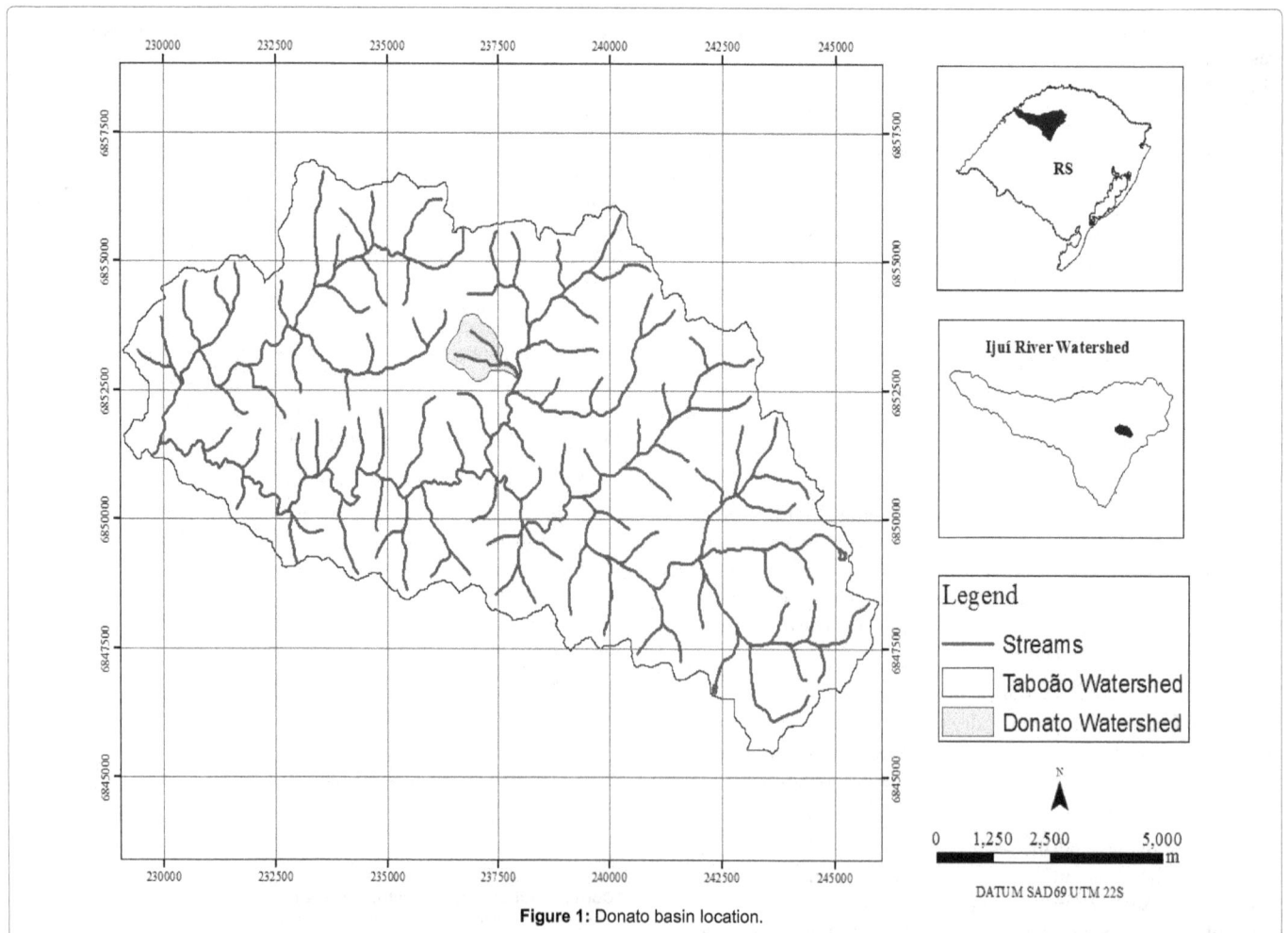

Figure 1: Donato basin location.

this region is above 22°C and the minimum temperature oscillates between -3 and 18°C. The average (1990-2001) annual rainfall is 1826 mm, October being the wettest month (216 mm) and August the driest (84 mm). The mean relative humidity is 74%, approximately [18]. The soils of the basin have high percentages of clay (> 60%), classified as Dystric Latosols and Eutric Nitosols.

The land use is primarily agricultural with the practice of no-till farming in the entire basin. Wheat, oats and soybeans are the most common crops cultivated. Soybean is sown in the summer and wheat or oats during the winter, all non-irrigated. From 1991 to 2012, soybean yields increased eight times and this is the tendency for the future [15]. The crops, soils and its use are representative of the northwest region of Rio Grande do Sul.

SWAP model

This one-dimensional model simulates the water and heat flow, solute transport, crop development and its interaction with surface water, based on the concepts and current techniques of modeling and simulation [19].

The water balance in the soil profile is the reference for all other modules, obtained by numerical solution of the nonlinear Richards equation [20]. In the SWAP model this equation has the general form:

$$\frac{\partial \theta}{\partial t} = \frac{\partial}{\partial z}\left[k(\theta)\left(\frac{\partial h}{\partial z} + 1\right)\right] - S_a(h) - S_d(h) - S_m(h) \qquad (1)$$

where θ is the volumetric water content (cm³ cm⁻³), z is depth along the soil profile (cm), k(θ) is the unsaturated hydraulic conductivity (cm d⁻¹), h is the soil pressure head (cm), t is time (d), $S_a(h)$ corresponds to the soil water extraction rate by plant roots (cm³ cm⁻³ d⁻¹), $S_d(h)$ represents the extraction rate by drain discharge in the saturated zone (d⁻¹) and $S_m(h)$ is the exchange rate of water between the soil matrix and macro pores (d⁻¹).

The numerical discretization of Richards equation is carried out using an implicit finite difference scheme, such that it allows the simultaneous simulation of the saturated and unsaturated zones. The SWAP model solves Equation (1) using known relationships between θ, h and K according to the combination of [21] and [22] models.

Soil samples

The SWAP model requires input on some soil physical and hydraulic properties such as the saturated hydraulic conductivity and the soil water retention curve. Thus, 55 points in the Donato basin were chosen, dispersed in a regular grid of approximately 140×140 m, with the most distant samples separated 200 m.

In each point, 4 samples were collected; two at 30 cm and two at 60 cm depth. For each depth, one sample was used to determine the soil water retention curve (RC) at pressures of 0.1, 0.3, 0.5, 0.7, 1.0, 1.5, 2.0, 2.5, 3.0 and 5.0 atm using the Richards chamber method. It was assumed that this pressure limit was sufficient to characterize the plant roots zone, since the first soil sample collected provided almost horizontal retention curves. The other sample was used to determine the saturated hydraulic conductivity (K_{sat}), obtained experimentally by the variable head permeameter method.

For each retention curve, the parameters of the model of [23] were adjusted according to Equation (2) and using the RETC software [23]. As some adjusted values of α were very large compared to the others, it was preferred to calculate the median of each of the parameters in the

retention curve instead of the average, since the latter would be affected by extreme values. In the case of K_{sat}, the averages at 30 and 60 cm were considered.

$$\theta(h) = \theta_{res} + (\theta_{sat} - \theta_{res})(1 + |\alpha h|^n)^{-m} \qquad (2)$$

where θ_{sat} is the saturated water content (cm³ cm⁻³), θ_{res} is the residual water content in the very dry range (cm³ cm⁻³) and α (cm⁻¹), n (-) and m (-) are empirical shape factors, with m = 1- (1/n). The final parameters used to characterize the physical and hydraulic properties of the soil are presented in Table 1.

Meteorological data

The SWAP model also requires input on a set of meteorological information concerning the daily time series of the variables: precipitation (P-mm), air vapor pressure (U-kPa), minimum and maximum temperatures (Tmin and Tmax-°C), wind speed (V-m/s) and solar radiation (R-KJ/m²). These variables were used to calculate the reference evapotranspiration (ET_{ref}).

In order to use meteorological data that were representative of the study area, climate projections from 7 different locations in the vicinity of the basin were considered (Figure 2), described in Table 2 and divided into the baseline (1961-1990), near term (2011-2040, 2025s), midterm (2041-2070, 2055s) and far term (2071-2100, 2085s). The GCMs considered are shown in Table 3, for which only monthly series of meteorological data in each location were initially available.

Besides these GCMs models, projections generated by the regional ETA model were also applied. This model is a descendant of the HIBU (Hydrometeorological Institute and Belgrade University) model, previously developed by [24]. It is a regional model that couples the HadCM3 general circulation model as lateral boundary condition for mesoscale simulations. A more detailed description of this method can be found in [25]. The coupling of this global model considers a set of regional members (CTRL, LOW, MID and HIGH), representing the climate sensitivity. The horizontal resolutions were 20 and 40 km, then providing 5 other projections: ETA 20-CTRL, ETA 40-CTRL, ETA 40-LOW, ETA 40-MID, ETA-40 HIGH. Yet, only the control member of each GCM was considered and hereafter they will be mention simply by their abbreviations, as shown in Table 3.

The daily series of meteorological data for GCMs were then generated by the Change Factor method [26]. This method of bias correction is based on the calculation of anomalies between the values predicted by climate models in the baseline and future time slices. The anomaly is then used to perturb the observed series of the variable in the baseline period generating the corrected series to be used in future periods. In this work, the time series to be disturbed were obtained after bias correction of the control member (CTRL) of the ETA-40 model. Thus, all the results generated by the use of global models should be compared with the baseline period related to the ETA 40-CTRL model.

Thus, for each of the 7 locations shown in Table 2, there are 10 sets

Parameter	Soil depths	
	30 cm	60 cm
Hydraulic conductivity, K_{sat} (cm/d)	25.4549	12.416
Parameter α (cm⁻¹)	0.1276	0.5046
Residual water content, θ_{res} (cm³ cm⁻³)	0.0000	0.0000
Saturated water content, θ_{sat} (cm³ cm⁻³)	0.4940	0.5088
Parameter n (-)	1.0652	1.0407

Table 1: Fitted soil water retention curve parameters.

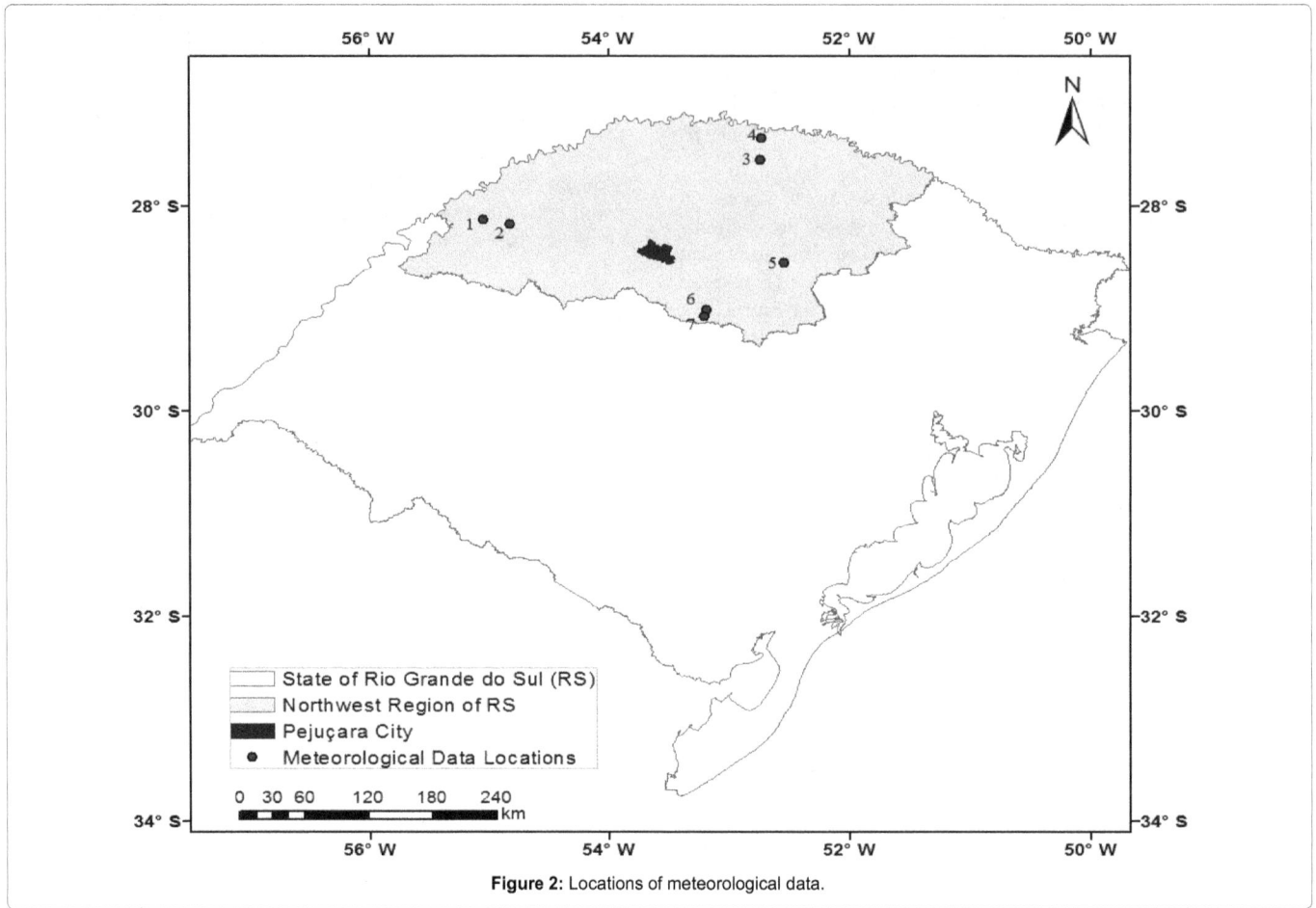

Figure 2: Locations of meteorological data.

Location	City	Latitude (°)	Longitude (°)	River	Altitude (m)
1	Roque Gonzales	-28.14	55.05	Ijuí	123
2	Salvador das Missões	-28.18	54.83	Ijuí	138
3	Entre Rios do Sul	-27.55	52.74	Passo Fundo	590
4	Nonoai/Faxinalzinho	-27.35	52.73	Passo Fundo	275
5	Tio Hugo	-28.56	52.55	Jacuí	470
6	Salto do Jacuí	-29.02	53.19	Jacuí	323
7	Salto do Jacuí	-29.07	53.21	Jacuí	283

Table 2: Meteorological data locations in Northwest region of Rio Grande do Sul.

Model	Abbreviation*	Modeling group
CCSM3, 2005	NCCCSM	National Center for Atmospheric Research, USA
ECHAM5/MPI-OM, 2005	MPEH5	Max Planck Institute for Meteorology, Germany
GFDL-CM2.1, 2005	GFCM21	U.S Dept. of Commerce/NOAAA/Geophysical Fluid Dynamics Laboratory, USA
MRI-CGCM2.3.2, 2003	MRCGCM	National Center for Atmospheric Research, USA
UKMO-HadCM3, 1997	HADCM3	Hadley Centre for Climate Prediction and Research/Met Office, United Kingdom

*Abbreviation adopt in this study.

Table 3: General circulation models applied in this study.

of meteorological data derived from these different models. All sets of data will be used in the uncertainty analysis of the climate variables.

Modelling process

The general form of the Richards equation (Eq. 1), when applied by SWAP model to achieve the objectives of this study, becomes:

$$\frac{\partial \theta}{\partial t} = \frac{\partial}{\partial z}\left[k(\theta)\left(\frac{\partial h}{\partial z} + 1 \right) \right] - S_a(h) \tag{3}$$

where only the term of water extraction by plant roots ($S_a(h)$) will be considered and it is assumed that this term is equal to the crop transpiration. The crops at Donato basin are non-irrigated. When calculating future irrigation demands it will be assumed that the total water requirement of the crop will be supplied, and therefore:

$$IWR = T_p - T_a \tag{4}$$

where IWR is the irrigation water requirement (cm), T_p is the potential transpiration (cm) when the plant has no water limitation and T_a is the actual transpiration (cm). Thus, any value lower than the potential transpiration means that the crop will suffer a water deficit and will need to be irrigated. The IWRs were evaluated for soybean only, since this is the most important crop in the region for which irrigation could be profitable. This is justified by the fact that in 2012 soybean production in this region was 3,585,710 tons compared to only 1,731,219 tons of corn. Additionally, soybean cultivated area (ha) is almost five times greater than for corn [15].

A different simulation for each of the ten climate projections and each of the seven data sources was performed, totaling 70 simulations (Figure 3 for methodology scheme illustration). Initially, the SWAP model calculates the evapotranspiration for a hypothetical reference crop based on the Penman-Monteith equation (not shown), assuming the crop height of 0.12 m, a fixed surface resistance of 70 s m^{-1} and an albedo of 0.23, as bellow:

$$ET_{ref} = \frac{0.408\Delta(R_n - G) + \gamma \dfrac{900}{T+273}u_2(e_s - e_a)}{\Delta + \gamma(1 + 0.34u_2)} \tag{5}$$

where ET_{ref} is the transpiration rate of the canopy (mm day^{-1}), Δ is the slope vapour pressure curve (kPa °C^{-1}), R_n is the net radiation flux at the canopy surface (J m^{-2} day^{-1}), G is the soil heat flux density (MJ m^{-2} day^{-1}), e_s is the saturation vapour pressure (kPa), e_a is the actual vapour pressure (kPa), γ is the psychrometric constant (kPa °C^{-1}), u_2 is the wind speed at 2 m height (m s^{-1}) and T is the mean daily air temperature at 2 m height (°C).

From ET_{ref} calculated in the previous step, the model then calculates the evapotranspiration for soybean (ETc) by using the coefficient of culture (k_c):

$$ET_c = k_c \cdot ET_{ref} \tag{6}$$

In very dry conditions, the evaporation rate declines much faster than transpiration, and for this reason the SWAP considers the calculation of these rates separately. At this stage, therefore, using the leaf area index (LAI), the model separates the evaluation of the potential transpiration (T_c) of a crop and the potential evaporation (Ep) of the soil. Yet, when the crop is wet due to interception, SWAP assumes that the energy available for evapotranspiration is entirely used to evaporate the intercepted water, independent of the soil cover fraction (W_{frac}). This is valid for higher values of W_{frac}. The potential evaporation is then calculated by:

$$E_p = E_{p0}(1 - W_{frac})e^{-\kappa_{gr} LAI} \tag{7}$$

where W_{frac} (-) is the fraction of the day that the crop is wet, P_i (cm d^{-1}) is the ratio of the daily amount of intercepted precipitation, ET_{w0} (cm d^{-1}) is the potential evapotranspiration rate for the wet canopy, E_{p0} (cm d^{-1}) is the evaporation rate of a wet bare soil and κ_{gr} (-) is the extinction coefficient for solar radiation. From equation 7, SWAP calculates T_c by the difference:

$$W_{frac} = \frac{P_i}{ET_{w0}} \tag{8}$$

$$T_c = ET_c(1 - W_{frac}) - E_p \tag{9}$$

where ET_c (cm d^{-1}) is the total evapotranspiration rate in periods with dry canopy.

The actual transpiration (T_a) is then calculated by taking into account only the water stress due to dry conditions. This is because the critical pressure heads reported for the model will force conditions in which the crop will not suffer a reduction in potential transpiration due to wet conditions, thus preventing it from being incorporated into IWR evaluation. When integrated over the entire depth of the roots, the maximum rate of water uptake by the plant roots (hence, potential) is given by [27]:

$$S_c(z) = \frac{\ell_{root}(z)}{\displaystyle\int_{-D_{root}}^{0} \ell_{root}(z)\,dz} T_c \tag{10}$$

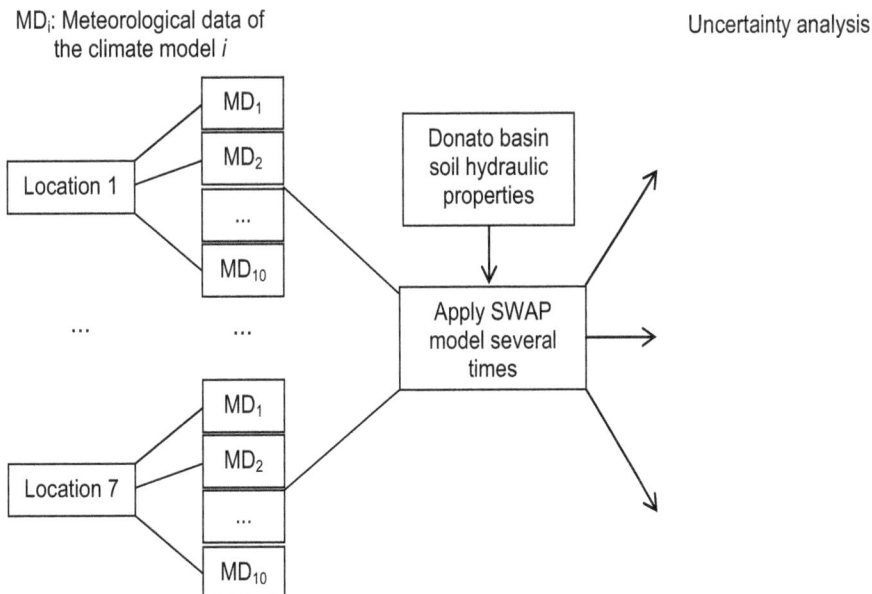

Figure 3: Methodology scheme illustration. IWR is the irrigation water requirement of 2025s, 2055s and 2085s time slices.

where $S_c(z)$ is the potential root water extraction rate at a certain depth (d^{-1}), D_{root} is the root layer thickness (cm), z is the root depth (cm), T_c is the crop transpiration (cm), $\ell_{root}(z)$ is the root length density distribution.

Thus, water stress of dry conditions will reduce the value of $S_c(z)$ to its actual value $S_a(z)$ by:

$$S_a(z) = \alpha \, S_c(z) \tag{11}$$

where α is the stress factor of water stress of dry conditions.

Integrating $S_a(z)$ along the root zone, then the crop actual transpiration is given by:

$$T_c = \int\limits_{D_{root}}^{0} S_a(z)dz \tag{12}$$

Results and Discussions

Meteorological data

The mean annual temperature (T) and cumulative precipitation (P) for the Northwest region of Rio Grande do Sul during the baseline and future periods are shown in Figure 4. The curves represent the average temperature and precipitation projections considering all 70 (7 locations *versus* 10 climate projections). The baseline period reveals that the temperature has already increased at about 1.71°C while the projections for 2025s indicate an even warmer climate, with an increase of over 3.12°C till the end of the century.

Climate projections also suggest an increase in precipitation, but lesser pronounced than for the temperature. Table 4 presents the average temperature and rainfall projected by GCMs and the RGM. All models predicted a gradual increase in temperature, while the projections of ETA 20 and ETA 40 models provide the strongest increases in precipitation compared to the baseline period.

Figure 5 presents the average annual temperature and precipitation anomalies for all different locations in the study area. The anomalies (Δ) refer to the difference between a variable in a future time slice (for example, 2011-2040) and the same variable in the baseline period (1961-1990). Notice that although the temperature anomalies in some locations are very similar, the same pattern does not repeat for precipitation anomalies, where the variability is much greater. The largest anomalies were predicted for the end of the century, and a reduction in T and P in the short term (2025s) is expected only in location 6. Based on these differences, this study proposes to assess the impact of climate change on future water demands for irrigation, considering these different possibilities for the future climate.

Irrigation Water Requirement (IWR)

Table 5 presents the average cumulative IWRs (cm) of all simulations obtained with the SWAP model for the baseline, 2025s, 2055s and 2085s periods for each location and climate model. The IWR anomalies are also presented in this table. As global models were generated from daily meteorological data of the ETA 40-CTRL model, IWRs from the formers were compared with the baseline period of this latter (Tables 4 and 5).

As can be seen (Table 5), increasing IWRs are expected for the end of the century (2085s), with the highest values being provided by

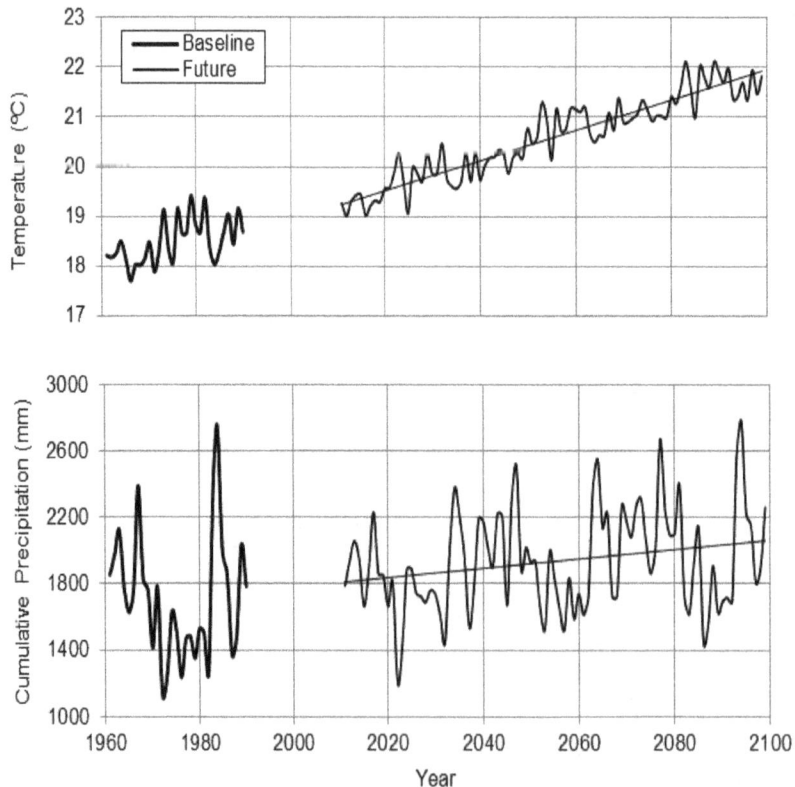

Figure 4: Historical and projected average temperature (°C) and average annual cumulative precipitation (mm) of all locations and models. The curves refer to the arithmetic mean of the 70 time series (7 locations and 10 models) of temperature and precipitation.

Variable	Time Slice	Models									
		ETA 20	ETA 40 CTRL	ETA 40 HIGH	ETA 40 LOW	ETA 40 MID	GFCM21	HADCM3	MPEH5	MRCGCM	NRCCCSM
Temperature (°C)	Baseline	18.98									
	2025s	19.9	19.7	20.0	19.3	20.0	19.6	19.8	19.4	19.2	19.7
	2055s	20.9	20.6	21.1	20.2	21.0	21.1	20.8	20.3	20.0	20.5
	2085s	21.2	21.3	22.5	20.6	22.1	21.7	22.2	21.6	20.3	21.0
Cumulative Precipitation (mm)	Baseline	1724.9									
	2025s	1953.7	1904.8	1840.3	1923.0	1995.5	1685.8	1701.4	1720.5	1785.5	1855.8
	2055s	2083.9	2086.2	1989.5	2089.8	2187.4	1429.2	1865.4	1851.2	1987.7	1921.8
	2085s	2266.5	2206.9	2151.5	2339.5	2065.1	1426.0	1834.2	1936.7	2093.1	1982.4

Table 4: Average annual temperature and precipitation projected by different climate models.

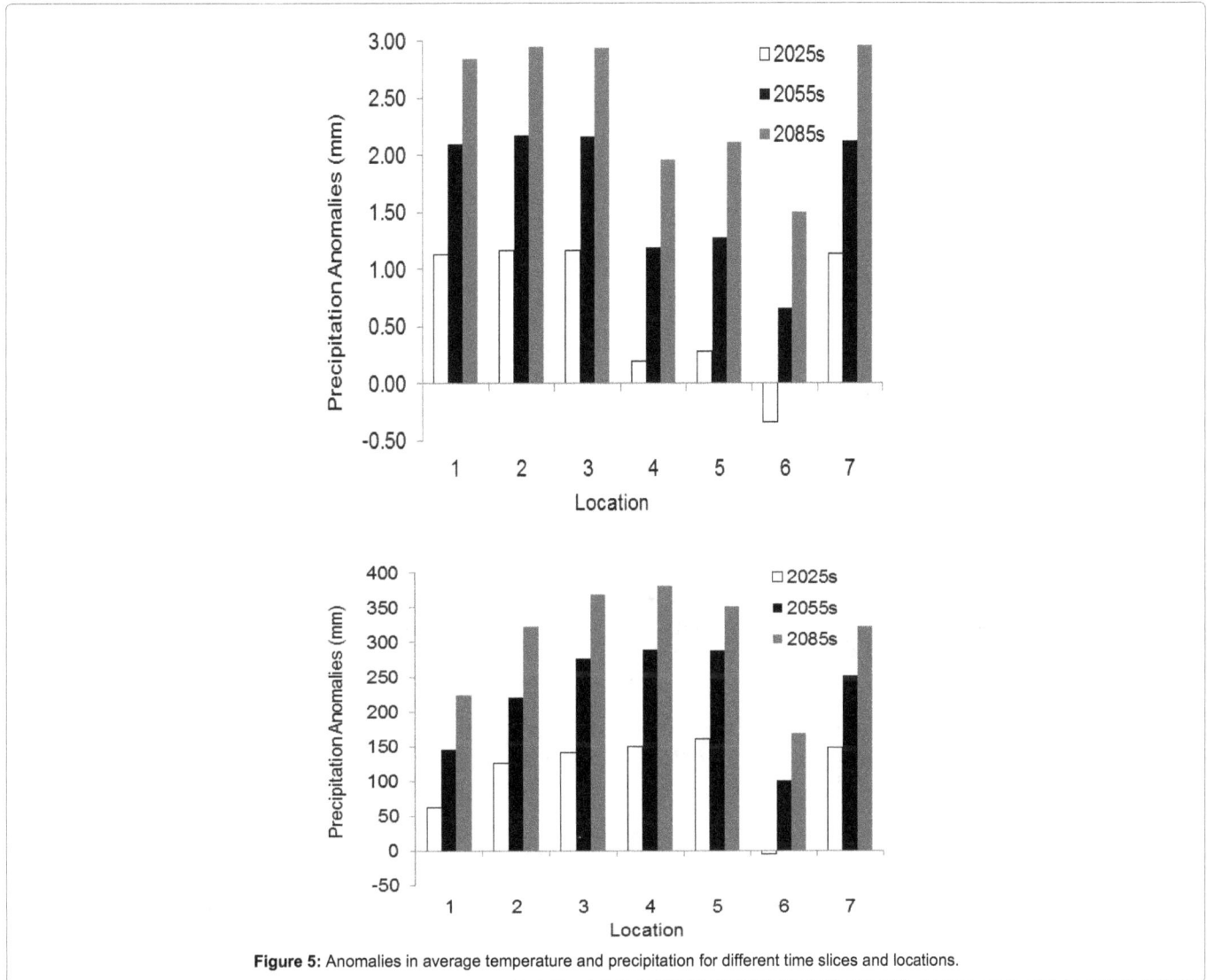

Figure 5: Anomalies in average temperature and precipitation for different time slices and locations.

global models. Notice also the differences among the locations, which demonstrated the degree of uncertainty of IWR forecasted for the region. In most cases, ETA model projections indicate a reduction in IWR, while global models suggest the opposite, except for MRCGCM and NRCCCSM models in some locations.

It could be concluded that models do not agree during 2025s, because some of them suggest a reduction in IWRs while others suggest an increasing in IWRs. In 2055s, it is observed that in most locations

prevail estimates of reductions, probably due to a higher cumulative precipitation, and this is the period when the models agree more often. Similar behavior can be observed during the end of the century (2085s).

The histograms in Figure 6 correspond to all 70 IWRs calculated for the study area. It is evident that the influences of climate change on water demands for irrigation will be less severe in the short term due to lower data departure around the mean of the baseline period (7.22 cm). In subsequent periods, higher frequencies of IWR simulated above this

Model	Baseline	IWR (2025s)	IWR (2055s)	IWR (2085s)	Δ' (2025s)	Δ (2055s)	Δ (2085s)
			Location 1				
ETA 20	4.11	4.91	3.32	1.32	0.80 (19)	-0.79 (-19)	-2.79 (-68)
ETA 40 CTRL	5.13	5.11	3.20	2.14	-0.02 (0)	-1.93 (-38)	-2.99 (-58)
ETA 40 HIGH	4.83	4.15	3.79	5.14	-0.68 (-14)	-1.04 (-22)	0.31 (6)
ETA 40 LOW	5.22	3.12	2.39	2.52	-2.10 (-40)	-2.83 (-54)	-2.70 (-52)
ETA 40 MID	5.61	8.36	8.41	10.50	2.75 (49)	2.80 (50)	4.89 (87)
GFCM21	5.13	5.64	10.34	8.91	0.51(10)	5.21 (102)	3.78 (74)
HADCM3	5.13	7.63	7.77	7.57	2.50 (49)	2.64 (51)	2.44 (48)
MPEH5	5.13	6.49	7.90	11.67	1.36 (27)	2.77 (54)	6.54 (127)
MRCGCM	5.13	5.43	5.30	6.13	0.30 (6)	0.17 (3)	1.00 (19)
NRCCCSM	5.13	4.19	4.13	5.33	-0.94 (-18)	-1.00 (-19)	0.20 (4)
			Location 2				
ETA 20	5.28	5.05	3.48	2.01	-0.23 (4)	-1.80 (-34)	-3.27 (-62)
ETA 40 CTRL	6.10	6.56	4.29	3.69	0.46 (8)	-1.81 (-30)	-2.41 (-40)
ETA 40 HIGH	5.98	5.38	4.72	6.11	-0.60 (-10)	-1.26 (-21)	0.13 (2)
ETA 40 LOW	6.22	4.20	3.57	2.86	-2.02 (-32)	-2.65 (-43)	-3.36 (-54)
ETA 40 MID	5.83	2.25	3.00	4.76	-3.58 (-61)	-2.83 (-49)	-1.07 (-18)
GFCM21	6.10	6.90	11.10	9.72	0.80 (13)	5.00 (82)	3.62 (59)
HADCM3	6.10	8.42	9.06	8.71	2.32 (38)	2.96 (49)	2.61 (43)
MPEH5	6.10	7.18	8.34	11.93	1.08 (18)	2.24 (37)	5.83 (96)
MRCGCM	6.10	6.33	5.91	6.54	0.23(4)	-0.19 (-3)	0.44 (7)
NRCCCSM	6.10	5.08	5.06	6.17	-1.02 (-17)	-1.04 (-17)	0.07 (1)
			Location 3				
ETA 20	7.41	8.47	6.51	3.91	1.06 (14)	-0.90 (-12)	-3.50 (-47)
ETA 40 CTRL	9.62	10.11	6.62	6.24	0.49 (5)	-3.00 (-31)	-3.38 (-35)
ETA 40 HIGH	9.17	8.36	8.25	8.33	-0.81 (-9)	-0.92 (-10)	-0.84 (-9)
ETA 40 LOW	8.04	6.75	5.39	3.44	-1.29 (-16)	-2.65 (-33)	-4.60 (-57)
ETA 40 MID	8.74	3.93	4.59	7.12	-4.81 (-55)	-4.15 (-47)	-1.62 (-19)
GFCM21	9.62	10.96	15.71	14.64	1.34 (14)	6.09 (63)	5.02 (52)
HADCM3	9.62	12.99	14.32	13.82	3.37 (35)	4.70 (49)	4.20 (44)
MPEH5	9.62	10.61	10.94	14.75	0.99 (10)	1.32 (14)	5.13 (53)
MRCGCM	9.62	10.13	9.42	10.54	0.51 (5)	-0.20 (-2)	0.92 (10)
NRCCCSM	9.62	7.33	7.25	9.59	-2.29 (-24)	-2.37 (-25)	-0.03 (0)
			Location 4				
ETA 20	4.21	4.18	3.00	2.94	-0.03 (-1)	-1.21 (-29)	-1.27 (-30)
ETA 40 CTRL	7.07	7.01	4.36	4.02	-0.06 (-1)	-2.71 (38)	-3.05 (-43)
ETA 40 HIGH	6.75	5.90	6.17	6.13	-0.85 (-13)	-0.58 (-9)	-0.62 (-9)
ETA 40 LOW	5.44	4.46	3.33	2.09	-0.98 (-18)	-2.11 (-39)	-3.35 (-62)
ETA 40 MID	6.38	2.18	3.04	5.04	-4.20 (-66)	-3.34 (-52)	-1.34 (-21)
GFCM21	7.07	8.41	12.92	12.18	1.34 (19)	5.85 (83)	5.11 (72)
HADCM3	7.07	10.15	11.51	10.93	3.08 (44)	4.44 (63)	3.86 (55)
MPEH5	7.07	8.13	8.37	12.07	1.06 (15)	1.30 (18)	5.00 (71)
MRCGCM	7.07	7.64	7.06	8.10	0.57 (8)	-0.01 (0)	1.03 (15)
NRCCCSM	7.07	4.89	4.84	6.91	-2.18 (-31)	-2.23 (-32)	-0.16 (-2)
			Location 5				
ETA 20	7.63	7.27	6.27	4.35	-0.36 (-5)	-1.36 (-18)	-3.28 (-43)
ETA 40 CTRL	9.17	9.40	5.44	5.47	0.23 (3)	-3.73 (-41)	-3.70 (-40)
ETA 40 HIGH	8.66	7.40	7.06	7.21	-1.26 (-15)	-1.60 (-18)	-1.45 (-17)
ETA 40 LOW	8.09	6.43	3.88	2.96	-1.66 (-21)	-4.21 (-52)	-5.13 (-63)
ETA 40 MID	8.46	3.59	3.81	6.28	-4.87 (-58)	-4.65 (-55)	-2.18 (-26)
GFCM21	9.17	10.08	14.83	13.37	0.91 (10)	5.66 (62)	4.20 (46)
HADCM3	9.17	11.70	13.17	12.64	2.53 (28)	4.00 (44)	3.47 (38)
MPEH5	9.17	10.19	10.62	14.36	1.02 (11)	1.45 (16)	5.19 (57)
MRCGCM	9.17	9.55	8.50	10.02	0.38 (4)	-0.67 (-7)	0.85 (9)
NRCCCSM	9.17	6.78	6.70	8.74	-2.39 (-26)	-2.47 (-27)	-0.43 (-5)
			Location 6				
ETA 20	6.35	6.43	5.68	5.17	0.08 (1)	-0.67 (-11)	-1.18 (-19)
ETA 40 CTRL	7.35	13.36	9.97	10.19	6.01 (82)	2.62 (36)	2.84 (39)
ETA 40 HIGH	6.89	12.93	12.13	12.91	6.04 (88)	5.24 (76)	6.02 (87)
ETA 40 LOW	7.17	11.25	8.99	8.07	4.08 (57)	1.82 (25)	0.90 (13)

ETA 40 MID	6.87	3.16	3.15	5.88	-3.71 (-54)	-3.72 (-54)	-0.99 (-14)
GFCM21	7.35	8.23	11.93	10.07	0.88 (12)	4.58 (62)	2.72 (37)
HADCM3	7.35	9.35	10.85	9.70	2.00 (27)	3.50 (48)	2.35 (32)
MPEH5	7.35	8.35	9.52	13.11	1.00 (14)	2.17 (30)	5.76 (78)
MRCGCM	7.35	7.85	6.99	8.39	0.50 (7)	-0.36 (-5)	1.04 (14)
NRCCCSM	7.35	5.53	5.55	6.88	-1.82 (-25)	-1.80 (-24)	-0.47 (-6)
Location 7							
ETA 20	7.48	6.76	5.45	4.35	-0.72 (-10)	-2.03 (-27)	-3.13 (-42)
ETA 40 CTRL	7.93	8.48	4.99	4.57	0.55 (7)	-2.94 (-37)	-3.36 (-42)
ETA 40 HIGH	7.95	6.90	6.25	7.52	-1.05 (-13)	-1.7 (-21)	-0.43 (-5)
ETA 40 LOW	8.41	6.18	3.83	2.54	-2.23 (-27)	-4.58 (-54)	-5.87 (-70)
ETA 40 MID	7.75	3.25	3.33	5.89	-4.5 (-58)	-4.42 (-57)	-1.86 (-24)
GFCM21	7.93	8.82	12.95	11.11	0.89 (11)	5.02 (63)	3.18 (40)
HADCM3	7.93	10.22	11.45	10.66	2.29 (29)	3.52 (44)	2.73 (34)
MPEH5	7.93	9.14	10.74	14.86	1.21 (15)	2.81 (35)	6.93 (87)
MRCGCM	7.93	8.55	7.52	9.18	0.62 (8)	-0.41 (-5)	1.25 (16)
NRCCCSM	7.93	6.08	5.99	7.46	-1.85 (-23)	-1.94 (-24)	-0.47 (-6)

* Δ: irrigation water requirement anomalies (cm); terms in parenthesis indicate percentage change from baseline period.

Table 5: Irrigation water requirements (IWRs) and anomalies (Δ) of all simulations and time slices (cm).

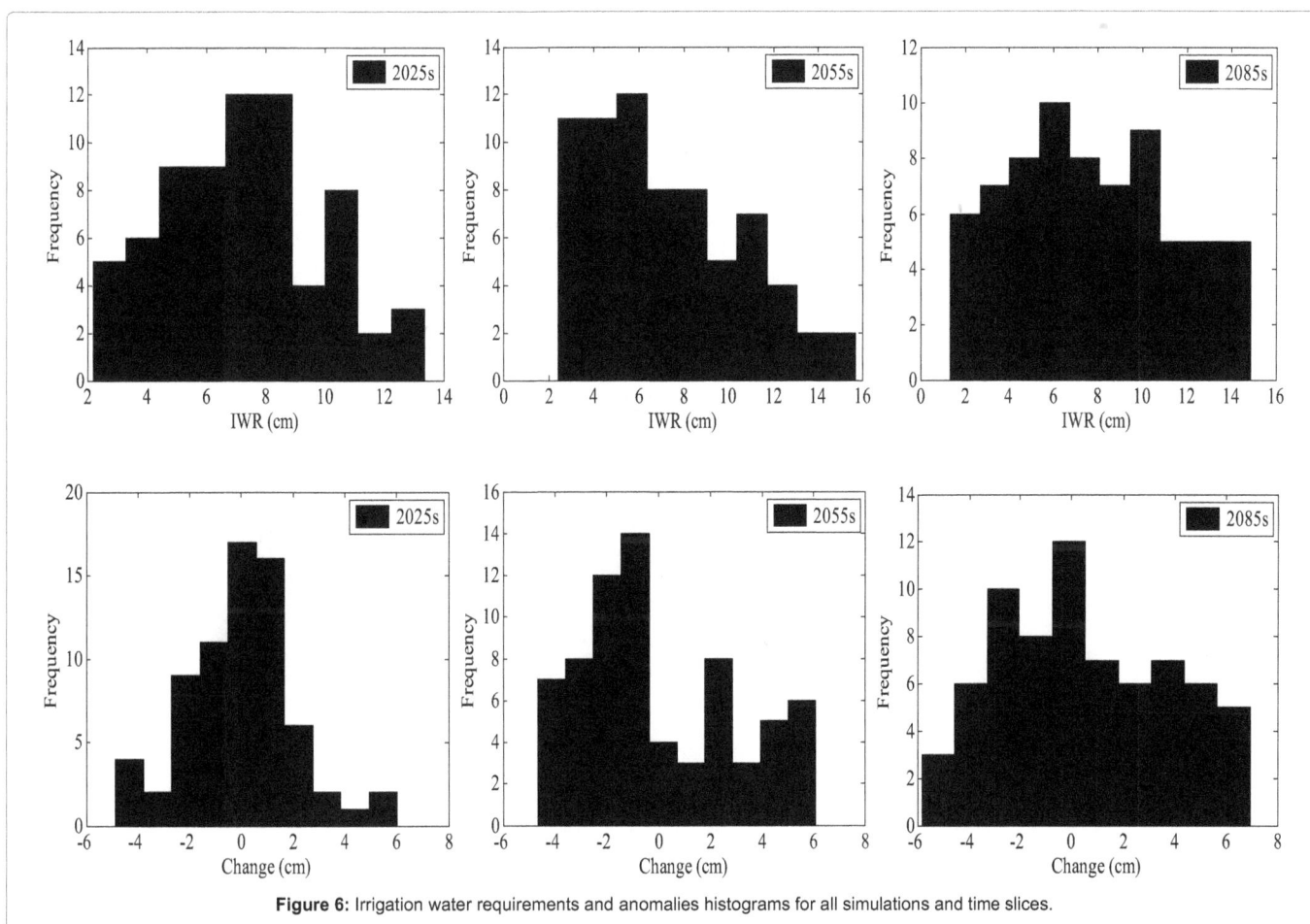

Figure 6: Irrigation water requirements and anomalies histograms for all simulations and time slices.

average were obtained. However, the most frequent values are in the range below 7.22 cm, confirming the possibility that the area will be positively influenced by lower irrigation requirements in the future.

Likewise, the analysis of the histograms of the anomalies also confirms that in the short term most of simulations fall on the range around zero, indicating little or no difference with respect to the

baseline period. During 2055s, there is a higher frequency of negative anomalies ranging from -4 to -1 cm. Yet, lower frequencies of null or positive anomalies were obtained.

These histograms can also be used to analyse the IWR uncertainties in terms of its anomalies. As expected, the uncertainty during 2085s is higher and is reflected by a more dispersed histogram. These results are in agreement with predictions of a population growth peak in the middle of the century and the intensification of negative effects on the Earth, such as an increase in temperature and increase or decrease in precipitation amounts.

In order to quantify the uncertainties of the simulations, the IWR cumulative distributions functions (cdfs) were plotted in Figure 7. Table 6 presents the respective percentiles for each of the cdfs to support in the interpretation of the graphs.

Table 6 shows the degree of uncertainty of the simulations. The uncertainties reflect how different all IWRs are, as a result of the use of different meteorological data in SWAP model. This evaluation was carried out for IRW and its anomalies separately. If we take the values between the percentiles of 5 and 95%, we may note that 90% of IWR simulations are between the ranges of 3.16-11.70, 3.04-13.17 and 2.14-14.36 cm, resulting in uncertainties of 8.54, 10.13 and 12.22 cm for 2025s, 2055s and 2085s, respectively. These values are significantly large, demonstrating the error that can be committed when defining which data will be used by the model.

Thus, it is essential that agricultural planning decisions that are based on climate projections should consider the greater number of possible models. For example, according to the projections employed in this study, it can be stated that in 90% of the simulations the average annual IWRs will not be greater than 10.42, 12.03 and 13.01 cm for the three future periods. However, many other forecasts were not included in this analysis, which could add more information about the uncertainties of climate models, or even the future will not match any of them. Only for an average analysis (50% percentile), the future periods are not different from baseline period.

It is worth mentioning that the uncertainties derived from model

parameters, among many other sources, were not considered in this study. In addition, these results represent one-dimensional simulations. When the area dimension is considered, the volumes of demanded water will reflect the impacts on water resources.

Hypothesis test

Although Table 5 presents the anomalies of future water demands compared to the period of 1961-1990, it does not reveal whether each future period is different from the baseline, since they refer to averages of 30 years period. Thus, the two-sample hypothesis test of Kolmogorov-Smirnov was performed to verify whether the probability distribution functions of future IWR were statistically different from the baseline period in each location. In this case, the null hypothesis H_o tests whether the compared cdfs are statistically equal. The results are shown in Table 7 for the significance level α=0.05.

Analyzing the values of h in this table, it can be seen that the vast majority of models agree that there are no significant differences between 2025s and the baseline IWRs, except for ETA 40 MID model, which rejects the null hypothesis (h=1) in all locations. During 2055s, the predictions generated by other models also indicate that this period is different from the baseline. The same is observed for the end of the century (2085s), coupled with some other rejections of the null hypothesis.

In general, one can say that there is not enough statistical evidence in most cases to admit that the cdfs differ to the period of comparison for the adopt level of significance. This fact is also reflected by the very high p-value, indicating that it is not safe to reject the null hypothesis without assuming a large error. Greater tendency to reject H_o was observed by simulations performed by ETA model.

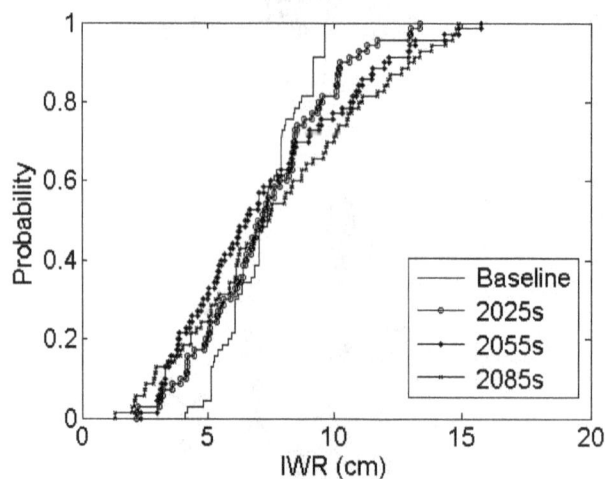

Figure 7: Irrigation water requirements and anomalies cumulative distribution functions.

Percentile	IWR (2025s)	IWR (2055s)	IWR (2085s)	Δ (2025s)	Δ (2055s)	Δ (2085s)
5	3.16	3.04	2.14	-4.20	-4.21	-3.70
25	5.38	4.36	4.76	-1.05	-2.23	-2.41
50	7.10	6.57	7.34	0.34	-0.85	0.10
75	8.82	9.52	10.50	1.06	2.64	3.18
90	10.42	12.03	13.01	2.52	4.85	5.12
95	11.70	13.17	14.36	3.37	5.24	5.83

Table 6: Cumulative distribution functions percentiles of irrigation water requirements and anomalies (cm).

	ETA 20	ETA 40 CTRL	ETA 40 HIGH	ETA 40 LOW	ETA 40 MID	GFCM21	HADCM3	MPEH5	MRCGCM	NRCCCSM
					Location 1					
h^a (2025s)	0	0	0	0	1	0	0	0	0	0
p^b	0.9360	0.9360	0.3420	0.1088	0.0017	0.9360	0.1088	0.5372	0.9970	0.5372
D^c	0.1333	0.1333	0.2333	0.3000	0.4667	0.1333	0.3000	0.2000	0.1000	0.2000
h (2055s)	0	0	0	0	1	1	0	0	0	0
p	0.2003	0.2003	0.5372	0.0550	0.0006	0.0113	0.0550	0.1088	0.9970	0.3420
D	0.2667	0.2667	0.2000	0.3333	0.5000	0.4000	0.3333	0.3000	0.1000	0.2333
h (2085s)	1	1	0	0	1	1	0	1	0	0
p	0.0010	0.0022	0.8383	0.0627	0.0008	0.0325	0.2468	0.0026	0.7240	0.9950
D	0.4897	0.4621	0.1552	0.3299	0.4977	0.3598	0.2563	0.4575	0.1736	0.1046
					Location 2					
h (2025s)	0	0	0	1	1	0	0	0	0	0
p	0.3420	0.9360	0.5372	0.0259	0.0046	0.7600	0.2003	0.7600	0.9970	0.9360
D	0.2333	0.1333	0.2000	0.3667	0.4333	0.1667	0.2667	0.1667	0.1000	0.1333
h (2055s)	1	0	0	1	1	1	0	0	0	0
p	0.0113	0.3420	0.2003	0.0259	0.0259	0.0113	0.1088	0.2003	0.9970	0.7600
D	0.4000	0.2333	0.2667	0.3667	0.3667	0.4000	0.3000	0.2667	0.1000	0.1667
h (2085s)	1	1	0	1	0	0	0	1	0	0
p	0.0001	0.0134	0.9670	0.0293	0.2657	0.2706	0.4474	0.0076	0.9977	0.9973
D	0.5598	0.3966	0.1241	0.3644	0.2517	0.2506	0.2161	0.4184	0.0989	0.1000
					Location 3					
h (2025s)	0	0	0	0	1	0	0	0	0	0
p	0.7600	0.9360	0.5372	0.3420	0.0017	0.3420	0.0550	0.3420	0.9360	0.1088
D	0.1667	0.1333	0.2000	0.2333	0.4667	0.2333	0.3333	0.2333	0.1333	0.3000
h (2055s)	0	0	0	0	1	1	1	0	0	0
p	0.3420	0.0550	0.3420	0.2003	0.0259	0.0046	0.0113	0.2003	1.0000	0.1088
D	0.2333	0.3333	0.2333	0.2667	0.3667	0.4333	0.4000	0.2667	0.0667	0.3000
h (2085s)	1	1	0	1	0	1	1	1	0	0
p	0.0027	0.0142	0.8045	0.0008	0.4206	0.0189	0.0444	0.0194	0.6858	0.9999
D	0.4563	0.3943	0.1609	0.4943	0.2207	0.3828	0.3460	0.3816	0.1793	0.0828
					Location 4					
h (2025s)	0	0	0	0	1	0	0	0	0	0
p	0.2003	0.7600	0.3420	0.3420	0.0001	0.3420	0.1088	0.5372	0.7600	0.1088
D	0.2667	0.1667	0.2333	0.2333	0.5667	0.2333	0.3000	0.2000	0.1667	0.3000
h (2055s)	0	0	0	0	1	1	1	0	0	0
p	0.1088	0.1088	0.5372	0.2003	0.0113	0.0046	0.0259	0.5372	1.0000	0.1088
D	0.3000	0.3000	0.2000	0.2667	0.4000	0.4333	0.3667	0.2000	0.0667	0.3000
h (2085s)	0	1	0	1	0	1	1	1	0	0
p	0.1356	0.0061	0.9030	0.0142	0.5856	0.0206	0.0422	0.0200	0.3698	0.9934
D	0.2908	0.4264	0.1425	0.3943	0.1943	0.3793	0.3483	0.3805	0.2299	0.1069
					Location 5					
h (2025s)	0	0	0	0	1	0	0	0	0	0
p	0.5372	0.9360	0.2003	0.2003	0.0006	0.9360	0.5372	0.9360	0.9970	0.2003
D	0.2000	0.1333	0.2667	0.2667	0.5000	0.1333	0.2000	0.1333	0.1000	0.2667
h (2055s)	0	1	0	1	1	1	1	0	0	0
p	0.3420	0.0259	0.3420	0.0017	0.0046	0.0046	0.0259	0.7600	0.9360	0.1088
D	0.2333	0.3667	0.2333	0.4667	0.4333	0.4333	0.3667	0.1667	0.1333	0.3000
h (2085s)	1	1	0	1	0	0	0	1	0	0
p	0.0009	0.0317	0.7975	0.0020	0.0813	0.0852	0.1603	0.0444	0.8869	0.9810
D	0.4920	0.3609	0.1621	0.4655	0.3172	0.3149	0.2816	0.3460	0.1460	0.1172
					Location 6					
h (2025s)	0	1	1	1	1	0	0	0	0	0
p	0.9360	0.0046	0.0006	0.0259	0.0017	0.9360	0.2003	0.7600	0.9360	0.2003
D	0.1333	0.4333	0.5000	0.3667	0.4667	0.1333	0.2667	0.1667	0.1333	0.2667
h (2055s)	0	1	1	0	1	1	0	0	0	0
p	0.3420	0.0113	0.0002	0.0550	0.0017	0.0259	0.0550	0.3420	0.9970	0.3420
D	0.2333	0.4000	0.5333	0.3333	0.4667	0.3667	0.3333	0.2333	0.1000	0.2333
h (2085s)	1	1	1	0	0	0	0	1	0	0
p	0.0010	0.0007	0.0001	0.1141	0.2468	0.1704	0.3947	0.0054	0.9030	0.9977
D	0.4897	0.5000	0.5598	0.3000	0.2563	0.2782	0.2253	0.4310	0.1425	0.0989

	Location 7									
h (2025s)	0	0	0	0	1	0	0	0	0	0
p	0.7600	0.9360	0.5372	0.0550	0.0017	0.7600	0.5372	0.7600	0.9970	0.5372
D	0.1667	0.1333	0.2000	0.3333	0.4667	0.1667	0.2000	0.1667	0.1000	0.2000
h (2055s)	0	0	0	1	1	0	0	0	0	0
p	0.2003	0.1088	0.5372	0.0046	0.0006	0.1088	0.2003	0.3420	0.9970	0.5372
D	0.2667	0.3000	0.2000	0.4333	0.5000	0.3000	0.2667	0.2333	0.1000	0.2000
h (2085s)	1	0	0	1	0	0	0	1	0	0
p	0.0174	0.0658	0.9977	0.0002	0.2514	0.4822	0.5780	0.0155	0.8574	0.9790
D	0.3862	0.3276	0.0989	0.5322	0.2552	0.2103	0.1954	0.3908	0.1517	0.1184

[a] h is the test output (H$_0$ = both samples come from the same distribution; h=1, reject H$_0$; h=0, no statistic evidence to reject H$_0$) ; p is the p-value and D is the test statistic.

Table 7: Two-sample Kolmogorov-Smirnov test for irrigation water requirements.

From the results shown in Table 7, it can be inferred that the rejection of the null hypothesis only occurred when differences between the cdfs were strongly large. This fact is because the test is quite conservative, since deviations of 20 to 40% from the baseline period were not considered statistically different in most cases.

For example, analyzing the ETA 40-MID model of location 1, we reject the equality of all future periods compared to the baseline, whereas the test does not reject H$_0$ for ETA 40-LOW model, which also showed significant differences from the comparison period (-40, -54 and -52% for 2025s, 2055s and 2085s, respectively). It should be emphasized that the data presented in Table 5 refer to annual averages for periods of 30 years, but reveal nothing about the behavior of the data distribution.

The statistic of the test (D) is the maximum distance between the distributions tested and when compared among future periods it also indicates which of them is more different from the baseline. It is evident from the results in Table 7 that the highest D values were obtained for simulations of the end of the century (2085s), when it is expected greater uncertainty on climate projections and when models differ more strongly.

The individual analysis of the hypothesis test served to reveal the differences generated by each model and location but does not allow general inferences about the region. Thus, the same test was performed for the distributions of all 70 simulations presented in Figure 7. The test revealed that there is not enough statistical evidence to assume that the 2025s period is different from the baseline, but the test rejects H$_0$ for the other two periods with p-values of 0.1065, 0.005 and 0.00073 and D of 0.2000, 0.2857 and 0.3286 for 2025s, 2055s and 2085s, respectively.

Conclusions

In this study, future irrigation water requirements in the most important agricultural region of Rio Grande do Sul were determined to analyze how climate change could affect agriculture. The use of different projections of meteorological data provided by global and regional models enabled to identify the degree of uncertainty associated with forecasts of water demand.

In general, what can be concluded is that if the predictions of global models are confirmed, the region is likely to be adversely affected, since the simulations with the SWAP model using these projections indicated an increase in irrigation requirements for soybean. On the other hand, regional models are more accurate to the scale of application and spatial resolution, and in this study they showed significant reductions in water demand for agriculture due to an increase in precipitation till the end of the century.

It should be noted that these projections are not able to satisfactorily reproduce the occurrence of extreme events, such as hail, heavy rains, frosts, which are the major causes of crop losses. However, it is possible that changes in climate conditions in the region will make these events more frequent and, therefore, studies such as this serve to support forms of adaptation even in the short term.

Although it cannot be concluded with certainty whether the impacts will be positive or negative when analyzing the models individually, the results of the hypothesis test performed for all simulations supports the premise that the water demand for irrigation in the near future (2025s) are not statistically different from the baseline (1960-1990), but the opposite was observed from the middle of the century on. These results may support water management studies to define the size of reservoirs, based on climate projections. In this study, we can conclude that global climate projections suggest that larger reservoirs will be needed in the future if it is economically possible to attend 100% of the demand. On the other hand, the projections of the regional climate model ETA suggest that smaller reservoirs will be needed, since it projects an increase in precipitation.

It was shown that the uncertainties are large, although many sources have not been considered. In this study, we assumed that cropping patterns are not affected by climate change, as well as agricultural or irrigation expansion scenarios were not considered. Furthermore, a more detailed analysis of the meteorological variables may indicate which of the weather variables influence the most the results.

However, unlike other sectors, agriculture adapts quickly in the face of atypical weather events. This possibility has not been included here, but it is known that the effects of climate change simulated in this work could have been reduced by assumptions such as genetic enhancements, fertilization management techniques and other choices of crops adapted to the new climate.

Acknowledgements

We are very thankful to the National Council of Scientific and Technological Development (CNPq) for sponsoring this research and acknowledge all anonymous reviewers for their suggestions and improvements in this paper.

References

1. World Bank (2009) World Development Report 2010: Development and Climate Change. World Bank, Washington.

2. Trenberth K, Jones P (2007) Observations: surface and atmospheric climate change. In: Solomon S, Qin D, Manning M, Chen Z, Marquis M et al. (eds), Climate Change 2007: The Physical Science Basis. Cambridge University Press, Cambridge, UK and New York, USA.

3. Intergovernmental Panel on Climate Change (IPCC) (2007) Climate Change 2007: Synthesis Report. Core Writing Team, Pachauri, R; Reisinger, A. (Eds). Geneva: IPCC.

4. Hwan Y, Yong C, Hyun L, Gyeong O, Koun Y (2013) Climate change impacts

on water storage requirements of an agricultural reservoir considering changes in land use and rice growing season in Korea. Agr Water Manage 117: 43-54.

5. Bates B, Kundzewicz Z, Wu S, Palutikof J (Eds) (2008) Climate Change and Water Technical Paper of the Intergovernmental Panel on climate Change, IPCC Secretariat, Geneva.

6. Hamada E, Gonçalves R, Orsini J, Ghini R (2008) Future climate scenarios for Brazil. In: Ghini R, Hamada E (eds) Climate Changes: impacts on plant diseases in Brasil. Brasília-DF, EMBRAPA Informação Tecnológica.

7. Harmsen E, Niller N, Schlegel N, Gonzalez J (2009) Seasonal climate change impacts on evapotranspiration, precipitation deficit and crop yield in Puerto Rico. Agr Water Manage 96: 1085-1095.

8. Nkomozepi T, Chung S (2012) Assessing the trends and uncertainty of maize net irrigation water requirement estimated from climate change projections for Zimbabwe. Agri Water Manage 111: 60-67.

9. Barry R, Chorley R (2009) Atmosphere, Weather and Climate, ninth ed. Routledge, New York.

10. 10. Islam A, Ahuja L, Garcia L, Ma L, Saseendran A, et al. (2012) Modeling the impacts of climate change on irrigated corn production in the Central Great Plains. Agri Water Manage 110: 94-108.

11. Jalota S, Kaur H, Kaur S, Vashisht B (2013) Impact of climate change scenarios on yield, water and nitrogen-balance and use efficiency of rice-wheat cropping system. Agri Water Manage 116: 29-38.

12. Bocchiola D, Nana E, Soncini A (2013) Impact of climate change scenarios on crop yield and water footprint of maize in the Po valley of Italy. Agr Water Manage 116: 50-61.

13. Yoo S, Choi J, Nam W, Hong E (2012) Analysis of design water requirement of paddy rice using frequency analysis affected by climate change in South Korea. Agri Water Manage 112: 33-42.

14. Mehta V, Haden V, Joyce B, Purkey D, Jackson L (2013) Irrigation demand and supply, given projections of climate and land-use change, in Yolo County, California. Agri Water Manage 117: 70-82.

15. http://dados.fee.tche.br/

16. Daccachea A, Weatherhead E, Stalhamb M, Knox J (2011) Impacts of climate change on irrigated potato production in a humid climate. Agr Forest Meteorol 151: 1641-1653.

17. Stone D, Knutti R (2010) Weather and Climate. In: Fung C, Lopez A, New M. (eds), Modelling the Impact of Climate Change on Water Resources, Willey-Blackwell, Oxford.

18. http://www.inmet.gov.br/portal/

19. Van Dam J (2000) Field scale water flow and solute transport. SWAP model concepts, parameter estimation and case studies. PhD thesis, Wageningen University.

20. Richards L A (1931) Capillary conduction of liquids through porous mediums. J Appl Phys 1: 318-333.

21. Van Genuchten M (1980) A closed-form equation for predicting the hydraulic conductivity of unsaturated soils. Soil Sci Soc Am J 44: 892-898.

22. Mualem Y (1976) A new model for predicting the hydraulic conductivity of unsaturated porous media. Water Resour Res 12: 513-522.

23. Van Genuchten, M, Simunek F, Leij F, Sejna M (2009) The RETC code (version 6.02) for quantifying the hydraulic functions of unsaturated soils.

24. Mesinger F, Janjic Z (1974) Noise due to time-dependent boundary conditions in limited area models. The GARP Programme on Numerical Experimentation, Rep. No. 4, WMO, Geneva.

25. Black T (1994) The new NMC mesoscale Eta Model: Description and forecast examples. Weather Forecasting 9: 265-278.

26. Gellens D, Roulin E (1998) Streamflow response of Belgian Catchments to IPCC climate change scenarios. J Hydrol 210: 242-258.

27. Kroes J, Van Dam J, Groenendijk P, Hendriks R, Jacobs C (2008) SWAP version 3.2. Theory description and user manual. Wageningen, Alterra, Alterra Report 1649.

6

Assessment of Natural Self Restoration of the Water of Al-Mahmoudia Canal, Western Part of Nile Delta, Egypt

Alaa F. Abukila*

Drainage Research Institute, National Water Research Center, El-Qanater El-Khairiya, Egypt

Abstract

Al-Mahmoudia canal in northern edge of Beheira Governorate, west part of Nile Delta, has important role in the economic development and prosperity of the people in Beheira and Alexandria Governorates. It has been exploited to support agriculture, fisheries, public water supply, industry, hydroelectric power and recreation. The continuing deterioration of water quality in the canal has become a routine water pollution case. Therefore, it is necessary to solve the canal pollution problems and upgrade the water quality. The objective of this study was to characterize and understand the water quality of Al-Mahmoudia canal. Samples of water were collected monthly from eleven locations for 12 month during 2010-2011. *In situ* measurements included; Temp, TDS, pH and DO, and laboratory determinations included TSS, BOD_5, COD, NO^-_3, NH_4^+, TC and FC, in addition to Cd, Cu, Fe, Mn, Ni, Pb and Zn. Natural self-purification model based on oxygen sag curve introduced by Streeter and Phelps was applied. The obtained results showed that the majority of water quality problems of Al-Mahmoudia canal are due to receive low grade water quality of Rosetta Branch. Natural self-purification is calculated and observed in two cases. The first is normal case, which no drainage water is discharging into Al-Mahmoudia canal; hence, Edko irrigation pump station is stopping lift drainage water of Zarkon drain into the canal. The result of this case showed that the deoxygenation rate is higher than the reoxygenation rate from km 14 to km 17.87 of Al-Mahmoudia canal. The second is simulated case, which simulated Edko irrigation pump station is lifting drainage water of Zarkon drain into the canal. The result of this case showed the deoxygenation rate is higher than the reoxygenation rate from km 14 to km 18.06 of the canal and the reach need 10.83 km to get rid of the influence of pollutants from Edko irrigation pump station discharge. The difference between conceptual and pragmatic approaches was used in identifying the most polluted reaches by non-point pollution sources along the canal. According to the obtained result the difference between observed and calculated values in the watercourse from south to north direction has been increased and contribution of the nonpoint pollution sources at Al-Mahmoudia canal is related to the four reaches.

Keywords: Al-Mahmoudia canal; Oxygen sag curve; Dissolved oxygen deficit; Natural self-purification

Introduction

Al-Mahmoudia canal is located at the northern edge of Beheira Governorate. The canal off-takes from Rosetta branch at km 194.200. The actual served area for the canal is 130,200 hectares. The total length of the canal is 77.170 km and there are seventy canals off-take from this canal. Al-Mahmoudia canal has three sources of water; two fresh water sources which are from Rosetta branch via El-Atf pump stations at the head of the canal, and Al-Khandaq Eastern canal at km 13.200 on Al-Mahmoudia canal, the third is drainage water from Zarkon drain at km 8.500 on Al-Mahmoudia canal *via* Edko irrigation pump station which lifting part of Zarkon drain water into Al-Mahmoudia canal. The canal receives pollutants from point and non-point sources [1,2]. These pollutants lead to significant deterioration of the quality of the water in the canal. The point source of pollutants is Edko drain in Beheira Governorate which supplies Al-Mahmoudia canal with water in order to cover irrigation needs along the canal and the drinking water for Alexandria city. The intake of the water treatment plant of Alexandria and many water treatment plants of Beheira Governorates are the upstream of these mixed three sources (Figure 1). The water treatment plants which feeding by Al-Mahmoudia canal are listed in the table 1. In Alexandria, water supply companies are producing various amounts of water in different seasons. In summer due to huge number of tourists, the water demands increases and thus the production too. Al-Mahmoudia canal suffered from the negative effects of nonpoint pollution sources [1,2].

The Dissolved Oxygen (DO) concentration is a primary measure of a stream's health, but the dissolved oxygen concentration responds to the Biochemical Oxygen Demand (BOD) load. Many streams in Egypt have suffered from DO deficit, which is very critical to aquatic life [3].

Water quality modeling in a river has developed from the pioneering work of Streeter and Phelps [4] who developed a balance between the dissolved oxygen supply rate from reaeration and the dissolved oxygen consumption rate from stabilization of an organic waste in which the Biochemical Oxygen Demand (BOD) deoxygenation rate was expressed as an empirical first order reaction, producing the classic dissolved oxygen sag model and it was really a great achievement when Streeter and Phelps [4], in 1925, were able to propose a mathematical equation that demonstrating how dissolved oxygen in the Ohio River decreased with downstream distance due to degradation of soluble organic biochemical oxygen demand. By considering a first order of degradation reaction, for a constant river velocity [5].

This paper presents calculations on stream sanitation and the main portion covers the evaluation of water assimilative capacities of Al-Mahmoudia canal. The procedures include classical conceptual approaches and pragmatic approaches; the conceptual approaches use simulation models based on Streeter-Phelps equation [4]. The pragmatica approaches uses observed Dissolved Oxygen (DO) and Biochemical Oxygen Demand (BOD_5) levels which are measured

*Corresponding author: Alaa F. Abukila, Drainage Research Institute, National Water Research Center, El-Qanater El-Khairiya, Post Code 13621/5, Egypt E-mail: Alaafg@gmail.com

Figure 1: Al-Mahmoudia canal.

Governorate	Water treatment plant	Production (m³/day)
Beheira	Algadih*	25,000
	Ficha*	25,000
	MonchatNassar	25,000
	AbouHommos	100,000
	Com Alkuenatur	250,000
	Kafr El-Dawar	100,000
Alexandria	Al-Sayouif	970,000
	Al-Mamoura	240,000
	Bab Sharki	630,000
	Al-Manshia	420,000
	Forn el garia	50,000
	Al-Nozha	200,000

Table 1: Water supply companies which feeding by Al-Mahmoudiacanal.

at several sampling points along Al-Mahmoudia canal reach. Both approaches are useful for estimating oxygen deficit and related dissolved oxygen with respect to time and space. The difference between conceptual and pragmatic approaches was used in identifying the most polluted reaches by non-point pollution sources along the canal.

Materials and Methods

Water sampling

Samples of water were collected monthly from eleven locations, during a period of 12 months starting from May 2010 to April 2011, nine from Al-Mahmoudia canal, one from Al-Khandaq Eastern canal and one from Zarkon drain (Figure 1). The collected water samples were transported preserved and the physical, chemical and biological analyses were determined by the procedures recommended in the Standard Methods for the Examination of Water and Wastewater [6].

In situ measurements

The *in situ* measured parameters: temperature, Total Dissolved Solid (TDS, mg/l), pH, Dissolved Oxygen (DO, mg/l) were carried out by WTW Multi 350i multimeter.

Laboratory measurements

Total Suspended Solid (TSS, mg/l), Biochemical Oxygen Demand (BOD$_5$, mg/l), Chemical Oxygen Demand (COD, mg/l), nitrate (NO$_3^-$,

mg/l), ammonium (NH$_4^+$, mg/l), Boron (B, mg/l) total coliform (TC, CFU/100ml) and fecal coliform (FC, CFU/100ml) were carried out according to APHA [6]. Metals were determined after preliminary treatment of water sample [6] and the concentrations of Cd, Cu, Fe, Mn, Ni, Pb and Zn were measured by atomic absorption spectrophotometer (Perkin Elmer 5300 DV).

Natural self-purification of Al-Mahmoudia canal

The Natural self-purification model consists of five measures. These five measures are described as follows:

Dissolved oxygen saturation, DO$_{sat}$: Which represents values for various water temperatures can be computed using the American Society of Civil Engineers formula [7] was calculated as equation (1).

$$DO_{Sat=14.652-0.41022T+0.0079910T^2-0.000077774T^3}$$ (1)

Where

DO$_{sat}$ = dissolved oxygen saturation concentration, mg/l

T = water temperature, °C

The DO$_{sat}$ concentrations generated by the formula must be corrected for differences in air pressure caused by air temperature changes and for elevation above the Mean Sea Level (MSL). The correction factor can be calculated as equation (2).

$$f = \frac{2116.8 - (0.08 - 0.000115A) \times E}{2116.8}$$ (2)

The corrected DO$_{sat}$ = outputequation$_1$ × outputequation$_2$ (3)

Where

f = correction factor for above MSL

A = air temperature, °C

E = elevation of the site, feet above MSL

Because elevation of Al-Mahmoudia canal is between 0 to less than 2 meter above the MSL, the equations 2 and equation 3 are neglected.

Ultimate BOD$_5$, L$_a$: The BOD test measures (1) the molecular oxygen consumed during a specific incubation period for the biochemical degradation of organic matter (carbonaceous BOD$_5$); (2) oxygen used to oxidize inorganic material such as sulfide and ferrous iron; and (3) reduced forms of nitrogen (nitrogenous BOD$_5$) with an inhibitor (trichloromethyl pyridine). If an inhibiting chemical is not used, the oxygen demand measured is the sum of carbonaceous and nitrogenous demands, so-called total BOD$_5$ or ultimate BOD$_5$. Ultimate BOD$_5$ can be computed according to Lee and Lin [8] which was calculated using equation (4).

$$L_a = BOD_5 \times 1.46$$ (4)

Where

L$_a$ = Ultimate BOD$_5$, mg/l

Streeter-Phelps oxygen sag formula: The method most widely used for assessing the oxygen resources in streams and rivers subjected to effluent discharges is the Streeter-Phelps oxygen sag formula that was developed for use on the Ohio River in 1914. The well-known formula is defined as follows [4] was calculated as equation (5).

$$D_t = \frac{k_d \times L_{aU}}{k_2 - K_d}\left(10^{-k_d t} - 10^{-k_2 t}\right) + D_a 10^{-K_2 t}$$ (5)

Where

D_t = DO saturation deficit downstream, mg/l (DO_{sat} - DOa) at time t

t = time of travel from two points, days

D_a = initial DO saturation deficit of upstream water, kg/day

L_{au} = ultimate upstream biochemical oxygen demand (BOD_5), kg/day

k_d = deoxygenation coefficient to the base 10, per day

k_2 = reoxygenation coefficient to the base 10, per day

Deoxygenationrate, k_d: The Streeter-Phelps oxygen sag equation is based on two assumptions: (1) at any instant the deoxygenation rate is directly proportional to the amount of oxidizable organic material present; and (2) the reoxygenation rate is directly proportional to the dissolved oxygen deficit. According to Lee and Shun Dar Lin [8] mathematical expressions for k_d can calculated as equation (6).

$$k_d = \frac{1}{\Delta t} \log \frac{L_{au}}{L_{ad}} \tag{6}$$

Where

k_d = Deoxygenation rate, day

Δt = time of travel from upstream to downstream, days

L_{ad} = ultimate downstream biochemical oxygen demand (BOD_5), mg/l

The K_d values are needed to correct for stream temperature according to the equation (7)

$$k_{d@T} = k_{d@20} \times (1.047)^{T-20} \tag{7}$$

k_d value at any temperature T °C and $k_{d@20}$ = k_d value at 20

Because BOD_5 has determined laboratory at 20°C so equation (7) not use.

Reoxygenation rate, k_2: According to Lee and Shun Dar Lin [8] mathematical expressions for k_d can calculated according to equation (8).

$$k_2 = k_d \frac{\overline{L}}{\overline{D}} - \frac{\Delta D}{2.303 \Delta t \overline{D}} \tag{8}$$

Where

k_2 = Reoxygenation rate, day

$\overline{L}$ = Average Ultimate BOD_5 load upstream and downstream (Kg/day)

$\overline{D}$ = Average Dissolved oxygen deficit load upstream and downstream (Kg/day)

ΔD = Difference Dissolved oxygen deficit upstream and downstream (Kg/day)

The k_2 values are needed to correct for stream temperature according to the equation (9)

$$k_{2@T} = k_{d@20} \times (1.02)^{T-20} \tag{9}$$

K_2 value at any temperature T °C and $k_{2@20}$ = k_2 value at 20

Because BOD_5 has determined laboratory at 20 °C so equation (9) not use.

Procedure of applied natural self-purification of Al-Mahmoudia canal and evaluated contribute of nonpoint source pollution

a) Natural self purification is calculated for the reach from km 14 to 24.2 subsequent to Edko irrigation pump station and Al-Khandaq Eastern canal discharge at km 8.5 & 13.20, respectively, (Figure 2).

b) Natural self purification is applied in two cases:

1. The first is normal case, which Edko irrigation pump station is stopped; hence, it has stopped from June, 2009.

2. The second is simulated case, which simulated data for DO, BOD_5 of Edko irrigation pump station dischargeat the same reach as equation (10).

$$C_d = \frac{Q_u \times C_u + Q_e \times C_e}{Q_u + Q_e} \tag{10}$$

Where

C_d = Completely mixed new constituent concentration at the test location, mg/l.

Q_u = Discharge at the test location, m³/month.

C_u = Constituent concentration at the test location, mg/l

Q_e = Edko irrigation pump station discharge, According to United States Agency for International Development [9] available drainage water at this pump station is 13,000,000 m³/month.

C_e = Constituent concentration of the effluent of Edko irrigation pump station, mg/l.

C) Compared calculated data with observed value along Al-Mahmoudia canal to assess the contribution of nonpoint sources pollution.

Statistical analysis

The obtained data is analyzed statistically using SPSS software Version 16.0.2 [10].

Results and Discussion

Assessment of water quantity which feeding Al-Mahmoudia canal

According to Egyptian Ministry of Water Recourses and Irrigation,

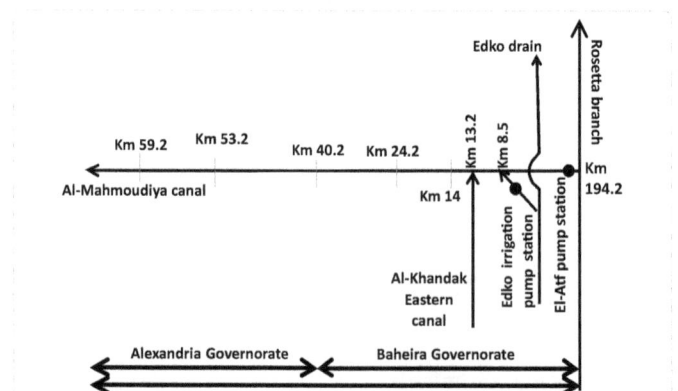

Figure 2: Schematic diagram of the measured water samples.

Assessment of Natural Self Restoration of the Water of Al-Mahmoudia Canal, Western Part...

45

a quantity of 3.360 billion m³/year have been discharged into Al-Mahmoudia canal (Table 2) from both Rosetta branch (km 1940) via El-Atf pump stations at the head of the canal and Al-Khandaq Eastern at km 13.200 on Al-Mahmoudia canal and the third water source is the drainage water from Zarkon drain at km 8.500 on Al-Mahmoudia canal via Edko irrigation pump station which lifts part of Zarkon drain into the canal. It is worthy to not that Edko Irrigation Pump Station has been stopped science June, 2009 due to water quality problems. This is because many drinking water intakes located on downstream of the mixing point.

Assessment of water quality which feeding Al-Mahmoudia canal

Table 3 represents the statistical analysis of the water quality of Al-Mahmoudia canal at km 0.0; out fall of Al-Khandaq Eastern at km 13.200 on Al-Mahmoudia canal and Zarkon drain. The variations in water quality can be summarized as follows:

Al-Mahmoudia canal (km 0.0):

- pH of water are within the permitted standard range (pH 6.5-8.4) according to FAO [11].

- The concentrations of TDS in water varied from 281 to 546 mg/l. No health-based guideline value for TDS has been proposed by WHO [12]. However, the palatability of water with a TDS level of less than 600 mg/litre is generally considered to be good since drinking-water becomes significantly and increasingly unpalatable at TDS levels greater than about 1000 mg/litre [12]. The quality of irrigation water is defined by the type and the concentrations of dissolved salts and substances. The most significations are the cations of calcium, magnesium, sodium and the anions of carbonate, sulfate, and chloride. They are apart from the absolute concentrations of ions [13]. The quality criteria of irrigation water have deducted from FAO regulations for three hazard categories: I) No problems, II) Gradual increasing problems from the continuous use of water, III) Immediate development of severe problems [11]. The water quality for irrigation use according to the criteria indicates that is no problem when Al-Mahmoudia canal water is used for irrigation.

- The concentrations of TSS in the waters varied from 1.10 to 5.80 mg/l.

- The median value of dissolved oxygen concentrationsis 5.62 mg O_2/L. This indicates that pollution loading is depleting oxygen levels.

- The median values of BOD_5 and COD concentrations are 17.60 mg BOD_5/L and 29.33 mg COD/L. which are reflecting the high organic load in water of Al-Mahmoudia canal which are from Rosetta branch.

- Nitrate and ammonia concentrations were within the permissible limits (<10 and<5, respectively) according to FAO [11].

- Fecal coliform counts exceeded the WHO Guidelines [14] of 1000 CFU/100 ml in almost all waterhence, the median is 3050 CFU/100ml. This is an indication of the discharge of human wastes in Al-Mahmoudia canal through Rosetta branch.

According to United States Agency for International Development [9] Rosetta Branch, starting downstream of Delta Barrage receives

relatively high concentrations of organic compounds, nutrients and oil & grease. The major sources of pollution are Rahawy drain (which receives part of Greater Cairo wastewater), Sabal drain, El-Tahrrer drain, Zawiet El-Bahr drain and Tala drain. At Kafr El-Zayat, Rosetta branch receives wastewater from Maleya and salt and soda companies. This indicates that the majority of water quality problems are occurring in the intake of Al-Mahmoudia canal due to receive low-grade water quality from Rosetta Branch.

Outfall of Al-Khandaq eastern canal: Al-Khandaq Eastern canal has discharged at km 13.200 on Al-Mahmoudia canal. The water quality of this canal can be summarized as follows:-

- pH of water are withinthe permitted standard.

- The concentrations of TDS are less than TDS concentrations in Al-Mahmoudia canal. The maximum concentration is 317mg TDS/L.

- The concentrations of TSS in the waters varied from 2.95 to 9.5mg/l.

- Dissolved oxygen concentrations ranged from 5.17 to 7.31mg/l.

- The median values of BOD_5 and COD concentrations are 11 mg BOD_5/L and 19 mg COD/L. which are reflecting the organic load received in Al-Khandaq Eastern canal.

- Nitrate and ammonia concentrations were within the permissible limits (<10 and<5, respectively) according to FAO [11].

- Fecal coliform counts exceeded WHO Guidelines [14] of 1000 CFU/100 ml in almost all water hence, the median is 2550 CFU/100ml. This is an indication of the discharge of human wastes intoAl-Khandaq Eastern canal.

The mixing of the drainage water at Etay El-Barud pump station in Al-Khandaq Eastern canal lowered water quality of Al-Khandaq Eastern canal downstream of the point of re-supply. More water with high pollution load results in worse water quality. This reproduces high concentration of BOD_5, COD, total coliform and fecal coliform. However,the concentration of contaminates in water of Al-Khandaq Eastern canal were less than Al-Mahmoudia canal.

ZarkonDrain: Zarkon drain discharges its water at km 8.500 of Al-Mahmoudia canal via Edko irrigation pump station which is lifting part of Zarkon drainage water into Al-Mahmoudia canal. As previously

Month	Discharge (billion m³/month)	
	El-Atf pump station	Al-Khandaq Eastern Canal
May, 2010	0.313024	0.0465
Jun, 2010	0.328546	0.0450
Jul, 2010	0.371786	0.0465
Aug, 2010	0.346819	0.0465
Sep, 2010	0.288056	0.0450
Oct, 2010	0.240332	0.0465
Nov, 2010	0.197236	0.0450
Dec, 2010	0.14797	0.0465
Jan, 2011	0.079902	0.0465
Feb, 2011	0.131224	0.0420
Mar, 2011	0.139648	0.0465
Apr, 2011	0.228372	0.0450
Total	2.812915	0.5475**3.360**

Table 2: Water Resources lifted by El-Atf pump station and Al-Khandaq Eastern Canal.

Location	Statistical analysis	pH	TDS (mg/l)	TSS (mg/l)	DO (mg/l)	BOD_5 (mg/l)	COD (mg/l)	NO_3-N (mg/l)	NH_4-N (mg/l)	Total Coliform (CFU/100ml)	Fecal Coliform (CFU/100ml)
Al-Mahmoudia canal (km 0.0)	Maximum	8.50	546	5.80	6.97	26	46	7.86	4.20	27000	17000
	Minimum	7.65	281	1.10	5.00	14	24	2.25	0.72	2400	1100
	Range	0.85	265	4.70	1.97	12	22	5.61	3.48	24600	15900
	Mean	8.03	383	3.13	5.70	18.10	30.61	3.38	2.08	8308	3975
	Median	8.03	384	3.07	5.62	18	29	2.54	2.12	7250	3050
	S.D.	0.25	77	1.29	0.58	3.09	5.96	1.88	1.04	6495	4238
	Skew	0.14	0.80	0.57	0.85	1.50	1.65	1.98	0.46	2.43	3.08
	Kurtosis	0.13	0.72	0.36	0.53	3.33	3.53	2.76	-0.05	7.01	10.11
Outfall of Al-Khandaq Eastern Canal	Maximum	8.47	317	9.50	7.31	14	24	4.31	1.56	15000	5000
	Minimum	7.80	227	2.95	5.17	7	12	1.54	0.04	1500	100
	Range	0.67	90	6.55	2.14	7	12	2.77	1.52	13500	4900
	Mean	8.20	260	4.95	6.23	10.67	18.25	2.51	0.47	8125	2525
	Median	8.21	254	4.90	6.20	11	19	2.11	0.29	9450	2550
	S.D.	0.18	30	1.66	0.52	1.97	3.52	0.97	0.43	3600	1128
	Skew	-0.73	0.57	1.98	0.06	-0.38	-0.44	1.01	1.85	-0.13	0.04
	Kurtosis	1.40	-0.81	5.32	1.84	-0.10	-0.30	-0.19	3.17	0.34	3.06
Zarkon Drain	Maximum	8.06	1074	7.00	3.80	36	60	11.00	4.20	82000	30000
	Minimum	7.45	537	5.50	0.37	24	40	5.11	1.44	9000	4000
	Range	0.61	537	1.50	3.43	12	20	5.89	2.76	73000	26000
	Mean	7.81	764	6.44	2.01	28.75	48.00	6.38	2.49	21558	9042
	Median	7.80	700	6.47	2.00	29	48	5.73	2.32	17400	7750
	S.D.	0.16	168	0.40	1.11	3.36	5.53	1.78	0.90	19279	6905
	Skew	-0.56	0.70	-0.84	0.39	0.50	0.56	2.15	0.55	3.30	2.94
	Kurtosis	1.13	-0.68	1.65	-0.71	0.64	0.73	4.05	-0.61	11.23	9.45

S.D. = Standard deviation

Table 3: Statistical analysis of water quality sources, which feeds Al-Mahmoudiacanal.

mentioned, Edko Irrigation Pump Station has been stopped from June, 2009 up till now. The water quality of this drain can be summarized as follows:

- pH of water are within the permitted standard.

- The concentrations of TDS in water varied from 537 to 1074 mg/l. It is less than the maximum limit (2000 mg/l) according to FAO [11].

- The concentrations of TSS in the waters varied from 7 to 5.5 mg/l.

- Dissolved oxygen concentrations ranged from 0.37 to 3.80 mg/l. This indicates that pollution loading is depleting oxygen levels.

- The median values of BOD_5 and COD were 29 mg BOD_5/l and 48 mg COD/l.

- Nitrate concentrations were within the permissible limits (<10) except at July, 2010 which was 11 mg NO_3-N/l.

- Ammonia concentrations were within the permissible limits (<5) according to FAO [11]

- The median count of fecal coliform was 7750 CFU/100ml. This is an indication of the discharge of human wastes into Zarkon drain.

According to United States Agency for International Development [9] Delta drains are mainly used for discharge of predominantly untreated or poorly treated wastewater (domestic and industrial), and for drainage of agricultural areas. Therefore, they contain high concentrations of various pollutants such as organic compounds (BOD_5, COD), nutrients, fecal bacteria, heavy metals and pesticides.

This explains increased concentrations of BOD_5, COD and fecal coliform.

Calculation procedures to evaluate natural self-purification in Al-Mahmoudia canal

Natural self-purification is calculated for the reach from km 14 to 24.2 subsequent to Edko irrigation pump station and Al-Khandaq Eastern canal discharge at km 8.5 & 13.20, respectively (Figure 2). Table 4-7 represent an example of calculation of natural self-purification using the data set from the reach from km 14 to 24.2

Natural self purification in Al-Mahmoudia canal according to case 1: Case 1 represents no drainage water is discharging into Al-Mahmoudia canal. Hence, Edko irrigation pump station is stopping lift drainage water of Zarkon drain into the canal.

Figure 3, illustrates the oxygen sag curve based on dissolved oxygen deficit. The data showed that dissolved oxygen deficit increased with distance and the lowest point of the oxygen sag curve (critical point) is at km 3.87. This point out that the deoxygenation rate is higher than the reoxygenation rate from km 0.0 to km 3.87, then dissolved oxygen deficit decreased with distance, consequently after km 3.87. Thus, the reoxygenation rate is higher than deoxygenation rate. It is clear that the first point on the oxygen sag curve where the oxygen deficit is less than the oxygen deficit at km 0.0 (restore activity point) is at km 10.31.

Figure 4 pointed out to the oxygen sag curve based on dissolved oxygen andas a result, the same trends for dissolved oxygen, the critical point is at km 3.87. Therefore, from km 0.0 to km 3.87 the deoxygenation rate is higher than the reoxygenation rate. Hence, dissolved oxygen was 5.59 and 5.44 mg/l, respectively. The restore activity point is at km 10.31and was 5.62 mg DO/l. According to Chapman, [15] the release into stream of untreated domestic or industrial wastes high in organic

Case	Q_b (m³/month)	Q_e (m³/month)	Q_b (m³/sec)	Q_e (m³/sec)	Q_a (m³/sec)	V (m/sec)	V (Km/h)
1	273,232,200	242,355,633	105.41	93.50	99.46	0.99	3.58
2	286,232,200	255,355,633	110.43	98.52	104.47	1.04	3.76

Q_b= Q at Km 14, Q_e= Q at Km 24.2, Q_a= 0.5(Q_b + Q_e), V=velocity

Table 4: Measured hydraulic data of the stream.

Case	Temp ºC	BOD_5 (mg/l)	BOD_5 (Kg/day)	L_a (Kg/day)	DO_{sat} (mg/l)	DO (mg/l)	DO deficit (mg/l)	DO deficit (Kg/day)
Station 1 (ZawyetGazal town at Km 14)								
1	24.44	16.67	151,826	221,666	8.2640	5.59	2.67	24,354
2	24.44	17.22	164,297	239,874	8.2640	5.43	2.83	27,039
Station 2 (AbouHommos city at km 24.2)								
1	24.47	14.52	117,300	171,258	8.2592	5.14	3.12	25,199
2	24.47	15.24	129,721	189,392	8.2592	4.98	3.28	27,912

BOD_5(Kg/day)= $Col_{3 \, of \, Table \, 5}$, X $Col_{4 \, of \, Table \, 4}$, X 60X60X24/1000, L_a (Kg/day)= $Col_{3 \, of \, Table \, 5}$, X 1.46 DO_{sat} compute from equation, DO deficit (mg/l) = $Col_{6 \, of \, Table \, 5}$, − $Col_{7 \, of \, Table \, 5}$,
DO deficit (Kg/day)= $Col_{8 \, of \, Table \, 5}$, X $Col_{4 \, of \, Table \, 4}$, X 60X60X24/1000

Table 5: Measured field and laboratory chemical characteristic data of the Stream.

Case	Δt	K_d (day)	$\overline{L}$ (Kg/day)	$\overline{D}$ (Kg/day)	ΔD (Kg/day)	K_2 (day)
1	0.12	0.94	196,462	24,776	845	7.36
2	0.11	0.91	214,633	27,476	873	6.97

Δt = (distance between station, &station₂/$Col_{8of \, Table4}$)/24 K_d compute from equation₆
$\overline{L}$= Average Ultimate BOD_5 load upstream and downstream
$\overline{D}$=Average Dissolved oxygen deficit load upstream and downstream
ΔD= Difference Dissolved oxygen deficit upstream and downstreamK_2 compute from equation₈

Table 6: Calculation procedures to estimate decay rates.

t (day)	Case 1				Case 2			
	Distance (km)	DO deficit (Kg/day)	DO deficit (mg/l)	DO (mg/l)	Distance (km)	DO deficit (Kg/day)	DO deficit (mg/l)	DO (mg/l)
0.000	0.000	24354	2.67	5.59	0.000	27039	5.43	5.43
0.015	1.289	25161	2.76	5.50	1.354	27831	5.35	5.35
0.030	2.578	25586	2.81	5.45	2.708	28251	5.30	5.30
0.045	3.867	25721	2.82	5.44	4.062	28383	5.29	5.29
				Etc..... until km 50				

t= Proposed time step to perform calculations (day) = 0.015
Distance= $Col_{1of \, Table7}$ X $Col_{8of \, Table4}$X 24
DO deficit (Kg/day) compute from equation₅
DO deficit (mg/l)=$Col_{3of \, Table7}$X 1000/($Col_{4of \, Table4}$ 60 X 60 X 24)
DO (mg/l)=$Col_{6of \, Table5}$− $Col_{8of \, Table7}$

Table 7: Estimated DO deficit and DO with respect to time (days) and space (km).

Case	Parameters		Distance (Km)								
		Al-Mahmoudiacanal	14	17.87	18.06	24.31	24.83	27.5	39.78	53.96	59.11
		Reach	0.0	3.87	4.06	10.31	10.83	13.5	25.78	39.96	45.11
Case 1	DO deficit (Kg/day)	Calculated	24354	25712	-	24043	-	-	16938	11866	10417
		Observed	24354	-	-	25199	-	-	19812	14916	13489
		Difference	0.0	-	-	1156	-	-	2874	3050	3072
	DO (mg/l)	Calculated	5.59	5.44	-	5.62	-	-	6.40	6.96	7.12
		Observed	5.59	-	-	5.14	-	-	5.01	4.97	4.92
		Difference	-	-	-	0.48	-	-	1.39	1.99	2.2
	Critical point		-	√	-	-	-	-	-	-	-
	Restore activity point		-	-	-	√	-	-	-	-	-
Case 2	DO deficit (Kg/day)	Calculated	27039	-	28383	-	26657	-	-	-	-
	DO (mg/l)	Calculated	5.43	-	5.29	-	5.47	5.59	-	-	-
	Critical point		-	-	√	-	-	-	-	-	-
	Restore activity point		-	-	-	-	√	-	-	-	-

Table 8: Natural self-purification along Al-Mahmoudiacanal.

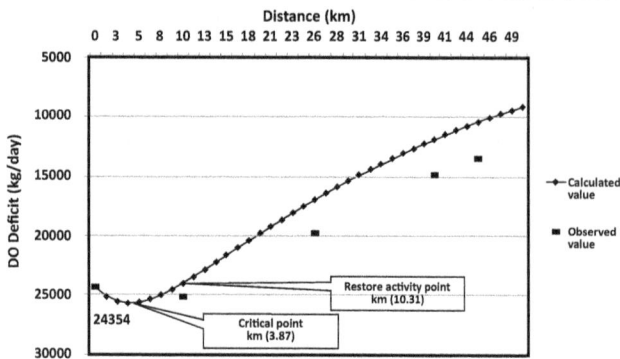

Figure 3: Calculated and observed value of DO deficit and related distance (Case 1).

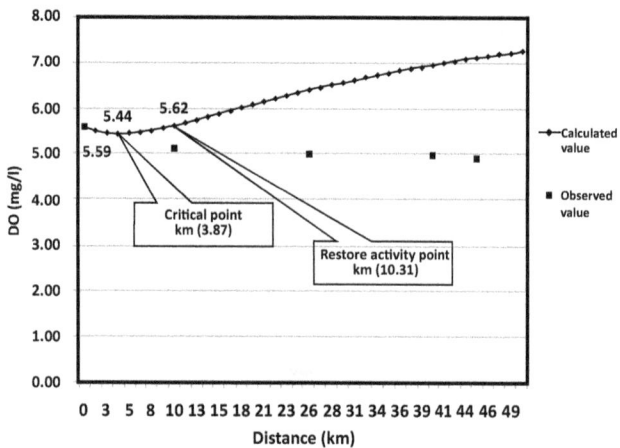

Figure 4: Calculated and observed value of DO and related distance (case1).

matter results in a marked decline in oxygen concentration (sometimes resulting in anoxia) and a release of ammonia and nitrite downstream of the effluent input. The effects on the river are directly linked to the ratio of effluent load to river water discharge. The most obvious effect of organic matter along the length of the river is the "oxygen-sag curve" which can be observed from a few kilometers to 100 km downstream of the input.

Table 8 compared the calculated data with the observed value to assess the impact of nonpoint pollution sources. The difference between observed and calculated values represents the contribution of nonpoint sources. The data showed that the difference between observed and calculated dissolved oxygen deficit was 1156, 2874, 3050 and 3072 Kg dissolve oxygen deficit /day, at km 10.31, 25.78, 39.96 and 45.11, respectively. As a result, the same trends can be reported for the dissolved oxygen hence, the difference between observed and calculated were 0.48, 1.39, 1.99 and 2.2mg DO/l, respectively. There is a progressive increase in the contribution of nonpoint pollution sources from south to north direction since the difference between observed and calculated values in the watercourse from south to north direction have been increased. Contribution of the nonpoint pollution source sat Al-Mahmoudia canal is related to the following four reaches:

- Reach No. 1: From Zawyet Gazal town at km 14to Abou Hommos city at km 24.2. Some degree of contribution of the nonpoint pollution sources.

- Reach No. 2: From Abou Hommos city at km 24.2to Kafr

El-Dawar city at km 42.0. Elevated degree of contribution of nonpoint pollution sources due to the effect of the former reach which presents residents and establishes the stables animals on the banks of Al-Mahmoudia canal, which are throwing the remnants of cattle, houses, and wash the cattle and the dumping of dead animals in Al-Mahmoudia canal.

- Reach No. 3: From Kafr El-Dawar city at km 42.0 to Khorshid city at km 55. High degree of contribution of the nonpoint pollution sources due to the effect of the previously reach and present many workshops and gasoline stations on the banks of Al-Mahmoudia canal.

- Reach No. 4: From Khorshid city at km 55 to Seiouf water treatment plant intake at km 61.3. Elevated degree of contribution of the nonpoint pollution sources due to the effect of the previously reach and effect of high population presents in Alexandria city.

Natural self-purification in Al-Mahmoudia canal according to case 2: Case 2 represents simulated case which simulated Edko irrigation pump station for lifting drainage water of Zarkon drain into the canal.

Figure 5 illustrated the oxygen sag curve based on dissolved oxygen deficit. The data showed that dissolved oxygen deficit had increased with distance until the critical point at km 4.06. Thus, from km 0.0 to km 4.06 the deoxygenation rate is higher than the reoxygenation rate. Then dissolved oxygen deficit had decreased, the restore activity point is at km 10.83.

Figure 6 showed the oxygen sag curve based on dissolved oxygen. As a result, the same trends for dissolved oxygen, the critical point is at km 4.06. Thus, before km 4.06 the deoxygenation rate was higher than the reoxygenation rate and after km 4.06 the deoxygenation rate is lower than the reoxygenation rate and the dissolved oxygen were 5.59 and 5.44 mg/l, for both respectively. The restore activity point is at km 10.31 and was 5.62 mg DO/l.

According to table 8, after 10.83 km the concentration of dissolved oxygen in the case 2 almost equal the initial concentration of dissolved oxygen in the same case. Therefore, the reach needs 10.83 km to get rid of the influence of pollutants from Edko irrigation pump station

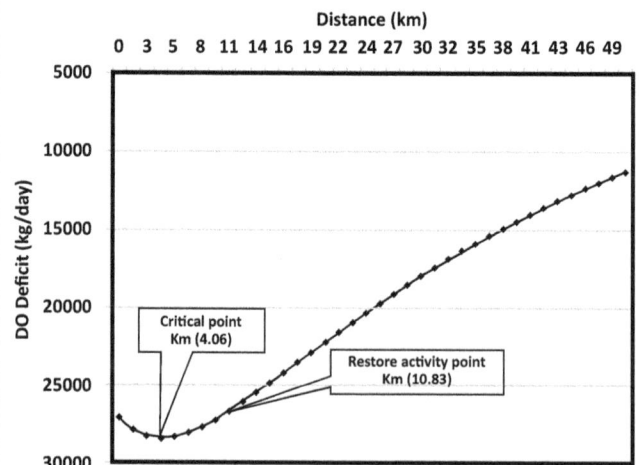

Figure 5: Calculated and observed value of DO deficit and related distance (case 2).

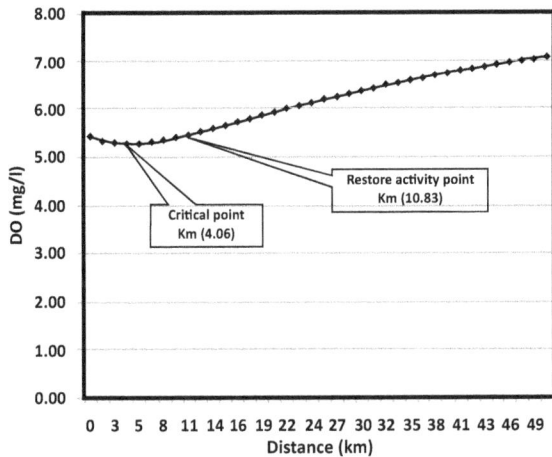

Figure 6: Calculated and observed value of DO and related distance (case2).

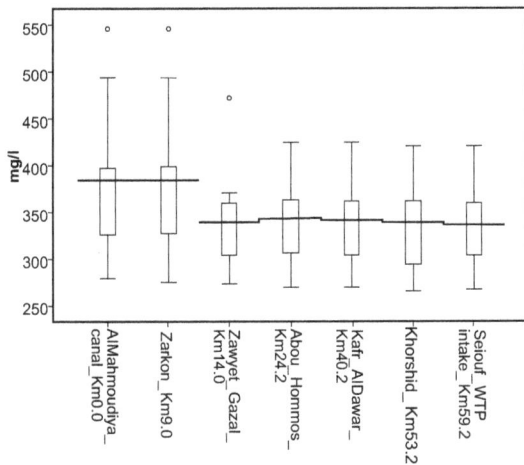

Figure 7: TDS along Al-Mahmoudia canal.

Assessment of water quality along Al-Mahmoudia canal

- There was a remarkable decrease in the levels of TDS in water samples collected from sites after Al-Khandaq Eastern canal outfall (Figure 7) due to the lower salinity of Al-Khandaq Eastern canal than that of Al-Mahmoudia canal.

- Figure 8 represents the pH values of the canal water during the study period the results indicated that pH values are within the standard pH permissible levels (pH 6.5-8.4) according to FAO [11].

- Figure 9-14 showed that there were great fluctuations in the levels of BOD_5, COD, NO_3, NH_4, total coliform and fecal coliform along Al-Mahmoudia canal. These variations can be summarized as follows:-

1. There is a marked decrease in the levels of BOD_5, COD, NO_3, NH_4, total coliform and fecal coliform in waters of sites located after the outfalls of Al-Khandaq Eastern canal. According to table 3, the pollutions load of Al-Khandaq Eastern canal is lower than that of Al-Mahmoudia canal.

2. Although there is no discharge from any drains into Al-Mahmoudia canal, the levels of tested parameters are

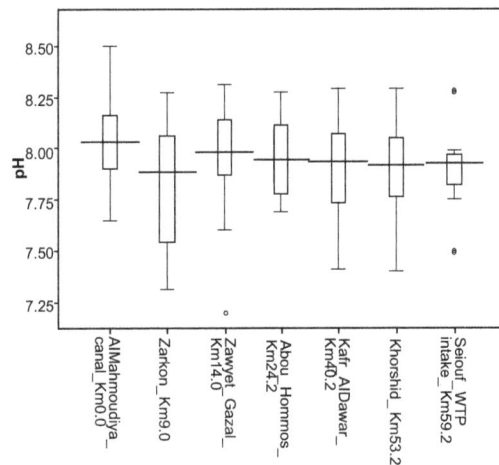

Figure 8: pH along Al-Mahmoudia canal.

discharge. As a result, most of water treatment plant in Beheira Governorate will be affected by Edko irrigation pump station discharge in a suit running,while all water treatment plant in Alexandria Governorate will not be affected by Edko irrigation pump station discharge in a suit running.

According to results obtained (Figures 3-6), the results of interplay of the biological oxidation and reaeration rates. Each is represented by first-order kinetics. In the early stages, oxidation greatly exceeds reaeration because of high CBOD concentrations and stream dissolved oxygen concentrations close to saturation (i.e., small deficit). Oxygen is used faster than it is resupplied, and stream dissolved oxygen concentrations decrease. As the wastes moves downstream, the consumption of oxygen decreases with the stabilization of wastes and the supply of oxygen from the atmosphere increases because of greater deficits. The driving force to replenish oxygen by atmospheric reaeration is directly proportional to the oxygen deficit, (i.e., low oxygen concentration). At some point downstream from the waste discharge, the decreasing utilization and the increasing supply are equal. This is the critical location, where the lowest concentration of dissolved oxygen occurs. Further downstream, the rate of supply exceeds the utilization rate, resulting in a full recovery of the dissolved oxygen concentration. This explanation is also supported by USEPA [16].

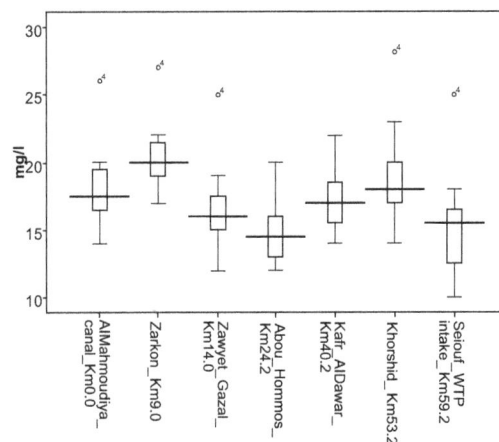

Figure 9: BOD_5 along Al-Mahmoudia canal.

flocculated along the canal. This is due to the impact of nonpoint pollution sources along the canal. These results agree with those previously mentioned for natural self-purification in Al-Mahmoudia canal.

3. The concentrations of the different metals in the waters along Al-Mahmoudia canal in addition to Al-Khandaq Eastern canal and Zarkon drain during the study period are within the normal range. As showing in table 9 there are no toxic levels of cadmium, copper, iron, manganese, nickel, lead and zinc since

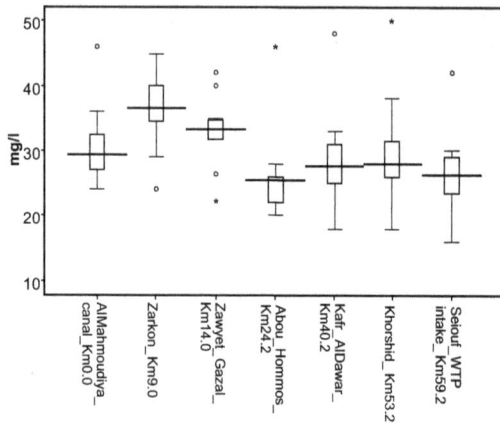

Figure 13: TC along Al-Mahmoudia canal.

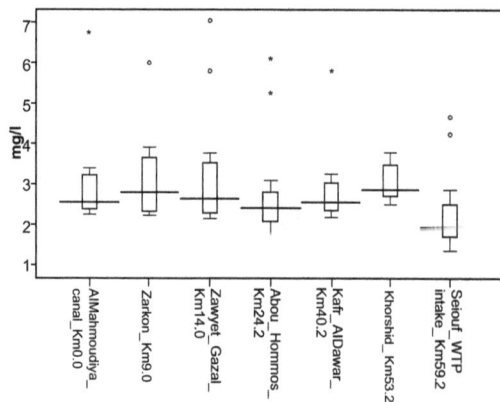

Figure 10: COD along Al-Mahmoudia canal.

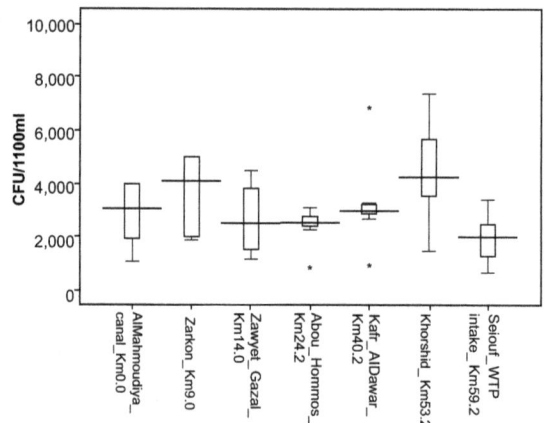

Figure 14: FC along Al-Mahmoudia canal.

these concentrations are less than the recommended maximum limits reported by the WHO [12].

Conclusion

Al-Mahmoudia canal is located at the northern edge of Beheira Governorate. It-off-takes Rosetta branch (km 194.200). It has three sources of water; two fresh water sources which are from Rosetta branch via El-Atf pump stations at the head of the canal, and Al-Khandaq Eastern canal and drainage water source which is from Zarkon drain at km 8.500 on Al-Mahmoudia canal via Edko irrigation pump station which lifting part of Zarkon drain into Mahmoudia canal. The majority of water quality problems are occurring in intake of Al-Mahmoudia canal due to receive low-grade water quality from Rosetta Branch. Edko Irrigation Pump Station has been stopped from June, 2009 due to water quality problems; especially many drinking water intakes are located on downstream of the mixing point. Contribution of the nonpoint pollution sources to Al-Mahmoudia canal is related to the following four reaches:

*   Reach No. 1: From Zawyet Gazal town at km 14 to Abou Hommos city at km 24.2. Some degree of contribution of the nonpoint pollution sources.

*   Reach No. 2: From Abou Hommos city at km 24.2 to Kafr El-Dawar city at km 42.0. Elevated degree of contribution of the nonpoint pollution sources.

*   Reach No. 3: From Kafr El-Dawar city at km 42.0 to Khorshid

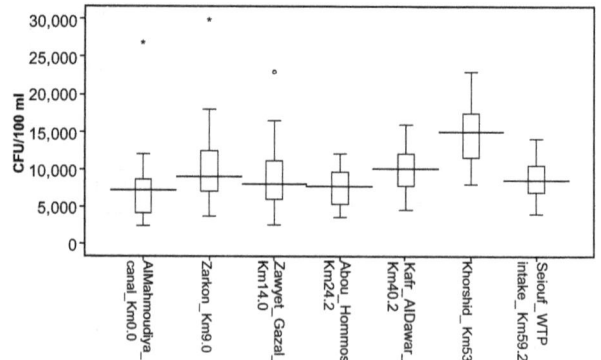

Figure 11: Nitrate along Al-Mahmoudia canal.

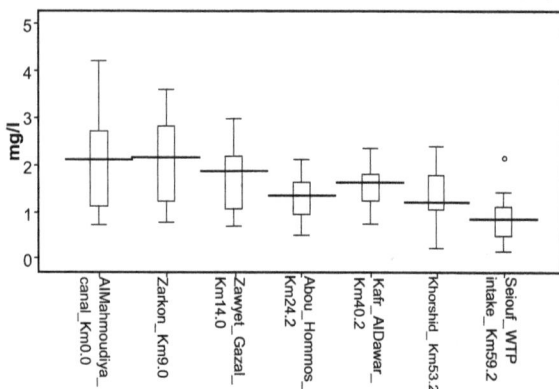

Figure 12: Ammonia along Al-Mahmoudia canal.

Location	Statistical	Cd	Cu	Fe	Mn	Ni	Pb	Zn
Al-Mahmoudia canal Km 0.0	max	L0.0001	0.007	0.081	0.335	0.048	L0.001	L0.0002
	min	L0.0001	L.0004	0.049	0.084	L0.0001	L0.001	L0.0002
	Median	L0.0001	L.0004	0.059	0.059	L0.0001	L0.001	L0.0002
Zarkon Km 9.0	max	L0.0001	0.006	0.083	0.338	0.052	L0.001	L0.0002
	min	L0.0001	L.0004	0.056	0.089	L0.0001	L0.001	L0.0002
	median	L0.0001	L.0004	0.069	0.065	L0.0001	L0.001	L0.0002
ZawyetGazal Km 14	max	L0.0001	0.006	0.077	0.270	0.04	L0.001	L0.0002
	min	L0.0001	L.0004	0.043	0.065	L0.0001	L0.001	L0.0002
	median	L0.0001	L.0004	0.052	0.038	L0.0001	L0.001	L0.0002
AbouHommos City Km 24.2	max	L0.0001	0.007	0.082	0.387	0.059	L0.001	L0.0002
	min	L0.0001	L.0004	0.053	0.089	L0.0001	L0.001	L0.0002
	median	L0.0001	L.0004	0.066	0.042	L0.0001	L0.001	L0.0002
Kafr El-Dawar city Km 42.0	max	L0.0001	0.006	0.079	0.131	0.018	0.004	0.002
	min	L0.0001	L.0004	0.055	0.057	L0.0001	L0.001	L0.0002
	median	L0.0001	L.0004	0.070	0.047	L0.0001	L0.001	L0.0002
Khorshid Km 55.0	max	L0.0001	0.007	0.077	0.343	0.052	0.001	L0.0002
	min	L0.0001	L.0004	0.045	0.073	L0.0001	L0.001	L0.0002
	median	L0.0001	L.0004	0.051	0.037	L0.0001	L0.001	L0.0002
Seiouf water treatment plant intake Km 61.3	max	L0.0001	0.007	0.062	0.273	0.043	L0.001	0.002
	min	L0.0001	L.0004	0.032	0.070	L0.0001	L0.001	L0.0002
	median	L0.0001	L.0004	0.030	0.044	L0.0001	L0.001	L0.0002
Outfall of Al-Khandaq Eastern canal	max	L0.0001	0.006	0.008	0.344	0.047	L0.001	L0.0002
	min	L0.0001	L.0004	0.004	0.046	L0.0001	L0.001	L0.0002
	median	L0.0001	L.0004	0.005	0.004	L0.0001	L0.001	L0.0002
Zarkon drain	max	L0.0001	0.006	0.074	0.340	0.052	L0.001	L0.0002
	min	L0.0001	L.0004	0.047	0.101	L0.0001	L0.001	L0.0002
	median	L0.0001	L.0004	0.063	0.066	L0.0001	L0.001	L0.0002

Detection limit for instrument are Cd 0.0001; Cu 0.0004; Fe 0.0001; Mn 0.0001; Ni 0.0005; Pb 0.001 and Zn 0.0002 mg/l.
WHO 2008 maximum contaminant limit for drinking water are Cd 0.0003; Cu 2; Fe 3 (PG); Mn 0.4; Ni 0.07; Pb 0.01 and Zn 3 mg/l (PG). PG means provisional guideline value

Table 9: Concentration of metals (mg/l) along Al-Mahmoudiacanal, outfall of Al-Khandaq Eastern canal and Zarkon drain.

city at km 55. High degree of contribution of the nonpoint pollution sources.

- Reach No. 4: From Khorshid city at km 55 to Seiouf water treatment plant intake at km 61.3. Elevated degree of contribution of the nonpoint pollution sources.

Natural self-purification of water of Al-Mahmoudia canal according to oxygen sag curve showed that the reach need 10.83 km to get rid of the influence of pollutants from Edko irrigation pump station discharge. As a result, most of water treatment plant in Beheira Governorate will be affected by Edko irrigation pump station discharge in a suit running, while all water treatment plant in Alexandria Governorate will not be affected by Edko irrigation pump station discharge in a suit running.

References

1. El-Gamal T, Meleha ME, Evelene SY (2009) The effect of main canal characteristics on irrigation improvement project. J Agric Sci Mansoura Univ 34: 1078-1079.

2. Hamdard M (2010) Fresh Water Swaps: Potential for Wastewater Reuse A Case Study of Alexandria, Egypt. UNESCO-IHE.

3. Elsokkary IH, AbuKila AF (2012) Prospective speculation for safe reuse of agricultural drainage water in irrigation. Alex Sci Exchange J 33: 134-152.

4. Streeter HW, Phelps EB (1958) A study of the pollution and natural purification of the Ohio River. Cincinatti: US Public Health Service.

5. Yudianto D, Xie Yuebo (2008) The development of simple dissolved oxygen sag curve in lowland non-tidal river by using matlab. Journal of Applied Science in Environmental Sanitation 3: 137-155.

6. American Public Health Assoc, American Water Works Assoc, Water Environment Federation (1998) Standard methods for the examination of water and wastewater. Washington DC, USA.

7. American Society of Civil Engineering Committee on Sanitary Engineering Research (1960) Solubility of atmospheric oxygen in water. J Sanitary Eng Div 86: 41-53.

8. Lee CC, Lin SD (2000) Handbook of environmental engineering calculation. McGraw-Hill.

9. US Agency for International Development (2003) Nile river water quality management study. Report NO.

10. SPSS software (2008) Version 16.0.2.

11. Ayers RS, Westcot DW (1985) Water quality for agriculture. FAO, Rome.

12. WHO (2008) Guidelines for drinking-water quality. (3rd edn), World Health Organization, Switzerland.

13. Loukas A (2010) Surface water quantity and quality assessment in Pinios River, Thessaly, Greece. Desalination 250: 266-273.

14. WHO Scientific Group (1989) Health guidelines for the use of wastewater in agriculture and aquaculture. Report of a WHO Scientific Group. World Health Organ Tech Rep Ser 778: 1-74.

15. Chapman D (1996) Water quality assessments: a guide to the use of biota, sediments and water in environmental monitoring. (2nd edn), Chapman and Hall, London.

16. USEPA (1997) Technical Guidance Manual for Developing Total Maximum Daily Loads, Book 2: Streams and Rivers, Part 1: Biochemical Oxygen Demand/ Dissolved Oxygen and Nutrients/ Eutrophication. United States Environmental Protection Agency. Office of Water, Washington DC, USA.

7

Energy Saving in a Variable-Inclination Archimedes Screw

Joaquim Monserrat*, Rubén García Ortiz, Lluis Cots and Javier Barragán

Department of Agroforestry Engineering, Universitat de Lleida, Lleida, Spain

Abstract

An analysis was undertaken of a new development of Archimedes screw consisting of a variable-inclination screw suitable for applications in which the downstream level may vary (e.g. sea level). As a result the energy consumption when the downstream level is low is less than that of a conventional, fixed Archimedes screw. In addition, investment costs are lower as a result of less expensive civil work being required. Moreover, this screw can also act as a check gate because it has a float at his downstream end. The flow rate for different rotational speeds was measured and a good fit was obtained with a developed graphical model. This paper analyses the operation of an Archimedes screw prototype pump of variable inclination and models its behavior. Its energy consumption is then measured and compared with that of conventional fixed inclination screw pumps.

Keywords: Archimedes screw; Discharge flow rate; Elevation height; Conventional pump; Variable inclination pump

Introduction

The Archimedes screw is a device for lifting water to low heads which dates back to the third century BC and is still in use today. Its main advantage is its effectiveness when lifting low-head debris-laden water. Though technological advances have logically seen various modifications to the device since its first use to the present day, its operating principle remains the same. Its principle applications are in drainage water pumping stations and water treatment plants. It has also proved valuable in installations where damage to aquatic life needs to be minimized. Given the antiquity of the device and the consequent empirical knowledge that has been acquired, little is to be found in the literature on its technical aspects. A manual was however written by Nagel [1] for the design of installations with this type of pump in which he explains a graphical method for calculating the flow rate in a screw pump according to the screw's geometry and rotational speed. Wijdieks and Bos [2] proposed a simple empirical equation to obtain the discharge from an Archimedes pump,

$$Q = k.n.D^3 \left[\frac{m^3}{s} \right] \tag{1}$$

Where: k is an empirical coefficient which depends on the shape of the screw, characterized by: S/D, d/D, and the inclination β [2], n is the turning speed (rev/s) and D is the outer diameter (m).

Rorres [3] developed computer software to find the volume of water lifted in one turn of the screw depending on the inner radius and pitch. From this he was able to obtain the optimum geometry values that maximize water volume per turn. However, the programme does not take into account the losses which occur in the gap between the screw and the casing. The screws are driven by electric motors the sizing of which depends on two basic factors: The discharge flow rate and the elevation height. Discharge can be regulated through advances in the control of motor turning speed, either through the use of variable frequency drives [4] or a gear reduction system. This paper analyses a new development of the conventional Archimedes screw. The modification involves variation of the inclination of the screw depending on the downstream water level, thereby obtaining a considerable energy saving in those applications in which the downstream level is variable (i.e. sea level).

Description

The pump used in this study was designed and manufactured

by B&G BUERA Ltd. under patent No. ES 1 051 874 U. It has a variable inclination capability through the incorporation of a shaft perpendicular to the input section and a float in the output section which additionally allows the pump to act as a check gate. The pump screw is comprised of a tubular carbon steel shaft onto which is welded a three-bladed helix of cold-rolled carbon steel sheet. Steel terminations are screwed onto the ends of the shaft and the support flanges are fixed onto these. The exterior cylinder of the device is also made from carbon steel sheet and is TIG welded. There is also a support frame where the pivotal shaft of the pump is mounted which can be fully removed from a sub-frame embedded in the concrete sidewalls which acts as a runner. The presence of neoprene perimeter strips which act as hydraulic seals impede any manner of filtration or contraflow as a result of the arrangement of these intermediate elements between the flow inlet and outlet. The pump dimensions are as follows: outer diameter 1500 mm; shaft diameter 750 mm; total length 4500 mm; pitch length 1500 mm; blades 3 and blade gap 5 mm (Figure 1). The pump's asynchronous motor is located in the pump input section, transmitting motion via a transmission connected to a reduction gearbox located near the input below the level of the water. The transmission system has no

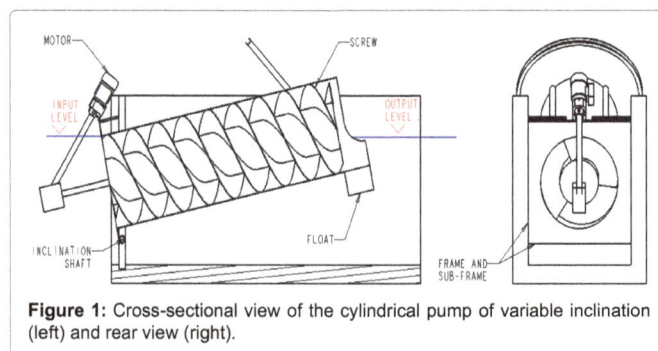

Figure 1: Cross-sectional view of the cylindrical pump of variable inclination (left) and rear view (right).

*Corresponding author: Joaquim Monserrat, Department of Agroforestry Engineering, Universitat de Lleida, Lleida, Spain
E-mail: monserrat@eagrof.udl.cat

intermediate pulley-based mechanism, thereby reducing mechanical losses.

Methodology

In the analysis of the device, a comparison was made of data measured in situ with data obtained through modelling. Therefore, once the model had been calibrated flow rate values could be obtained without the necessity of making a reduced-scale model. Analysis of the screw involved measuring the flow rate and energy transferred for different rotational speeds. Measurement of the difference in head between the inlet and outlet of the pump was performed using a piezometric tube. Flow rate was measured using an Acoustic Doppler Current Profiler (ADCP). The ADCP emits sound signals below the audible range which strike the water particles below. The sound waves are scattered back from particles suspended in the water producing an echo which is detected by the ADCP enabling it to determine the water speed and depth. Using these data, the ADCP calculates the flow rate. The amount of energy transferred was calculated by adding together the potential and kinetic energy of the water at the input and output of the device. Evidently, in this calculation the potential energy term has greater weight than the kinetic energy term given that the increase in flow speed is almost zero. Nonetheless, it was possible to calculate the flow speed at the input and output of the device by measuring the wet cross-section at the two ends of the device. A graphical method was used [1] to calculate the volume of the water trapped in one of the screw chutes, the length of which is equal to the pitch. This was done by representing the intersection points of the water surface and the blade. In this method, the projection of the screw helix is represented onto a plane which perpendicularly intercepts the axis, giving two lines which correspond to the upper edge of the blade and the contact point of blade and axis. The screw is represented horizontally, so the water surface is drawn with an inclination equal to the inclination of the screw which facilitates graphical representation of the projection lines of the intersection of the surface of the water with the blade. The idea is to make a projection of this surface onto the cross-section perpendicular to the screw axis. The starting point with this method is to project onto the cross-section perpendicular to the screw axis the intersection points of the water surface with the screw blade (A, B,...H). Following this, in order to draw the upper and lower limits of the water surface in each chute in the plane perpendicular to the axis, the segments bounded by the intersection points are divided into equal parts and this division is taken to the circular cross-section. When doing this, it should be remembered that the displacement is angular not linear, as can be seen in Figure 2. Once all the intersection points have been found, they need to be joined in such a way that the whole area that lies below the lines C-D and G-H represents that part of the chute which is completely full, in contact with the two delimiting walls (Figure 3). Following this, the portion of this volume which is not maximally connected with the walls of a chute is calculated. For this, the shaded area of Figure 4 is divided into equal parts, calculating for each of these parts the trapped water volume (Figure 5).

$$l1 + l2 = \frac{\alpha_1 + \alpha_2}{2.360} \cdot 2 \cdot \pi \cdot \left(r + \frac{R-r}{z} \cdot N \right); \quad x = \frac{R-r}{z}; \qquad (2)$$

Where:

l_n: Arc length of each portion (m).

α: Angle of the perimeter of the portion analysed (°).

x: Width of each portion (m).

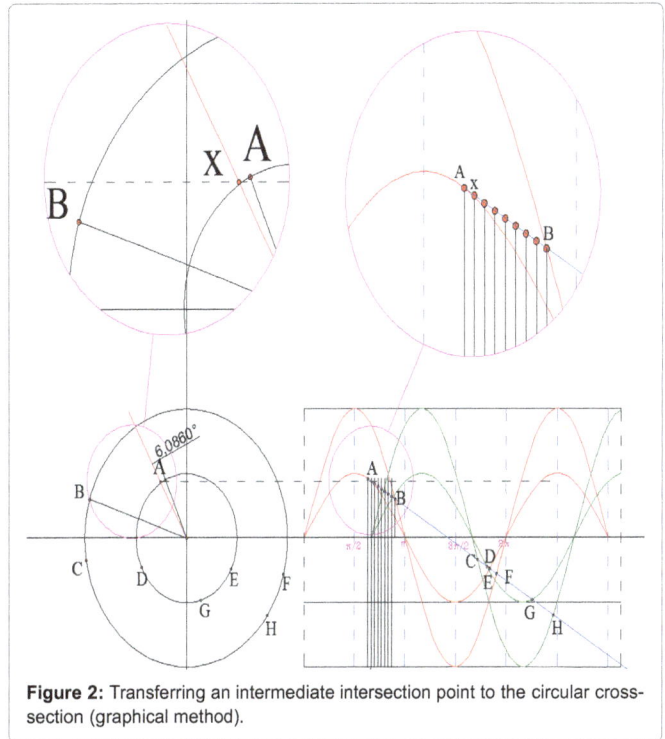

Figure 2: Transferring an intermediate intersection point to the circular cross-section (graphical method).

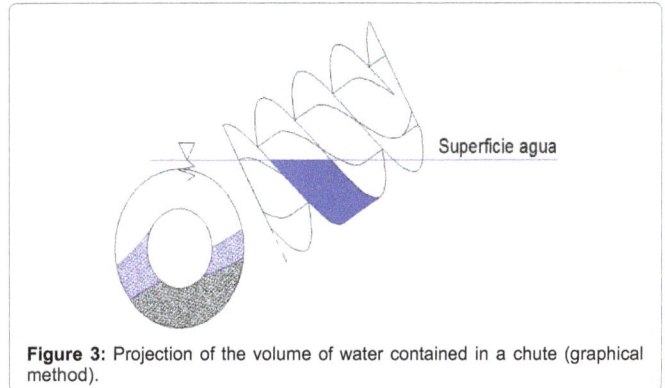

Figure 3: Projection of the volume of water contained in a chute (graphical method).

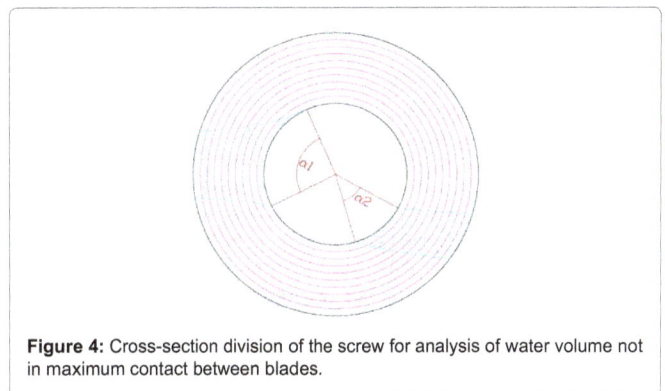

Figure 4: Cross-section division of the screw for analysis of water volume not in maximum contact between blades.

R: Outer screw radius (m).

r: Inner screw radius (m).

z: Number of divisions made (dimensionless).

N: Position of the portion under study, ordered from innermost to outermost (dimensionless).

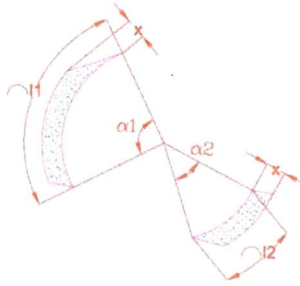

Figure 5: Dimensions considered for calculation of the unit volumes of water of the areas not maximally connected between blades.

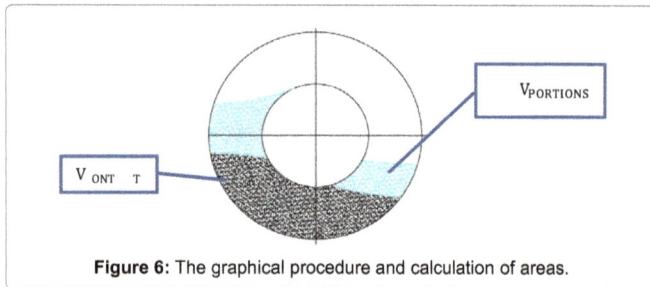

Figure 6: The graphical procedure and calculation of areas.

So, the volume of each portion can be calculated as:

$$l1 + l2 = \frac{\alpha_1 + \alpha_2}{2.360} \cdot 2 \cdot \pi \cdot \left(r + \frac{R - r}{z} . N \right); x = \frac{R - r}{z} . h \left[m^3 \right] \qquad (3)$$

So, the volume discharged in each turn of the screw will be:

$$V_c = a . V_u \left[m^3 \right] \qquad (4)$$

Where "a" is the number of screw blades and V_U the unit volume present between two screw blades, with V_U being:

$$V_u = \sum V_{PORTIONS} + V_{CONTACT} \qquad (5)$$

The graphical procedure and calculation of areas was performed using Autocad software [5] which allows us to know the shaded area through the use of a specific function (Figure 6). From volume V_c, the discharge can be calculated with the following equation, by applying an empirical coefficient [1]:

$$Q_n = 1.15 \frac{V_c \, n}{60} \qquad (6)$$

Finally the leakage loss (Q_1) is calculated according to blade gap (S_{sp}) and pump diameter (D) [1]:

$$Q_l = 2, 5 . S_{sp} . D . \sqrt{D} \left[\frac{m^3}{s} \right] \qquad (7)$$

The lifted discharge would therefore be:

$$Q_c = Q_n - Q_l \qquad (8)$$

Field Results

After processing the data, the flow rate and transferred energy were calculated for two rotational speeds (Table 1). It can be seen that when increasing the rotational speed the flow rate, logically, also increases. It is also the case that when increasing the volume of water in the pump and, therefore, its weight, the angle of inclination decreases slightly. The k coefficient values (eq. 1) are relatively low compared

to those presented in Wijdieks and Bos [2]. If the data taken in situ are compared with the theoretical calculation of the flow rate using the graphical method described above, it can be seen that the values obtained are very similar. Global efficiencies, i.e. the ratio between hydraulic power output and electrical power input were measured, but have not been presented due to a problem of an oversized motor in the prototype which produced abnormally low efficiency values. Wijdieks and Bos [2] reported values of between 0.65, for small diameter screws, and 0.75, for large diameter ones.

Comparison of the Variable Inclination and Conventional Pump

To show the energy saving that the installation of a variable inclination pump entails in comparison with a conventional Archimedes screw, modelling using the graphical method was performed of an Archimedes screw of the same geometric characteristics as the variable inclination pump analysed in this paper, but with a fixed inclination of 30° (conventional pump). The same pitch, number of blades and outer and inner diameter of each blade were considered. It was also considered that both pumps work under the same rotational regime, taking into account for each flow rate calculation the same flow loss factor for leakage. For comparison purposes, the energy saving of the variable inclination pump vs. the conventional pump is represented in Figure 7. The energy savings attained when the pump is of variable inclination can be clearly seen. The energy saving is greatest when the head difference is smallest and falls as the head difference between the inlet and outlet of the pump increases.

Economic Study

Table 2 shows the costs of the pump and civil works of the two cases studied. It can be seen how the total cost of the conventional

$n(rpm)$	$Q_m \left(\frac{m^3}{s} \right)$	$\beta m(^0)$	$H_m(m)$	$Q_c \left(\frac{m^3}{s} \right)$	$\Delta \frac{Q}{Q_m}$	K_c (eq 1)
20	0.396	11.67	0.334	0.410	0.035	0.352
40	0.802	10.59	0.321	0.900	0.021	0.392

Table 1: Results of the measured and calculated variables. Q_m: measured flow rate, β: pump inclination, H: transferred energy, Q_c calculated flow rate.

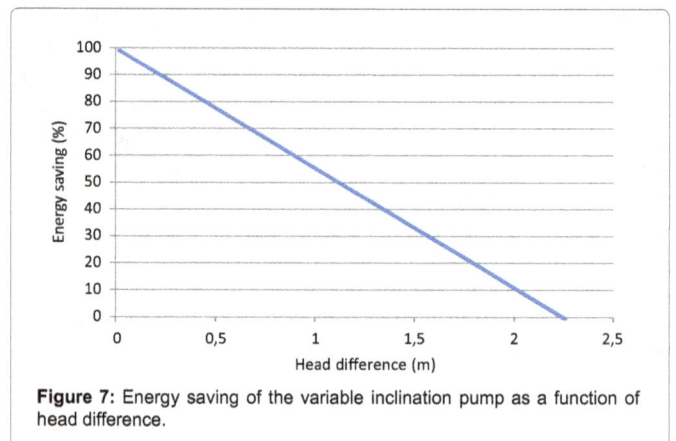

Figure 7: Energy saving of the variable inclination pump as a function of head difference.

Concept	Conventional pump	Variable inclination pump
Pump	€ 80,000	€ 110,000
Civil works	€ 105,000	€ 47,000
Total	€ 185,000	€ 157,000

Table 2: Partial and total costs of the two pump types.

pump is greater than that of the variable inclination pump as a result of the higher cost of the civil works. The reason for this is that in a conventional pump, it is necessary to build a semi-circular casing and to anchor the shaft, while in the case of the variable inclination pump, it is only necessary to construct a frame to which to attach the pump. It is therefore concluded that the variable inclination pump is more economic from both an investment and energy consumption point of view.

Conclusions

The energy consumption of the variable inclination pump was compared with that of a conventional fixed inclination pump and it was observed that significant energy savings can be attained when the downstream water level is low. The cost of the device and civil works is also lower for the variable inclination pump, making it a more recommendable option. The method of calculating the volume of water raised by the pump according to its geometry and rotational speed gave good results when compared with data taken in situ. This allowed us modeling of the performance of the pump for other conditions.

References

1. Nagel G (1968) Archimedean Screw Pump Handbook. RITZ Pumpenfabrik OHG, Schwäbisch Gmünd.

2. Wijdieks J, Bos MG (1972) Drainage principles and applications. International Institute for Land Reclamation and Improvement. Wageningen, The Netherlands.

3. Rorres C (2000) The turn of the screw: Optimal design of an Archimedes screw. Journal of Hydraulic Engineering 126: 72-80.

4. Burt C, Piao X, Gaudi F, Busch B,Taufik N (2008) Electric Motor Efficiency under Variable Frequencies and Loads. Journal of Irrigation and Drainage Engineering 134: 129-136.

5. Autocad (2012) Drawing software. AutoDesk Inc.

8

Support of Drainage for Management of Nitrogen Cycle in Africa

Yadav RC*

Ex Head of Research Centre, Soil and Water Conservation, Agra 282006, Uttar Pradesh, India

Abstract

Drainage engineering known to provide mere relief from excess surface water, water logging, rise in ground water table, and removal of accumulated toxic salts from the fields. However, since 1994 its scope was broadened and poised to cater functions of bringing social goods by enhancing food security and protection of environment. The continent Africa has the minimum area under irrigation so drainage has not been promoted in general, beyond the small areas under irrigation. The useful transformations of stages of nitrogen cycle occur under aerobic conditions. This study presents how good drainage will help bring aerobic condition. The study presents innovative technologies to enhance productivity of agriculture, grasslands and forest and reduce contribution of nitrous oxide. Provision of drainage on entire land uses has great potential in both irrigated and rainfed agriculture, natural or manmade forest and grasslands that will promote reduction of insect, pest and diseases in Africa. An innovative technology of auto drainage devised by author is to go long way in bringing drainage condition both under irrigated as well as the rainfed agriculture. This study while elaborating support of drainage in bringing management of nitrogen cycle depicts lack of appreciation by the professionals that limit application and cope in utilisation of potential of drainage. The study presented sufficient Justification for the change of names of existing departments of water engineering to production of professionals for execution and that for using the water related facilities for both engineering and agriculture disciplines of research and academic developments. Good scientific knowledge in the course curriculum should be broad based to bring food security and reduce global environmental problems of Greenhouse gases. Convergence of services for nitrogen and water use efficiency amply supports the need for the broadened curriculum of drainage for professional developments.

Keywords: Decomposition; Drainage engineering; Environment; Nitrogen cycle; Nitrous oxide; Productivity

Introduction

Drainage, the activity of removal of excess water, from surface, in ponded or saturated, lowering of ground water table below the root zone, removal of salts accumulated in the root zone soil [1,2]. Because of large volume of water and salts etc, it involves engineering skill and activity to carryout work with convenience and feasibility. In Africa irrigated agriculture has been mere 5 percent [3,4]. Provision of drainage has been limited to such low hactarage of irrigated agriculture. The remaining 95% of land uses have been subjects of natural drainage and water logging and ponding, salinization and rise in ground water table causing large scale anaerobic decompositions to continue to go in their own pace. This means nitrogen cycle largely operated through the bad path and produced N_2O and nitrogen use efficiency remained far low. The drainage technology brings relief by modifying the process of anaerobic decomposition which causes denitrification and in turn production of nitrous oxide (N_2O), one of the green house trace gases (GHGs) that depletes ozone layer.

Nitrogen is a primary nutrient essential for living organisms and development of primary productivity. Pastures, grassland, agriculture, and forestry land uses are the primary producers enabling browsers, the primary consumers in the ecosystem to produce secondary produces *viz* milk, meet, hides, dung, urine and energy etc. The nitrogen cycle (Figure 1) depicts transformation of N in various forms. The microbial transformations are organic nitrogen RNO_2, Ammonia NH_3, Ammonium NH_4, Nitrite NO_2, Nitrate NO_3 Nitric acid NO, nitrous oxide N_2O and reformed nitrogen gas N_2 that goes back to the atmosphere.

The transformation of nitrogen into its many oxidation stages is key to productivity in the biosphere and highly dependent on the activities of a diverse assemblage of microorganisms such as bacteria, algae and fungi. Nitrogen fertilizer is expensive and losses can be detrimental to the environment. The health issues include respiratory ailments, heart disease and several cancers. In any situation all the phases of

decompositions and transformations continue simultaneously, of course on different organic wastes and so are the emission of nitrous oxide is continuous. The N_2O, important minor GHG is produced during the formation of nitrite NO_2 (nitrification) and again *during* reformation of nitrite (denitrification) [4,5]. The flux of release will vary with time and situation of the material under decomposition.

The situation of surrounding environment and water interactions brings good and bad transformation stages in any decomposition process. The decomposition under aerobic condition is different from that under anaerobic condition. The decomposers work differently and produce lot of environmental hazards. The decomposition forms the detritus chains and decomposers such as bacteria, algae, and fungi play their dominant roles. The insects such as flies and mosquitoes, and fungus thrive on the detritus food chain and produce nuisance for the environment. These detritus consumer become food chain for creeping, crawling and flying insects and further add to the problem of environmental concern. The example of such detritus food chain consumers and secondary consumers those bring bad environmental impacts such as malaria, Tse Tse fly responsible for yellow fever, known in Africa. Thus, if by any means, the decomposition process can be controlled, it will improve the environmental quality. Yadav [6] established that provision of good drainage will bring large scale transformation of the decomposition process from anaerobic to aerobic. Drainage which used to be taken as mere rectifier of bad developments of misdeeds in the irrigated agriculture such as water ponding, water logging, rise in ground water table and salinization,

---

*Corresponding author: Yadav RC, Ex Head of Research Centre, Soil and Water Conservation, Agra 282006, Uttar Pradesh, India, E-mail: ramcyadav@rediffmail.com

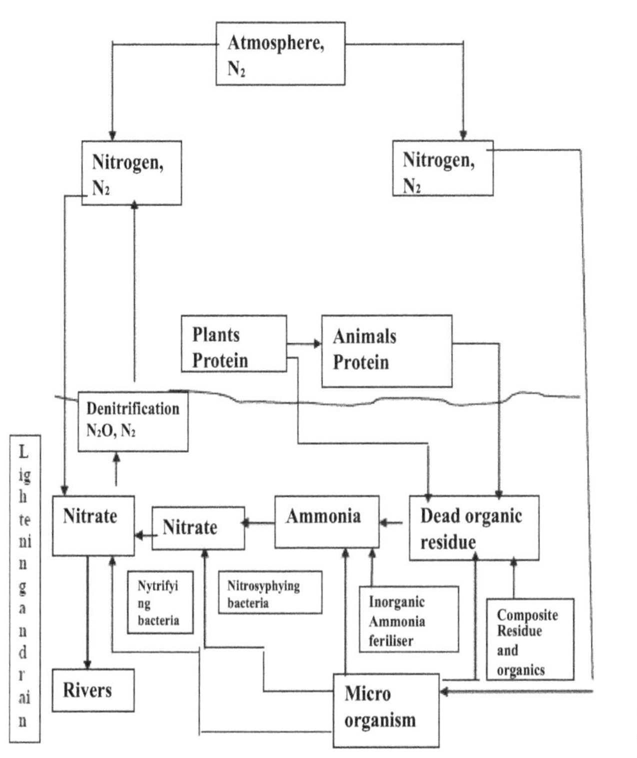

Figure 1: The Nitrogen Cycle [20] and updates.

alkalinisation and removal of salts by leaching, will play significant role in improving environmental condition of all land forms and land uses. Thus, implication of the nitrogen cycle has to be looked at in association with drainage (aerobic condition) and no drainage (anaerobic condition). The misdeeds in nitrogen cycle caused dis balance and brought several undesirable results of local as well as global domain. Clear understanding of the nitrogen cycle opens avenues for enhancing productivity and development of entrepreneurship to generate employment and become people supported environment protective self driven process. This automation will enable enhance stabilised sustainable productivity and protection of environment for the present and posterity.

The agriculture, forest and grasslands remained as emitter of N_2O gas. As the domain of drainage has enlarged since 1994 [7] to cover the society and environment, its applicability has broadened. Thus, drainage has become a saviour of enhancement of productivity and environment [6]. It has wider applicability in vast land stretch in Africa having grassland, forest, rainfed agriculture and large cattle and wild animal population as depicted by presence of many insect, diseases, and low productivity. The innovative nature agriculture and allied technology can be demonstrated to become saviour for Africa too. Thus, objective of this study was to present aspects where drainage has prospect of managing nitrogen cycle for tackling productivity and environmental problems in Africa. These technologies will find better appreciation if the academic teaching and research endeavours are rightly supplemented and complemented to understand the natural process. Thus, this study was carried out to make beneficial use of Nitrogen cycle to enhance productivity and protect environment.

The manuscript comprises after introduction materials and method; the nitrogen cycle, nitrogen cycle in soil, strategy of management of nitrogen cycle, and technology capsule. The results section comprises subheads; Broadened scope of drainage to cover aspects of environment and society, Biological nitrogen fixation in different land uses are viz agriculture under sub head of intra row banding, and inter cropping, nitrogen fixation in grassland and pasture, with annual leguminous crops, that by perennial crops, and innovative cropping practice in the green fodder fields and N for forestry again with annual and with perennial leguminous crops. Reformation of cropping and agro forestry practices, Innovative universally applicable green technology -Racy nature agriculture–racy combo, Sanitary, sewage and waste water treatment, Industrial process enabling transfer from N excess to N deficit sites and Convergence of other sciences for promoting activities of nitrogen cycle are included. Following results under these sub heads, is the SWOT analyses, ie strength, weakness, opportunity and threat. The action initiatives and conclusion and research needs are described. Thus, the manuscript presents a unique scenario of subject of management of nitrogen cycle in Africa trough the support of drainage engineering.

Materials and Method

The nitrogen cycle

Nitrous oxide (N_2O) is third trace gas in the order of dominance of GHGs. Anthropogenic ally induced mainly originates from soil and agricultural activities. GHG-N_2O is also responsible for ozone depletion. The nitrogen containing compositions *viz.* ammonium, nitrite and nitrate resulting in reaction of nitrogen with water due to hydrolysis are the radicals [8]. Chemistry related to nitrogen N cycle in the troposphere, terrestrial, hydro-ecosystems and ground water etc is highly researched upon science. There are four main ways by which nitrogen can naturally be made available for use in ecosystems. Firstly, by plants themselves, when bacteria, most notably those associated with leguminous (bean or pea) plants, trap nitrogen from the air and combine it with hydrogen to form ammonia (NH_3). The second way is by thunder and rains that bring down nitrogen in the nitrate form (Figure 1). Third way is the decomposition of plants and animals that release organic nitrogen in to the soil as ammonia. Fourth way of nitrogen build up in the soil is by addition of nitrogenous fertilisers. Bacteria and fungi in the soil then convert this ammonia into ammonium (NH_4), which can be used by plants. Further, chemical reactions by nitrosomonas bacteria transform the NH_4 into nitrite -NO_2^-. The nitrobacters then convert the nitrite NO_2- to NO_3- nitrate. This nitrate is very soluble and used by plants. The cycle is concluded when denitrifying bacteria in soil convert nitrates in anaerobic soil to either nitrogen gas (N_2) or nitrous oxide (N_2O) and these gasses then return to the atmosphere. In order to reach at scientific interventions in the nitrogen cycle process knowledge would be necessary. Hence the processes are included in the following section.

The Nitrogen cycle in soil

The Nitrogen Cycle in soils is a very dynamic process and has great practical implications on modern agriculture and protection of nitrate pollution of water courses and pollution of air by nitrous oxide. Most plants can absorb nitrogen in both the nitrate and ammonium forms. However, ammonium-N is rapidly nitrified to nitrate-N which is the main form absorbed by agricultural crops. Nitrous oxide can be produced during nitrification, denitrification, dissimilatory reduction of NO_3^- to NH_4^+ and chemo-denitrification. Since soils are a mosaic of aerobic and anaerobic zones, it is likely that multiple processes are contributing simultaneously to N_2O production in a soil profile. The

N_2O produced by all processes may mix to form one pool before being reduced to N_2 by denitrification.

Understanding the N Cycle is important in managing the nitrogen needs of the future. The N from atmosphere by the action of microbial fixation is well known since ancient time. During the primitive time many crops were sown as mixed crop by broadcasting. The association of nitrogen fixing crop and nitrogen using crops were closer. Application of scientific method of line sowing and mechanization kept both the types of nitrogen fixing and fixed nitrogen using crops at distance, thereby weakening association by the distance between sole lines of mixed crops/intercrops. Thus, the transfer of fixed organic RNO_2 got reduced.

Nitrogen transformation by bacteria: Nitrogen cycle depicts entire domain of nitrogen circulation. Bacterial transformation is main route and the other route being the thundering and sparking as depicted in (Figure 1). The bacterial transformation is by following four reactions [9] is given in the following.

Nitrogen Fixation by *Rhizobium*

The atmospheric N_2 is fixed as organic nitrogen.

$$3\{CH_2O\} + 2N_2 + 3H_2O + 4H^+ \longrightarrow 3CO_2 + 4NH_4^+ \qquad (1)$$

Nitrification is the process of *nitrosomnas* and *nitrobacter* bacteria

$$NH_3 + 3/2O_2^+ \longrightarrow H^+ + NO_2 + H_2O \qquad (2)$$

$$NO2- + 1/2O2 \longrightarrow NO3 \qquad (3)$$

Nitrate reduction

$$1/2NO_3^- + ¼\{CH_2O\} \longrightarrow ½NO_2 + ¼H_2O + 1/4CO_2 \quad (4)$$

Denitrification which involves the reduction of NO_3 and NO_3^- to NO_2^- followed by recycling of N_2 to the atmosphere

$$4NO_3^- + 5\{CH_2O\} + 4H^+ \longrightarrow 2N_2 + 5CO_2 + 7H_2O \qquad (5)$$

It is evident from the five reactions that transformations of nitrogen

in the nitrogen cycle require presence of oxygen which can be brought in by provision of good drainage. This fact implicates that aerobic condition can be brought in all land uses which comprise land terrain, crops/vegetation. The eq (5) reveals that reaction under anaerobic condition brought about by water logging forms denitrification, which will get transformed to release nitrous oxide, a green house gas.

Strategy of management of Nitrogen cycle

Management of any aspect of things, behaviour and situation requires certain strategy with a view to bring change. In case of the nitrogen cycle where all process are aimed at to function under aerobic reactions, the entire cycle is discretised and each segment is manoeuvred to function in the desired manner.

Discretisation of the nitrogen cycle: The nitrogen cycle discretised in various functions of stages are listed in (Table 1). The steps are reduced to single function in the cycle for better understanding and isolation of aspects for management. The discretised sub sections of nitrogen cycle are to be looked at so that it functions in its best possible manner.

Strategy of management: Strategy was to devise ways and means so that nitrogen fixation is the maximum by the nitrogen fixing bacteria. Further, strategy was to maintain, as far as possible, the aerobic condition so that the denitrification is the minimum. Incorporation of crops that nitrogen fixing crops remain in close association and for long duration crop is able to utilise the biologically fixed nitrogen to the maximum extent possible. The fact that presence of sulphate in low concentration enhances nitrogen and water use efficiency demands chartering of sulphur cycle in the crop to function by creating decomposition of cellulose in sulphate which occurs during aerobic decomposition. Thus, it becomes scientific need of addition of organic cellulose to decompose under aerobic condition and produce sulphate. The sulphate is good remedy of overcoming of salinity is one of the benefits of drainage or management of saline soil and water. Any agricultural system where water and crops are to have good interactive actions, drainage becomes a precondition, Thus, management of nitrogen cycle is aimed bring good effect on yield and reduce imbalance

Stages	Type of transformations	From	To	Problems	Needed engineering improvements
I	BNF	Nitrogen gas N_2	Organic nitrogen RNO_2	By microbial nodulation	Enhancing efficiency of fixation
		R NH_2	Ammonia NH_3	By natural process	-
II	Ammonification	NH_3	Ammonium NH_4	Receptacle for quick acceptance and mixing	Restrict escape from the site by modular transformation of decay process.
	Mineralization	NH_3	NH4	Conserving N during intervening period	By repeated ploughing
III	Nitrification	NH_4	Nitrite NO_2	By nitrosyfying bacteria	Provide aerobic condition
	Volatisation	NH_3	Capture and mixing	It is loss of soil stored N	Quick incorporation by ploughing
IV	Nitrification	NO_2	NO_3^-	Nitrate leaching is problem	Avoidance of nitrate leaching by Conservation by arresting in soil root zone
	Natural fall from atmosphere	Thunder, Storm and rain	NO_3		
V	Denitrification	NO_2	Nitric acid NO		Arrest N_2O at critical peak times of release by biochar
	Urea addition	CO NH_4	From industrial sources	Loss as volatile ammonia	Quick incorporation and split doze application
VI	Immobilisation	NH_4 and/ or NH_3	Organic RNH_2	-	Building N in the intervening season.
VII	Denitrification	NO_3	Nitrous oxide, N_2O	GHG emission	Reduce quantum of balanced Nitrate that may get under denitrification.
		N_2O	N_2	Nitrogen gas back to atmosphere	Arrest N_2O
Combo	Racy nature agriculture	Accommodates all stages and encompass all processes to reduce adverse effects and combine integrated nutrient management.			

Table 1: Transformations in the Nitrogen cycles, gain and loss and effective measures to enhance utility and reduced GHG burdens.

require integral support of drainage. That means nitrogen cycle can not be managed to bring good function without support of drainage, Any academic documentation on land water management should give cognizance to this effect. These facts support the importance of drainage beyond the irrigated agriculture. In the rainfed areas also drainage will help enhance productivity even without requirement of excess water, high ground water table or excess of salt needing removal by washing or leaching. For the management of various sub components in (Table 1), the measures should eliminate release of nitrous oxide, nitric acid and nitrogen oxide from the cycle. These harmful gases release reducing measures are situational with respect to utilisation of NH_4 and NO_3 from the agriculture and plantations. It is necessary to have minimum reserve balance, hence the quantum of release of the nitrogen based harmful gases will be the minimum. Presence of well drained condition is basic for the aforesaid situation to develop that will eliminate release of the GHGs.

Further, the N load in waste water should be low so as to eliminate the eutrophication to develop in the water bodies. The different approaches and strategies are necessary to bring down the net result of balance in N cycle. The excessive N under any situation will result in dis balance that will produce bad effect on environment and food commodity.

The technology module

From the fore mentioned facts it is clear that for management of nitrogen cycle to bring beneficial effects under any plant production system the aerobic condition has to be maintained by providing good drainage provision of the land. The nitrogen cycle should always remain and function under aerobic condition. Nitrogen fixations should continue to function all the time so that other plants are in position to make use during the company of the companion crop as well as after the cessation of the growth of short duration crop, as well. This situation will keep the emission of nitrous oxide in low magnitude. Vegetation will be able to extract nitrogen to the maximum extent. During the period of no crop in the fields, nitrogen present in the soil should be mineralised and immobilised to eliminate the loss from the root zone. With these aspects the new technologies were devised which are presented under different subheads in the results.

Results

Broadened scope of drainage

Drainage for protection of environment and welfare of society: Earlier belief was that drainage is required as an accompanying component in irrigated field to remove water ponding, remove water logging, control rise in the ground water table and control salinization and removal of salts from the root zones by leaching. But, since the year 1994 the scope of drainage has been broadened to cover aspects related to society and environment [7]. However, this realisation came late. Innovative application of scientific facts devised by the author enables one to realise that, as a phenomena, excess water triggers anaerobic decomposition which leads to emission of green house gases namely, methane and nitrous oxide. Thus, water ponding, water logging and salinisation do occur in the lands under the rainfed situation as well. Thus, drainage becomes the need of rainfed land under various land uses such as agriculture, grasslands and forests etc. Application of drainage for creating aerobic condition should be global activity for enhancing productivity and enhancement of environment. This broadened scope of drainage emphasised that tropical countries such as Brazil and Asia with intense rainfall during rainy season, which

bring intensive water ponding, water logging and rise in water table contribute more GHG percentage contribution of GHG build up in the environment. In the temperate regions receiving winter rainfall, the anaerobic condition might exist but the due to low temperature the emission of GHGs such as methane and the nitrous oxide remains low. Thus, contribution of agricultural GHG is high in the tropics and low in the temperate region. It means that the need of drainage to account for environment protection is more than that for irrigated agriculture. Hence, the domain of drainage in consideration of environmental aspect i.e. emission of GHG is more broadened. This chemical phenomenon has existed all the time, but late realisation came due to lack of perception of application of environmental sciences and environmental engineering principles. The late appraisal of application of process based knowledge lead to damage to environment; set back of GHGs and accumulation and cause of global warming. This fact was sufficiently substantiated by [10]. Thus, it is evidently clear that need of drainage exists in the global land percentage far more than percentage of land under agriculture. In Africa land under irrigated agriculture is mere 5%, but drainage requirement is to cover almost all land of the continent i.e. 100%.

Drainage in Ethiopia- a Highland: Ethiopia is a high land country at altitude ranging from 2000-3000 m, maintaining about one third height of troposphere (11 km). That means Ethiopian climate will depict climatic feature of stratosphere above mean sea level at height of 5-6 km. Natural green house gases contributions are H_2O vapour (62) CO_2 (22), O_3 (7), N_2O (4) and CH_3 (4) percent. Likewise, contribution of anthropogenic green house gases are CO_2 (61), CH_4 (15), CFC (11), N_2O (4) and ozone O_3 (unknown) [3]. The ultraviolet radiation is intensive as sky is largely clear of atmospheric pollutants. Contribution of methane from low land rice (lack of rice cultivation) fields is very low, but it can be compensated by many lakes and large cattle population. Since the country is not intensively industrialised, contribution to green house gases would be largely from grasses, upland agriculture, rainfed grains namely wheat, sorghum, pearl millet and *teff* (a famous cereal produce) etc. Due to low temperature contribution green house gas through CFC is also to be the minimum because of minimum use of fridges and air conditioners. However, specific data on these factors are not available.

Because of altitude, the problem of drainage is not visible, but there are marshy lands stretches close to the water bodies such as lakes and rivers, River Nile originates from the Ethiopian highlands and traverses long route. Thus, there does exist local problem of drainage. These local problems of drainage at far distance do not come to attention of the local gentry. There seems to be no example of built drainage as the irrigated area itself is little and all soil salts get washed by the runoff to join the River Nile upland branches.

Now the role drainage has to be looked from enhancing productivity of grass lands and upland agriculture. Management of nitrogen cycle for enhancing productivity is very relevant. The cereal crops when grown with intra row banding of leguminous crops [11] or as intercrops in fine grain cereals, it will make sure way of nitrogen fixation in the field to enhance yield and reduce N_2O emission [9]. The nitrogen content of grasses and cereal food grains will increase that will improve health of animals and people. General health of cattle in Bale Robe, (South Western) Ethiopia, a grassland valley with good rainfall, is good and animals keep heavy weight due to availability of ample grasses to graze.

The implications of nitrogen cycle have further to be seen. When animals are adding nitrogen by way of dung and urine the urea and ammonia are added and in due course of decomposition will encompass ammonification, nitrification, to convert ammonium to nitrite and nitrate and then denitrification to again convert nitrate to nitric oxide and nitrous oxide. Thus, there is likely build up of nitrous oxide from the grass land in the wetland to the atmosphere and nitrate movement to rivers. The nitrate content of water of the river Nile is high. Similarly the nitrate poisoning may happen. On the other hand movement of nitrogen and phosphorus rich runoff may cause profuse weed infestation and decomposition of cellulose under anaerobic condition (in deep lakes water) will make ground water acidic. Therefore, as a security measure, the wetlands what so ever may be, application of drainage is warranted in Ethiopian highland in particular and any high land in general.

Drainage need in the low land countries: The low lands are more susceptible to water logging due to physiographic setting to retain more excess water for longer duration. The gravity of drainage is going to increase as a result of increase in the extremes and frequency of its occurrence during the ongoing phenomena of global warming and climate change.. The utility and role of drainage will be gaining still greater importance under the lowland condition than any highland situation. In the lowland situation sub surface drainage will be additional burden towards bringing the aerobic condition.

Biological nitrogen fixation in different land uses

The results of high esteem value with regard to management of nitrogen cycle are presented under different sub heads of phases (Table 1) as accomplished in the present study listed in Table 2.

Agriculture: In order to avoid losses due to volatisation the organic manure as well as the inorganic nitrogen fertiliser, the intervening period free from crops or when the nitrogen extraction is low during initial stage of crop developments, the field should be as far as possible, kept tilled or inter tilled, as may be the case, to keep process of

mineralisation and immobilisation to work. The nitrogen present in the fields will be saved from the losses by nitrification and denitrification. This aspect of tillage requirement in reducing the GHGs was covered in detail in study [10], that indicated need of tillage in winter season cropping in lieu of zero tillage. The zero tillage is justified to conserve soil and nutrient losses by surface runoff, but no thought was given to the reduction of emission of GHG nitrous oxide.

Cyno bacterial leguminous crop intra-row banding: The situation of crop cultivation with tillage facilitated mineralisation and immobilization in the soil hence build-up of N in such fields. Since the process is continuous, devising ways and means of N fixation of crop by the leguminous crop in the closest vicinity of the user crop will enable enhanced N transfer efficiency [11] devised an innovative method of leguminous crop banding right in the lines of cereals or oil seeds. The referred to study established that sowing of leguminous crops such as green gram and lentil @50% of normal seed rate right in the line of cereals and oilseeds produced contrasting visible effect of N fixation. The protein content was highest in this banding. Further, efficiency of N fixation and transfer can be enhanced by microbial inoculation of the legumes N before sowing [12].

Inter cropping: Inter cropping of leguminous crop in pure band of cereals and oil seed has been known practice worldwide. Researches [11] have established that intercropping of short duration legume crop in the long duration legume crop also extracts advantage of nitrogen fixation. Thus, it became a practice of growing green gram or black gram in pigeon pea and castor. Thus, natural process of biological nitrogen fixation has been extended to enhance yield of legume with another legume. It needs to extend this nitrogen fixing characteristics of leguminous crops for roots and tuber crops such as sweet potato and potato and vegetable production. In addition to enhancement in yield, the quality of produce will also be better than that without nitrogen fixing crops, as prevalent worldwide.

Agro-ecosystems	Innovative technology	Resulting benefits after implementation
Biological Nitrogen fixation BNF		
Agriculture	Intra row banding	Fix nitrogen in cereal and oilseeds by intra row banding of inoculated leguminous crop.
	Inter cropping	Carry out Annual legume in another annual legume of longer growing period
	Racy nature agriculture	Both the two systems can be applied in the new green technology of racy nature agriculture, described in the following part of the results.
Grasslands	Legume intercropping	Intercropping of legume in pasture transfers about 50% of the nitrogen fixed bay leguminous crops
	Annual legume	Annual legume in grassland
	Perennial legume	Perennial legume in grassland
Forest	Understory plantation of legumes	Annual legume in tree basin
	Perennial legume inter tree plantation	Perennial legume in tree basin.
Green technology combo		
All terrestrial ecosystems	Racy nature agriculture	It encompasses all needed good effect for green technology for enhancing food productivity and protection of environment.
Reduction of N and P in waste water disposal		
Waste Water disposal	Grow Reeds to conserve biodiversity by saving wood and forest.	Protection from eutrophication. Enhancement in ecosystem services.
Transfer from N excess to N deficit places		
Industrial N transfer technology	Preparation of Diammonium sulphophosphate (DASP)	The local excess n resources used to prepare DASP which will be transferable from N excess to N deficit places
Convergence of other sciences		
Agriculture, grassland and water resources	Biological control for eradication of aphids and other secondary detritus consumer Biological control of malaria	By promoting parasitic action of feeding wasp on aphids Inactivation of mosquito larva by spreading mustard seeds.

Table 2: Innovative technologies for management of nitrogen cycle.

Nitrogen fixation in the grassland and pastures

Provision of drainage: The grass and pasture land are naturally drained and at places there may be pot holes. Since there is no use of mechanisation, the grass and pasture lands should be drained by providing random field ditches and water led to suitable site where it can be stored for further use by the animals.

Incorporation of Annual legume bands at regular interval: In the established grasslands and forest annual legume should be sown that will fix the nitrogen and keep the grasses active in utilising the nitrate by the plants, to eliminate losses in runoff water. Nitrate and phosphorus movement in runoff cause eutrophication, a process undesirable for the water bodies. Leguminous crop will solubilise the phosphorus added by animal dung and urine. The role of leguminous crop in phosphate solubilisation is available in research presented by [13]. The production of organic acid is related to phosphate solubilisation of $Ca HPO_4$ and $Fe PO_4.2H_2O$. Thus, in addition to nitrogen fixation some effect of phosphate solubilisation is brought in the rangelands and pastures. The combined effect will be that reduction of release of nitrous oxide to the atmosphere. The cultivation in the grass and pasture will bring some effect of sulphur cycle and enhance grass yield and its palatability for the animals. Since the grasslands and pastures are largely not externally supplemented with fertilisers, incorporation of legume will supplement them by biological nitrogen fixation.

Incorporation of Perennial legume in rangelands and pastures: The extensive root system and permanent ground cover tend to provide greater nutrient retention, erosion reduction and carbon sequestration relative to annual plant covers [14]. A central tenet of ecological theory is that increasing biodiversity increases the stability of ecosystem properties, such as net primary productivity (NPP). The species composition can shift in response to environmental changes, buffering the sensitivities of species to environmental changes. The transfer of N from the legume to the grasses facilitates grass NPP in grass legume mixture in grasslands where N is limiting. Therefore, one potential benefits of grass legume mixture is their ability to optimise N retention and supply across a range of environmental condition. Red clover is known as one perennial legume suitable as mixture in the grasslands and pastures. In this study way as how to integrate the grass land and the red clover and bring nitrogen fixation in the grassland is new addition to the existing knowledge of nitrogen cycle management. The band of red clover at interval can be sown to enhance N land equivalent ratio (NLER), and land equivalent ratio (LER). Red clover is a tap root species while grasses have fibrous root system, their association increases total biomass, particularly, root biomass, resulting in greater soil C and N accumulation relative to corresponding monoculture. In order to increase this complementary association one or two rows of red clover should be banded in the entire grasslands. Studies [11] indicated that incorporation of perennial legume in grasses fixed 15-37 kg N/ha in soil and apparent transfer of N to the grasses was 8-16 kg/ha (almost 50%). However, it needs selecting suitable local perennial legumes to be inter sown in the grasslands and pastures. Grass legume mixture improve synchrony through slower decomposition of litter with higher C:N resulting in short term N immobilisation and retention in microbial biomass and soil organic matter pool. These results highlight the need to apply ecological theories to research in using plant functional diversity to enhance ecosystem services in ecosystems. It is utmost necessary to remember that irrigation and drainage systems are created and management to bring sustainable productivity and that can not be possible without application of nitrogen. Thus, management of nitrogen is essential function involved in drainage. This is covered in the broadened scope of drainage.

New cropping pattern for cultivated green fodder for animals

For animals especially dairy cattle green fodder are cultivated in some fields depending on the need of the cattle to be fed. The green fodders are harvested from one side continuously and irrigated. The fields get cultured and the fodder sown once is kept for many year after year. There is strong build up of nitrogen in such fields. In order to create biodiversity and utilise the nitrogen reserve sowing in bands other crops such vegetable green vegetables, leguminous and non leguminous crops which require high doses of nitrogen. The reserve of nitrogen will be utilised and scope for denitrification will get reduced. Thus, it will be new aspect of capitalising the N equivalent and LER of the field under fodder crops. This new practice will attract dairy owners and farmers around city make fodder cultivation to sell the green fodder and generate their income by cultivating vegetables. The net result will be capitalising that nitrogen existing at some sites. It will be new resource from ecosystem services involving concept of plant biodiversity.

Incorporation of annual and perennial legumes in forest: Sowing of leguminous annual and perennial crops in establishes forest enables nitrogen cycle to continue to function in fixation, nitrification, immobilisation, and volatilization and denitrification, whereby useful products are produced. This strategy will enable crop/plants use it the field. The scientific facts and experiences enable establishment of the fact that lead to development of practices that convenes aerobic condition under the crop bed alike appropriate technology. The reduction in denitrification was found to reduce N_2O emission [15]. N losses during the decomposition period could be controlled by synchronising the culture in legume based system. In permanent legume plantations, the availability of N to the forest species will be for longer duration than that for annual crops.

Reformation of cropping and agro forestry practices: Cropping of legume in legume of different growth durations e.g. pigeon pea and green gram, sowing of leguminous vegetable such as ladies finger in fodder clover /alfafa or *barseem*, cultivation of legumes in the pastures and plantation of legumes in lieu of sole crop enhance land equivalent ratio (LER) by enhancing nitrification and use by the crops leaving low N reserve to undergoes denitrification leading to N_2O emissions. Plantation of perennial legume surrounding the trees will be innovative way of reducing nitrous oxide. This will enhance scope of forest and trees in reducing global warming by adding new dimension of controlling nitrous oxide.

The most important for life the nitrogen use and nitrogen cycle are vitiated by human efforts to fulfil present and future needs of sustainable food security and environments. Earlier practices had focus to enhance yield but no emphasis on environment protections that generated problems of environment, disruption of ecosystem services, price rise and socio-political unrest in many countries, health hazards, resources constraints etc leading to worry for the world at large [9]. The policies and issues formulations were focussing problems and hoping emergence of scientific technologies to alleviate this global problem related to N use. The scientific innovative developments presented for N and all resources by development of racy nature agriculture, which has capability to overcome all misdeeds in agricultural practices and work as universally applicable panacea green technology for agriculture and environment.

Developments of N_2O induced pollution control measures in to entrepreneurships and business attract peoples' participation to undertake difficult jobs in feasible and easy way for application. Example cases of nitrogen fertiliser plants, Ireland and northern European countries facing misdeeds of vitiations of nitrogen cycle can be easily alleviated by application of technologies developed in this study. Thus, developments go long way in scrapping global worry concerning future sustainable food security and protection of environment. The racy nature agriculture has raised niche of low sounding pro-genetic world opinions to raise harvest index of crops by land, water and crop husbandry management practices. Many terrifying estimates and foreseen dangers will get slowly diminished by application of technologies developed in this study.

The auto drainage technology -Racy nature agriculture–racy combo: The racy nature agriculture which has acquired status of sun technology [16] meaning thereby a green innovative universally applicable technology which suitable for all ecological reasons, all soils, all crops, all cropping patterns and for both irrigated and rain-fed agriculture be it under large scale or small scale peri-urban as well as control environment agriculture. The details of the technology are available in other research articles published by the author(s). The (Figure 2) gives self explanatory depiction of the technology particularly the land forms under the field.

The technology capsule special component of nutrient management

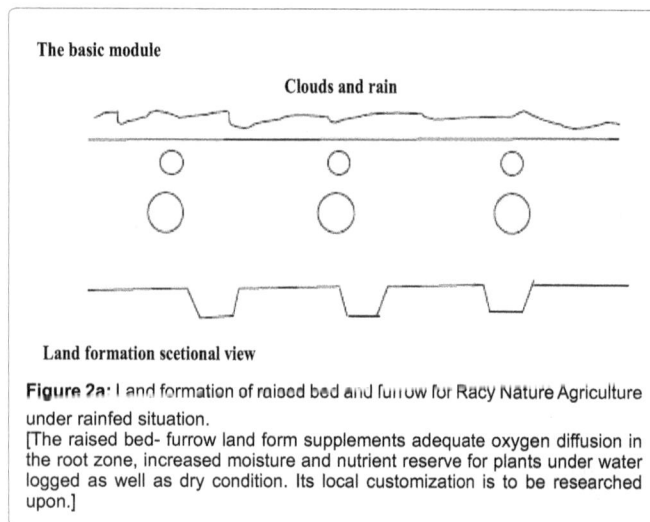

The basic module

Clouds and rain

Land formation scetional view

Figure 2a: l and formation of roised bed and furrow for Racy Nature Agriculture under rainfed situation.
[The raised bed- furrow land form supplements adequate oxygen diffusion in the root zone, increased moisture and nutrient reserve for plants under water logged as well as dry condition. Its local customization is to be researched upon.]

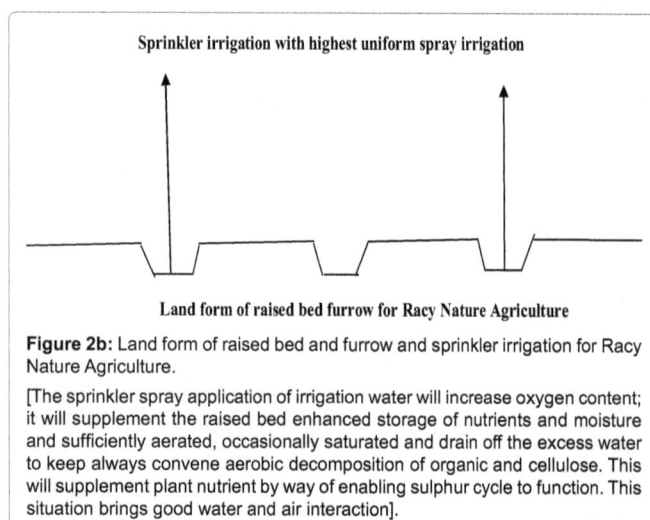

Sprinkler irrigation with highest uniform spray irrigation

Land form of raised bed furrow for Racy Nature Agriculture

Figure 2b: Land form of raised bed and furrow and sprinkler irrigation for Racy Nature Agriculture.

[The sprinkler spray application of irrigation water will increase oxygen content; it will supplement the raised bed enhanced storage of nutrients and moisture and sufficiently aerated, occasionally saturated and drain off the excess water to keep always convene aerobic decomposition of organic and cellulose. This will supplement plant nutrient by way of enabling sulphur cycle to function. This situation brings good water and air interaction].

by primary and secondary natural resources that develop by chemical reactions [11,16,17]. The racy nature agriculture promotes productivity with existing situation and conserves resources for posterity. The technology has capacity to endure adverse impacts of droughts and floods that are likely to become severe due to global warming and climate change in future. The racy nature agriculture focuses and meets world over challenge in the use of natural and fixed resources for agriculture and environment conservation, which were not found in the existing scientific ventures.

Local optimizations of the technology will take care of customization accuracy to account for existing roles of agro-eco-regions, man- machine and socio economic status. The alteration of decomposition process, arrest of GHG gases and heavy metals will reduce GHG s load in atmosphere, reduce load of heavy metals that will reduce global warming and avert climate change. This aspect, totally new application in agriculture will produce food better than so called organic food. Thus, in lieu of some high profile accessible limited organic food, a better quality and accessible to all surpassed solution for all is developed. Further scope for refinements for the second generation research has opened so as to bring technology refinement. The lag in the situation and makeup in the shortfall in present day agriculture can be made by recognition of motivation by oriental saying i.e. late is better than never. Therefore, it requires to makeup of mind, without further delay, and come in action for implementation of the racy nature agriculture. The implementation will revamp all to join in mission to create Manson of global sustainable food sufficiency for present and posterity.

Sanitary, sewage and waste water treatment: Treatment of sewage, waste water and solid waste has been largely inadequate that has been causing problem due to excess nitrogen in its surrounding. One glaring example of excess nitrogen imbalance in the nitrogen cycle is the Ire Land. Most of Ise Lands have become oxygen deficient due to excess growth of water planktons which consume oxygen for respiration. Contrary to this, is a situation of Nitrogen deficiency in the sub Saharan region where excessive withdrawal of N through the crop is causing deficiency in the soil.

The waste water treatments are carried out up to secondary treatments where carbon is removed but nitrogen and Phosphorus rich content in waste water are released in the lower places, usually lakes, rivers and even sea. The excessive N and P result in eutrophication in the water bodies, leading to slow death water bodies. On one hand eutrophication impairs water quality and fish kills and on the other extinction of water resources, thus surrounding to bear adverse impact of drought. Some people believe denitrification as promoter of Nitrate conversion in to atmosphere instead of discharging the nitrate to the water bodies in the process of disposal. The land treatment of the N and P rich waste water by land application is known practice but that too is not being adopted as a popular practice. Utilisation of the N and P waste water through specially prepared bed and planting reed is simple and effective way of containing excessive nitrogen thus, reducing cause of building excessive N build-up. The fast renewable canes become source of alternative wood for manufacturing of house hold furniture and conserving biodiversity. The collect toxic materials will find use in preparation of bio bricks and bio char for further use in controlling reducing bioavailability of heavy metals and producing quality food. Thus, the problem can be converted as an opportunity for making use of the resources.

Industrial process enabling transfer from N excess to N deficit sites

A new formulation of Diammonium sulphophosphate is designed as green DASP which will use nutrient recovered from waste water streams and adding cellulose and their decomposition in aerobic decomposition. The product can be physically transferable from N excess to N deficit place to make balance in N cycle.

Convergence of other sciences for promoting activities of nitrogen cycle

Crop protection from aphids: There are many scientific areas where their contributions are needed either to promote or to protect nitrogen cycle activities bringing inimical effects. For example, when water is conserved in the water bodies in the grasslands and forest, malaria becomes a normal problems. Simple biological control of malaria spread suggested by [18] is spread of mustard seed that gets attached to the larva of malaria at the water surface and sinks in water where the larva do not further grow and die. Thus, this or similar material can be used for controlling spread of diseases of animals in the African grasslands and parasitism on these animals reduced.

Harnessing biological nitrogen from the bird droppings

Bird droppings (known as guano) [18,19], contain high amount of nitrogen. The birds collectively sit and wait for long hours in search of their feed of small fish on the bank of rivers and on the boulders existing in the river bed having elevation more than the flow depths. These bird droppings when get washed away and accumulate in the water bodies cause eutrophication. The nitrogen is one of the major nutrients for the plants. The profuse growth of water hyacinth absorbs oxygen present in the water that leads of fish kill and deterioration in the water quality. This nitrogen can be manoeuvred by the application of nitrogen cycle to enhance productivity of crops. Therefore, whatever quantity of nitrogen that gets mixed up in the long stretch of rivers and brings bad effect to the quality of water on can be manoeuvred to eliminate the bad effect on one hand and produce useful good effect on the enhancement of productivity of nitrogen and water, on the other.

The collection of the bird droppings can be easily carried out from the trees where the birds sit for their night stay. Suitable litter or plastic sheets can be spread to collect the droppings, which can be replaced by the new ones at certain intervals at convenience say, 15 to 30 days. The collection of the bird droppings along strategic points (where bird cluster along the bank) can be done by erecting bird sitting stands/ raft for sitting of birds. The stands equipped with provision of collecting tray type channel as an when it is dropped or when the accumulated droppings get washed during rains in a jerry can attached to it. Further, in the river reaches of boulder zones, birds sit on boulder extruding above water surface and keep waiting for the fish for their feed. The protruding stones on which birds sit can be covered with a plastic cap with bottom edge formed as channel and having provision of a collection bottle to collect the washed down droppings as and when rain or mist occur. This cap and collection bottles can be again changed at intervals. Thus, the rare natural resource which produces undesirable effects to the water resources can be augmented to produce beneficial effects.

Discussion

From the reactions that occur in transforming the nitrogen cycle, it is clear that drainage is utmost necessary to bring the aerobic condition that implicates need of drainage in Africa. The study has devised

innovative management of nitrogen cycle for enhancing productivity and protection of environment. The new crops and cropping managements were devised based on innovative application of scientific facts of nitrogen cycle, and sulphur cycle. Since these concepts are new it will require producing data that will show the degree of effectiveness of application. In this study innovative application of the scientific facts has established the correct approach and direction of research work. It remains application and customisation of universal technology and establishment of its efficacy, and finding the shortfall, which can be made up by correction under the refinements.

The broadened scope of drainage still had not been realised by the drainage engineers as they failed to appreciate the role that drainage can play in bringing productivity and protecting the environment. The scope of drainage is limited for removal of excess water, lowering of ground water table and removal of excess salts etc. The conservation of drainage water for re use to reduce water stress due to water shortages, management of nitrogen and sulphur and waste products, reduction of eutrophication etc are not coming to the minds of drainage functionaries. The research endeavours remain limited to conducting conventional field experiments and monitoring the response to the enhancement in yield only. The extended role of drainage on protection of environment and welfare of the society is still not coming to the minds of drainage engineers. These aspects suffer lack of appreciation of the role that good drainage can play. The lack of appreciation culminated in pointing out the aspects towards more elaboration, and conventional field experimental results. The innovative concept and developments brought out in this study still need more appreciation and adopt to derive benefit from drainage. These aspects suffer set back of lack of appreciation due to unawareness of the existence on new concepts of the aspects presented in the article. It requires redesign of course curriculum of water resources facility creating professionals and the users of the facility. These aspects are dealt with in detail in authors other publication [17]. This study had already substantiated that drainage engineering is the best water resources engineering, because of its cost effectiveness, no risk of over drainage by any means, easily applicable to varied problems and field conditions.

In the present study importance of drainage in enhancing productivity and protection environment by reducing the emission of GHGs are brought in with sufficient justifications. The auto drainage technology will eliminate water logging, reduce salts and reduce adverse effect of rise in water table. These aspects are different and new from the earlier functions of drainage. Since these are new researches of the author, various references cited in the text are unique. As there have been lack of support of such references in the existing literature, large number of publications cited by the author is in support of innovativeness of this study on the extended utility and importance of drainage. Therefore, the fact of having many references from the author should be accepted as new development by the author. Acceptance of these facts in the scientific community will enhance utilisation of natural resources viz nitrogen, oxygen, carbon and water to some extent. Thus, drainage will bring lot of social good of food, environment in Africa.

The innovative developments on drainage and practices supprted by the drainage for enhancing productivity and protection of environment are scale insensitive. These measures can be applied at any location and can be up scaled to any extent. In its new domain it is beyond the field experiment where conventional way of researched are carried out by executing site experiment, observe response and formulate recommendations. The other approach towards accomplishment of

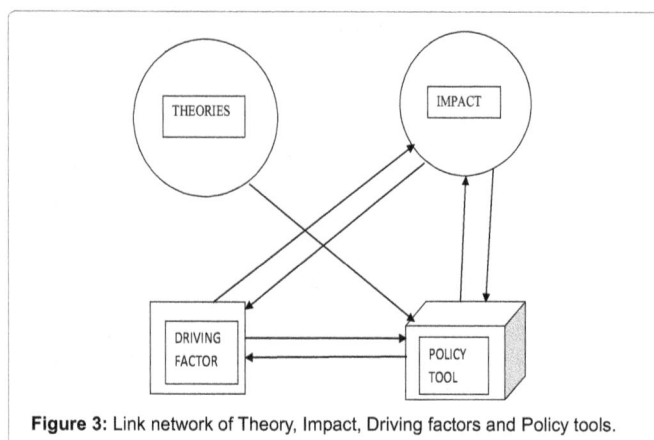

Figure 3: Link network of Theory, Impact, Driving factors and Policy tools.

similar objective is formulation of recommendations based on theory which of course can be modified by customisation, as depicted by (Figure 3). This philosophy of research promotes think global and act local, as the most practical and feasible way of universalisation of researches. This study has excelled in the sphere water resources. It is no way contradicting the known and existing practices of drainage systems. In addition, new auto drainage technology ie racy (alive, smart and enthusiastic) nature agriculture practice is an advancement in the existing types of drainage systems.

SWOT Analyses

Strength

The study has substantiated that with drainage aerobic condition will get developed that will conduct nitrogen cycle to function by bacteria nitrogen fixation from the atmosphere to the soil and transform ammonia in to nitrate and from nitrate to nitrite and then to nitrate. The denitrification converts nitrate again in to nitrite and then to nitrous oxide. The nitrous oxide reserve is kept to minimum to reduce the emission of nitrous oxide to the atmosphere which remains as pollutant and also causes ozone layer depletion. The process based knowledge was applied to devise cropping pattern so as to capitalise the nitrogen build up and reduce excessive and surplus situation to exist. Such cropping patterns are not existence. The new cropping patterns, although suffer setback from the lack of appreciation of drainage experts, have large potential to increase productivity with betterment of quality and protection of environment by reduction of emission of the GHG N_2O. The cropping systems elaborated in this study add very good scientific strength in development of cropping practices to make good use of natural resources. Therefore, there is direct gain from nitrogen cycle management to the technology practising people and indirect gain to the environment. When people carry out their business there is reduction in the emission of nitrous oxide of which about 60% is contributed by agriculture. Methane is largely reported to be contributed by submerged paddy fields and cattle population. By providing drainage aerobic condition can be maintained and new practice of aerobic rice can be use. However, there is difference of opinion that for good yield of rice submergence of paddy fields is necessary. Study by [16] has substantiated that by good practice methane emission from the submerged paddy fields can be cut down to zero methane emission level.

The emission of green house gases such as methane and nitrous oxide will get eliminated to a great extent when drainage is provided to create condition for aerobic decomposition and eliminate anaerobic

decomposition. The study has capacity to transfer excess nitrogen from its place where it is in surplus to the place where it is in deficit. Thus it will be a new method of management of nitrogen cycle. The methods presented here have strong scientific strength and have capacity to produce ecosystem services on sustainable bases.

The scientifically reasoned expected effects of global warming are being experienced in the recent years that convince one to believe the fact of occurrence of global warming. Thus, it is clear that management of nitrogen is an important and appropriate approach to reduce emission of nitrous oxide and capitalise the processes of nitrogen transformations. The new technology i.e. racy nature agriculture is very strong in its formulation and strong in coping up all situations of soil, climate, crops, irrigated or rainfed agriculture. In general, in the rainfed agriculture always emphasis has been to enhance yields, but no consideration on the environmental aspects involved. This study has substantiated that how the role of nitrogen cycle is important in plant production systems. Appropriate management of N improves the services of ecosystems. Therefore, extension of provision of drainage facility in the rainfed agro- ecosystem is very important.

Weakness

All the aspects of management of nitrogen cycle are devised on sound scientific basis and there is no any scientific flaw. Thus, the study has no weakness. Most of results presented here are well established and substantiated by the author in the present study by application of scientific facts or by other supporting results. This situation results incitation of several research articles of the author..

Opportunity

The study makes very strong case that application of drainage engineering will come to alleviate environmental problem of emission of GHG such as methane and nitrous oxide. It has opened opportunities to harness the process for enhancing the ecosystem services by cropping management for nitrogen fixation from the atmosphere. This is a good example of secondary natural resources management. Development of enterprise from the methods presented here will make the methane and nitrous oxide get controlled by the indirect effects from the work that people would undertake for their own benefits. Thus, this is a case of bringing automation in exercising control of reducing emission of GHGs. The process of creating aerobic condition by application of drainage engineering work will be the foundation on which success stories can be transcripted. Thus, this study provides great opportunity for management of nitrogen cycle and produce global benefit to the atmosphere. Therefore, agricultural development in Africa, be it irrigated or rainfed condition, has to be based on provision of land drainage. Drained condition is necessary for bringing direct and indirect social good in Africa.

Threat

There is no threat in the implementation of the results of the study. All the threats of adverse condition that promote deleterious effects on the environment in Africa are removed. The present study has added strength and created tremendous opportunities that will go long way for improving food and environmental situation in Africa.

It will be difficult to realise the discrepancies in knowledge of science of agriculture that did not care for environment, which remained discrepant in always bringing good effects, to accept the facts presented in the present study. It is expected that real realisations will develop soon that will attract attentions of those with the cons views and go in favour of pros of the facts presented in this study.

Action Initiatives

All agricultural practices involving in the management of nitrogen and productivity have been, largely, on the basis of carry out study, feel the benefit and again go for further verification on the pretext of change in agro ecological condition. There had been lack of universally applicable innovative technology. The green technology encompassing all necessary scientifically justified components was developed. However, appreciation and application of the technologies requires acquisition of process based knowledge of chemistry, physics, biological sciences, environmental sciences and environmental engineering. This can be supplemented by adding broadened scope syllabus of drainage engineering to the civil and water resources and irrigation engineering, which produces professionals for infrastructure developments. Equally important aspect is adding syllabus in the study curriculum of environmental engineering in the Agricultural Engineering in particular and Agriculture in general world over, who use the facilities of the infrastructure. Thus, in order to make it simple and justifiable to bring human resources development in drainage engineering, it is very necessary to change names of civil engineering department of Water Resources and Irrigation Engineering to Department of Water Resources and Irrigation and Drainage Engineering. Likewise, in Agricultural Engineering colleges, the existing departments of Irrigation Engineering should be changed to Department of Irrigation and Drainage Engineering. These academic supplementations to bring tuning in thinking will produce knowledgeable professionals in the domain of infrastructure executers and users of the facilities so created to appreciate needful aspects for enhancement in productivity and protection of environment. Thus, such initiatives will develop solid foundations on which mansions of food security and environment can be created. Development of human resources in this direction is possible by bringing slight strategic changes in the academic institutions. In any developmental programme role of human resources is very crucial that will all development pro productive or counterproductive, which has been the case in agriculture, irrigation and drainage and related infrastructure developments. The cases of convergence of services presented in the present support the need of process based broad knowledge to bring some innovative applications enhancing use of water and nitrogen.

The improvement in nitrogen fixation along with the enhancement in yield will be brought by the racy nature agriculture. Further, testing of food quality with respect to amino acid balance needs to be examined. Studies have shown that absorption and utilisation of protein are optimised when adults evenly distribute their protein uptakes throughout the day [20]. All technologies presented in the present study are applicable at global scale. Their application for local condition will indicate efficiency level to emerge from customisation. This study has opened opportunity for managing agriculture and environment scientifically at global scale as well as at local scale. This study enables capture under grip for manoeuvre the global worrisome problem emerging due to distortions in the nitrogen cycle. The scientific management of agriculture and environment will replace pseudo managements in vogue science of agriculture.

Conclusion and Future Research Needs

The study presents prospects of drainage in creating condition favourable to apply and implement innovative technologies of management of nitrogen cycle in Africa. Creation of aerobic condition by providing drainage in all land uses is a foundation development

to conduct by itself necessary aspects of reduction of GHGs. Technological application will involve actions for harnessing direct benefits of enhancement in quantity and quality of primary produces and protection of environment will emerge as indirect automatic benefits. Technologies of nitrogen cycle management described are pertinent to situation in Africa. Technologies for biological nitrogen fixation in agriculture, grasslands and pastures and forestry will enhance N use efficiency and reduce emission of nitrous oxide, physical transfers of excess to deficit site and convergence of services that go in promoting management of N are developed. Being universally applicable, these technologies will be beneficial for global perspective developments. Needful action initiatives are suggested to be brought in human resources development. The universally applicable technologies need be applied in different regions to promote customisation of the technologies.

Acknowledgement

The author acknowledges use of references cited in support of statements of academic values included in the study. Certified that no institutional support was availed for preparation of the manuscript.

References

1. Schwab GO, Fangmeier DD, Elliot WJ, Frevert RK (2005) Soil and Water Conservation Engineering. IV edition, John Wiley and Sons, Asia.

2. USDA USSL (1958) Diagnosys and Improvement of saline and alkali soils. US. Soil salinity laboratory, USA.

3. Ehlers, Wilfried, Goss, Michael (2003) Water dynamics in plant production; CABI Publication, Walling Ford, U.K.

4. Kuenen JG, Robertson LA (1994) Combined nitrification-denitrification processes. FEMS Microbiol Rev 15: 109-117.

5. Smith KA, Arah, JRM (1990) Losses of nitrogen by denitrification and emission on nitrogen oxides from soils; In Proceedings of the Fertilizer Society No 299. London.

6. Yadav RC (2014) Drainage Engineering: A savvier for sustainable resources use, protection of environment and professional development. J of Civil Engineering, Photon 107: 200-213.

7. FAO (2002) Agricultural drainage water management in arid and semi-arid areas. Fao, Rome, Italy.

8. Hammer Mark J, Hammer Mark Jr (2005) Water and Waste Water Technology; Printice Hall of India. New Delhi

9. Yadav RC (2014) Innovative application of scientific facts for arresting GHG-N2O and improvising lucrative ventures with enhanced land, water and nutrient use efficiency. J of Energy and Environment 128: 486-520.

10. Yadav RC (2012) Innovative application of scientific facts for nutrient recovery from waste water Streams for sustainable agriculture and protection of environment. Hydrology: Current Research 5: 1000142.

11. Yadav RC, Om Prakash, Deshwal JS (2013) Biotechnology of intra-row banding of cyno-bacterial leguminous crops for raising yield plateaus of cereals and oil seeds. Intern. J. Agronomy and Plant Production 4: 3330-3336.

12. Catherine H, Revellin, Cecile (2011) Inoculants of leguminous crops for mitigating soil emissions green house gas nitrous oxide. Plant Soil 34: 289-296.

13. Marra LM, Soares CRFS, de Oliveira SM, Ferreira PAA, Moreira et al. (2012) Biological nitrogen fixation and solubilization by bacteria isolated from tropical soils. Plant Soil. 357: 289-307.

14. Glover JD, Culman SW, DuPont ST, Brousand W, Young L, et al. (2010) Harvested perennial grassland provides ecological bench mark for agricultural sustainability. Agr Eco Sys Environ 137: 3-12.

15. Catherine H, Revellin Cecile (2011) Inoculants of leguminous crops for mitigating soil emissions greenhouse gas nitrous oxide. Plant Soil 34: 289-296.

16. Yadav RC (2013) Racy nature agriculture versus other allied technologies: A technologies contrast. Middle East Journal of Scientific Research, Dubai, Published on line.

17. Yadav RC (2014) Innovative application of scientific facts for natural resources management for enhancing productivity and protection of environment. International J of Agronomy and Plant Production.

18. Wendy H (2003) Drainage. How it Works, Science and Technology 5: 673-675.

19. Gustafson AF (2010) Hand Book of Fertilisers. Agrobios; Jodhpur, India.

20. Gupta US (2000) Crop Improvement: Quality Characters. Science publishers, USA.

9

New Approaches to Agricultural Land Drainage

Luis Gurovich* and Patricio Oyarce

Departamento de Fruticultura y Enología, Pontificia Universidad Católica de Chile. Santiago, Chile

Abstract

A review on agricultural effects of restricted soil drainage conditions is presented, related to soil physical, chemical and biological properties, soil water availability to crops and its effects on crop development and yield, soil salinization hazards, and the differences on drainage design main objectives in soils under tropical and semi-arid water regime conditions. The extent and relative importance of restricted drainage conditions in Agriculture, due to poor irrigation management is discussed, and comprehensive studies for efficient drainage design and operation required are outlined, as related to data gathering, revision and analysis about geology, soil science, topography, wells, underground water dynamics under field conditions, the amount, intensity and frequency of precipitations, superficial flow over the area to be drained, climatic characteristics, irrigation management and the phenology of crop productive development stages. These studies enable determining areas affected by drainage restrictions, as well as defining the optimal drainage net design and performance, in order to sustain soil conditions suitable to crops development.

Keywords: Restricted drainage; Drainage studies; Drainage design parameters

Introduction

Agricultural land drainage consists of a set of technical strategies and hydraulic structures allowing the removal of water and/or salts excesses present in the soil volume occupied by crop roots, to provide an adequately oxygenated environment suitable for root normal development, keeping adequate water and air relative proportions according to crop physiological needs, to enable soil sustainability for crop productive conditions [1-4]. Under deficient drainage conditions, resulting from excessive water stored in the soil pore space, oxygen vapor pressure is very limited for root crop normal biological activity, as well as for the microflora and microfauna activity in soil [5]. This condition induces multiple physiological disorders in plants, such as stomata closure induction processes in leaves, due to the increase in the ABA (abscisic acid) concentration, as well as to a lower permeability of root exodermis cell membranes to water and nutrient absorption [6-8].

Anoxia conditions in soils inhibit root tissue respiration rates and energetic processes at the cellular level [9], thus affecting important metabolic processes in plants, which react to the oxidative stress using an unique substrate for the ADP phosphorylation to ATP, and thus generating metabolic energy disorders, resulting in fermentative glycolysis instead of oxidative respiration [10,11]. As a consequence, plant photosynthetic rate is inhibited [11,12] and the transport of solutes among plant organs, through its conductive phloematic and xylematic tissues, is significantly reduced, modifying transport and storage mechanisms of photosynthates and minerals in the whole plant [13,14]. It has been also demonstrated that oxygen deficiencies induce photo-oxidative damage in leaves, due to the generation of reactive oxygen species (ROS) such as superoxide (O_2^-), hydrogen peroxide (H_2O_2) and hydroxyl radicals (OH^-), which affect chloroplast fundamental functions, resulting in significant levels of chlorosis and premature senescence induction in plant tissues [15]. Another important effect of oxygen deficiency is the generation of plant chemical signals, stimulating ABA production in leaves, which activate stomata closure mechanisms, further decreasing photosynthetic rates and transpiration [16]. For soils under limited aeration conditions, an increase in CO_2 concentration is produced inside the soil porous volume, with a transient increment of pH around the root absorbing system [10,12]. Also, accumulation of others gases in the soil has been reported, such as methane and hydrogen sulfide, generated during anaerobic degradation of soil organic matter, which may have phytotoxic properties, causing environmental damage in flooded soils [17].

As a result of plant metabolic disorders, caused by conditions of total or partial anoxia near the roots, resulting from deficient drainage conditions, a decrease in crop production often occurs [8,15,16,18]. In anoxic or hypoxic soils, one of the earliest detectable changes is the decline in net CO_2 assimilation rate, as reported for diverse crops such as avocado (Persea americana Mill) [8], sunflower (Helianthum annus L) [19] beans (Vigna radiate) [9]. This reduction is often coupled to decreases in stomata conductance, transpiration rate and intercellular CO_2 partial pressure in leaves [20]. In poorly drained soils, the intrinsic dependence between water content, apparent density and porosity also affect soil physical and hydrodynamic properties, such as major changes in soil structure, reducing its permeability and temperature fluctuations, with significant impacts on crop development and yield [5,6,21,22]; under low soil temperature, chemical and biological reactions rates decrease drastically [23,24]. Severe restrictions on the uptake of major plant nutrients, like nitrogen, phosphor, sulfur and calcium, have been reported in cold, poorly drained soils [25]. Microorganisms responsible of soil organic matter degradation are influenced by changes in the soil environment; low temperature and anaerobic conditions restrict microorganism activity and abundance. A decrease in soil temperature, resulting from waterlogging, produces a proportional decrease in organic matter decomposition rates [26,27].

In poorly aerated soils, high concentrations of mineral elements in reduced forms (Fe^{+2}, Al^{+3}, and NH_4^+) are common [17]. Oxygen diffusion into saturated soil is extremely low, therefore, as depth in the saturated soil profile increases, the presence of O_2 diminishes and the redox potential turns extremely low. Moreover, the soil may lose some of its soluble nitrogen species, due to an oxidation of ammonia nitrogen to both nitric and nitrous species, followed afterwards by a reduction of these N-species to various forms of gaseous nitrogen, producing a release of gaseous nitrogen to the atmosphere [25]. Reduction-

---

*Corresponding author: Luis Gurovich, Departamento de Fruticultura y Enología, Pontificia Universidad Católica de Chile. Santiago, Chile
E-mail: lgurovic@puc.cl

reactions predominate over oxidation-reactions in soils presenting drainage restrictions, so when iron is present in its Fe^{+2} form (reduced), a gray-bluish tint can be observed in the soil; the opposite occurs when oxidation reactions prevail, with changes in soil color, from reddish to brown tints. When alternate periods of poor and adequate drainage develop throughout the year, the soil profile exhibits some very characteristic stains of yellowish–orange appearance; for permanent saturation conditions the dominant soil color is gray and these soils are generically known as *gley* soils [28,29].

Waterlogging in soils is the major factor influencing soil compaction process [30,31], which is defined as the process in which soil particles organize themselves spatially, diminishing void spaces in soils, thus increasing its apparent density; in other words, the increase in soil moisture content reduces its capacity to support loading and decreases the permissible pressure due to mechanical work [31]. High levels of soil compaction are related to soil penetration resistance [32]. Soil compaction is mainly produced by tillage activities and machinery weight, that is transmitted to soil either through the wheels or others contact elements; agricultural machinery traffic intensity is determinant for soil compaction problems in poorly drained soils [33]. Soil tillage is often needed to maintain optimal aeration within the soil rooting profile [30,31,34]; in clay soils, intensive plough use at a specific depth generate a compacted, almost impermeable soil layer located immediately below plough depth, which restricts both crop root development, as well as the free water and air flow within the profile [35,36]. Soil profiles accumulate dissolved salts present in phreatic layers; in arid and semiarid regions, these salts may ascend by capillarity from the phreatic layers into the soil volume occupied by crop roots; the existence of a saline phreatic mantle is frequent at soil depressions [37]. Gradual soil salinization resulting from deficient drainage conditions can occur [3,38-41]; main salts present in poorly drained soil profiles correspond to sulphates [42], nitrates [43], chlorides [44], carbonates and bicarbonates [45]. Gradual soil salinization, and the eventual crystallization of salts [40,46], when its concentrations exceed the pK values for specific saline solutions, is produced as the result of water absorption by crops, which normally excludes most of the salts, by means of its selective root permeability [47,48], as well as by salt free water direct evaporation from the soil surface [1,49]. Soil salinization processes also modify soil structure, leading to a gradual impairment in its agricultural productivity, because high salt concentrations can disperse clay soil aggregates, reduce soil porosity and permeability [38,39], reduce water availability to crops and can be phytotoxic. These phenomena originate crusting and hardening of the soil surface, incrementing soil resistance to penetration and reducing soil gaseous exchange with the atmosphere, as well as reducing soil water infiltration rate [50]. Saline and sodic soils not only are physical and chemically degraded, but also can be biologically affected, due to reduced concentration and activity of heterotrophic microorganisms. Organic matter decomposition rate in poorly drained soils is considerably lower, as compared to soils with normal aeration conditions [51]. A decrease in soil fertility and nutrient supply to crops occurs with deficient drainage conditions, and larger rates of crop fertilizer needs are needed to reach profitable crop yield [52,53].

Removal of waterlogging conditions in soils though agricultural drainage, is crucial for efficient and sustainable crop production in irrigated areas, both in tropical zones as well as in rain-fed zones, were crops satisfy its hydric requirements exclusively from precipitations [39,54,55]. In some areas, soil drainage is produced naturally, but in other areas, soil intervention is necessary to create conditions for an efficient artificial drainage [3,56]. In tropical regions, precipitation

generally is larger than evapotranspiration; for low hydraulic conductivity soils, superficial drainage problems are common. In these areas, the excess of precipitations nearly always guarantees that saline balance in soil is kept, so drainage main purpose in these areas is to evacuate water from the soil profile and to provide the ideal aeration conditions in the soil profile [57]. On arid and semi-arid zones, atmospheric evaporative demand is high and precipitations are scarce; generally the yearly total precipitation is significantly lower than crop evapotranspiration. Agricultural activity is only possible by implementing irrigation, to acknowledge for hydraulic balanced conditions, enabling crops reach profitable yields. Excessive irrigation determines saturation conditions to develop, thus raising phreatic levels in the soil profile [1,46]. Moreover, in arid and semiarid areas, irrigation water with significant salt concentrations can generate soil productivity degradation and crop reduced water availability, by a decrease of soil water potentials [7,46,58]. Therefore, artificial drainage in this areas is aimed towards reducing phreatic levels and eliminate salinity from the soil profile occupied by crop rooting systems [38,39,59].

Artificial agricultural drainage design in tropical and arid zones share common principles, but there are important differences in drainage system design and planning objectives. The main goal on tropical zones is soil profile waterlogging control, to provide ideal aeration conditions for crops; in arid zones and semi - arid zones, drainage not only includes soil aeration considerations, but also it must consider salinity control and soil moisture deficit avoidance [60].

Agricultural drainage plays an important role on world food production, providing a safeguard for investments in irrigation, as well as the conservation of soil resources. During the XX century second half, large drainage projects were built worldwide, on about 150 million hectares affected by flooding and salinity problems, thus contributing to world food production significant improvements, by intensifying and diversifying competitive and financially sustainable agricultural activities. Soil profile artificial intervention, associated to the annexed hydraulic structures needed to eliminate the drained water, represent significant public and private investments, oriented to solve deficient agricultural drainage situations [2,54,57]. Therefore, it is essential that drainage systems should be properly designed, installed, operated and maintained. Planning and design of drainage networks determines the efficiency of water and salinity removal from agricultural soil profiles [4,53,54,57]. In developing countries, most drainage facilities are far from being adequate or sufficient; only 14% of the 1.500 million hectares cultivated worldwide, including both irrigated and non-irrigated lands, have been implemented with some kind of drainage structures (Figure 1) [60]. Agricultural soils subsurface drainage systems must include quantitative information on relevant aspects of hydraulic engineering,

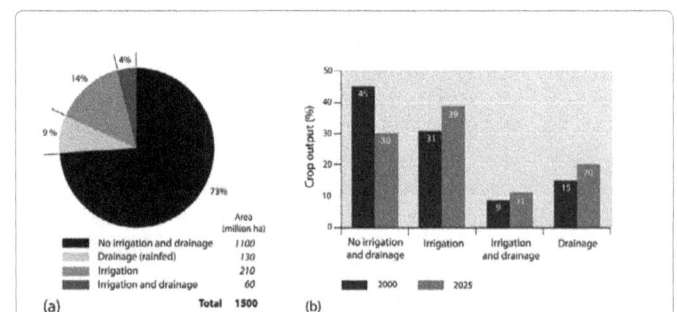

Figure 1: Drainage plays an essential part to sustainable food production: (a) Worldwide total agricultural areas. (b) Relative agricultural output from agricultural lands provided with irrigation and drainage systems in 2000 and in 2025 [60].

1. Topographic Studies	i.	Slope constraints
	ii.	Landscape
2. Investigation of soils	i.	Soil maps
	ii.	Soil salinity and alkalinity data
	iii.	Oxidation - reduction potential data
	iv.	Soil composition
	v.	Hydraulic conductivity measurements
3. Origin of the water present in soil profile	i.	Precipitations
	ii.	Deficient irrigation
	iii.	Underground water (aquifer)
4. Studies of underground waters	i.	Water table relative location in the soil profile
	ii.	Water table level and flow fluctuations
	iii.	Water table salinity fluctuations
5. Irrigation practices and requirements	i.	Irrigation water quality
	ii.	Irrigation type and frequency
	iii.	Irrigation water depths applied
	iv.	Water depth requirements for salinity control
	v.	Water losses due to percolation
	vi.	Irrigation techniques

Table 1: Main studies relative to the design and planning of drainage systems [53].

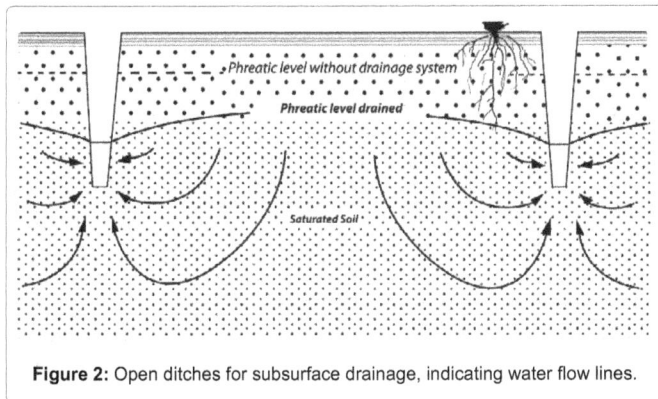

Figure 2: Open ditches for subsurface drainage, indicating water flow lines.

agriculture, environment and hydrology, as well as economic, social and socio-political aspects, in order to establish suitable criteria for optimal drainage [57]; however, the main purpose of drainage design is to ameliorate soils presenting water saturated layers. This improvement will be oriented towards soil conservation, as well as to optimize crop yield, to enable crops diversification and optimization of agricultural tillage operations using specific agricultural machinery. Adequate drainage system planning and design necessarily requires soil accurate studies, underground water origin and recharge rates, phreatic layer depth and time evolution, vertical water flow rates and crops characteristics related to soil profile water and salt contents [53] (Table 1).

Agricultural drainage studies require comprehensive data gathering, revision and analysis about geology, soil science, topography, wells, water level and its fluctuations, the amount, intensity and frequency of precipitations, superficial flow over the area to be drained, climatic characteristics and the phenology of crop productive development stages. A quantitative study relative to underground water dynamics under field conditions is also required, including information on water table positioning and its fluctuations throughout the crop production season, at various points within the region to be drained. These studies enable determining areas affected by drainage restrictions,

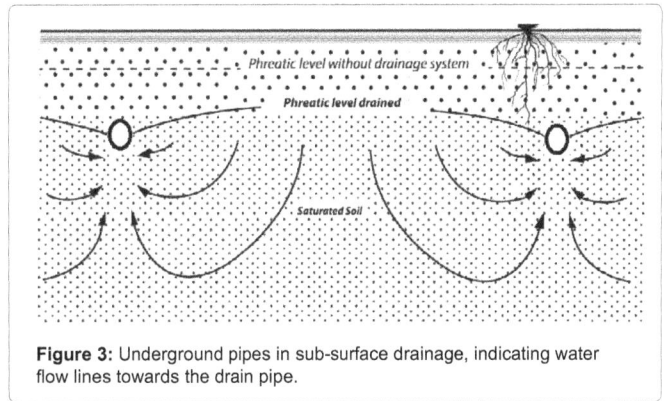

Figure 3: Underground pipes in sub-surface drainage, indicating water flow lines towards the drain pipe.

defining the optimal distance and depth for lateral drains, and the flow to be removed, in order to sustain soil conditions suitable to crops development. There are two major drainage systems for controlling underground waters: open ditches (Figure 2) and subsurface piping (Figure 3). Open ditches systems consist of excavations in the soil that collect the water stored at existing saturated layers; it also can be used to remove surface run-off; this system can account for significant land farming losses, smaller soil units for farm machinery operation and interference with irrigation systems, making agricultural tasks more expensive [61,62]. Subsurface pipe drainage systems consists on plastic tubes, either smooth or corrugated, provided with perforations, that are placed buried within the soil, at specified distances and depths; this system is used mainly to lower the water table in the unconfined aquifers [54,63,64]. These drainage systems in most cases consist on a main drain, a collector drain and a network of field drain pipes; the position of the main drain depends on the field slope and the location of the lowest field level, through which the collected water is removed from the drained area. The collector drain and the network of field drains are usually located in parallel to each other; field drains are perforated pipes along their extension and its function is phreatic level control, by receiving water excesses present in the soil profile and convey this effluent towards the collector drain. Secondary drains and main drain main functions are water conductions from the drain pipes to the site of water discharge. These conductive drains are either open ditch type or underground pipes, the selected option will depend on costs and dimensions of piping [53, 54, 57, 60, 63, 64]. Subsurface drain design corresponds to a set of agronomic, hydraulic and engineering characteristics that a lateral drainage system must fulfill, to eliminate the volume of soil water required to satisfy crop optimal growth and production [1, 3, 65, 66]. In general, design features must define the proper criteria and parameters relevant to spacing among lateral drains, its depth placement inside the soil profile and the hydraulic characteristics of the hydraulic net, required to transport the water volume to be collected and removing it from the cultivated area. In relation to construction aspects, it must include definitions about drain hydraulic net layout, the materials to be used, the density and kind of perforations, as well as building techniques and network installation and maintenance.

Optimal distance calculations between consecutive lateral drains are closely related to water flow towards the drains. The development of a mathematical model for quantitative description of the sub-surface flow towards lateral drains is possible only based on mathematical simplifications, deduced from the theory of underground water saturated flow, with pre-established initial and border conditions.

Conclusion

The normative and protocols established for agricultural land drainage in countries having expertise in the subject, have not been validated for the specific conditions of soils and situations of deficient drainage existing in local agricultural conditions. International standards for drainage networks specify the required properties for drainage materials (concrete, plastics and ceramics), as well as raw materials specifications, in terms of its chemical composition and the recommended additives. Also, very seldom specifications for drainage pipe resistance are available, as well as the proportions of recycled plastic as raw material allowance.

The existing norms for drainage materials proceeding from countries with a long drainage experiences might be used as a reference to define national standards, specifically needed for local circumstances. Optimization of perforation density and shape for PVC drainage pipe, allowing to increment water extraction efficiency and reducing pipe costs, is needed to define design and evaluation techniques of new components. A continuous, applied research program, carried on jointly by Universities, Research Institutes and Industry, can provide technologies to develop efficient and low cost drainage systems, adapted to local conditions.

Acknowledgement

This study was supported by Conicyt (The Chilean National Commission for Science and Technology, under the PAI PROGRAM (Industry involvement in research) 2013. Project Folio Code 7813110015.

References

1. Ayars J, Christen E, Hornbuckle J (2006) Controlled drainage for improved water management in arid regions irrigated agriculture. Agricultural Water Management 86: 128-139.

2. Ritzema H, Nijland H, Croon (2006) Subsurface drainage practices: From manual installation to large-scale implementation. Agricultural Water Management 86: 60-71.

3. Ritzema H, Satyanarayana T, Raman S, Boonstra J (2008) Subsurface drainage to combat waterlogging and salinity in irrigated lands in India: Lessons learned in farmers' fields. Agricultural Water Management 95: 179-189.

4. Naz B, Ale S, Bowling L (2009) Detecting subsurface drainage systems and estimating drain spacing in intensively managed agricultural landscapes. Agricultural Water Management 96 (4): 627-637.

5. Saqib M, Akhtar J, Qureshi R (2004) Pot study on wheat growth in saline and waterlogged compacted soil: I. Grain yield and yield components. Soil and Tillage Research 77: 169-177

6. Palta J, Ganjeali A, Turner N, Siddique K (2010) Effects of transient subsurface waterlogging on root growth, plant biomass and yield of chickpea. Agricultural Water Management 97: 1469-1476.

7. Askri B, Ahmed A, Abichou T, Bouhlila R (2014) Effects of shallow water table, salinity and frequency of irrigation water on the date palm water use. Journal of Hydrology 513: 81-90.

8. Schaffer B (2006) Efectos de la deficiencia de oxígeno del suelo en Paltos (Persea Americana Mill). Seminario internacional; Manejo del agua de riego y suelo en cultivo del palto. Serie Actas INIA 41: 9-21.

9. Sairam R, Dharmar K, Chinnusamy V, Meena R (2009) Waterlogging-induced increase in sugar mobilization, fermentation, and related gene expression in the roots of mung bean (Vigna radiata). Journal of Plant Physiology 166: 602-616.

10. Kumutha D, Sairam R, Ezhilmathi K, Chinnusamy V, Meena R, et al. (2008) Effect of waterlogging on carbohydrate metabolism in pigeon pea (Cajanus cajan L.): Upregulation of sucrose synthase and alcohol dehydrogenase. Plant Science 175: 706-716.

11. Yin D, Chen S, Chen F, Guan Z, Fang W, et al. (2009) Morphological and physiological responses of two chrysanthemum cultivars differing in their tolerance to waterlogging. Environmental and Experimental Botany 67: 87-93.

12. Ozcubukcu S, Ergun N (2013) Effects of waterlogging and nitric oxide on chlorophyll and carotenoid pigments of wheat. Food, Agriculture and Environment 11: 2319-2323.

13. Pang J, Zhou M, Mendham N, Shabala S (2004) Growth and physiological response of six barley genotypes to water logging and subsequent recovery Aust J Agr Res 55: 895-906.

14. Maryam A, Nasreen S (2012) A Review: Water Logging Effects on Morphological, Anatomical, Physiological and Biochemical Attributes of Food and Cash Crops. International Journal of Water Resources and Environmental Sciences 1: 113-120.

15. Yordonova R, Christov K, Popova L (2004) Antioxidative enzymes in barley plants subjected to soil flooding. Environ. Exp. Bot 51: 93-101.

16. Ahmed S, Nawata E, Sakuratani T (2006) Changes of endogenous ABA and ACC, and their correlations to photosynthesis and water relations in mungbean (Vigna radiata (L.) Wilczak cv. KPS1) during waterlogging. Environmental and Experimental Botany 57: 278-284.

17. Virtanen S, Simojoki A, Hartikainen H, Yli-Halla M (2014) Response of pore water Al, Fe and S concentrations to waterlogging in a boreal acid sulphate soil. Science of the Total Environment 485: 130-142.

18. Sauter M (2013) Root responses to flooding. Current Opinion in Plant Biology 16(3): 282-286.

19. Grassini P, Indaco G, Pereira M, Hall A, Trápani N, et al. (2007) Responses to short-term waterlogging during grain filling in sunflower. Field Crops Research 101: 352-363

20. Voesenek L, Bailey-Serres J (2013) Flooding tolerance: O2 sensing and survival strategies. Current Opinion in Plant Biology 16: 647-653.

21. Rodríguez-Gamir J, Ancillo G, González-Mas M, Primo-Millo E, Iglesias D (2011) Root signalling and modulation of stomatal closure in flooded citrus seedlings. Plant Physiology and Biochemistry: PPB. Société Française de Physiologie Végétale 49: 636-45.

22. Li H, Bielder C, Payne WA, Li T (2014) Spatial characterization of scaled hydraulic conductivity functions in the internal drainage process leading to tropical semiarid soil management. Journal of Arid Environment 105: 64-74.

23. Zanchi F, Meesters A, Waterloo M, Kruijt B, Kesselmeier J, et al. (2014) Soil CO2 exchange in seven pristine Amazonian rain forest sites in relation to soil temperature. Agricultural and Forest Meteorology 192: 96-107.

24. Zheng C, Jiang D, Liu F, Dai T, Jing Q, et al. (2009) Effects of salt and waterlogging stresses and their combination on leaf photosynthesis, chloroplast ATP synthesis, and antioxidant capacity in wheat. Plant Science 176: 575-582.

25. Steffens D, Hütsch B, Eschholz T, Lošák T, Schubert S, et al. (2004) Water logging may inhibit plant growth primarily by nutrient deficiency rather than nutrient toxicity. Plant Soil Enviroment 2: 545-552.

26. Kwon M, Haraguchi A, Kang H (2013) Long-term water regime differentiates changes in decomposition and microbial properties in tropical peat soils exposed to the short-term drought. Soil Biology and Biochemistry 60: 33-44.

27. Wang H, Yang J, Yang S, Yang, Lu Y, et al. (2014) Effect of a 10 °C-elevated temperature under different water contents on the microbial community in a tea orchard soil. European Journal of Soil Biology 62: 113-120.

28. Fiedler S, Sommer M (2004) Water and Redox Conditions in Wetland Soils-Their Influence on Pedogenic Oxides and Morphology. Soil Science Society of America Journal 8: 326-335.

29. Startsev D, McNabb D (2009) Effects of compaction on aeration and morphology of boreal forest soils in Alberta, Canada. Canadian Journal of Soil Science 89: 45-56.

30. Strudley M, Green T, Ascoughii J (2008) Tillage effects on soil hydraulic properties in space and time: State of the science. Soil and Tillage Research 99: 4-48.

31. Hamza M, Anderson W (2005) Soil compaction in cropping systems: A review of the nature, causes and possible solutions. Soil and Tillage Research 82: 121-145.

32. Govaerts B, Fuentes M, Mezzalama M, Nicol J, Deckers J, et al. (2007) Infiltration, soil moisture, root rot and nematode populations after 12 years of different tillage, residue and crop rotation managements. Soil and Tillage Research 94: 209-219.

33. Afzalinia S, Zabihi J (2014) Soil compaction variation during corn growing season under conservation tillage. Soil and Tillage Research 137: 1-6.

34. Dani O, Ghezzehei T (2002) Modeling post-tillage soil structural dynamics: a review. Soil and Tillage Research 64: 41-59.

35. Botta G, Jorajuria D, Balbuena R, Ressia M, Ferrero C, et al. (2006) Deep tillage and traffic effects on subsoil compaction and sunflower (Helianthus annus L.) yields. Soil and Tillage Research 91(1-2): 164-172.

36. Raper R (2005) Subsoiling: Encylopedia of soil in the environmental. Hill D, Rosenzweig C, Powlson D, Scow K, Singer M, Sparks, D. Academic Press. Columbia University New York, NY. USA

37. Nosetto M, Acosta M, Jayawickreme D, Ballesteros S, Jackson R (2013) Land-use and topography shape soil and groundwater salinity in central Argentina. Agricultural Water Management 129: 120-129.

38. Bahçeci I, Dinç N, Tarı A, Ağar A, Sönmez B, et al. (2006) Water and salt balance studies, using SaltMod, to improve subsurface drainage design in the Konya-Çumra Plain, Turkey. Agricultural Water Management 85: 261-271

39. Konukcu F, Gowing J, Rose D (2006) Dry drainage: A sustainable solution to waterlogging and salinity problems in irrigation areas. Agricultural Water Management 83: 1-12

40. Qureshi A, McCornick P, Qadir M, Aslam Z (2008). Managing salinity and waterlogging in the Indus Basin of Pakistan. Agricultural Water Management 95: 1-10.

41. Yao R, Yang J, Zhang T, Hong L, Wang M (2014) Studies on soil water and salt balances and scenarios simulation using SaltMod in a coastal reclaimed farming area of eastern China. Agricultural Water Management 131: 115-123.

42. Green R, Macdonald B, Melville M, Waite T (2006) Hydrochemistry of episodic drainage waters discharged from an acid sulfate soil affected catchment. Journal of Hydrology 325(1-4): 356-375.

43. Karpuzcu M, Stringfellow W (2012) Kinetics of nitrate removal in wetlands receiving agricultural drainage. Ecological Engineering 42: 295-303.

44. Kennedy C, Bataille C, Liu Z, Ale S, Van de Velde J, et al. (2012) Dynamics of nitrate and chloride during storm events in agricultural catchments with different subsurface drainage intensity (Indiana, USA). Journal of Hydrology 466-467: 1-10.

45. Zuo Y, Ren L, Zhang F, Jiang R (2007) Bicarbonate concentration as affected by soil water content controls iron nutrition of peanut plants in a calcareous soil. Plant Physiology and Biochemistry : PPB. Société Française de Physiologie Végétale 45: 357-64.

46. Aragüés R, Urdanoz V, Çetin M, Kirda C, Daghari H, et al. (2011) Soil salinity related to physical soil characteristics and irrigation management in four Mediterranean irrigation districts. Agricultural Water Management 98: 959-966.

47. Deinlein U, Stephan A, Horie T, Luo W, Xu G, et al. (2014) Plant salt-tolerance mechanisms. Trends in Plant Science 19: 1-9.

48. Roy S, Negrão S, Tester M (2014) Salt resistant crop plants. Current Opinion in Biotechnology 26: 115-24.

49. Hornbuckle J, Christen E, Ayars J, Faulkner R (2005) Controlled water table management as a strategy for reducing salt loads from subsurface drainage under perennial agriculture in semi-arid Australia. Irrigation and Drainage Systems 19: 145-159.

50. Rengasamy P, Smith A (2010) Diagnosis and management of sodicity and salinity in soil and water in the Murray irrigation region. The University of Adelaide South Australia.

51. Bennett S, Barrett-Lennard E, Colmer T (2009) Salinity and waterlogging as constraints to saltland pasture production: A review. Agriculture, Ecosystems & Environment 129: 349-360.

52. Barrios E (2007) Soil biota, ecosystem services and land productivity. Ecological Economics 64: 269-285.

53. Skaggs R, Van Shilfgaarde J (1999) Agricultural Drainage. American Society of Agronomy. Agronomy, madison, Wisconsin, USA.

54. Rimidis A, Dierickx W (2004) Field research on the performance of various drainage materials in Lithuania. Agricultural Water Management 68: 151-175.

55. Darzi-Naftchali A, Shahnazari A (2014) Influence of subsurface drainage on the productivity of poorly drained paddy fields. European Journal of Agronomy 56: 1-8.

56. Singh R, Helmers M, Crumpton, W, Lemke D (2007) Predicting effects of drainage water management in Iowa ' s subsurface drained landscapes 92: 162-170.

57. Vander Molen W, Martínez B, Ochs W (2007) Guidelines and computer programs for the planning and design of land drainage systems. FAO, Roma.

58. Sun J, Kang Y, Wan S, Hu W, Jiang S, et al. (2012) Soil salinity management with drip irrigation and its effects on soil hydraulic properties in north China coastal saline soils. Agricultural Water Management 115: 10-19.

59. Houk E, Frasier M, Schuck E (2006) The agricultural impacts of irrigation induced waterlogging and soil salinity in the Arkansas Basin. Agricultural Water Management 85: 175-183.

60. Nijland B, Croon F, Ritzema H (2005) Subsurface Drainage Practices Guidelines for the implementation and operation. Wageningen, Alterra, ILRI Publication.

61. Scholz M, Trepel M (2004) Hydraulic characteristics of groundwater-fed open ditches in a peatland. Ecological Engineering 23: 29-45.

62. Kröger R, Cooper C M, Moore M (2008) A preliminary study of an alternative controlled drainage strategy in surface drainage ditches: Low-grade weirs. Agricultural Water Management 95: 678-684.

63. Stuyt L, Dierickx W (2006) Design and performance of materials for subsurface drainage systems in agriculture. Agricultural Water Management 86: 50-59.

64. Stuyt L, Dierickx W, Martínez J (2009) Materiales para sistemas de drenaje subterráneo. Estudios FAO, Riego y drenaje. Roma.

65. Welderufael W, Woyessa Y (2009) Evaluation of surface water drainage systems for cropping in the Central Highlands of Ethiopia Agricultural Water Management 96: 1667-1672.

66. Rahman M, Lin Z, Jia X, Steele D, DeSutter T, et al. (2014) Impact of subsurface drainage on streamflows in the Red River of the North basin. Journal of Hydrology 511: 474-483

10

Transportation Module Determination for the Urban Landscapes with Linear Programming Pattern in the Urmia, North-West Iran

Solmaz Javanbakht* and Reza Dadmehr

Department of Water Engineering, Urmia University, Urmia, Iran

Abstract

Urban landscapes are the crucial key factors in natural human life stability in modern urban civilization. However, despite of the importance of urban landscape, common suitable urban landscape per capita in cities of Iran is between 7 to 12 square meters and the average urban landscape per capita in Urmia metropolis is 6.9 square meters, which indicates a serious gap for 20 to 25 square meters as global standards. On the other hand, the urban landscape growth for achieving global standards causes an increase in vegetation which by itself results in greater demand for water resources. Consideration of arid and semi-arid climate of Iran and limitation in water resources makes urgent need for planning and water allocation. The main purpose of the this study is putting forward a linear programming pattern in the form of transportation model in-order to allocate water optimally from the existing and future water resources (i.e. surface water, ground water and drinking water) to Urmia urban landscape pieces, considering minimization of the cost of supplying water. To achieve the above mentioned model, North-West Corner method, Least Cost method and Vogal Approximation method have been applied and the obtained results have been compared. According to the obtained results, in summary, it can be claimed that Vogal Approximation method, has the higher capacity for optimal allocation of water resources in Urmia urban landscape than that of Least Cost method as well as North-West Corner methods With regard to the present availability of water resources for every seven months of irrigation, the amount of optimal allocation of water from drinking water is 5400 cubic meters per day for boulevards, 1400 cubic meters per day for nurseries and 1200 cubic meters per day for other landscapes. The allocated optimal amount of water from surface water resources is 6500 cubic meters per day for forest parks. The allocated amount of ground water resources is 5000 cubic meters per day for parks in urban areas and 1600 cubic meters per day for boulevards and 900 cubic meters per day for forest parks. During the hottest month of each year (June 22nd to July 14th), with respect to irrigation, given the above variables in optimal allocation of water resources for urban landscape in the city of Urmia in comparison with that of seven month irrigation, are the same. However from quantity point of view, drinking water and ground water quantity, in some landscapes are less. Also, concerning the optimal allocation of water from future water resources (increment water supply), given variables such as mentioned above are present in the existing conditions. However, the quantity of groundwater usage for some landscapes is more. Finally, through the aforesaid allocation, only a portion of water demand for the pieces of Urmia landscapes has been partially met and the existing water resources would not be sufficient to bridge the gap.

Keywords: Optimal allocation of water; Landscape; Urmia city; Water resources; Transportation model; Linear programming

Introduction

Urmia has many parks and touristic coastal villages in the shore of Urmia Lake. The oldest park in Urmia, called Park-e Saat, was established in the first Pahlavi's era. Urmia's largest park is Ellar Bagi Park (Azerbaijani "People`s Garden") along the Shahar Chayi, or the "City River". In most private landscapes in Urmia, water used for irrigation is potable water. As a consequence, poor landscape irrigation performance results in high economic and environmental costs. In addition, the Urmia water act gives the highest priority to urban uses in the case of drought. As a consequence the characterization of landscape water use is a valuable tool to rationalize water consumption in urban environments and in whole river basins. Landscape irrigation can become a key local water use in the presence of water shortages [1,2].

Water is a unique material in nature. It is capable of almost complete return of light waves from its surface. In addition to the water surface being seen, images of surrounding objects may also be reflected. When the surface is calm, extremely clear images of mountains, rocks, trees, wildlife, and at times, the observer him/herself are displayed. If the surface is ruffled by a breeze or by the flow of the water, the reflections lose their sharpness and detail, producing an impressionist's image of the surrounding world. Water requirements for landscapes are calculated taking into account different factors, the two most important being the local climate and the type of species present in the landscape. Other factors include the coexistence of two or more species

in the same area (i.e., turf, trees or shrubs) and factors modifying the climate, such wind exposure. Research work determining landscape water requirements (LWR) usually follows one of three methodological approaches: The first option is to put landscape water requirements at the level of ET0 values [3]. This comparison is logical if most of the landscape area is turf. The second option is based on direct estimation of landscape water requirements through the use of instruments such as volumetric soil water sensors [4,5] or weighing lysimeters [6]. The last group of authors [7] follows the methodology proposed by Costello et al., developers of the WUCOLS method for determining landscape water requirements. The WUCOLS method is based on ET0, and uses an ad hoc procedure to estimate the coefficients that replace the crop coefficient by a landscape coefficient [8].

Recently, scientists in hydrology, ecology, geography, pedology, environmental sciences are concerned about the changes in landscape

*Corresponding author: Solmaz Javanbakht, Department of Water Engineering, Urmia University, Urmia, Iran
E-mail: javanbakht.solmaz@yahoo.com

composition of watershed, their cumulative impaction on water quality, and emerges hundreds of water quality models on non-point source pollution mechanism and nutrient migration and transformation. Of particular concern is the degree to which landscape conditions at watershed scales influence nitrogen, phosphorus, and sediment loadings to surface waters [9,10]. High levels of nutrients and sediment in water can pose significant human health and ecological risks [11]. In watershed scales models, there are the mechanism models based on hydrological processes and empirical models based on the correlative regression analysis between landscape and water quality [12-18]. However, it often occurs that parameters of these models have undefined ecological significance when we models landscape and water quality using these methods, which is the one reason that leads to the limited applying of empirical models.

In arid environment, water is the most limiting factor to plant growth, and the spatio-temporal dynamics of vegetation are therefore largely determined by water availability [19] Understanding the relationship between water supply and spatio-temporal variations of vegetation and landscape pattern is critically important to developing and implementing strategies for biodiversity conservation and maintenance of ecosystem structure and function in arid regions [20].

Population growth, subsequent urban development, changing lifestyle patterns and increasingly unreliable rainfall have all placed unprecedented pressure on individuals and governance institutions to contend with potable water scarcity throughout the world. In particular, securing water supply for growing residential areas has prompted many countries to investigate and invest in alternative water supply schemes, such as desalination plants, recycled water schemes, and the utilization of existing groundwater supplies [21,22]. Successful and unsuccessful attempts to incorporate alternative schemes have occurred in diverse locations such as Singapore, California, Namibia and Australia, and have led to the recognition that the fate of such projects is largely determined by the local communities [23,24].

Water dramatically influences and shapes the landscape; it can create `monumental sculptured environments [25]. The nature of the ground materials, the quantity of water, its duration of flow and type and amount of particulates carried determine the sculpturing effects. Where water meets the most resistance it works the hardest. The harder the material, the narrower the area carved. The larger the vertical height differences and the shorter the horizontal distance between the point of origin and the point of termination, the greater the carving forces of water.

Designers have long taken advantage of the many attractive visual and non-visual qualities of water in the landscape. Water can be still or move at various speeds. It can be shallow or deep, reflect the sky, sun, vegetation, and other objects surrounding it. Water can gain various colors, create sounds, and, when touched, cause cool sensation. Water color is associated with other perceptual and experiential characteristics as well. Blue water is associated with coolness, and white water with power and roaring sound [22-27].

The main purpose of the this study is putting forward a linear programming pattern in the form of transportation model in-order to allocate water optimally from the existing and future water resources (i.e. surface water, ground water and drinking water) to Urmia urban landscape pieces, considering minimization of the cost of supplying water.

Materials and Methods

Introduction of urmia

Urmia is a city in capital of West Azerbaijan Province, Iran. Urmia is located on a vast and verdurous plain, which is 70 km long and 30 km wide. The plains of rivers, rich deposits Barandoezchay, Sharchay, Nazloochaei tea and drink it regularly every year are covered. The cities geographical position 37 degrees 34 minutes north latitude and 44 degrees longitude is located 58 minutes. The area of 7764 hectares and its population according to the 1385 census, 583,255 people. Figures 1-3 shows a view of Urmia.

Analysis of green space per capita address

The source of information for green space Urmia has been divided into four regional divisions of the city. Green area of the city with the last changes in 2010, 4/4006860 obtained m this area contains a variety of green spaces (Parks, Squares, Boulevard, Delta, Trees, Streets, Nurseries and Forests) and 15/5 percent the area includes the entire city. Green spaces in the city include the percentage of the entire city. However, their distribution is such that regions 2 and 3, the lowest level of zone 1 have the highest percentage of green space. Green spaces that do not have the proper distribution of their distribution is as follows Zone 1 with an area of 2774 hectares and a population of 1,434,689 of whom 172,407 square meters of green space. The per capita area of 3/8 sq m has been calculated that a total of 84/1% of the city area and 8/35

Figure 1: Shows a view of Urmia.

Figure 2: Shows of green space area of Urmia.

Figure 3: Shows of green space area of Urmia.

of the total green area of the city, and two fifths of the area are included. Images (2) selection of green space area 1 is shown.

Identification of green spaces urmia

In order to increase the accuracy and efficiency studies, the green spaces of the city identify and profile should be studied and developed. Based on the identification of urban green spaces studied by organizations to identify and develop parks and green space Urmia was prepared in order to come to the table that contains information and specifications of the green space areas are separated Urmia.

Studies on water supply for landscape urmia

Overview: The plain subsurface layers of juicy collector of a vast repository of natural and regulator inlet water from large drainage area, which both operate, provides the storage capacity of water. Regional hydrogeological studies indicate that these reservoirs are located in an area of approximately 868 square kilometers.

Hydrated surface layer comprises an area of 764 square kilometers in area, but with deep moist layer over an area of 868 square kilometers, respectively. Groundwater within the aquifer layers are scattered artesian aquifer water (water contribution Consulting Engineers, 1381).

Optimization of water resources, water resources currently available green space Address

Optimization of irrigation water for an average of 7 months: First, the initial optimal solution using triple northwest corner, at least cost and Vogel approximation provided and the optimality condition using each method (MODI) evaluation and optimization methods is presented.

Results and Discussion

Initial optimal solution using the northwest corner

Table 1 of the optimal solution using the northwest corner of the current landscape (Median 7 months, irrigation) shows. As this table is a fundamental variable allocation of water to urban parks and boulevards, the basic variables are the boulevards and parks and forest allocation of surface water as a basic variable allocation of ground water parks, forest and other the green space of estimated nurseries become at the same time Ghyrasasy variables are estimated by assigning zero. Optimal solution in this case is 15675000 Rials per day.

Optimal solution of the minimum cost method

Table 2 optimal solution of the minimum cost for current green method (Average 7 month's irrigation) shows. This solution differs from the solution obtained from the northwest corner (Table 1). In this method, according to the Table 2 fundamental variables allocation of water to urban parks , boulevards, nurseries and other green spaces, the basic parameters of surface water allocation and forest parks just basic variables allocation of groundwater and boulevards forest parks was estimated at the same time non-fundamental variables are estimated by assigning zero. Optimal solution in this case is 15035000 Rials per day.

Optimal solution of Vogel approximation method

Table 3 of the optimal solution (Initial justification) Vogel's approximation method to the current status of green space (Median 7 months, irrigation) shows. The different responses with the reply obtained from the northwest corner method in the Tables 1 and 2 indicated. In this method, according to the Table 3, the basic parameters of water allocation Boulevard, nurseries and other green spaces, the basic parameters of surface water allocation and forest parks just basic variables allocation of water to urban parks, boulevards parks and forests was estimated at the same time Ghyrasasy variable assignment, are estimated to be zero. Optimal solution in this case is 14035000 Rials per day. Therefore, the optimal solution of the Vogel approximation method with a value of £ 14035000 = Z and the worst of the northwest corner of ways 15675000 Z = is the value of the Rial. But since this is the optimal solution, is not the final completion of the optimization method (MODI) optimality condition for each of the above three methods were tested and the results are presented in the following topics.

Initial feasible solution optimality study using three methods (MODI)

Optimality test procedure included three procedures, respectively. For the northwest corner of the Tables 4-10, for the least expensive method of the Tables 11-13 and the method Vogel approximation of the Tables 4-14 with the average irrigation season (7 months) utilized. In the method northwest corner after three iterations to reach the final optimal solution Tables 4-10, also in the least -cost method of the repetition he obtained their optimal solutions Tables 11-13 and Vogel 's approximation method is finally beginning to provide optimal solutions Table 14. This means that the optimal allocation of water resources availability in the Vogel approximation method for green spaces, more than capable of Urmia least expensive and also the northwest corner of potential methods for water in the current situation mean during. According to the results of the same basic variables Vogel 's approximation method and cost method , the optimal values for green spaces Urmia, according to the basic parameters of drinking water , 5,400 cubic meters per day Blvd 1400 cubic meters per day to 1,200 cubic meters per day nurseries other green spaces and the basic parameters of surface water , 6,500 cubic meters per day to park and forest due to the fundamental variables of water , 5,000 cubic meters per day in urban parks , boulevards , and 900 cubic meters to 1600 cubic meters per day forest Parks on the optimal cost £ 14035000 per day is recommended Tables 13 and 14.

It is showing that high accuracy. Thus, the final optimum solution for water allocation model using three methods above for irrigation 14035000Z = IRR is an average of 7 months. However, how to allocate the northwest corner method is different from the other two methods.

Supply sources	Urban parks		Boulevards		Forest parks		Nurseries		Other green spaces		a_i
		Demand destination									
	1700	5000	1500		1700		1300		1200		8000
Drinking water	*1		3000		*2						
	100		100	4000	100	2500	100		100		6500
Surface water											
	270		270		270	4900	270	1400	270	1200	7500
Ground water											
b_j	5000		7000		7400		1400		1200		22000

1*- basic variables
2*- nonbasic variables

$X_{11}=5000$ $X_{21}=0$ $X_{31}=0$
$X_{12}=3000$ $X_{22}=4000$ $X_{32}=0$
$X_{13}=0$ $X_{23}=2500$ $X_{33}=4900$
$X_{14}=0$ $X_{24}=0$ $X_{34}=1400$
$X_{25}=0$ $X_{35}=1200$
$X_{15}=0$

O.F=Z=15675000 Rial

Table 1: The initial optimal solution using the northwest corner to the current situation (mean 7 months, irrigation).

Supply sources	Urban parks		Boulevards		Forest parks		Nurseries		Other green spaces		a_i
		Demand destination									
	1700	5000	1500	400	1700		1300	1400	1200	1200	8000
Drinking water	*1				*2						
	100		100		100	6500	100		100		6500
Surface water											
	270		270	6600	270	900	270		270		7500
Ground water											
b_j	5000		7000		7400		1400		1200		22000

1*- basic variables
2*- non basic variables

$X_{11}=5000$ $X_{21}=0$ $X_{31}=0$
$X_{12}=400$ $X_{22}=0$ $X_{32}=6600$
$X_{13}=0$ $X_{23}=6500$ $X_{33}=900$
$X_{14}=1400$ $X_{24}=0$ $X_{34}=0$
$X_{15}=1200$ $X_{25}=0$ $X_{35}=0$

O.F=Z=15035000 Rial

Table 2: The initial optimal solution of the least-cost method to the current situation (median 7 months, irrigation).

Supply sources	Urban parks		Boulevards		Forest parks		Nurseries		Other green spaces		a_i
		Demand destination									
	1700		1500	5400	1700		1300	1400	1200	1200	8000
Drinking water			*1		*2						
	100		100		100	6500	100		100		6500
Surface water											
	270	5000	270	1600	270	900	270		270		7500
Ground water											
b_j	5000		7000		7400		1400		1200		22000

1*- basic variables
2*- non basic variables

$X_{11}=0$ $X_{21}=0$ $X_{31}=5000$
$X_{12}=5400$ $X_{22}=0$ $X_{32}=1600$
$X_{13}=0$ $X_{23}=6500$ $X_{33}=900$
$X_{14}=1400$ $X_{24}=0$ $X_{34}=0$
$X_{15}=1200$ $X_{25}=0$ $X_{35}=0$

O.F=Z=14035000 Rial

Table 3: The initial optimal solution of the vogel approximation method to the current situation (median 7 months, irrigation).

Optimization of supplying irrigation water, the warmest month (July)

For July, the first response early optimization is done using three methods. Repeat steps similar to the steps for evaluating the optimality condition with an average of 7 months, irrigation frequency on the 4-2-1, here is only to provide the tables were filled, and a final stop.

Furthermore, the initial optimal solution by employing three methods northwest corner, least cost method and Vogel's approximation method for allocating water resources, green spaces in the Urmia city at hottest months of the year (such as July) represent. Here optimality test indicates greater Vogel approximation method is compared to two other methods, successfully tested on its optimality, the optimal allocation of water from the boulevards of 5100 cubic meters of drinking water per day, 1,600 cubic meters per day to 1,300 cubic meters per day nurseries and other green spaces. The optimal allocation of water from surface water sources, 6,500 cubic meters per day to park and forest allocation of water from groundwater sources,

Supply sources	Urban parks	Boulevards	Forest parks	Nurseries	Other green spaces	a_i	U_i
	1700	1500	1700	1300	1200		
Drinking water	5000	3000				8000	U_1
	100	100	100	100	100		
Surface water		4000	2500			6500	U_2
	270	270	270	270	270		
Ground water			4900	1400	1200	7500	U_3
b_j	5000	7000	7400	1400	1200	22000	
V_j	V_A	V_B	V_C	V_D	V_E	-	-

$C_{ij} - U_i - V_j = \bar{C}_{ij}$ $C_{3B} - U_3 - V_B = 0$

$U_i + V_j = C_{ij}$ $U_1 = 0$
$U_1 + V_A = 1700$ $V_C = 1500$ $C_{1c} - U_1 - V_C = 200 > 0$
$U_1 + V_B = 1500$ $V_A = 1700$ $C_{1D} - U_1 - V_D = -200 < 0$
$U_2 + V_B = 100$ $V_B = 1500$ $C_{1E} - U_1 - V_E = -300 < 0$
$U_2 + V_C = 100$ $U_2 = -1400$ $C_{2A} - U_2 - V_A = -200 < 0$
$U_3 + V_C = 270$ $U_3 = -1230$ $C_{2D} - U_2 - V_D = 0$
$U_3 + V_D = 270$ $V_D = -1500$ $C_{2E} - U_2 - V_E = 0$
$U_3 + V_E = 270$ $V_E = 1500$ $C_{3A} - U_3 - V_A = -200 < 0$

Table 4: The initial optimal solution using the northwest corner to the U_i column and row V_j.

Supply sources	Urban parks	Boulevards	Forest parks	Nurseries	Other green spaces	a_i	U_i
	1700	1500	1700	1300	1200		
Drinking water	*1 5000	-0 ←3000				8000	$U_1 = 0$
	100	100	100	100	100		
Surface water	*2	+0	-0 2500			6500	-1400
	270	270	270	270	270		
Ground water			+0 4900	1400	-0 1200	7500	-1230
b_j	5000	7000	7400	1400	1200	22000	-
V_j	1700	1500	1500	1500	1500	-	-

Table 5: Stepping stone path input variable X_{15}.

Supply sources	Urban parks	Boulevards	Forest parks	Nurseries	Other green spaces	a_i	U_i
	1700	1500	1700	1300	1200		
Drinking water	5000	1800			1200	8000	U_1
	100	100	100	100	100		
Surface water		5200	1300			6500	U_2
	270	270	270	270	270		
Ground water			6100	1400		7500	U_3
b_j	5000	7000	7400	1400	1200	22000	-
V_j	V_A	V_B	V_C	V_D	V_E	-	-

1*- basic variables
2*- non basic variables

$C_{ij} - U_i - V_j = \bar{C}_{ij}$ $C_{3E} - U_3 - V_E = 300 > 01$

$U_i + V_j = C_{ij}$ $U_1 = 0$
$U_1 + V_A = 1700$ $V_A = 1700$ $C_{1C} - U_1 - V_C = 200 > 0$
$U_1 + V_B = 1500$ $V_B = 1500$ $C_{1D} - U_1 - V_D = -200 < 0$
$U_1 + V_E = 1200$ $V_E = 1200$ $C_{2A} - U_2 - V_A = -200 < 0$
$U_2 + V_B = 100$ $U_2 = -1400$ $C_{3A} - U_3 - V_A = -200 < 0$
$U_2 + V_C = 100$ $V_C = 1500$ $C_{2D} - U_2 - V_D = 0$
$U_3 + V_C = 270$ $U_3 = -1230$ $C_{2E} - U_2 - V_E = 300 > 0$
$U_3 + V_D = 270$ $V_D = 1500$ $C_{3B} - U_3 - V_B = 0$

Table 6: Answer northwest corner of the first iteration method.

Demand destination											
Supply sources	Urban parks		Boulevards		Forest parks		Nurseries		Other green spaces		a_i	U_i
	1700		1500		1700		1300		1200		8000	0
Drinking water	-0	5000	1800		*2				*1	1200		
			+0									
	100		100		100		100		100		6500	-1400
Surface water				5200	300							
			-0		+0							
	270		270		270		270		270		7500	-1230
Ground water	0					6100						
					-0		1400					
b_j	5000		7000		7400		1400		1200		22000	-
V_j	1700		1500		1500		1500		1200		-	-

1*- basic variables
2*- non basic variables

Table 7: Stepping stone path input variable X_{31}.

Demand destination							
Supply sources	Urban parks	Boulevards	Forest parks	Nurseries	Other green spaces	a_i	U_i
	1700	1500	1700	1300	1200	8000	U_1
Drinking water	*2	6800 *1			1200		
	100	100	100	100	100	6500	U_2
Surface water		200	6300				
	270	270	270	270	270	7500	U_3
Ground water	500		1100	1400			
b_j	5000	7000	7400	1400	1200	22000	-
V_j	V_A	V_B	V_C	V_D	V_E	-	-

1*- basic variables
2*- non basic variables

$$C_{ij} - U_i - V_j = \overline{C}_{ij}$$

$U_i + V_j = C_{ij}$ $U_1 = 0$

$U_1 + V_B = 1500$ $V_B = 1500$ $C_{1A} - U_1 - V_A = 200 > 0$

$U_1 + V_E = 1200$ $V_E = 1200$ $C_{1C} - U_1 - V_C = 200 > 0$

$U_2 + V_B = 100$ $U_2 = -1400$ $C_{1D} - U_1 - V_D = -200 < 0$

$U_2 + V_C = 100$ $V_C = 1500$ $C_{2A} - U_2 - V_A = 0$

$U_3 + V_A = 270$ $U_3 = -1230$ $C_{2D} - U_2 - V_D = 0$

$U_3 + V_C = 270$ $V_A = 1500$ $C_{2E} - U_2 - V_E = 300 > 0$

$U_3 + V_D = 270$ $V_D = 1500$ $C_{3B} - U_3 - V_B = 0$

$C_{3E} - U_3 - V_E = 300 > 0$

Table 8: Answer northwest corner of the second iteration method.

Demand destination											
Supply sources	Urban parks		Boulevards		Forest parks		Nurseries		Other green spaces		a_i	U_i
	1700		1500		1700		1300	0	1200		8000	$U_1=0$
Drinking water		*2	6800						*1	1200		
			-0									
	100		100	200	100		100		100		6500	-1400
Surface water			+0		-0	630						
	270		270		270		270	-0	270		7500	-1230
Ground water		5000			1100		1400					
b_j	5000		7000		7400		1400		1200		22000	-
V_j	1500		1500		1500		1300		1200		-	-

Table 9: Stepping stone path input variable X_{14}.

Supply sources	Demand destination										a_i	U_i
	Urban parks		Boulevards		Forest parks		Nurseries		Other green spaces			
	1700		1500		1700		1300		1200		8000	U_1
Drinking water	*1		5400	*2				1400		1200		
	100		100		100		100		100		6500	U_2
Surface water				1600		4900						
	270		270		270		270		270		7500	U_3
Ground water		500				2500						
b_i	5000		7000		7400		1400		1200		22000	-
V_j	V_A		V_B		V_C		V_D		V_E		-	-

1*- basic variables
2*- non basic variables

$C_{ij} - U_i - V_j = \bar{C}_{ij}$

$U_i + V_j = C_{ij} \quad U_1 = 0$

$U_1+V_B=1500$	$V_B=1500$	$C_{1A}-U_1-V_A=200>0$
$U_2+V_D=1300$	$V_D=1300$	$C_{1C}-U_1-V_C=200>0$
$U_1+V_E=1200$	$V_E=1200$	$C_{2D}-U_2-V_D=200>0$
$U_2+V_B=100$	$U_2=-1400$	$C_{2A}-U_2-V_A=0$
$U_2+V_C=100$	$V_C=1500$	$C_{2D}-U_2-V_D=200>0$
$U_3+V_A=270$	$U_3=-1230$	$C_{2E}-U_2-V_E=300>0$
$U_3+V_C=270$	$V_A=1500$	$C_{3B}-U_3-V_B=0$
$C_{3D}-U_3-V_D=300>0$		$C_{3E}-U_3-V_E=300>0$

Since all values of $\bar{C}_{ij}$ are non-negative. The result in table 9 in the third iteration is optimal. Therefore, the optimal solution values are extracted as follows:

$X_{11}=0 \quad X_{21}=0 \quad X_{31}=5000$
$X_{12}=5400 \quad X_{22}=1600 \quad X_{32}=0$
$X_{13}=0 \quad X_{23}=4900 \quad X_{33}=2500 \quad$ O.F=Z=14035000 Rial
$X_{14}=1400 \quad X_{24}=0 \quad X_{34}=0$
$X_{15}=1200 \quad X_{25}=0 \quad X_{35}=0$

Comparison of the optimal solution and improved the northwest corner method, it is concluded that there is a difference between these two amounts of £ 1640000.

Table 10: Answer northwest corner of the third iteration method.

Supply sources	Demand destination										a_i	U_i
	Urban parks		Boulevards		Forest parks		Nurseries		Other green spaces			
	1700		1500		1700		1300		1200		8000	U_1
Drinking water	*1	5000	400		*2			1400		1200		
	100		100		100		100		100		6500	U_2
Surface water						6500						
	270		270		270		270		270		7500	U_3
Ground water				6600		900		1400		1200		
b_i	5000		7000		7400		1400		1200		22000	-
V_j	V_A		V_B		V_C		V_D		V_E		-	-

1*- basic variables
2*- non basic variables

$C_{ij} - U_i - V_j = \bar{C}_{ij}$

$U_i + V_j = C_{ij} \quad U_1 = 0$

$U_1+V_A=1700$	$V_A=1700$	$C_{1C}-U_1-V_C=200>0$
$U_1+V_B=1500$	$V_B=1500$	$C_{2D}-U_2-V_D=200>0$
$U_1+V_D=1300$	$V_D=1300$	$C_{2A}-U_2-V==-200<0$
$U_1+V_E=1200$	$V_E=1200$	$C_{2B}-U_2-V_B=0$
$U_2+V_C=100$	$V_C=1500$	$C_{2E}-U_2-V_E=300>0$
$U_3+V_B=270$	$U_3=-1230$	$C_{3A}-U_3-V_A=-200<0$
$U_3+V_C=270$	$U_2=-1400$	$C_{3E}-U_3-V_E=300>0$
$C_{3D}-U_3-V_D=200>0$		

Table 11: The initial optimal solution of the minimum cost method with column and row $U_i V_j$.

5,700 cubic meters per day in urban parks, boulevards and 1800 cubic meters per day to 500 cubic meters per hectare is in forest park.

According to the results obtained from the solution of the water allocation model in July, our data suggested that as least-cost allocation model to allocate irrigation during the middle of the flowers and also with the approximation method in proportion to the supply in the July with average from 7 months to irrigation but does not affect the rate of increase in costs.

The final optimal solution (improved) July

Vogel approximation method for evaluating the optimality condition for this conclusion was estimated that the final optimal solution is the same answer to the same basic principles in order to express the optimal solution to the same table method.

Initial feasible solution, this method is limited. However, to achieve the optimal solution in the northwest corner of techniques and treatments and to evaluate the optimality of a repeat procedure cost

Demand destination								
Supply sources	Urban parks	Boulevards	Forest parks	Nurseries	Other green spaces		a_i	U_i
	1700	1500	1700	1300	1200		8000	U_1=0
Drinking water	-0 ← 500	400 +0	*2	*1	1400	1200		
	100	100	100	100	100		6500	-1400
Surface water			650		-0			
	270	270	270	270	270		7500	-1230
Ground water	0	-0	900					
b_j	5000	7000	7400	1400	1200		22000	-
V_j	1700	1500	1500	1300	1200		-	-

Table 12: Stepping stone path input variable X_{31}.

Demand destination							
Supply sources	Urban parks	Boulevards	Forest parks	Nurseries	Other green spaces	a_i	U_i
	1700	1500	1700	1300	1200	8000	U_1=0
Drinking water		*1 5400	*2	1400			
	100	100	100	100	100	6500	-1400
Surface water			650				
	270	270	270	270	270	7500	-1230
Ground water	5000	1600	900				
b_j	5000	7000	7400	1400	1200	22000	-
V_j	1700	1500	1500	1300	1200	-	-

1*- basic variables
2*- non basic variables

$C_{ij}-U_i-V_j=\overline{C}_{ij}$

$U_1+V_B=1500$	$V_B=1500$	$C_{1C}-U_1-V_C=200>0$
$U_1+V_D=1300$	$V_D=1300$	$C_{2D}-U_2-V_D=200>0$
$U_1+V_E=1200$	$V_E=1200$	$C_{2A}-U_2-V_A=0$
$U_2+V_C=100$	$V_C=1500$	$C_{2B}-U_2-V_B=0$
$U_3+V_A=270$	$V_A=1500$	$C_{2E}-U_2-V_E=300>0$
$U_3+V_B=270$	$U_3=-1230$	$C_{1A}-U_1-V_A=200>0$
$U_3+V_C=270$	$U_2=-1400$	$C_{3E}-U_3-V_E=300>0$
$C_{3D}-U_3-V_D=200>0$		

Since all values of $\overline{C}_{ij}$ are non-negative. Therefore, the optimal solution values are extracted as follows:

$X_{11}=0$	$X_{21}=0$	$X_{31}=5000$
$X_{12}=5400$	$X_{22}=0$	$X_{32}=1600$
$X_{13}=0$	$X_{23}=6500$	$X_{33}=900$ O.F=Z=14035000 Rial
$X_{14}=1400$	$X_{24}=0$	$X_{34}=0$
$X_{15}=1200$	$X_{25}=0$	$X_{35}=0$

Comparison of the optimal solution and improved the northwest corner method, it is concluded that there is a difference between these two amounts of £ 1000000.

Table 13: The essential response of least expensive method of the first iteration.

Jam and final solution to optimize the cost of the solution is obtained by Vogel refused to repeat them here, but due to different allocation northwest corner results in is given.

Conclusions

In conclusion, Urmia city is separated into four regional divisions of the city. Green area of the city with the last changes in 2010, 69/400 hectare park that includes a variety of green space, square, boulevard, street trees, plantations and forest parks and 15/5 % of the total area of the city covers. Figures 1-3 shows the four regions in Urmia city. By linear programming, formulation, modeling and solving models considered and the following results in ensuring optimum water green Urmia identified, assessed and will be supplied. Green spaces of the Urmia city in terms of distribution and area under the international standards and national level not. The current green spaces of city and develop a global standard requires optimal allocation of water. Moreover, irrigation of green spaces Urmia through groundwater resources of 180 liters per second, a rate of 150 liters per second of

water resources and the water resources of 185 liters per second and the supply takes place. And also, green spaces irrigation application efficiency Urmia through gravity irrigation by sprinkler irrigation by 50 percent and 70 percent. Urmia irrigate green spaces in accordance with Table 2 respectively. Ability Vogel approximation method in optimal allocation of water resources for green spaces of Urmia and its availability at low cost way more than the northwest corner of the capability approach. To 7 -month irrigation, optimal water allocation amounts of water, 5,400 cubic meters per day to the boulevards, 1,400 cubic meters per day to 1,200 cubic meters per day nurseries and other green spaces. The optimal allocation of water from surface water sources, 6,500 cubic meters per day to park in the forest. Values for optimal allocation of water from groundwater sources , 5,000 cubic meters per day in urban parks , boulevards , and 900 cubic meters to 1600 cubic meters per day in the park is forested.

The hottest month of July, irrigation, optimal water allocation amounts of water, 5,100 cubic meters per day to the boulevards, 1,600 cubic meters per day nurseries, and 1,300 cubic meters per day to other

Supply sources	Urban parks		Boulevards		Forest parks		Nurseries		Other green spaces		a_i	U_i
					Demand destination							
Drinking water	1700		1500		1700		1300		1200		8000	U_1
			5400	*1	*2		1400		1200			
Surface water	100		100		100		100		100		6500	U_2
						6500						
Ground water	270		270		270		270		270		7500	U_3
		5000		1600		900						
b_j	5000		7000		7400		1400		1200		22000	-
V_j	V_A		V_B		V_C		V_D		V_E		-	-

1*- basic variables
2*- non basic variables

$C_{ij} - U_i - V_j = \bar{C}_{ij}$

$U_i + V_j = C_{ij}$ $U1=0$
$U_1 + V_B = 1500$ $V_B = 1500$ $C_{1C} - U_1 - V_C = 200 > 0$
$U_1 + V_D = 1300$ $V_D = 1300$ $C_{2D} - U_2 - V_D = 200 > 0$
$U_1 + V_E = 1200$ $V_E = 1200$ $C_{2A} - U_2 - V_A = 0$
$U_2 + V_C = 100$ $V_C = 1500$ $C_{2B} - U_2 - V_B = 0$
$U_3 + V_A = 270$ $V_A = 1500$ $C_{2E} - U_2 - V_E = 300 > 0$
$U_3 + V_B = 270$ $U_3 = -1230$ $C_{1A} - U_1 - V_A = 200 > 0$
$U_3 + V_C = 270$ $U_2 = -1400$ $C_{3E} - U_3 - V_E = 300 > 0$
$C_{3D} - U_3 - V_D = 200 > 0$

Table 14: The initial optimal solution of the vogel approximation method to the U_i column and row V_j.

green spaces is estimated. The optimal allocation of water from surface water sources, 6,500 cubic meters per day to park in the forest. Values for optimal allocation of water from groundwater sources , 5,700 cubic meters per day in urban parks , boulevards , and 500 cubic meters to 1800 cubic meters per day in the park is forested.

Results 7 and 8 state that the basic variables in the optimal allocation of water resources for irrigation of green spaces Urmia in during 7 months compared to the average of the warmest month (July), but from the point of view of the same quantity of water for drinking water and, below are some green spaces. The optimal allocation of water resources, drinking water, surface water and groundwater for the future of green spaces in Urmia are present and also, the increased supply of essential variables entered into the basic variables for the average warmest months of the year 7 months and Irrigation (July) in conditions are present. The quantity of groundwater and some have greener spaces. The adequacy or inadequacy of the daily allocation of water for irrigation in the seventh month of each of the various components of green space in accordance with Tables 4 and 5 showed, respectively.

Suggestions

i. Water delivered to the green spaces of the city of Urmia, requires a particular database and organized in terms of water volume delivered to the green spaces as well as financial costs them. This can be organized in the future development and management of green spaces in Urmia is very useful.

ii. The physical constraints of allocation of surface water, groundwater and drinking water in relation to space and time in the range of Urmia, in providing the optimal allocation of a water management plan for green spaces for future research are suggested.

iii. For a population of 583,255 people, Urmia, landscaping standards require an average of 1300 ha.be added to the existing water resources.

iv. Work and forested areas as possible, rather than gravity irrigation method used.

v. Use plant species need less water than grass to green space for future expansion Urmia recommended.

Acknowledgements

The authors thank staff of Dr.Javad Javanbakht, for help this manuscript.

References

1. Salman AZ, Al-Karablieh EK, Fisher FM (2001) An Inter–seasonal Agricultural Water Allocation System (SAWAS). Agr Syst 68: 233-252.

2. Dadmehr R (1996) Optimal Water Resources Hydraulic Modeling and Management in Lower Macquarie valley. University of Technology, Sydney, NSW, Australia.

3. Haley MB, Dukes MD, Miller GL (2007) Residential irrigation water use in central Florida. J Irrig Drain E-ASCE 133: 427-434.

4. Morari Г, Giardin L (2001) Estimating evapotranspiration in the Padova botanical garden. Irrigation Sci 20: 127-137.

5. White R, Havalak R, Nations J, Pannkuk T, Thomas J, et al. (2004) How much Water is Enough? Using Pet to Develop Water Budgets for Residential Landscapes Texas Water Resources Institute. College Station, Texas, USA.

6. Brown PW, Mancino CF, Young MH, Thompson TL, Wierenga PJ, et al. (2001) Penman monteith crop coefficients for use with desert turf systems .Crop Sci 41: 1197-1206.

7. Domene E, Sauri D (2003) New Urban Lifestyles and Welfare: Water Consumption In The Suburbs Of Barcelona. Geogr Res 32: 5–17.

8. Costello LR, Matheny NP, Clark Jr. A (2000) A Guide to Estimating Irrigation Water Needs of Landscape Plantings in California. The Landscape Coefficient Method and WUCOLS III. University of California, USA.

9. Ator SW, Ferrari MJ (1997) Nitrate and selected pesticides in ground water of the Mid-Atlantic Region. US Geological Survey.

10. Jones KB, Neale AC, Nash MS, Van Remortel RD, Wickham JD, et al. (2001) Predicting nutrient and sediment loadings to streams from landscape metrics: A multiple watershed study from the United States Mid-Atlantic Region. Landscape Ecol 16: 301-312.

11. Bouraoui F, Benabdallah S, Jrad A, Bidoglio G (2005) Application of the SWAT model on the Medjerda river basin (Tunisia). Phys Chem Earth 30: 497-507.

12. Fohrer N, Moller D, Steiner N (2002) An interdisciplinary modeling approach to evaluate the effects of land use change. Phys Chem Earth 27: 655-662.

13. Nasr A, Bruen M, Jordan P, Moles R, Kiely G, et al. (2007) A comparison of SWAT, HSPF and SHETRAN/GOPC for modeling phosphorus export from three catchments in Ireland. Water Res 41: 1065-1073.

14. Mander U, Kull A, Kuusemets V (2000) Nutrient flows and land use change in a rural catchment: A modelling approach. Landscape Ecol 15: 187-199.

15. Eugene Turner R, Rabalais NN (2003) Linking landscape and water quality in the Mississippi river basin for 200 years. Bioscience 53: 563-572.

16. Johnson LB, Richard C, Host GE, Arthur JW (1997) Landscape influences on water chemistry in Midwestern stream ecosystems. Freshwater Biol 37: 193-208.

17. Basnyat P, Teeter LD, Flynn KM, Graeme Lockaby B (1999) Relationships between landscape characteristics and non-point source pollution inputs to coastal estuaries. Environ Manage 23: 539-549.

18. Elmore AJ, Manning SJ, Mustard JF, Craine JM (2006) Decline in alkali meadow vegetation cover in California: the effects of groundwater extraction and drought. J Appl Ecol 43: 770-779.

19. Wu JG, Hobbs R (2002) Key issues and research priorities in landscape ecology: an idiosyncratic synthesis. Landscape Ecol 17: 355-365.

20. Kennewell C (2008) Perth, Western Australia. Cities 25: 243-255.

21. Ingram PC, Young VJ, Millan M, Chang C, Tabucchi T (2006) From controversy to consensus: The Redwood City recycled water experience. Desalination 187: 179-190.

22. Hurlimann A, Dolnicar S (2010) When public opposition defeats alternative water projects – The case of Toowoomba Australia. Water Res 44: 287-297.

23. Campbell CS (1978) Water in Landscape Architecture. Van Nostrand Reinhold Company, New York.

24. Litton RB (1977) River landscape quality and its assessment. Proceedings river recreation management and research symposium, USDA Forest Service 46-54.

25. Valipour M (2013) Necessity of Irrigated and Rainfed Agriculture in the World. Irrigat Drainage Sys Eng S9: e001.

26. Valipour M (2013) Evolution of Irrigation-Equipped Areas as Share of Cultivated Areas. Irrigat Drainage Sys Eng 2: e114.

27. Valipour M (2013) Need to Update of Irrigation and Water Resources Information According to the Progresses of Agricultural Knowledge. Agrotechnol S10: e001.

11

Computational Fluid Dynamics (CFD) Picture of Water Droplet Evaporation in Air

Giulio Lorenzini[1], Alessandra Conti[2] and Daniele De Wrachien[3]*

[1]University of Parma, Department of Industrial Engineering, viale G.P. Usberti no.181/A, Parma 43124, Italy
[2]Alma Mater Studiorum-University of Bologna, Department of Energetic Nuclear and Environmental Control Engineering, viale Risorgimento no. 2, Bologna 40136, Italy
[3]Department of Agricultural Hydraulics, University of Milan, via Celoria no.2, Milan 20133, Italy

Abstract

The study of droplet evaporation is applied to many and varied fields: the present approach is oriented to sprinkler irrigation. This paper examines a parametric study on the evaporation in air of a single droplet, with the aim of highlighting the influence of each parameter alone on the evaporative process. Four parameters are investigated: air temperature, droplet initial velocity, droplet initial diameter, diffusion coefficient of vapour in air. Droplet evaporation is studied through numerical-CFD simulation employing STAR-CCM+ version 5.04.012 software, which treats the evaporative phenomenon hypothesizing quasi-steady conditions, given the interface low liquid-gas vapour concentration gradients. The results are provided as time- and space-dependent in-percentage evaporation rates, the latter ones after defining a specific distance, from the injection point, to be covered. Apart from a qualitatively predictable effect of air temperature and diffusion coefficient of vapour in air, droplet initial velocity and above all droplet initial diameter prove not at all to be negligible when managing an irrigation process, the latter being inversely proportional to droplet mass evaporation. These results prove that droplet evaporation is a complicate fluid dynamic effect and cannot be simply regarded as a diffusive process. The final discussion provides some practical remarks useful to irrigation operators.

Keywords: Droplet evaporation in air; Sprinkler irrigation; Computational fluid dynamics; Numerical modelling; Parametric study

Introduction

Droplet evaporation is thoroughly investigated [1-4] as it is applied to multivarious engineering fields like automotive [5-15], air-conditioning [16,17], fluidized beds [18], fire suppression [19], geophysics [20], meteorology [21] and agriculture [3,4,22-28]. In the latter case the focus is on irrigation systems, whose design for both civil and agricultural use must consider, within a sustainable economy context, the worrying depletion of an important resource: water. As a matter of fact, during irrigation the single water particles come in contact with air and evaporate: this is to be attributed to many parameters, among which are those that will be here considered, i.e. air-water temperature difference, droplet initial velocity, droplet initial diameter and diffusion coefficient of vapour in air. Several studies, both experimental and theoretical [3,4,26,27,29-37], were carried out in the past in order to quantify evaporation loss. As regards the second type, a "classic" approach was that by Kinzer and Gunn [29], who modelled evaporation for falling droplets but neglected the dynamics acting upon the in-flight droplets; they arrived at the following expression, suitable for a limited Reynolds number range:

$$m^{\wedge} = 4 \pi a^2 D_{va} G$$

where: $m^{\wedge}$ [kg s^{-1}] is droplet mass evaporated with time, a [m] is its radius; D_{va} [m^2 s^{-1}] is diffusion coefficient of vapour in air and G [kg m^{-4}] is the vapour-density gradient established at the surface of the droplet. Also Ranz and Marshall [30,31] described in-flight droplet evaporation with an equation for molecular transfer rate, which was modified by Goering et al. [38], who used empirical formulae even from other authors:

$$D^{\wedge} = -2 M^{\wedge} K^{\wedge} R P Nu^*$$

where: $D^{\wedge}$ [m s^{-1}] is droplet diameter variation with time; $M^{\wedge}$ [-] is the ratio between molecular weights of vapour and air; $K^{\wedge}$ [m s^{-1}] is the ratio between D_{va} [m^2 s^{-1}] and d_p [m]; R [-] is the ratio between air and

droplet density; P [-] is the ratio: difference between saturation pressure at wet bulb air temperature and vapour pressure at dry bulb temperature / partial pressure of air; and Nu* [-] is a specially defined Nusselt number for mass transfer. Still on theoretical approaches, a simplified mathematical model, validated by experimental data, was developed by Lorenzini and applied to the case studies examined by Edling [22] and Thompson et al. [23] for comparison purposes [3,4,27,28]. There is a strong correlation among the chemical-physical processes that characterize the evolution of sprays and an analytical model attempting at this description may be strongly conditioned in arriving at a closed-form solution by the non-linear nature of the partial differential equations arising. A further step in literature was taken thanks to CFD implementation [13,39-43], even though most of the researches were applied to the field of combustion rather than to agricultural sprays and this makes the temperature and chemical context quite far from that here faced, so considering even high values of the Spalding Number which, on the contrary, did not happen in this study. The present paper analyses the evolution, alternatively with time or with space (i.e. for a simulation time equal to 4 s or for a simulation path of 20 m from the injection point, where a suitable check-plane was located), of the evaporative phenomenon in air of a single water droplet. Phenomenon modelling and solving were performed by means of the CFD control volume code STAR-CCM+ version 5.04.012. Treatment included a straight droplet trajectory defined downwards along the vertical axis

---

*Corresponding author: Daniele De Wrachien, University of Milan, Department of Agricultural Hydraulics, via Celoria no.2, Milan 20133, Italy
E-mail: daniele.dewrachien@unimi.it

and not affected neither by wind action nor by solar radiation. The present parametric study included the following variables: droplet initial velocity; droplet initial diameter; air temperature; and diffusion coefficient of vapour in air. Air relative humidity, a parameter which is indeed relevant to the whole phenomenological picture, was instead not considered here due to technical limits, because the solver implemented within STAR CCM+ could not tackle that issue due to a water-air interface modelling limit: this, anyway, does not affect the generality of the present study as the "classical" parametric approach adopted (one variable varying at a time for every simulation; all the others kept as constants) assures the independence of any parametric result from another. Droplet evaporation was described by means of an Eulerian-Lagrangian approach, typical of this kind of processes [44,45], even if fully Lagrangian approaches [46] and other approaches [47] are also reported in literature. A range of twenty cases study was faced, five for each analysis parameter.

Method

The simulations performed involve the evaporation of a single water droplet into the air, considering the influence of four physical parameters (air temperature, diffusion coefficient of vapour in air; droplet initial velocity and droplet initial diameter): the aim is to examine a dynamically realistic process, also accounting for the contribution of air friction to evaporation [4]. Simulations involve the investigation and determination of the evaporation rate of a liquid particle moving vertically downwards; such particle, located within the control volume, starts its path at a given set of initial conditions. Friction and gravity are the forces influencing the system and are opposite to one another, directed along the x axis as Figure 1 displays together with a detail of the meshed domain. The numerical modelling of the phenomenon was carried out adopting an Eulerian-Lagrangian approach, in which a Lagrangian phase (water droplet) moves within a continuous Eulerian phase (air). Liquid particles are rigid spheres, hence the simulations do not consider the degree of deformation caused by air friction and undergone by a droplet along its path: previous researches demonstrated this assumption to be satisfactory and realistic [48,49]. The numerical approach employed by STAR-CCM+ version 5.04.012 does not model the liquid-gas interface directly, but solves it defining the diffusion law at the interface (Fick's law). All the simulations were unsteady, given the intrinsic dependence of unsaturated evaporation upon time [50]. Physical properties of air and water were taken by the STAR-CCM+ version 5.04.012 code library; the same for the values of the constants in the relations presented below (for example, the Sutherland's or Antoine's laws). Sutherland's law was adopted for air viscosity μ_a [Pa s] [51,52]:

$$\mu_a = \mu_0 \cdot \frac{T_0 + 111}{T_a + 111} \cdot \left(\frac{T_a}{T_0}\right)^{\frac{3}{2}} \tag{1}$$

in which μ_0 (dynamic viscosity at the temperature T_0) is equal to 1.716 $\times 10^{-5}$ Pa s, $T_0 = 273.15$ K, and T_a [K] is the air temperature. Antoine's law was used for computing the saturation pressure and hence the vapour tension [52,53]:

$$p_{sat} = p_{atm} \exp\left(A - \frac{B}{T - C}\right) \tag{2}$$

in which T [K] is temperature, p_{atm} is equal to atmospheric pressure (= 1 atm), A = 11.949; B = 3978.205 and C = -39.801 [52]. Friction force has the following expression, valid just in case of a no-wind condition:

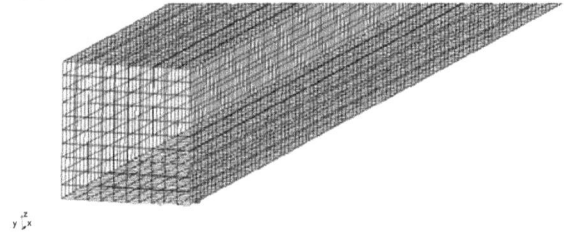

Figure 1: Detail of the trimmed domain mesh.

$$\vec{F}_d = \frac{1}{2} \cdot C_d \cdot \rho_a \cdot A_p \cdot \vec{v}_p \cdot |v_p| \tag{3}$$

Where C_d [-] is the friction factor, ρ_a [kg m^{-3}] is air density, A_p [m^2] is droplet cross sectional area and v_p [m s^{-1}] is velocity of the Lagrangian phase. The code employs Schiller-Naumann law for calculating the friction factor [54]:

$$C_d = \begin{cases} \dfrac{24}{Re_p} \cdot \left(1 + 0.15 \cdot Re_p^{0.687}\right) & Re_p \leq 1000 \\ 0.44 & Re_p \geq 1000 \end{cases} \tag{4}$$

Evaporation is a phenomenon directly connected with energy, hence it involves the heat exchange coefficients, namely the Nusselt number (here called Nu_p [-] as referred to the water particle). The Ranz-Marshall correlation was used for calculating the heat transfer coefficients during evaporation and has to be retained valid for spherical droplets up to $Re_p \approx 5000$ [30,31]:

$$Nu_p = 2 + 0.6 \cdot Re_p^{\frac{1}{2}} \cdot Pr_a^{\frac{1}{3}} \tag{5}$$

Where the Prandtl number Pr_a [-] refers to the gas phase (air) and the Reynolds number of the droplet Re_p [-] is defined as follows:

$$Re_p = \frac{\rho_a \cdot v_p \cdot d_p}{\mu_a} \tag{6}$$

in which ρ_a [kg m^{-3}] is air density, μ_a [Pa s] is air viscosity, d_p [m] is droplet diameter and v_p [m s^{-1}] is water particle velocity. Air density and viscosity are calculated at the mean temperature between those of air (T_a [K]) and water (T_w [K]) as: $[(T_a + T_w)/2]$ [55]. It has to be specified that temperature gradients within a droplet have not been analysed here, as Biot numbers throughout the investigation proved to be always very close to zero thus making it reasonable, in a just-evaporative investigation, to consider the droplet itself as isothermal. About recirculation in droplets: this is an important issue but we feel it more as a possible further development of the current research (based on comprehensive droplet evaporation assessment and not on a punctual thermal fluid dynamic analysis of the transient process) rather than a detail of the present work.

In this study the diffusion coefficient of vapour in air D_{va} [m^2 s^{-1}], which essentially depends upon many factors (e.g.: temperature, air pressure, vapour pressure gradient), is kept constant throughout the simulation duration; its values can be calculated in different ways: through both theoretical [56] and semi-empirical formulas [57-60], and through tabulated values found in literature [53,55,61]. In this investigation the diffusion coefficients were first calculated according to the former way (formulas) and then compared to the latter one (tabulated values), identifying an interval between 0.2×10^{-4} m^2 s^{-1} and 0.3×10^{-4} m^2 s^{-1} as realistic within the analysis parameters range tested. Schmidt number Sc [-] and Sherwood number Sh [-] are both connected to the diffusion coefficient D_{va} [m^2 s^{-1}]. Sh number can be expressed in function of Re [-] and Sc [-]. In the case studies tackled here, being the droplet spherical, Ranz-Marshall expression can be employed [30,31]:

$$Sh = 2 + 0.6 \cdot Re_p^{\frac{1}{2}} \cdot Sc^{\frac{1}{3}} \tag{7}$$

being this expression analogous to equation (5), with Re_p [-] calculated according to equation (6). In-percentage droplet mass evaporation rate Δm [%] is defined as follows:

$$\ddot{A}m = \frac{m_0 - m_1}{m_0} \cdot 100 \tag{8}$$

where m_0 [kg] represents the droplet initial mass and m_1 [kg] the mass observed at a dimensionless time t* defined as:

$$t^* = \frac{t}{t_{max}} \tag{9}$$

in which t [s] is the variable time and t_{max} [s] is the maximum simulation time (4 s). Alternatively, a target location placed at set distance (20 m) from the droplet injection point is considered and Δm [%] is checked there. This choice will add significant insight to the present study, as the Discussion section will explain. To add generality to the approach it has been defined a dimensionless travel distance L* [-]:

$$L^* = \frac{L}{L_{max}} \tag{10}$$

Where L [m] is travel distance and L_{max} = 20 m is the target location just above defined, reached after a simulation time of t_{Lmax} [s] which is made dimensionless (t^*_{Lmax} [-]) as it follows:

$$t^*_{Lmax} = \frac{t_{Lmax}}{t_{max}} \tag{11}$$

Parameters and Boundary Conditions

The study is a parametric one: all the parameters, except one kept constant during each test, were kept constant throughout the simulation. The following analysis parameters were considered:

air temperature (T_a [K]);

droplet initial velocity (v_i [m s^{-1}]);

diffusion coefficient of vapour in air (D_{va} [m^2 s^{-1}]);

droplet initial diameter (d_i [m]).

Table 1 reports the whole set of analysis parameters and cases study faced in the present investigation. In relation to T_a [K], five values were chosen within an interval range compatible with the climatic conditions of a hot-arid environment: 300 K, 305 K, 310 K, 315 K and 320 K. Whereas, droplet initial velocities and diameters were selected among typical values of a wide range of sprinkler systems [62,63]. The diffusion coefficient of vapour in air D_{va} [m^2 s^{-1}] is a rather important parameter in the dynamics of evaporation as it is directly connected to the vapour film which envelopes water droplets: its range of variation was determined, as explained in the previous section of this paper, comparing computed to tabulated data in relation to the general conditions of the tests performed. As explained in the previous

section, after comparing computed to tabulated data the range of D_{va} [m^2 s^{-1}] investigated was $0.2 \times 10^{-4} \div 0.3 \times 10^{-4}$ m^2 s^{-1} [53,55,56-61], choosing in detail the following test values: 0.2×10^{-4} m^2 s^{-1}, 0.225×10^{-4} m^2 s^{-1}, 0.25×10^{-4} m^2 s^{-1}, 0.275×10^{-4} m^2 s^{-1} and 0.3×10^{-4} m^2 s^{-1}. The droplet velocity is directly connected to the friction force (a quadratic function of it) [55], which plays a significant role in the process under exam [4,27]. To gather direct information on its effect five ascending values of v_i [m s^{-1}] were investigated: 1 m s^{-1}, 5 m s^{-1}, 15 m s^{-1}, 25 m s^{-1} and 30 m s^{-1}. Finally, as droplet evaporation also depends on the droplet diameter, five values of d_i [m] were tested, within the typical range $0.001 \div 0.003$ m [62-66]: 0.001 m, 0.0015 m, 0.002 m, 0.0025 m and 0.003 m. Water temperature T_w [K] was a constant and kept equal to 288 K throughout the investigation. The simulations were performed in unsteady state, given the intrinsic nature of evaporation under conditions far from saturation, and the path followed by the liquid particle within the control volume was alternatively observed at t* = 1 or L* = 1, as explained above.

Numerical Settings

The numerical domains based on control volume approach are geometrically simple: they are in fact represented by parallelepiped volumes and in this study their dimensions are identical in all cases study but those where the droplet initial diameter was the parameter tested (Table 2). In the latter cases the geometric dimensions of the computational domain were variable from one case to another along the droplet path direction (x-axis in Figure 1), while dimensions along z- and y-axes remained unchanged. Table 2 summarises the geometrical dimensions of all the domains tested, in addition to the main numerical settings. Given the extremely simple geometry, the mesh for the numerical domain is of the "trimmed" type (see detail

Cases study	Analysis parameters				Constant
	v_i [m s^{-1}]	d_i [m]	D_{va} [m^2 s^{-1}]	T_a [K]	T_w [K]
Case 1	1	0.001	0.3×10^{-4}	320	288
Case 2	5	0.001	0.3×10^{-4}	320	288
Case 3	15	0.001	0.3×10^{-4}	320	288
Case 4	25	0.001	0.3×10^{-4}	320	288
Case 5 (*)	30	0.001	0.3×10^{-4}	320	288
Case 6 (*)	30	0.001	0.3×10^{-4}	320	288
Case 7	30	0.0015	0.3×10^{-4}	320	288
Case 8	30	0.002	0.3×10^{-4}	320	288
Case 9	30	0.0025	0.3×10^{-4}	320	288
Case 10	30	0.003	0.3×10^{-4}	320	288
Case 11	30	0.001	0.2×10^{-4}	320	288
Case 12	30	0.001	0.225×10^{-4}	320	288
Case 13	30	0.001	0.25×10^{-4}	320	288
Case 14	30	0.001	0.275×10^{-4}	320	288
Case 15 (*)	30	0.001	0.3×10^{-4}	320	288
Case 16	30	0.001	0.3×10^{-4}	300	288
Case 17	30	0.001	0.3×10^{-4}	305	288
Case 18	30	0.001	0.3×10^{-4}	310	288
Case 19	30	0.001	0.3×10^{-4}	315	288
Case 20 (*)	30	0.001	0.3×10^{-4}	320	288

Table 1: Analysis parameters and cases study (those marked with * are duplicated). The numbers reported are input data and they consequently represent exact numbers.

Cases study	Time step [s]	x-axis [m]	y-axis [m]	z-axis [m]	Mesh [m]	Number of cells	Number of faces
Case 1	1×10^{-2}	25	1×10^{-1}	1×10^{-1}	5×10^{-3}	845 952	2 472 540
Case 2	5×10^{-3}	25	1×10^{-1}	1×10^{-1}	5×10^{-3}	845 952	2 472 540
Case 3	1.5×10^{-3}	25	1×10^{-1}	1×10^{-1}	5×10^{-3}	845 952	2 472 540
Case 4	5×10^{-4}	25	1×10^{-1}	1×10^{-1}	5×10^{-3}	845 952	2 472 540
Case 5 (*)	3×10^{-4}	25	1×10^{-1}	1×10^{-1}	5×10^{-3}	845 952	2 472 540
Case 6 (*)	3×10^{-4}	25	1×10^{-1}	1×10^{-1}	5×10^{-3}	845 952	2 472 540
Case 7	7.5×10^{-4}	30	1×10^{-1}	1×10^{-1}	1×10^{-2}	300 000	839 900
Case 8	7.5×10^{-4}	35	1×10^{-1}	1×10^{-1}	1×10^{-2}	350 000	979 900
Case 9	7.5×10^{-4}	40	1×10^{-1}	1×10^{-1}	1×10^{-2}	400 000	1 119 900
Case 10	1×10^{-3}	45	1×10^{-1}	1×10^{-1}	1.2×10^{-2}	247 442	674 817
Case 11	3×10^{-4}	25	1×10^{-1}	1×10^{-1}	5×10^{-3}	845 952	2 472 540
Case 12	3×10^{-4}	25	1×10^{-1}	1×10^{-1}	5×10^{-3}	845 952	2 472 540
Case 13	3×10^{-4}	25	1×10^{-1}	1×10^{-1}	5×10^{-3}	845 952	2 472 540
Case 14	3×10^{-4}	25	1×10^{-1}	1×10^{-1}	5×10^{-3}	845 952	2 472 540
Case 15 (*)	3×10^{-4}	25	1×10^{-1}	1×10^{-1}	5×10^{-3}	845 952	2 472 540
Case 16	3×10^{-4}	25	1×10^{-1}	1×10^{-1}	5×10^{-3}	845 952	2 472 540
Case 17	3×10^{-4}	25	1×10^{-1}	1×10^{-1}	5×10^{-3}	845 952	2 472 540
Case 18	3×10^{-4}	25	1×10^{-1}	1×10^{-1}	5×10^{-3}	845 952	2 472 540
Case 19	3×10^{-4}	25	1×10^{-1}	1×10^{-1}	5×10^{-3}	845 952	2 472 540
Case 20 (*)	3×10^{-4}	25	1×10^{-1}	1×10^{-1}	5×10^{-3}	845 952	2 472 540

Table 2: Computational domain details and numerical settings (the cases marked with * are those duplicated). The numbers reported are input data and they consequently represent exact numbers.

Cases study	Δm [%]	m_0 [kg]	m_1 [kg]
Case 1	7.8116%	5.2232×10^{-7}	4.8152×10^{-7}
Case 2	8.0567%	5.2232×10^{-7}	4.8024×10^{-7}
Case 3	8.3400%	5.2232×10^{-7}	4.7876×10^{-7}
Case 4	8.4874%	5.2232×10^{-7}	4.7799×10^{-7}
Case 5 (*)	8.5755%	5.2232×10^{-7}	4.7753×10^{-7}
Case 6 (*)	8.5755%	5.2232×10^{-7}	4.7753×10^{-7}
Case 7	5.2412%	1.7628×10^{-6}	1.6704×10^{-6}
Case 8	3.6905%	4.1785×10^{-6}	4.0244×10^{-6}
Case 9	2.7776%	8.1612×10^{-6}	7.9346×10^{-6}
Case 10	2.1841%	1.4102×10^{-5}	1.3795×10^{-5}
Case 11	7.4287%	5.2232×10^{-7}	4.8352×10^{-7}
Case 12	7.7431%	5.2232×10^{-7}	4.8188×10^{-7}
Case 13	8.0301%	5.2232×10^{-7}	4.8038×10^{-7}
Case 14	8.2925%	5.2232×10^{-7}	4.7901×10^{-7}
Case 15 (*)	8.5755%	5.2232×10^{-7}	4.7753×10^{-7}
Case 16	6.1574%	5.2232×10^{-7}	4.9016×10^{-7}
Case 17	6.7168%	5.2232×10^{-7}	4.8724×10^{-7}
Case 18	7.3015%	5.2232×10^{-7}	4.8418×10^{-7}
Case 19	7.9084%	5.2232×10^{-7}	4.8101×10^{-7}
Case 20 (*)	8.5755%	5.2232×10^{-7}	4.7753×10^{-7}

Table 3: Results computed at t* = 1 (the cases marked with * are those duplicated). The numbers reported in the third and fourth columns are numerical results and their representation respect the criterion of significant figures homogeneity; the numbers in the second column are computed values and respect the criterion of decimal figures homogeneity.

in Figure 1): the hexagonal three-dimensional elements are arranged (cut and connected) to form a mesh made up of cubic elements. Their side was kept constant and equal to 0.005 m (apart from those case studies where the droplet initial diameter effect was assessed), while the time step varied in function of v_i [m s^{-1}], as highlighted in Table 2, decreasing as velocity increases (in order to achieve a suitable "resolution" of the physical phenomenon) and increasing with d_i [m] for numerical reasons related to convergence of the solution. By adopting an Eulerian-Lagrangian approach, the code STAR-CCM+

version 5.04.012 requires some bonds to be respected, for the internal subsistence of such approach [52]. In detail, the volume fraction occupied by the Lagrangian phase in a single cell must be lower than 0.01. For such reason in Cases from 7 to 10, in which droplet diameter d_i [m] varied from 0.0015 to 0.003 m (see Table 1), the mesh dimensions were increased and the cells were assigned a side of 0.01 or 0.012 m.

Results

The evaporation process involving a water droplet moving through the air is influenced by a wealth of factors. This paper deals with the importance of four among the most important influencing factors: droplet initial velocity, air temperature, diffusion coefficient of vapour in air and droplet initial diameter. In the figures displaying the evaporation rate trend, calculated according to eq. 8, it was made possible an analysis comparing the temporal-dependent and spatial-dependent homologous results, as already stated, setting a simulation time t* = 1, on the one hand, and a dimensionless travel distance L* = 1, on the other. Hereafter each case study (except those duplicated) is considered one at a time to reach a clear picture of the results arrived at. Tables 3 and 4 showcase the main evaporation results at t* = 1 and L* = 1, respectively.

Velocity is the first analysis parameter considered (Case 1 to Case 5).

Case 1 (v_i = 1 m s^{-1})

This case studies evaporation of a single water droplet which leaves an irrigation sprinkler nozzle with velocity of 1 m s^{-1}. The thermophysical conditions of water and air are reported in Table 1. At t* = 1 one gets an in-percentage mass evaporation rate of 7.8116% (Table 3); while fixing an ideal plane perpendicular to the droplet direction and located at L* = 1 gives an in-percentage mass evaporation rate of 10.4416% (Table 4) as the droplet reaches that set position at t*$_{L,max}$ = 1.3375 (Table 4).

Case 2 (v_i = 5 m s^{-1})

Here it is tackled the case of a single water droplet (evaporating) characterised by an initial velocity of 5 m s^{-1}. The thermophysical conditions of water and air are reported in Table 1. Considering t* = 1 one computes an in-percentage evaporation rate of 8.0567% (+0.2451% with respect to Case 1), as reported in Table 3. This datum gives the first signal that higher velocity may cause bigger evaporation rates: drag force, which depends on velocity according to eq. 3, affects the evaporation phenomenon. By fixing a target plane perpendicular to droplet path and located at L* = 1, the droplet reaches that ideal plane at t*$_{Lmax}$ = 1.2594 (Table 4), where an evaporation rate of 10.1278% is computed (-0.3138% with respect to Case 1, as t*$_{Lmax}$ [-] is now smaller because of a higher velocity).

Case 3 (v_i = 15 m s^{-1})

This case studies the evaporation of a single water droplet leaving an irrigation sprinkler nozzle with initial velocity of 15 m s^{-1}. The thermophysical conditions of water and air are reported in Table 1. If one considers a dimensionless time t* = 1, the related in-percentage mass evaporation rate is equal to 8.3400% (Table 3). With respect to Case 1 (+0.5284%) and Case 2 (+0.2833%) this result confirms proportionality between evaporation and velocity, which may be ascribed to the drag force as strictly related to velocity (eq. 3). Checking the space-dependent evaporation result (i.e. for a dimensionless travel distance of L* = 1), the droplet flight lasts t*$_{Lmax}$ = 1.1511 (Table 4) and determines an evaporation rate equal to 9.6572% (Table 4). The latter datum is proves smaller than those in both Case 2 (-0.4706%) and Case 1 (-0.7844%), being t*$_{Lmax}$ lower in this case due to an augmented droplet initial velocity.

Case 4 (v_i = 25 m s^{-1})

This case relates to the evaporation process occurring to a water droplet with velocity of 25 m s^{-1}. The thermophysical conditions of

water and air are reported in Table 1. After a simulation time of t* = 1 the in-percentage evaporation rate of 8.4874% is arrived at (Table 3). With respect to Case 1 (+0.6758 %), Case 2 (+0.4307%) and Case 3 (+0.1474%) the higher velocity – higher evaporation trend proves its consistency and confirms the air friction effect in affecting evaporation. After a dimensionless travel distance of L* = 1 (corresponding to a simulation time t*$_{Lmax}$ = 1.0728, see Table 4) in-percentage evaporation rate becomes 9.0927% (Table 4): this value, as it may now be expected, is less than those of Case 1 (-1.3489%), Case 2 (-1.0351%) and Case 3 (-0.5645%) because of t*$_{Lmax}$ decrease.

Case 5 (v_i = 30 m s^{-1})

In this case the initial droplet velocity is the highest tested: 30 m s^{-1}. The thermophysical conditions of water and air are reported in Table 1. Once reached a time dimensionless value of t* = 1, the in-percentage evaporation rate of 8.5755% is reached (Table 3). With respect to Case 1 (+0.7639%), Case 2 (+0.5188%), Case 3 (+0.2355%) and Case 4 (+0.0881%) the same parametrical trend is confirmed: higher velocities cause higher evaporation as air friction characterises the process (see also eq.3). If the spatial effect, instead, is considered: once reached a distance L* = 1 (which happens at t*$_{Lmax}$ = 1.0399, see Table 4), an evaporation rate equal to 8.8496% (Table 4) is arrived at. This figure is lower than those of Case 1 (-1.5920%), Case 2 (-1.2782%), Case 3 (-0.8076%) and Case 4 (-0.2431%), again because of the role played by t*$_{Lmax}$ as related to velocity.

Initial droplet diameter is the second analysis parameter considered (Case 6 to Case 10).

Case 6 (d_i = 0.001 m)

See Case 5 (duplicated case: the combination of the analysis parameters considered, leads case 6 to be coincident with case 5).

Case 7 (d_i = 0.0015 m)

In this case study it is investigated the evaporation of a single water droplet with initial diameter of 0.0015 m. The thermophysical conditions of water and air are reported in Table 1. At the dimensionless instant of time t* = 1 the corresponding in-percentage evaporation rate is 5.2412% (-3.3343% with respect to Case 6), as shown in Table 3. This datum shows that a bigger droplet diameter results in a lower evaporation rate: this may be credited to an augmented thermal inertia which tends to limit evaporation and to a decreased surface over volume ratio, reducing the diffusion of vapour in air in relation to the volume of the drop. At the dimensionless travel distance L* = 1 from inlet, which the droplet covers after a dimensionless time t*$_{Lmax}$ = 0.6746 (Table 4), the evaporation rate is 3.6569% (-5.1927% with respect to Case 6, due to a reduction in t*$_{Lmax}$ creditable to an increase in gravity), as Table 4 shows.

Case 8 (d_i = 0.002 m)

The present case tackles in-flight droplet evaporation when an initial diameter of 0.002 m is set. The thermophysical conditions of water and air are reported in Table 1. An in-percentage evaporation rate of 3.6905% (Table 3) is arrived at after the time dimensionless co-ordinate t* has reached a value equal to 1. With respect to Case 6 (-4.8850%) and Case 7 (-1.5507%) it may be evicted that increasing the initial diameter means decreasing the evaporation rate: increased droplet thermal inertia and decreased surface over volume ratio may be pointed out as responsible for that. Parallel, an in-percentage evaporation rate of 2.0061% (Table 4) is computed considering a dimensionless covered

Cases study	Δm [%]	m_0 [kg]	m_1 [kg]	t*$_{Lmax}$ [-]
Case 1	10.4416	5.2232 × 10^{-7}	4.6795 × 10^{-7}	1.3375
Case 2	10.1278	5.2232 × 10^{-7}	4.6942 × 10^{-7}	1.2594
Case 3	9.6572	5.2232 × 10^{-7}	4.7188 × 10^{-7}	1.1511
Case 4	9.0927	5.2232 × 10^{-7}	4.7483 × 10^{-7}	1.0728
Case 5 (*)	8.8496	5.2232 × 10^{-7}	4.7610 × 10^{-7}	1.0399
Case 6 (*)	8.8496	5.2232 × 10^{-7}	4.7610 × 10^{-7}	1.0399
Case 7	3.6569	1.7628 × 10^{-6}	1.6984 × 10^{-6}	0.6746
Case 8	2.0061	4.1785 × 10^{-6}	4.0947 × 10^{-6}	0.5111
Case 9	1.2975	8.1612 × 10^{-6}	8.0554 × 10^{-6}	0.4336
Case 10	0.9165	1.4102 × 10^{-5}	1.3973 × 10^{-5}	0.3824
Case 11	7.6981	5.2232 × 10^{-7}	4.8211 × 10^{-7}	1.0375
Case 12	8.0237	5.2232 × 10^{-7}	4.8041 × 10^{-7}	1.0381
Case 13	8.3221	5.2232 × 10^{-7}	4.7885 × 10^{-7}	1.0387
Case 14	8.5954	5.2232 × 10^{-7}	4.7743 × 10^{-7}	1.0393
Case 15 (*)	8.8496	5.2232 × 10^{-7}	4.7610 × 10^{-7}	1.0399
Case 16	6.5243	5.2232 × 10^{-7}	4.8824 × 10^{-7}	1.0728
Case 17	7.0826	5.2232 × 10^{-7}	4.8533 × 10^{-7}	1.0641
Case 18	7.6578	5.2232 × 10^{-7}	4.8232 × 10^{-7}	1.0558
Case 19	8.2470	5.2232 × 10^{-7}	4.7925 × 10^{-7}	1.0476
Case 20 (*)	8.8496	5.2232 × 10^{-7}	4.7610 × 10^{-7}	1.0399

Table 4: Results computed at L* = 1 (the cases marked with * are those duplicated). The numbers reported in the third and fourth columns are numerical results and their representation respect the criterion of significant figures homogeneity; the numbers in the second column are computed values and respect the criterion of decimal figures homogeneity.

distance $L^* = 1$ (at $t^*_{Lmax} = 0.5111$, see Table 4): such value, smaller than those of both Case 6 (-6.8435%) and Case 7 (-1.6508%) confirms the trend previously highlighted.

Case 9 ($d_i = 0.0025$ m)

This case takes into account the aerial evaporation of a single water droplet with initial diameter of 0.0025 m. The thermophysical conditions of water and air are reported in Table 1. If one considers a dimensionless time interval $t^* = 1$, the simulation provides an in-percentage evaporation rate of 2.7776% (Table 3). With respect to Case 6 (-5.7979%), Case 7 (-2.4636%) and Case 8 (-0.9129%) the general trend is confirmed, i.e. an increased droplet diameter causes a decrement in the evaporation rate. Augmented droplet thermal inertia and diminished surface over volume ratio are the reasons for that. Considering instead a dimensionless travel distance $L^* = 1$, covered in a dimensionless time $t^*_{Lmax} = 0.4336$ (Table 4), it is obtained an in-percentage evaporation rate of 1.2975% (Table 4), smaller than that of Case 6 (-7.5521%), Case 7 (-2.3594%) and Case 8 (-0.7086%) for the same physical and mechanical reasons explained above.

Case 10 ($d_i = 0.003$ m)

The present case study is about an evaporating droplet with initial diameter of 0.003 m. The thermophysical conditions of water and air are reported in Table 1. Considering $t^* = 1$: in-percentage evaporation rate is 2.1841% (Table 3). Comparing such figure to the previous cases one has: Case 6 (-6.3914%); Case 7 (-3.0571%); Case 8 (-1.5064%); and Case 9 (-0.5935%). This completes the whole picture examined proving that augmenting a diameter acts upon two significant parameters: thermal inertia (increasing it) and surface over volume ratio (decreasing it). Both these variations tend to limit evaporation. Same considerations may be made considering a spatial targeting of the analysis: $L^* = 1$ (for $t^*_{Lmax} = 0.3824$, see Table 4). In-percentage evaporation rate becomes 0.9165% (Table 4), being the figure smaller than in Case 6 (-7.9331%), Case 7 (-2.7405%), Case 8 (-1.0896%) and Case 9 (-0.3810%).

Diffusion coefficient of water vapour in the air is the third analysis parameter considered (Case 11 to Case 15).

Case 11 ($D_{va} = 0.2 \times 10^{-4}$ m² s⁻¹)

In this case study it is investigated the evaporation of a single water droplet when D_{va} is set equal to 0.2×10^{-4} m² s⁻¹. The thermophysical conditions of water and air are reported in Table 1. At the dimensionless instant of time $t^* = 1$ the corresponding in-percentage evaporation rate is 7.4287%, as shown in Table 3. At the dimensionless travel distance $L^* = 1$ from inlet, which the droplet covers after a dimensionless time $t^*_{Lmax} = 1.0375$ (Table 4), the evaporation rate is 7.6981%, as Table 4 shows.

Case 12 ($D_{va} = 0.225 \times 10^{-4}$ m² s⁻¹)

This test studies single droplet evaporation in case of a diffusion coefficient value of 0.225×10^{-4} m² s⁻¹. The thermophysical conditions of water and air are reported in Table 1. When $t^* = 1$, the in-percentage evaporation rate becomes equal to 7.7431%, as shown in Table 3. Compared to the previous case (Case 11) evaporation results to be increased by a 0.3144 percentage, easily (from a qualitative point of view) explainable as increasing D_{va} tends to favour evaporation (when all the other parameters are kept constant). Instead, when $L^* = 1$ (i.e. at $t^*_{Lmax} = 1.0381$, see Table 4) in-percentage droplet mass evaporation is equal to 8.0237% (Table 4). This figure is higher than in Case 11 (0.3256%), confirming, on the one hand, a direct proportionality

between diffusion coefficient, on the other, the correct predictions obtainable by the code.

Case 13 ($D_{va} = 0.25 \times 10^{-4}$ m² s⁻¹)

A D_{va} value of 0.25×10^{-4} m² s⁻¹ is here considered to check its effect on a single droplet evaporation. Again, the thermophysical conditions of water and air are reported in Table 1. In the present case the in-percentage mass evaporation rate at dimensionless time $t^* = 1$ is equal to 8.0301% (Table 3). If compared to Case 11 and Case 12, as qualitatively expectable, there is an evaporation augmentation of +0.6014% and +0.2870%, respectively, due to the modified diffusion coefficient value which enhances the process. By considering evaporation after a dimensionless travel distance equal to $L^* = 1$ (covered in $t^*_{Lmax} = 1.0387$, see Table 4), the evaporation result is equal to 8.3221% (Table 4): this datum is higher than in both Case 11 (+0.6240%) and Case 12 (+0.2984%), as the time of flight is higher in relation to the effect of gravity.

Case 14 ($D_{va} = 0.275 \times 10^{-4}$ m² s⁻¹)

Case 14 faces droplet evaporation when the diffusion coefficient of vapour in air is set equal to 0.275×10^{-4} m² s⁻¹ (thermophysical conditions of water and air are available in Table 1). The mass evaporation in-percentage figure results 8.2925% at $t^* = 1$ (Table 3). Compared to the previous cases (Case 11, Case 12, Case 13) the same trend is confirmed, showing an augmentation of +0.8638%, +0.5494% and +0.2624, respectively, being an increase of D_{va} [m² s⁻¹] in favour of a more intense evaporation. Moreover, if a dimensionless travel distance $L^* = 1$ is covered by the droplet after a time of $t^*_{Lmax} = 1.0393$ (Table 4), then a mass droplet evaporation of 8.5954% is arrived at (Table 4). The latter datum is, again, part of the increasing trend examined in the previous cases: it boasts a +0.8973%, +0.5717% and +0.2733% with respect to Cases 11, 12, and 13, respectively.

Case 15 ($D_{va} = 0.3 \times 10^{-4}$ m² s⁻¹)

See Case 5 (duplicated case: the combination of the analysis parameters considered, leads case 15 to be coincident with case 5).

Air temperature is the last parameter here investigated (Case 16 to Case 20).

Case 16 ($T_a = 300$ K)

This case study is about droplet evaporation in case of surrounding air at a temperature of 300 K. The thermophysical conditions of water and air are, again, reported in Table 1. As Table 3 displays, the in-percentage mass evaporation rate after a dimensionless time of $t^* = 1$ is equal to 6.1574%; while after a dimesionless distance $L^* = 1$ (covered after a time interval $t^*_{Lmax} = 1.0728$, see Table 4) a droplet mass percentage of 6.5243% is evaporated (Table 4).

Case 17 ($T_a = 305$ K)

It is here considered an air temperature of 305 K (thermophysical conditions of water and air are in Table 1) and its effect on a single droplet aerial evaporation. At $t^* = 1$ the mass evaporation figure is of 6.7168% (Table 3), that is +0.5594% with respect to Case 16 as an augmented air temperature, keeping all the other parameters constant, plays in favour of an increased evaporation. By setting a dimensionless travel distance of $L^* = 1$ (covered after $t^*_{Lmax} = 1.0641$, see Table 4), droplet mass evaporation becomes equal to 7.0826% (Table 4), i.e. higher than in Case 16 (+0.5583%).

Case 18 (T$_a$ = 310 K)

This case studies droplet evaporation when the surrounding air is at 310 K. The thermophysical conditions of water and air are reported in Table 1. At t* = 1 the computed mass evaporation is equal to 7.3015% (Table 3): +1.1441% and +0.5847% compared to Case 16 and Case 17, respectively, which can be explained as increasing the temperature of the medium into which the droplet flows makes its evaporation more intense. Considering instead the spatial effect of air temperature, i.e. L* = 1 (at t*$_{Lmax}$ = 1.0558, see Table 4), mass droplet evaporation becomes 7.6578% (Table 4). As one can easily see the rate of mass evaporation is higher than in both Case 16 (+1.1335%) and Case 17 (+0.5752%) depending on the droplet shrinkage increased by temperature.

Case 19 (T$_a$ = 315 K)

The present case study relates to an air temperature of 315 K and its effect on droplet evaporation. Table 1 displays the thermophysical conditions of water and air. In this case the rate of evaporation in mass at t* = 1 is equal to 7.9084% (Table 3). Compared to the previous cases (Case 16, Case 17, Case 18) there is here an increased evaporation rate (+1.7510% with respect to Case 16, +1.1916% with respect to Case 17, +0.6069% with respect to Case 18), confirming what is qualitatively a reasonable trend: a raised air temperature value enhances aerial evaporation, keeping constant all the other parameters. Again, checking the process at a dimensionless travel distance L* = 1, reached after a time t*$_{Lmax}$ = 1.0476 (Table 4), droplet evaporation becomes equal to 8.2470% (Table 4). As one can easily see, this figure confirms the trend highlighted in the previous three cases: Case 16 (+1.7227%), Case 17 (+1.1644%), and Case 18 (+0.5892%) which, again, may be attributed to increased droplet shrinkage due to air temperature.

Case 20 (T$_a$ = 320 K)

See Case 5 (duplicated case: the combination of the analysis parameters considered, leads case 20 to be coincident with case 5).

Discussion

The previous section showcased the main results that were arrived at numerically, together with a few preliminary comments that are here to be expanded for each analysis parameter. The influence of droplet initial velocity on evaporation may be inferred by comparing the results obtained in cases from 1 to 5 (Table 3). In general it may be deduced that the lower the initial velocity, the lower the evaporated mass. In detail, the evaporation rate percentage shifts from 7.8116% in Case 1 to 8.5755% in Case 5. The variation of the evaporation rate percentage against velocity shows a logarithmic trend displayed in Figure 2. Under a physical point of view such result is interpreted in the light of air friction effect both on the dynamics of vapour film surrounding the droplet and directly on evaporation (Lorenzini, 2004). Velocity enhances both convection and friction force but the latter depends on velocity raised at the second power and this makes its effect on droplet evaporation significantly more remarkable, especially for a so limited time interval as that here investigated. These comments, however, hold true just if time-dependence is analysed while if space-dependence is considered the situation changes significantly, as Figures 3a and 3b demonstrate. The following analysis is not negligible given that the present investigation has to do with a particular application, i.e. sprinkler irrigation, which is mainly interested in the spatial distribution of water for agricultural purposes. Figure 3a shows how, completing the same path, a slower droplet evaporates more than a faster one, because of a higher time of flight which acts upon water temperature

and friction. This also highlights the overlap between the effects of temperature and air friction, especially evident for faster droplets: in the first part of the flight the elevate contribution of the friction force (enhancing evaporation) does not balance the lower convective heat flux entering the droplet due to the quicker covering of the same path length. This effect becomes less significant for slower droplets, showing a linear trend, as Figure 3a displays. So, keeping the other parameters fixed, faster droplets tend to evaporate less than slower ones and consequently less water is wasted. On the contrary, Figure 3b shows that faster droplet evaporates more (especially in the first moments of flight) than slower droplet, because if the focus is on time than the friction force effect becomes predominant on ruling evaporation. From these comments it may be evicted that the phenomenon under exam is due to a dynamic and to a convective affection. The effect of droplet initial diameter on the evaporation rate may be deduced checking the cases from 6 to 10 (Table 3). In general: the bigger the diameter, the lower the evaporation mass. Evaporation rate reduction according to droplet diameter increase is remarkable, with exponential trend and results from 8.5755% (Case 6) for a 1 mm droplet diameter, to 2.1841% in the case of a 0.003 m droplet diameter (Case 10) (Figure 4). This result may be attributed to the higher thermal inertia which characterises bigger droplets, which is also bond to a less favourable surface-volume ratio (decreasing when the diameter augments). Such result may be displayed in function of the path covered (Figure 5a)

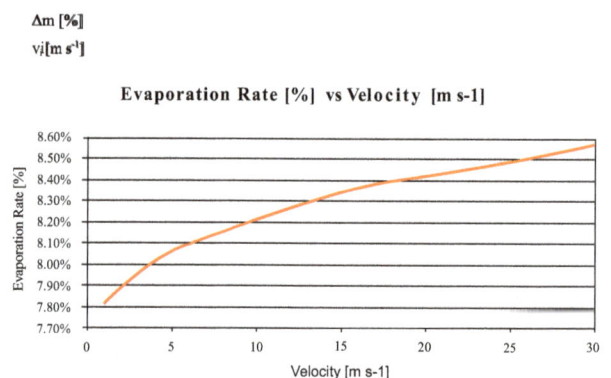

Figure 2: Evaporation rate (t* = 1) versus droplet initial velocity.

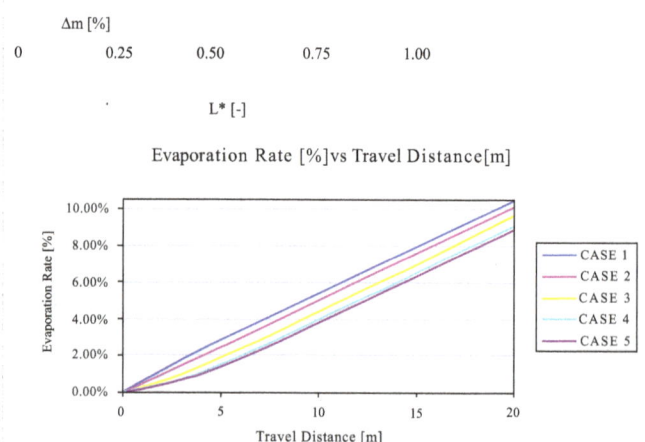

Figure 3a: Evaporation rate versus dimensionless travel distance (parameter: v$_i$ [m s^{-1}]).

or of the simulation time (Figure 5b) which both confirm the same trend and consequent analysis. The effect of the diffusion coefficient of vapour in air on aerial water droplet evaporation was faced from Case 11 to Case 15 (Table 3). It evidently appears that, augmenting D_{va} [m^2 s^{-1}], droplet evaporation raises, as the contribution of the diffusion term is enhanced, letting evaporation rate to shift from 7.4287% (Case 11, D_{va} = 0.2 × 10^{-4} m^2 s^{-1}), to 8.5755% (Case 15, D_{va} = 0.3 × 10^{-4} m^2 s^{-1}). The trend (Figure 6) proves to be linear, in accordance to Fick's law of diffusion, keeping the coefficient constant with temperature, given the low liquid-gas interface vapour concentration gradients, typical of the phenomenon here considered. Further parametric considerations may be performed checking carefully the spatial and temporal phenomenological effects of D_{va} [m^2 s^{-1}], in Figure 7a and 7b respectively, both confirming the general trend of Figure 6. Figure 7a shows how mass evaporation raises with D_{va} [m^2 s^{-1}] for a same L* [-] value: such trend may be interpreted in relation to the necessity of keeping a constant water vapour film at the air-droplet interface by drawing it from the droplet liquid water. The curves in Figure 7a are also very close from one another, especially at the beginning of the path, i.e. for small values of L* [-]: for bigger values of L* [-] the trends tend to become more separated, as a consequence of the cumulative effects of a higher diffusion coefficient. The temporal variations displayed

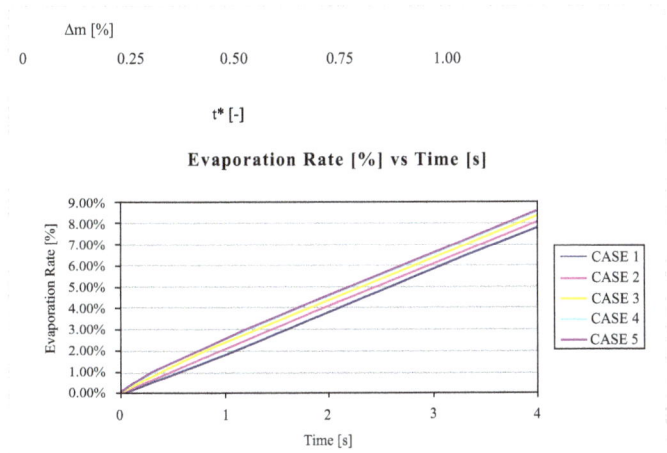

Figure 5a: Evaporation rate versus dimensionless travel distance (parameter: d_i [m]).

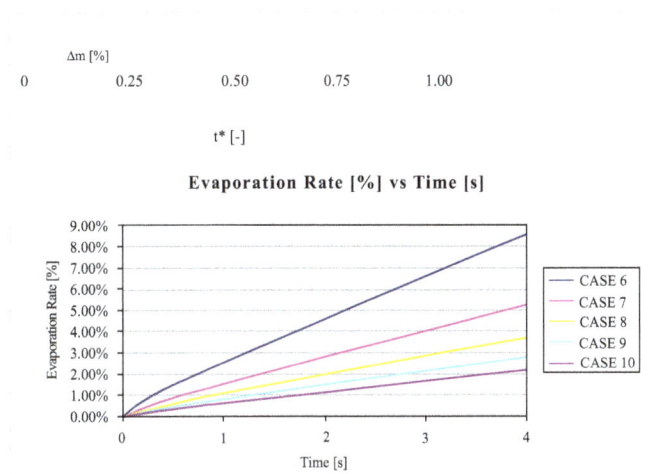

Figure 5b: Evaporation rate versus dimensionless simulation time (parameter: d_i [m]).

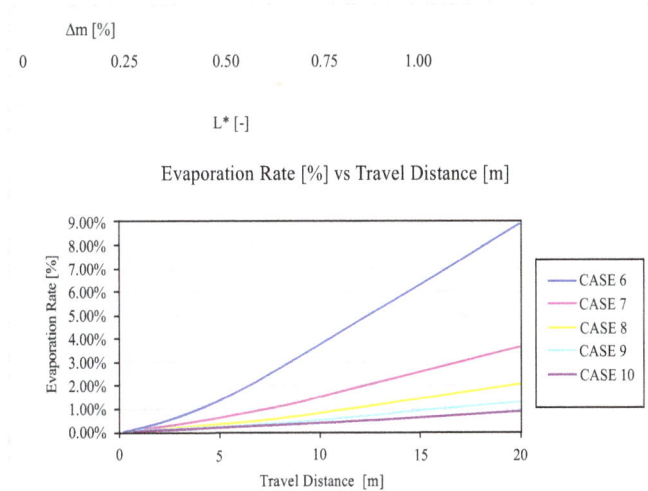

Figure 3b: Evaporation rate versus dimensionless simulation time (parameter: v_i [m s^{-1}]).

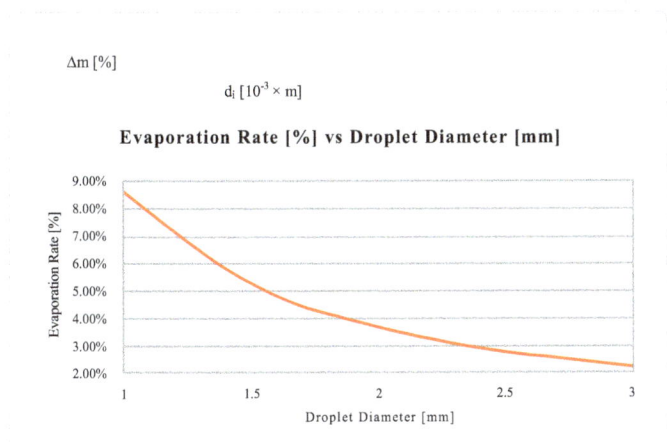

Figure 4: Evaporation rate (t* = 1) versus droplet initial diameter.

in Figure 7b prove similar, confirming the previously made analysis. The last parameter tested (in cases from 16 to 20) was air temperature (Table 3). When keeping all the other analysis parameters constant, droplet evaporation is promoted by air-water temperature gradient, which influences the convective contribution. In fact, varying T_a from 300 K to 320 K, evaporation augments from 6.1574% to 8.5755%, respectively. The evaporation vs. air temperature relation proves to be nearly linear (see Figure 8), which is to be attributed to the linear dependence between the temperature and the convective term in an evaporation process, in addition to the linear effect that temperature has on the entire thermal flux. As a matter of fact, as it may be deduced considering the cases study from 1 to 5, convection is a phenomenon which depends upon temperature and velocity, once the geometrical and physical features of the problem are set. The same general effect is confirmed when facing a spatial (Figure 9a) and temporal (Figure 9b) study of the parameter which is currently under examination. In particular: Figure 9a shows droplet mass evaporation rate versus the dimensionless path covered. It is confirmed a directly proportional relation between the two variables while the curves, initially very close from one another, tend to open when getting closer to $L^* = 1$ as a consequence of the ongoing cumulative convective effect. If time dependence is investigated (Figure 9b), the general trend is the same just described even if the first moments of the path prove to be more intensely affected (with respect to the space-dependent trend of Figure 9a) by the water transient heating due to the air temperature as affecting the diffusion laws.

Conclusions

Aerial droplet evaporation is a phenomenon that applies in several technical fields related to civil, mechanical and agricultural engineering (automotive, refrigeration and conditioning, fire safety, irrigation, water saving, etc.). This study is specifically referring to irrigation and water saving, even though its results would be suitable, at least as a first step, for many other applications among those just quoted, provided that pressures and temperatures analogous to those here tested were involved. A numerical approach was employed, based on Computational Fluid Dynamics (CFD) software called STAR-CCM+ version 5.04.012: such code adopts the control volume method. The system investigated is a single spherical droplet travelling within air and evaporating during its aerial path between the nozzle and the

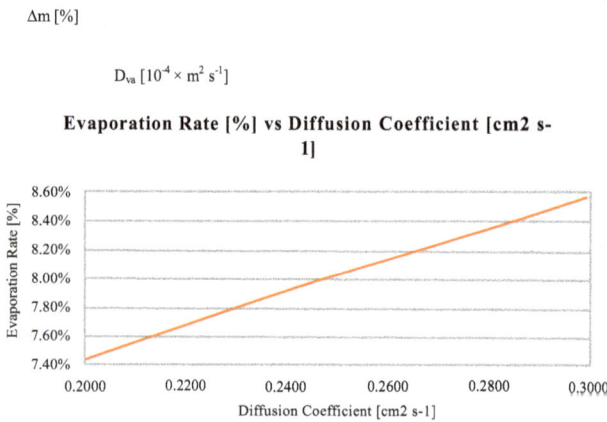

Δm [%]

D_{va} [$10^{-4} \times$ m^2 s^{-1}]

Figure 6: Evaporation rate (t* = 1) versus diffusion coefficient of vapour in air.

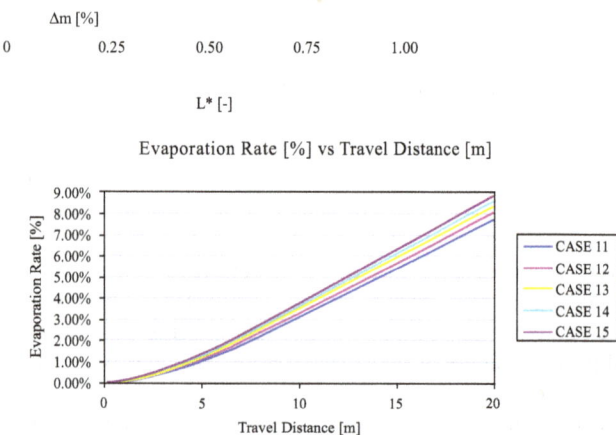

Δm [%]

L* [-]

Figure 7a: Evaporation rate versus dimensionless travel distance (parameter: D_{va} [m^2 s^{-1}]).

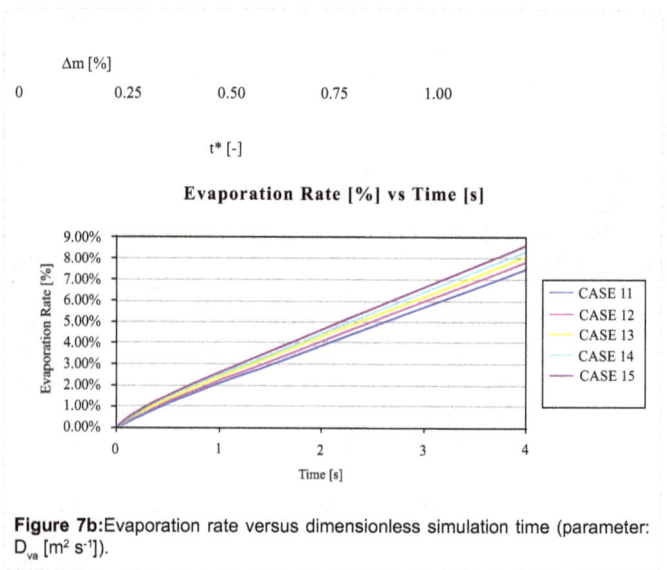

Δm [%]

t* [-]

Figure 7b: Evaporation rate versus dimensionless simulation time (parameter: D_{va} [m^2 s^{-1}]).

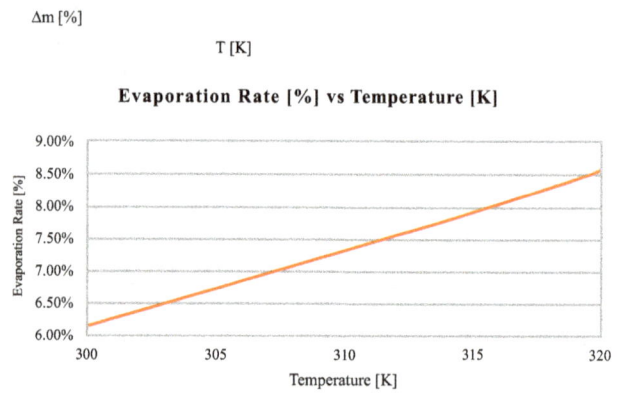

Δm [%]

T [K]

Figure 8: Evaporation rate (t* = 1) versus air temperature.

ground. The analysis parameters were: droplet initial velocity; droplet initial diameter; air temperature; and diffusion coefficient of vapour in air. Air relative humidity was not a parameter here because the code proved not acceptably reliable in managing such kind of water-air interface: such limitation, anyway, does not affect the generality of the present study because of the "classical" parametric approach adopted, one variable being entirely independent on the others. Twenty cases study, five for each analysis parameter, were faced.

The results obtained show that, at the conditions tested, the parameter which affects droplet evaporation more significantly is droplet initial diameter which, varying from 0.001 to 0.003 m (a range typical in sprinkler irrigation practice), determines a droplet mass evaporation decrement of 6.3914%, considering time dependence, and 7.9331% considering space dependence: this highlights the role played in the process by the dynamic components (air friction) and by the droplet thermal inertia. Air temperature (which initially may have been suspected as the most affecting parameter) also proves significant, with a 2.4181% (time dependence considered) and 2.3253% (space dependence considered) augmentation when passing from 300 to 320 K. Droplet initial velocity (varying from 1 to 30 m s^{-1}) and diffusion coefficient of vapour in air (varying from 0.2 × 10^{-4} to 0.3 × 10^{-4} m^2 s^{-1}) are instead interested by an evaporation variation within the

Δm [%]

0   0.25   0.50   0.75   1.00

L* [-]

Evaporation Rate [%] vs Travel Distance [m]

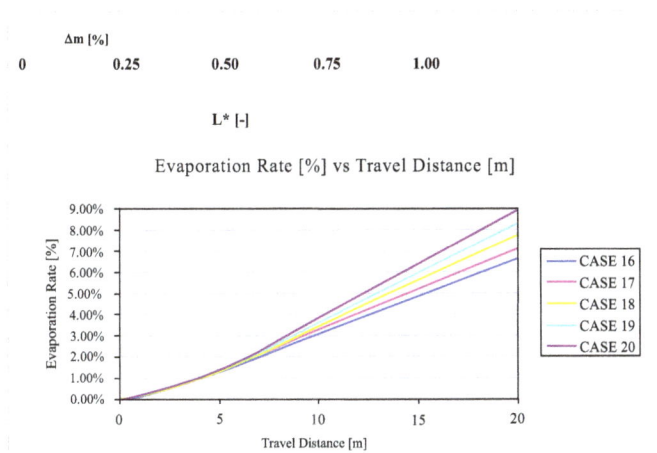

Figure 9a: Evaporation rate versus dimensionless travel distance (parameter: T_a [K]).

Δm [%]

0   0.25   0.50   0.75   1.00

t* [-]

Evaporation Rate [%] vs Time [s]

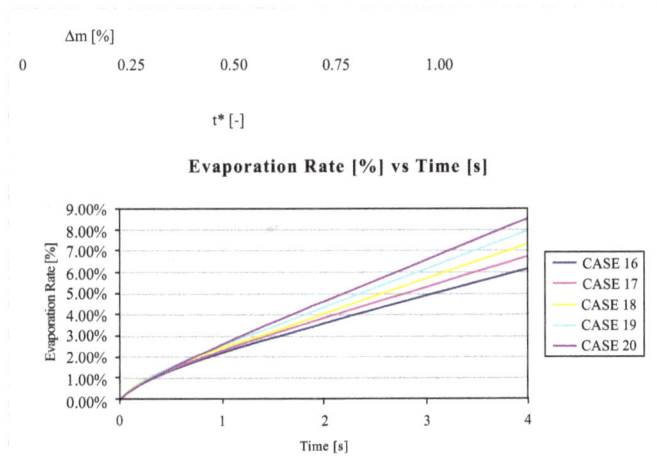

Figure 9b: Evaporation rate versus dimensionless simulation time (parameter: T_a [K]).

range investigated of 0.7639% (time dependence considered; 1.5920% when space dependence considered) and 1.1468% (time dependence considered; 1.1515% when space dependence considered), which are still significant data, anyway. Like in the case of air temperature, probably an increase of 50% in D_{va} [m^2 s^{-1}] would have suggested a much higher effect on droplet evaporation than that actually computed but the result correctly show that the droplet evaporation is a very complicate fluid dynamic process which cannot be reduced to a simply diffusive matter. Finally, for what relates to water saving in irrigation practice, one can conclude that, apart from a low air temperature and high diffusion coefficient of vapour in air condition (which, on the one hand, may have been somehow predictable but, on the other, serve to practically prove and validate the reliability of the model created and of the approach adopted) the conditions that help save water are: big droplet initial diameter and low droplet initial velocity, if the aim is that water reaches a specific location and watering covers a certain field, this (and not time) being generally the main goal which in-field irrigation wishes to achieve. Both these latter parameters, the influence of which was not predictable a priori of the present investigation, may be controlled and conditioned acting on the sprinkler operating conditions, and this makes of a numerical study a highly applicable

source to many practical issues.

Nomenclature

Symbols	Name	Unit
A:	droplet radius (Kinzer and Gunn, 1951; see Introduction)	m
A	coefficient = 11.949, (eq. 2)	-
A_p	droplet cross sectional area (eq.3)	m^2
B	coefficient = 3978.205, (eq. 2)	-
C	coefficient = -39.801, (eq. 2)	-
C_d	friction factor (eq.3)	-
CFD	Computational Fluid Dynamics	-
d_i	droplet initial diameter	m
d_p	droplet diameter, (eq.6)	m
D_{va}	diffusion coefficient of vapour in air	m^2 s^{-1}
D^	droplet diameter variation with time (Goering, 1972; see Introduction)	m s^{-1}
F_d	friction force, (eq.3)	N
G	vapour-density gradient at the droplet surface (Kinzer and Gunn, 1951; see Introduction)	kg m^{-4}
K^	ratio between D_{va} and d_p (Goering, 1972; see Introduction)	m s^{-1}
L	travel distance (eq.10)	m
L_{max}	target location (eq.10)	m
L*	dimensionless travel distance (eq.10)	-
m^	droplet mass evaporated with time (Kinzer and Gunn, 1951; see Introduction)	kg s^{-1}
m_0	droplet initial mass, (eq. 8)	kg
m_1	droplet final mass, (eq.8)	kg
M^	ratio between molecular weights of vapour and air (Goering, 1972; see Introduction)	-
Nu_p	Nusselt number (droplet), (eq. 5)	-
Nu*	specially defined Nusselt number for mass transfer (Goering, 1972; see Introduction)	-
P_{atm}	atmospheric pressure, (eq.2)	Pa
P_{sat}	saturation pressure, (eq.2)	Pa
P	difference between saturation pressure at wet bulb air temperature and vapour pressure at dry bulb temperature / partial pressure of air (Goering, 1972; see Introduction)	-
Pr_a	Prandtl number (air), (eq.5)	-
R	ratio between air and droplet density (Goering, 1972; see Introduction)	-

Re_p	Reynolds number (droplet), (eq.6)	-
Sc	Schmidt number (eq.7)	-
Sh	Sherwood number (eq.7)	-
T	time, (eq.9)	s
t^*	dimensionless time, (eq.9)	-
t_{Lmax}	simulation time after a path $L = L_{max}$, (eq. 11)	s
t^*_{Lmax}	dimensionless simulation time after a path $L = L_{max}$, (eq. 11)	-
t_{max}	maximum simulation time, (eq.9)	s
v_i	droplet initial velocity	m s^{-1}
v_p	velocity (droplet), (eq.3)	m s^{-1}
T_0	reference temperature (273.15 K), (eq.1)	K
T_a	air temperature, (eq. 1)	K
T_w	water temperature	K

Greek Symbols	Name	Unit
Δm	droplet mass evaporation rate, (eq.8)	%
μ_0	air viscosity at 273.15 K, (eq.1)	Pa s
μ_a	air viscosity (eq.1)	Pa s
ρ_a	air density (eq.3)	kg m^{-3}

References

1. Lefebvre AH (1989) Atomization and Sprays. Taylor & Francis: London.

2. Sirignano WA (1999) Fluid Dynamics and Transport of Droplet and Sprays. Cambridge University Press: Cambridge.

3. Lorenzini G (2002) Air temperature effect on spray evaporation in sprinkler irrigation. Irrigation and Drainage 51: 301-309.

4. Lorenzini G (2004) Simplified modelling of sprinkler droplet dynamics. Biosystems Engineering 87: 1-11.

5. Park TW, Aggarwal SK, Katta VR (1996) A numerical study of droplet-vortex interactions in an evaporating spray. Int J Heat Mass Transf 39: 2205-2219.

6. Abramzon B, Sirignano W (1989) Droplet vaporization model for spray combustion calculations. Int J Heat Mass Transf 32: 1605–1618.

7. Bertoli C, Migliaccio na M (1999) A finite conductivity model for diesel spray evaporation computations. International Journal of Heat and Fluid Flow 20: 552–56.

8. Gogos G, Soh S, Pope DN (2003) Effects of gravity and ambient pressure on liquid fuel droplet evaporation. Int J Heat Mass Transf 46: 283–296.

9. Sazhin SS, Krutitskii PA, Abdelghaffar WA, Sazhina EM, Mikhalovsky SV, et al. (2004) Transient heating of diesel fuel droplets. Int J Heat Mass Transf 47: 3327-3340.

10. Birouk M, Gokalp I (2006) Current status of droplet evaporation in turbulent flows. Progress in Energy and Combustion Science 32: 408-423.

11. Chen YC, Staerner SH, Masri AR (2006) A detailed experimental investigation of well-defined, turbulent evaporating spray jets of acetone. International Journal of Multiphase Flow 32: 389-412.

12. Qureshi M, Zhu C (2006) Crossflow evaporating sprays in gas–solid flows: Effect of aspect ratio of rectangular nozzle. Powder Technology 166: 60-71.

13. Sazhin SS (2006) Advanced models of fuel droplet heating and evaporation. Progress in Energy and Combustion Science 32: 162-214.

14. Sazhin SS, Kristyadi T, Abdelghaffar WA, Heikal MR (2006) Models for fuel droplet heating and evaporation: Comparative analysis. Fuel 85:1613-1630.

15. Shusser M (2007) The influence of thermal expansion flow on droplet evaporation. Heat Transfer 2: 443-449.

16. Belarbi R, Ghiaus C, Allard F (2006) Modeling of water spray evaporation: Application to passive cooling of buildings. Solar Energy 80: 1540-1552.

17. Barrow H, Pope CW (2007) Droplet evaporation with reference to the effectiveness of water-mist cooling. Applied Energy 84: 404-412.

18. Qureshi M, Zhu C (2006) Gas entrainment in an evaporating spray jet. Int J Heat Mass Transf 49: 3417-3428.

19. Chen YC, Peters N, Schneemann GA, Wruck N, Renz U, et al. (1996) The detailed flame structure of highly stretched turbulent premixed methane-air flames. Combustion and Flame 107: 223-244.

20. Rouault M, Mestayer PG, Schiestel R (1991) A Model of Evaporating Spray Droplet Dispersion. J Geophys Res 96: 7181-7200.

21. Nurnberger FV, Merva GE, Harrington JB Jr (1976) Microenviromental modification by small water droplet evaporation. Journal of Applied Meteorology 15: 858-867.

22. Edling RJ (1985) Kinetic energy, evaporation and wind drift of droplets from low pressure irrigation nozzle. Transaction of the ASABE 28: 1543-1550.

23. Thompson AL, Gilley JR, Norman JM (1993) A sprinkler water droplet evaporation and plant canopy model: II. Model application. Transaction of the ASABE 36: 743-750.

24. Lorenzini G, De Wrachien D (2003) Phenomenological analysis of sprinkling spray evaporation: the air friction effect. Rivista di ingegneria Agraria 49-54.

25. Playan E, Garrido S, Faci JM, Galan A (2004) Characterizing pivot sprinklers using an experimental irrigation machine. Agricultural Water Management 70: 177-193.

26. Lorenzini G, De Wrachien D (2004) Theoretical and Experimental analysis of spray flow and evaporation in sprinkler irrigation. Irrigation and Drainage Systems 918: 155-166.

27. Lorenzini G (2006) Water droplet dynamics and evaporation in an irrigation spray. Transaction of the ASABE 94: 545-549.

28. De Wrachien D, Lorenzini G (2006) Modeling jet flow and losses in sprinkler irrigation: overwiew and perspective of a new approach. Biosystems Engineering 94. 297-309.

29. Kinzer GD, Gunn R (1951) The evaporation, temperature and thermal relaxation-time of freely falling waterdrops. Journal of Meteorology 8: 71-83.

30. RanzWE, Marshall WR (1952a) Evaporation from drops Part I. Chem Eng Progr 48: 141-146.

31. RanzWE, Marshall WR (1952b) Evaporation from drops Part II. Chem Eng Progr 48: 173-180.

32. Dzumbova L, Schwarz J, Smolik J (1999) Evaporation of water droplet in the humid atmosphere. J Aerosol Sci 30: S337–S338.

33. McLean RK, Sri Ranjan R, Klassen G (2000) Spray evaporation losses from sprinkler irrigation systems. Canadian Agricultural Engineering 42: 1.1-1.15.

34. Carrion P, Montero J, Tarjuelo JM (2001) Applying simulation on sprinkling irrigatuion systems design: SIRIAS model. Revista Internacional de Métodos Numérico para Calculo y Diseno en Ingeneria 17: 347–362.

35. Qu X, Davis EJ, Swanson BD (2001) Non-isothermal droplet evaporation and condensation in the near-continuum regime. J Aerosol Sci 32: 1315–1339.

36. Moyle AM, Smidansky PM, Lamb D (2006) Laboratory studies of water droplet evaporation kinetics. In: Proceeding of 12th Conference on Cloud Physics, and Proceeding of 12th Conference on Atmospheric Radiation. American Meteorological Society.

37. Yan HJ, Bai G, He, JQ, Li YJ (2010) Model of droplet dynamics and evaporation for sprinkler irrigation. Biosystems Engineering 106: 440-447.

38. Goering CE, Bode LE, Gebhard MR (1972) Mathematical modeling of spray

droplet deceleration and evaporation. Transactions of the ASAE 15: 220-225.

39. Sirignano WA (1993) Fluid dynamics of sprays—1992 Freeman scholar lecture. Transaction of the ASME- J Fluids Eng 115: 345-378.

40. Kulmala M, Vesala T, Schwarz J, Smolik J (1995) Mass transfer from a drop-II. Theoretical analysis of temperature dependent mass flux correlation. Int J Heat Mass Transf 38: 1705-1708.

41. Gouesbet G, Berlemont A (1999) Eulerian and Lagrangian approaches for predicting the behaviour of discrete particles in turbulent flows. Progress in Energy and Combustion Science 25: 133-159.

42. Dombrovsky LA, Sazhin SS (2003) A simplified non-isothermal model for droplet heating and evaporation. International Communications in Heat and Mass Transfer 30: 787-796.

43. Kolaitis DI, Katsourinis DI, Founti MA (2009) Droplet evaporation assisted by "stabilized cool flames": scrutinizing alternative CFD modelling approaches. Seventh International Conference on CFD in the Minerals and Process Industries (CSIRO, Melbourne, Australia, 9-11 December).

44. Edson JB, Anquetin S, Mestayer PG, Sini, JF (1996) Spray droplet modelling 2. An interactive Eulerian-Lagrangian model of evaporating spray droplets. J Geophys Res 101: 1279-1293.

45. Burger M, Rottenkolber G, Schmehl R, Giebert D, Schaefer O, et al. (2002) A Combined Eulerian and Lagrangian Method for Prediction of Evaporating Sprays. J Eng Gas Turbines and Power 124: 481-488.

46. Salman H, Soteriou M (2004) Lagrangian simulation of evaporating droplet sprays. Physics of Fluids 16: 4601–4622.

47. Beck JP, Watkins AP (2003) The droplet number moments approach to spray modelling: The development of heat and mass transfer sub-models. International Journal of Heat and Fluid Flow 24: 242-259.

48. Okamura S, Nakanishi K (1969) Theoretical study on sprinkler sprays (part four) geometric pattern form of single sprayer under wind conditions. Transactions of the Japanese Society of Irrigation Drainage and Reclamation Engineering 29: 35-43.

49. de Lima JLMP, Torfs PJJF, Singh VP (2002) A mathematical model for evaluating the effect of wind on downward-spraying rainfall simulators. Catena 46 : 221-241.

50. Miliauskas G, Sabanas V (2006) Interaction of transfer processes during unsteady evaporation of water droplets. Int J Heat Mass Transf 49: 1790-1803.

51. Sutherland W (1893) The viscosity of gases and molecular force. Philosophical Magazine 5 36: 507-531.

52. Star-CCM+, version 5.04.012, (2010) Users' Guide.

53. Reid RC, Prausnitz JM, Poling BE (1988) The Properties of Gases and Fluids. (2ndedn), Mc-Graw Hill: New York.

54. Schiller L, Naumann A (1933) Ueber die grundlegenden Berechnungen bei der Schwerkraftaufbereitung. VDI Zeits 77: 318-320.

55. Bird RB, Stewart WE, Lightfoot EN (2006) Transport Phenomena. (4thedn), Wiley and Sons: New York.

56. Carmichhael LT, Sage BH, Lacey WN, (1955) Diffusion coefficients in hydrocarbon systems: n-Hexane in the gas phase of the methane—, ethane—, and propane—n-hexane systems. AIChE 1: 385-390.

57. Wilke CR, Lee CY (1955) Estimation of Diffusion Coefficients for Gases and Vapors. Ind Eng Chem 47: 1253-1257.

58. Fuller EN, Giddings JC (1965) A comparison of methods for predicting gaseous diffusion coefficients. Journal of Gas Chromatography 3: 222-227.

59. Fuller EN, Schettler PD, Giddings J C (1966) A new method for prediction coefficients of binary gas-phase diffusion Ind Eng Chem 58: 18-27.

60. Fuller EN, Ensley K, Giddings J C (1969) Diffusion of halogenated hydrocarbons in helium. The effect of structure on collision cross sections. J Phys Chem 73: 3679-3685.

61. Green DW, Perry RH (2008) Perry's Chemical Engineer's Handbook. (8thedn), McGraw-Hill: New York.

62. Keller J, Bliesner RD (1990).Sprinkler and Trickle Irrigation. Van Nostrand Reinhold: USA.

63. Salvador R, Bautista-Capetillo C, Burguete J, Zapata N, Serreta A, et al. (2009) A photographic method for drop characterization in agricultural sprinklers. Irrigation Science 27: 307-317.

64. Babinsky E, Sojka PE (2002) Modeling droplet size distribution. Progress in Energy and Combustion Science 28: 303-329.

65. DeBoer DW, Monnens MJ, Kincaid DC (2001) Measurement of sprinkler droplet size. Appl Eng Agric 17: 11-15.

66. DeBoer DW (2002) Drop and Energy Characteristics of a Rotating Spray-Plate Sprinkler. J Irrig Drain Eng 128: 137-146.

12

An Evaluation of the SRI on Increasing Yield, Water Productivity and Profitability; Experiences from TN-IAMWARM Project

Ravichandran VK[1]*, Vibhu Nayar[2], Prakash KC[1]

[1]Agronomy, Tamil Nadu Agricultural University, Chennai, India
[2]IAMWARM, Chennai, India

Abstract

The increasing dependence on groundwater for growing staple food crops like paddy causes lowering of water table and serious depletion of groundwater storage in many parts of India including Tamil Nadu, a southern Indian state. This study is based on field research conducted during the Kharif seasons in 2011-2013 in Villupuram district, Tamil Nadu, to evaluate the impact of TN-IAMWARM Project's adoption of SRI on agronomic productivity and irrigation water use efficiency. The adoption of SRI method in paddy cultivation has resulted in increased by 20% in paddy yield and net income 44.50% over the conventional cultivation. This has been achieved with substantial reduction in irrigation water application (42.33%), labor input (17.46%) and seed cost (87.47%). The economic attractiveness of SRI cultivation is very high, giving farmers a strong incentive to accept water-saving techniques as a new norm for irrigated paddy production. Hence, the cultivation of Kharif paddy (which is a high water consuming crop) through SRI practices promises to be a significant alternative for not only increasing paddy productivity, but also for savings on irrigation water and energy costs in the resource-starved regions of India.

Keywords: SRI; Kharif paddy; Water use efficiency; Ground water potential

Introduction

Tamil Nadu covers 4% of the geographical area (13.01 Million ha) and caters to 5.96% of the population of the country with 7.21 crore people living along the 17 river basins. More than 95% of the surface water potential and 80% of groundwater potential have been put into use. The total water potential of the State including ground water is 47,125 MCM (1664 TMC ft.). The total surface water potential of the State is 24,160 M cum (853 TMC ft) including the contribution (7391 MCM or 261 TMC ft.) from the neighboring States, viz., Kerala, Karnataka and Andhra Pradesh [1]. The annual per capita water availability in India is about 2200 M^3 whereas it is about 750 M^3 in Tamil Nadu. As per World standards (per capita availability - 1000 M^3), our State is under severe water scarcity. It has been assessed that the against the water potential of 47,125 MCM, the agriculture demand alone works out to 49,000 MCM indicating the overdrawal of the ground water resulting in the increase of overexploited blocks (Figure 1). This trend needs to be arrested which is possible only with adoption of new innovative technologies in agriculture practices and diversification of less water intensive crops.

Area irrigated and sources of irrigation

The State's irrigation potential in per capita terms is 0.08 ha when compared to the all-India average of 0.15 ha. The three main sources of irrigation in the State are rivers, tanks and wells [2]. There are about 41,127 tanks, 2239 Kms irrigation main canals and 18.26 lakh irrigation wells in the state. The area irrigated by various sources is furnished in the Figure 2. Among the production constraints, availability of irrigation water is a major one, with consumption of 70% of the water for paddy alone in agriculture. The gap between demand and supply for irrigated crops in Tamil Nadu is projected to reach 21,000 Mm^3 by 2025 [3]. Tamil Nadu State's policy of supplying free electricity to agriculture prompting farmers highly reluctant towards water conservation techniques for agriculture with consequent irrigation inefficiencies.

Under traditional methods of rice cultivation, 3000-5000 liters of water was used to produce one kilogram of rice [3]. A significant portion of the total water requirement for rice production is used for land preparation alone. Farmers have an urgent need for irrigated rice-based systems with technologies that save water by improving water productivity. The intensified efforts to improve both crop and water productivity and subsequently the farmers' income have resulted in many efficient water management practices in wetland rice [4]. The System of Rice Intensification (SRI) is a holistic agro-ecological crop management technique seeking alternatives to the high-input oriented agriculture and also a climatic resilient activity in the change in climate scenario.

Materials and Methods

Farm level investigation was conducted by the researchers in Villupuram district located in the North Eastern part of Tamil Nadu during 2011-2013. From a list of farmers in this district, a random sample of 60 SRI (IAMWARM) farmers and also an equal number of farmers cultivating paddy through conventional method under tube-well irrigation system were drawn from the same blocks (Kandamangalam, Olakkur and Kanai) and interviewed in-depth and collected primary data. The selected farmers were cultivating paddy using both SRI method and conventional practices on fields' side-by-side on similar type of Clay Loam soil [5]. The sampling location was purposively selected due to prevalence of SRI adopters in this area and for ease of monitoring as in this block, SRI cultivation has been promoted for over 6 years by World Bank assisted Tamil Nadu-Irrigated Agriculture Modernization and Water Bodies Restoration and Management (TN-IAMWARM) project.

The mean area under paddy cultivation for the total sample was 0.72 ha. This indicated that most of the farmers in the sample belonged

*Corresponding author: Ravichandran VK, Professor, Agronomy, Tamil Nadu Agricultural University, Chennai, India, E-mail: vkr9999@ yahoo.com

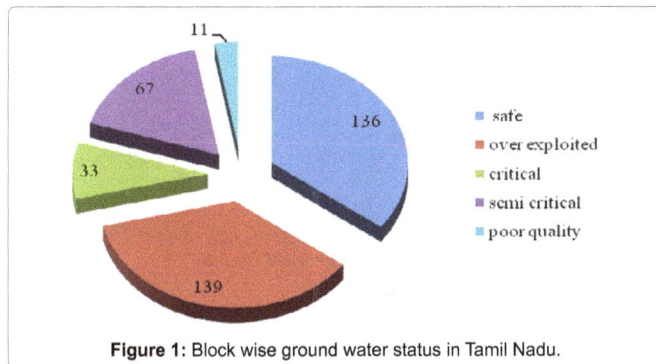

Figure 1: Block wise ground water status in Tamil Nadu.

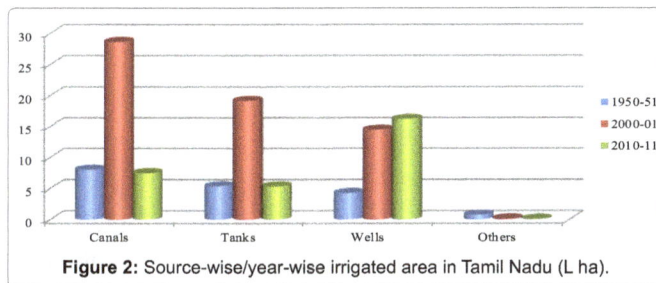

Figure 2: Source-wise/year-wise irrigated area in Tamil Nadu (L ha).

to small and marginal category and there was no significant difference seen in sampled land area between the two sets of farmers (Table 1). At the end of the season, four SRI farmers had to be dropped from the analysis because their data were incomplete. Thus, there were full sets of data for 56 SRI and 60 conventional farmers. The data was collected at an interval of 8-10 days, which coincided with the interval of activities performed by different farmers [6]. The data for labor and other inputs were collected throughout the crop period, starting from seed treatment to measuring the final yield. However, it was difficult to measure irrigation water volumetrically in the field situations. So, to derive a first approximation of the impact of SRI methods on irrigation water use, the number of pumping hours used for growing paddy was selected as a proxy variable to understand the differences in water consumption between SRI and conventional paddy cultivation. Thirty farmers under tube-well irrigation were selected for in-depth study and data generation [7-11]. Of these, 17 were practicing SRI and 13 were conventional paddy farmers. These sample farmers were selected in such a way that their tube-wells represented similar conditions in terms of water pump discharge and power rating. The data was collected from individual farmers on their method of watering, number of irrigations and the pumping hours for each irrigation required in different crop growth stages. Initially the water discharge from each tube well was calculated for a minute using physical measurement [3]. Observations were also recorded for various agronomic practices under the two methods as difference in agronomic practices is hypothesized to generate difference in output [12-16].

Observations on plant growth attributes, yield components and yield were recorded over the study period at regular intervals for both SRI and conventional methods. Field level activities and data measurement were supervised and monitored by the qualified technical staff and researchers of TN-IAMWARM. For comparing various observations taken on water parameters as well as yield under both SRI and conventional methods, mean values of the last three year demonstrations were taken into consideration and standard statistical tools were used to analyze the data for interpretation between two methods of paddy cultivation.

Results

Comparisons of Productivity and Profitability

The overall performance of SRI introduced at the project area of IAMWARM indicated an increase in rice productivity in SRI over the conventional cultivation in all the year. It could be observed from the Table 2. SRI methods significantly showed higher grain yield of 1069 kg/ha than conventional method. The reduction in straw yield was noticed in SRI method.

Irrespective of years, the yield contributing parameters viz., number of tillers per hill, number of productive tillers per hill and number grains per panicle were also higher in SRI than conventional method (Table 3).

It could be observed from the Table 4, that the seed rate was much lower (7.5 kg/ha) in SRI compared to conventional methods (60 kg/ha), which enabled savings of Rs. 1837/-per hectare on seed input, assuming that farmers purchased seed at Rs. 35/kg. The total labor requirement of SRI was less (880 hour/ha) than the conventional method (1000 hour/ha), enabling a considerable saving of Rs. 2359/ha on labor cost (considering the existing daily wage of Rs. 150 /labor in Villupuram district). The major contributor to labor cost in paddy cultivation is transplanting and weeding operations [17,18]. The estimated time for transplanting per hectare in SRI against conventional practices was 378 versus 435 hour/ha. Similarly, SRI also took 100 hour/ha for weeding compared to 122 hour/ha for conventional. Farmers using SRI practices had a 17.46% reduction in their overall labor costs, with disaggregated cost reductions shown in Figure 3. A comparison of net return per hectare, as shown Table 5, indicated that the increase in grain yield had a marked impact on farmers' net return from SRI cultivation. The net return per ha with SRI was significantly higher by Rs. 15548 (paddy grain valued at a constant price of Rs.11/kg), which was 44.5% higher than the conventional method of paddy cultivation.

However, the relative results are consistent with farmers' observations during the study period. Figure 4 summarizes in graphic form of the relative differences between expenditure, yield and economic returns for SRI and conventional.

Water use in SRI and conventional paddy cultivation

An attempt was made to evaluate the differences in water use between SRI and conventional paddy production. The data pertaining to 2011-2013, the years with relatively good rainfall and substantial groundwater recharge. The number of pumping hours was treated as a proxy for actual water use. The selected tube-wells with an electric submersible motor of 10 HP power rating were similar in average discharge of 78,000 litre/hour based on the physical tube-well discharge measurements done by our researchers [14,19,20]. This was an approximate estimation of water use at field level, but even reveled it reasonable relative differences between SRI and conventional practices. Figure 5 presents comparisons of the number of irrigations and the number of pumping hours at each stage of the crop growth, under conventional and SRI management [5,21].

There was not much difference during the land preparation stage, as no special water-saving tillage methods were employed in SRI for the season. Table 6 presents the number of irrigations and number of pumping hours per ha in SRI fields, which were 36.72% less than the conventional paddy. Ravindra et al. [22] and Mahendra Kumar et al. [23] have reported an observed reduction in water use on SRI over conventional paddy by observing through this method. The major irrigation savings with SRI irrigation management was during the

Particulars	Conventional		SRI		Total		Comparison of means F statistic (5%)
	Mean	SD	Mean	SD	Mean	SD	
Total land area	1.83	0.65	1.69	1.11	1.76	0.89	2.65
Total area under Paddy	0.62	0.31	0.78	0.46	0.72	0.41	2.17
Sampled area	0.63	0.32	0.59	0.30	0.60	0.31	N.S

N.S: Not Significant

Table 1: Landholding pattern of the sampled farmers (ha).

Cropping year	Grain yield (kg/ha)		Straw yield (kg./ha)	
	Conventional	SRI	Conventional	SRI
2011	5238	6285	2749	2347
2012	5415	6437	2823	2410
2013	5381	6519	2989	2491
Mean	5345	6414	2853	2416
SRI over conventional	1069 (20%)		-437 (-18%)	
SD	0.93	1.18	1.22	0.97
F statistic (5%)	2.30		N.S	

Notes: Figures in brackets are the differences in per cent between SRI and conventional results.

Table 2: Average productivity of paddy under conventional and SRI methods.

Year	No. of tillers per hill		No. of productive tillers per hill		No. of grains per panicle	
	SRI	Conventional	SRI	Conventional	SRI	Conventional
2011	23.5	11.2	19.3	8.4	138	127
2012	25.8	12.5	21.2	9.4	152	134
2013	24.6	11.8	20.2	8.9	145	128
Average	24.6	11.8	20.2	8.9	145	129

Table 3: Performance of yield components in SRI.

Input	Conventional (Mean)	SRI (Mean)	SRI over conventional	
			Amount	%
Land preparation	2005	1955	50	2.49
Seed	2100	263	-1837	- 87.47
Fertilizer	6996	7254	258	3.68
Pesticides	680	660	-20	-2.94
Labour	13,505	11,146	-2359	-17.46
Total expenditure	25,286	21,270	-4008	-15.85

Table 4: Input expenditures (Rs./ha).

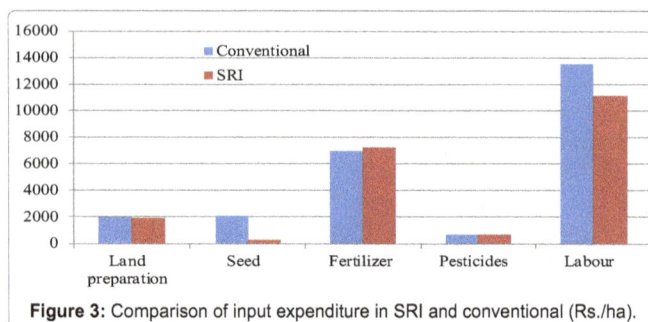

Figure 3: Comparison of input expenditure in SRI and conventional (Rs./ha).

nursery, weeding and upto panicle initiation stages. Adherence to all of the principles of SRI, with only slight variations, by giving higher yields and better returns would reinforce motivation for controlling irrigation and making savings in irrigation water. The cumulative irrigation water application (Figure 6) showed a large potential with SRI for reducing the quantum of water use and also to bring in purposeful management in irrigation water usage. Our study was also noted that each ha of SRI saved about 571.4 pumping hours in one season compared to the conventional method. This amounted to 3028 kWh of savings in electricity consumption, which was currently fully subsidized by the state. The state will be able to save about Rs. 12,112 on every ha of paddy (cost of power- Rs. 4 per unit kWh) converting to SRI management. The preliminary results of the study were very promising, but irrigation management with SRI needed more detailed systematic study to determine how to make use of SRI as a lever for introducing management reforms in groundwater irrigation system.

Discussion

SRI versus conventional

Evidence from the sample of farmers showed 20% yield advantage in SRI over conventional method, indicating better adoption of standard practices by SRI among farmers. In selected blocks of Villupuram district, prospective farmers were initially trained and demonstrations were conducted for adoption SRI under direct supervision of the scientists and researchers of IAMWARM project. Efficient utilization of externally applied nutrients with more foraging area of root volume along with intermittent irrigation in SRI plots enhanced the growth of tillers, root development, number of productive tillers and % of grain filling over conventional practices, which ultimately reflected on the grain yield of paddy [11,21]. Similarly from the field observations,

Parameter	Conventional (Mean)	SRI (Mean)	SRI over Conventional Difference	Comparison of means:Conventional and SRI F statistic (5%)
Total expenditure (Rs./ha)	25,286	21,278	-4008 (-18.83)	N.S
Grain value (Rs/ha)	58, 795	70,554	11,759 (20.00)	3.13
Straw value (Rs./ha)	1427	1208	-219 (-18.13)	N.S
Gross return (Rs./ha)	60,222	71,762	7264 (19.16)	2.43
Net return (Rs./ha)	34,936	50,484	15,548 (44.50)	2.75

Notes: Figures in brackets are the differences in per cent SRI over conventional difference.

Table 5: Differences in expenditure, yield and economic returns for SRI and conventional methods.

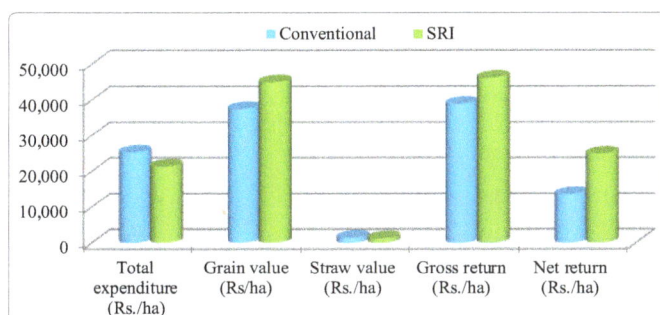

Figure 4: Comparison of expenditure and returns in SRI and conventional method.

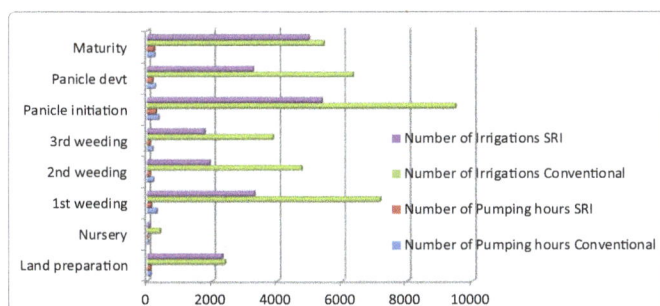

Figure 5: Comparisons of Tubewell irrigation between SRI and conventional method, according to different crop growth stage.

the innovations in the method of using invigorated younger seedling have better crop establishment and positive yield components over conventional practices (Table 3). This favorable influence might be due to efficient utilization of resources and less inter-and-intra space competition under SRI management, which may be responsible for yield attributes of rice and consequently increased yield [4].

SRI system of rice cultivation enhanced the labor productivity substantially with far higher net income than traditional cultivation of rice [6]. Present study suggested 17.46% reduction in labor cost than conventional method (Table 4). The direct economic benefits of SRI included lower seed rates, lesser nursery area lower expenses on labor and higher output of paddy grain. High reduction of seed cost (87.47%) in SRI gave farmers enough economic incentive to adopt this method. SRI allows each plant to be better exposed to sunlight and circulating air and this promotes the 'edge effect' besides dropping canopy humidity with change in micro climate an added advantage for disease reduction. It also includes the mutual/synergistic effects of all the components of SRI such as younger seedling age, wider spacing, better soil aeration through the use of weeder etc. [10,17]. Field study

also evaluated that, lowering cost of pesticides due to less incidence of disease in SRI practices might be attributed to wider plant spacing, less stagnation of water in fields and better aeration and light penetration. Besides the rodent nuisance is also reduced due to more land areas exposed to sun because of line planting.

Water saving effects of SRI

In the study area, the SRI farmers followed intermittent irrigation with Alternate wetting and drying cycle compared to continuous flooding of fields (>10 cm standing water) by conventional paddy farmers. This practice has led to a substantial reduction of irrigation numbers, pumping hours and overall water usage in paddy under SRI management (Table 5). The water consumption in conventional paddy fields was nearly two times greater than the normally suggested crop-water requirement of 12,500 m^3/ha in paddy. Zhao et al. [21] found 40- 47% reduction in water-use with SRI, 68-94% increase in water use efficiency (WUE) and 100-130% increase in irrigation WUE compared to traditional flooding.

Water saving during nursery preparation

The size of nursery bed and duration of seedling in nursery were relatively more in conventional than SRI. For conventional, it was necessary to supply more irrigation water continuously to large size nursery beds (0.032 ha nursery required for transplanting 1ha land) for a period of one month (as seedlings transplanted at 25-30 days age). The total amount of irrigation water supplied during nursery operation was estimated to about 390 m^3/ha. However, only 16% of this amount of water was required for SRI nursery management.

Water saving during weeding

In conventional paddy cultivation, the flooding is being practiced to suppress the weed growth up to 45–50 DAP. However, with SRI cultivation, weeds are incorporated into the soil by way of mechanical weeder which helps to build the organic matter in the soil and subsequently the large and diverse microbial population in the soil. Thus mechanical weeding operation facilitates the process of aeration in the soil and provides soil churning effect and pruning the older roots facilitating plants to produce new roots which help in the uptake of enhanced nutrients. This in turn mobilizes the micro nutrients required for the healthy growth of the rice plant. Under intermittent irrigation more availability of soil Phosphorus [15] favors root growth. Hence, SRI cultivation saved water by lowering 42% water consumption requirements and applying AWD without affecting the crop yield. This was supported by previous studies of Uphoff et al. [19] Ravindra et al. [22]. It had a great significance particularly in paddy cultivation which caused high absorption of groundwater and a serious depletion of groundwater storage in many locations of the state. This has helped

Crop stage	Number of pumping hours		Water application (m³/ha)		Cumulative water application(m³/ha	
	Conventional	SRI	Conventional	SRI	Conventional	SRI
Land preparation	98.7	89.3	2388	2298	2388	2298
Nursery	14.1	3.4	390	62	2778	2360
1st weeding	289.7	111.3	7166	3294	9944	5654
2nd weeding	178.7	78.9	4738	1925	14682	7579
3rd weeding	163.8	72.4	3862	1763	18544	9342
Panicle initiation	343.7	262.3	9458	5358	28002	14700
Panicle development	239.4	158.6	6325	3254	34327	17954
Maturity	227.8	208.3	5428	4969	39755	22923
Total	1555.9	984.5	39755	22923		
Reduction between SRI and conventional	571.4 (36.72%)		16832 (42.33%)			
Water productivity					0.13	0.27

Table 6: Estimated water applications in SRI and conventional paddy cultivation.

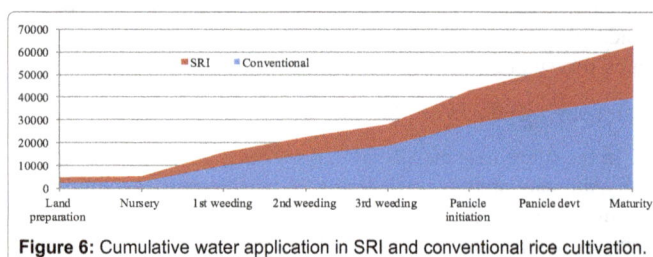

Figure 6: Cumulative water application in SRI and conventional rice cultivation.

farmers to reduce 42% of their irrigation during the crop season and still get higher yield and economic return. Moreover, the burden of power subsidy of the state would be reduced by saving pumping hours with SRI.

Adherence to all of the principles of SRI, with only slight variations guarantees higher yield with better returns reinforcing motivation for controlling irrigation and saving irrigation water. This further encourages farmers to extend their area under paddy with the savings made in water with SRI. Therefore there is more scope for expansion of SRI method, as it addresses the issues of excessive groundwater exploitation which resulted in lowering the water table and polluting water with salinity and arsenic [10]. But a majority of farmers in the tube-well commands are still wasting water by practicing continuous flooding. In this context, there is an urgent need for local institutions to take charge and initiate attitudinal and informational changes among the farmers at the village level for effective water management and more efficient use of valuable water resources.

Formation of effective more functional root system with abundant foraging area resulted in efficient absorption of externally applied nutrients and through positive physiological factors favoring yield components and yield in SRI paddy compared to conventional practice. The reduced straw yield (Table 2) under SRI reflects that efficient translocation of photosynthates to economic produce i.e grain from the vegetative portion.

Conclusion and Policy Implication

It is widely believed that one of the world's major staple foods, rice, is also one of the larger contributors to water scarcity. The search for alternative ways of growing rice, in a manner that substantially reduces water use resulted in the identification of SRI as one of the important alternative.

Firstly, SRI methods can produce significantly higher paddy yield with lower production costs (seeds, pesticides, labors) than

conventional practices, therefore, generating higher profits to farmers. Higher paddy yield obtained with SRI cultivation is the result of the combined effects of

(a) SRI transplanting methods,

(b) Mechanical weeding operation and

(c) Intermittent irrigation.

Secondly, the considerable savings of water with SRI at field level results in substantial savings of electric energy. The reduced extraction of groundwater and increased water productivity in SRI would be an additional benefit, having long-term implications for maintaining groundwater reserves which are declining faster due to over use, especially for the cultivation of paddy. The opportunity for making savings in electricity, groundwater and lower production costs (seeds, pesticides, labors) can justify more robust planning efforts for the promotion of SRI to engender systemic improvements in paddy production and efficient water resource management at the macro level. Such a policy measure is also necessary for improving the food security situation of India as a combination of irrigation and paddy production systems improvement. By reviewing the results of some of the studies across the globe and the experience in Villupuram district of Tamil Nadu, India, we find that SRI uses less water and fewer inputs including energy; reduces costs substantially and results in higher yields compared with conventional cultivation practices [11-16]. There is substantial net reduction in water use in SRI rice cultivation under a controlled water regime as compared to conventional practice [18].

In spite of these outstanding positive findings, not only validated at the field level in our own research which corroborates that of other scientists, but also widely recognized by national, state and local governments, civil society organizations and small-marginal farmers, the spread of SRI to newer rice growing areas is extremely slow. It has failed to make any significant effect on conventional practices and technologies.

Obstacles like the need to follow rigid, time-bound practices, the shift to relatively monotonous isolated work like mechanical weeding, are shown to be not insurmountable. Ingenious modifications to tools and practices have to be invented. But a further array of factors such as:

- The lack of resources for research and development in breeding appropriate varieties to overcome the rigid short-duration transplanting schedule,

- Facilitating the availability of Laser leveling machines at nominal cost to maintain uniform thin film of water,

- The lack of the appropriate type of weeder including simple mechanized ones that would remove the psychological strain from using the current designs of weeders,

- Proper IEC (Information, Education and Communication) measures for farmer to farmer exchange through FFS, exposure visits and using media as a tool for propagating success stories to local communities

- Reduced subsidies on irrigation water will be important cost savings for government and provides a real incentive for farmers to use SRI principles. Investments in training and extension for SRI could be covered in part by reducing present subsidies on water, fertilizer, and electricity for pumping water and

- Political resistance to adopt a framework to integrate training in SRI practices with TN-IAMWARM scientists so as to overcome certain perceived skill deficiencies

- Initially practicing of SRI should not be forced on farmers by the way of target to Agricultural ministry officials.

- SRI should not be advocated to all paddy growing areas. The extensions personnels should identify the right area highly suited for SRI

- The water resources department engineers and the Agriculture Officers shall mutually discuss and adopt water saving technique by way of effecting turn system of water delivery, in the canal commend area.

- During initial days of introduction of SRI critical inputs were given at free of cost to enable the farmer to apply necessary inputs in time to reap more benefit and to reduce the risk factor or fear of adoption. Now it is the right time to withdraw in phases the subsidy for SRI.

It is evident now that only 20% of adopters of SRI take to all core practices of SRI and the balance 80% are either partial or low adopters [14]. So, a commitment to adopt SRI on a large scale, if done, can save money for governments by reducing budget allocations needed for food imports and agricultural subsidies. For governments to maximize cost savings and efficiencies in national water use and energy, they will need to invest in better control over the delivery of irrigation water. This includes the design and management of water resources. It may also require institutional reforms and capacity building for water users to readjust water delivery mechanisms and schedules for more precise water allocation within irrigation schemes.

References

1. Rama Rao IVY (2011) Estimation of Efficiency. Sustainability and Constraints in SRI (System of Rice Intensification) vis-a-vis Traditional Methods of Paddy Cultivation in North Coastal Zone of Andhra Pradesh. Agricultural Economics Research Review 24: 325-331.

2. Palanisami K, karunakaran, Upali Amarasinghe, Ranganathan CR (2013) Doing Different Things or Doing It Differently: Rice Intensification Practices in 13 States of India. Economic and Political Weekly 8: 51-67.

3. ICRISAT-WWF Project (2009) SRI Fact Sheet, International Crops Research Institute for Semi-Arid Tropics. India.

4. Palanisami K, Paramasivam P (2000) Water scenario in Tamil Nadu-present and future. Tamil Nadu Agricultural University, Coimbatore: 36-46.

5. Dass Anchal, Chandra S (2013) Effect of different components of SRI on yield, quality, nutrient accumulation and economics of rice (Oryza sativa) in tarai belt of northern, India. Ind J Agron 57: 250-254.

6. Kassam Amir, Willem Stoop, Norman Uphoff (2011) Review of SRI modifications in rice crop and water managementand research issues for making further improvementsin agricultural and water productivity. Paddy and Water Environment 9: 163-180.

7. Bruno A (2002) Evaluation of the system of rice intensification in Fianarantso Province of Madagascar. Cornnel International Institute for Food, Agriculture and Development (CIIFAD). Ithaca.

8. Lin Xianqing, Defeng Zhu, Xinjun Lin (2011) Effects of water management and organic fertilization with SRI crop practices on hybrid rice performance and rhizospheredynamics. Paddy and Water Environment 9: 33-39.

9. Mishra A, salokhe VM (2008) Seedling characteristics and the early growth of transplanted rice under different water regimes. Exp Agri 44: 1-19.

10. Mishra A, Salokhe V (2010) The effects of planting pattern and water regime on root morphology, physiology and grain yield of rice. J Agron Crop Sci 196: 368-378.

11. Pandian BJ, Rajkumar D, Chellamuthu (2011) System of Rice Intensification: A Synthesis of Scientific Experiments and Experiences. system of Rice Intensification.

12. Pasuquin E, Lafarge T, Tubana B (2008) Transplanting young seedlings in irrigated rice fields: early and high tillering production enhanced grain yield. Field Crops Res 105: 141-155.

13. Quin Yanmei, Shuwei Liu, YanqinGuo, Qiaohui Liu, JianwenZou, et al. (2011) Methane and nitrous oxide emissions from organic and conventional rice cropping systems in Southeast China. Biology and Fertility of Soils 46: 825-834.

14. San-oh Y, Sugiyama T, Yoshita D, Ookawa T, Hirasawa T, et al. (2006) The effect of planting pattern on the rate of photosynthesis and related processes during ripening in rice plants. Fields Crops Res 96: 113-124.

15. Turner, Haygarlh (2001) Phosphorus solubilization in rewetted soils. Nature 411: 258.

16. Udyakumar (2005) Studies on system of rice intensification (SRI) for seed yield and seed quality. M.Sc. (Agri.) Thesis, Acharya N.G. Ranga Agricultural University, Hyderabad, India.

17. Uphoff N (2004) System of rice intensification responds to 21stcentury needs. Rice Today 42.

18. Uphoff, Norman (2007) Envisioning 'Post-Modern Agriculture': A Thematic Research Paper. Cornell Univerisity, Ithacca: Accessed through WAASAN Website.

19. Uphoff N, Kassam A, Thakur A (2013) Challenges of increasing water saving and water productivity in the rice sector: introduction to the system of rice intensification (SRI) and This Issue. Taiwan Water Conservancy 61: 1-13.

20. Yang J, Liu K, Wang Z, Du Y, Yang J, et al. (2007) Water-saving and high-yielding irrigation for lowland rice by controlling limiting values of soil water potential. J Integr Plant Biol 49: 1445-1454.

21. Zhao LM, Wu LH, Li YS, Lu XH, Zhu DF, et al. (2009) Influence of the system of rice intensification on rice yield and nitrogen and water use efficiency with different N application rates. Exp Agric 45: 275-286.

22. Ravindra, Asusumilli, Bhagya Laxmi S (2010) Potential of the System of Rice Intensification for Systemic Improvement in Rice Production and Water Use: The Case of Andhra Pradesh. India. Paddy and Water Environment 9: 89-97.

23. Mahendra Kumar R, Surekha K, Padmavathi CH, Subba Rao LV Latha PC, et al. (2010) Research experiences on system of rice intensification and future directions. J Rice Res 2: 61-71.

13

Comparative Analysis of Water Saving Techniques for Irrigating More Land with Less Water in Nguruman Scheme, Kenya: Design Principles and Practices

Muya EM[1]*, Sijali IV[1], Radiro M[1] and Okoth PFZ[2]

[1]Food Crop Research Institute, KALRO-Kabete, Kenya
[2]International Institute for Rural Technologies, Nairobi, Kenya

Abstract

A review of the past studies has demonstrated that rehabilitation of an irrigation scheme through infrastructural development and agricultural production value chains alone will not sustain the targeted agricultural production unless scientific principles, aimed at enhanced water use efficiency are incorporated in the development agenda to avoid future water losses and shortages. In Nguruman, the available water supply has the capacity to meet irrigation water requirement of 5,332,809 m^3 to irrigate the targeted 800 ha per year. However, with increasing number of farmers engaging in irrigated agriculture, more area is expected to be brought under irrigation to meet the increasing food demands, thus requiring significant amount of water savings through enhanced water use efficiency. For this reason, analysis and review of the irrigation design challenges, operational realities and technological options for improved water use efficiency was carried out in terms of the quantities of water saved and additional area that can be irrigated by each of the technological options considered in relation to the farmers' irrigation methods. The technological options reviewed were: drip plus digital instruments; drip plus analog instruments; sprinkler plus digital instruments; sprinkler plus analog instruments and farmers' irrigation method. It was deduced from literature review that drip plus digital instruments, drip plus analog instruments, sprinkler plus digital instruments and sprinkler plus analog instruments could save 465,600, 433,132, 365,520 and 323,637 m^3 of irrigation water respectively relative to the farmers' method. The additional area to be irrigated using water saved from drip plus digital instrument, drip plus analog instrument, sprinkler plus digital instrument and sprinkler plus analog instrument was 146.3, 136.7, 115.4 and 104.1 ha respectively. Since drip irrigation plus digital instrument had the highest water saving potential, it was identified as the most appropriate technology to be used tested, validated and applied to irrigate more land with less water in Nguruman Irrigation Scheme.

Keywords: Water savings; Irrigation technologies; Efficiency

Introduction

Study background

Nguruman Irrigation Scheme is one of the nine schemes targeted for development by the Small Scale Horticultural Development Project (SHDP), supported by African Development Bank. One of the main objectives of the project is rehabilitation and infrastructural development of the schemes through appropriate design followed by construction [1]. However, design, construction and entire infrastructural development will sustain the agricultural production with the limited water supply in the long-run if important scientific principles are incorporated in the irrigation development agenda [2]. The scientific approach is based on the production objective of optimizing crop yields through improved irrigation efficiency, focusing mainly on efforts to maintain appropriate soil-water-plant relationships [3]. As a result of extensive research into these relationships, water requirements by crops at various stages of growth under different soil and climatic conditions can be determined with great accuracy [4,5]. This development can enable the designers, planners or irrigation development agencies to assess the scheme water supply requirement in relation to available water from the source, the capacities of water distribution systems and design the operation of water supply to the irrigated field, based on accurate information on the field irrigation scheduling, aimed at ensuring the use of limited water with the designed irrigation efficiencies [6]. In order to understand the relationships between the design assumptions, irrigation schedules and operation reality and their effects on water use efficiency, Professional Training Consultants suggested an intimate interactions between all parties involved in irrigation development and identified four main parties, namely: planners/designers (e.g. consultants, donors or development

agencies); technical and operational unit (District Agricultural and/ or Irrigation Office); Scientists, equipped with scientific principles (e.g. Kenya Agricultural and Livestock Research Organization) and operational field staff and farmers at tertiary and secondary levels [7]. The study focuses on water savings to irrigate more land with less water through the study of the interactions between these parties during planning, design and operational stage of irrigation development.

Study focus

Water saving for irrigating more land is to be achieved by planning, designing and operating an irrigation system using adequate information and data input into each of the three important stages of irrigation development. The underlining objective of sound irrigation development and practices is to attain the quantity and quality of agricultural products desired in the market, while, at the same time, maintaining or enhancing the quality of soils and environment to ensure the biophysical sustainability of the designed system [8]. Planning stage entails determination of seasonal and peak scheme water supply requirements to use in estimating how much actual acreage that can be irrigated by water available from the source

*Corresponding author: Muya EM, Food Crop Research Institute, KALRO-Kabete, Kenya, E-mail: edwardmuya2011@gmail.com

(irrigation potential), followed by determination of the irrigable area to be matched with irrigation potential determined. To derive data for the design and operation of irrigation system, a detailed evaluation of water supply schedules is done, starting from the lowest irrigation unit and subsequently include the block of fields, served by tertiary system, area served by laterals, and eventually the entire project area, served by the sub-mains and mains. Detailed design followed by the construction of an irrigation system is not yet a conclusion of the irrigation development. No irrigation system will function perfectly the day it is commissioned and becomes operational [4]. The operation criteria must be developed immediately after the construction, based on the field irrigation supply schedules, i.e. size, duration and interval of supply, and method of supply (rotation, on demand or continuous). Development of operation criteria considers the supply schedules for individual fields and subsequently blocks of fields, the area served by lateral and the main canals Muchangi et al., [9]. It is a common knowledge that variations may occur in cropping systems and area being served, hence the need for the supply to be regulated accordingly. Therefore, based on the supply schedules, the capacities of water supply and regulating structures should be determined together with organizational framework for operating and maintaining the system through involvement of the key stakeholders, including the Irrigation Water User Association, Extension workers, farmers and Soil and Water Engineers. Against this background, the objective of this paper is to explore ways and techniques for enhanced water savings through examination of the data input, challenges and technological options in different phases of irrigation development, namely: planning, design and operation with specific reference to Nguruman Irrigation Scheme, Kenya.

Conceptual Framework

How efficient an irrigation system would be or how much water is saved by improving the irrigation efficiency depends not only on the type and sufficiency of the data collected at each stage of development, but also the accuracy and timeliness of the collected data [10]. The data collected and put into each stage of development are mainly on water, soils and crops. Based on these data, average and peak water supply, field irrigation schedules and field water budgets and balances are derived for the planning, designing and operation stage respectively [4]. This involves analysis of climatic data, water availability at the proposed intake and soil quality at planning stage (National Irrigation Board of Kenya, field water balance and permissible soil water deficit at design stage and operational parameters at the implementation level [3,11].

Stages of Irrigation Development, Challenges and their Redress

Planning stage, requirements and constraints

At the planning stage, a comprehensive inventory of available resources is done, involving the determination of the physical, chemical, hydrological characteristics of soils as well as climatic attributes and preliminary cropping patterns for the estimation of the irrigation potential of the project area. Two important results of these studies as far as water saving is concerned is how much acreage can be irrigated by the available water and how much is the irrigable area [8]. Crop water requirement is an essential component of planning and developing an irrigation system because of its application in the preliminary, design and operation stage [12]. At the preliminary planning stage, the crop water requirement is applied to calculate seasonal and peak water demands for a given cropping pattern which are needed to provide

the basis of determining the total acreage that can be irrigated by the available water supply from the source.

In Nguruman Irrigation Scheme, Small Scale Horticultural Development Project targets 800 ha (40% of total irrigable land) to be irrigated [13]. The annual water requirement for irrigating 800 ha is 5,333,809 m³, which can be met by the available water from the source (28 million m³) for the proposed cropping pattern comprising pasture, beans, maize, onions, water melon, and banana in phase one [14]. However, more area is likely to be brought under irrigation in the near future, considering the growing population, and the fact that 60% of the area is still available for irrigation. This would lead to a serious water deficit unless appropriate water saving techniques are identified to improve the irrigation efficiency from 55% to over 95% [15]. According to Muya et al. low irrigation efficiency of the current system, determination of net irrigation requirements without measured data on field water balance, and using climatic data derived from meteorological station which is Fourty kilometer away from the scheme, resulted into erroneous calculation of the irrigation water supply requirement, thereby, causing, not only reduction in the quantity and quality of agricultural products (Box 1), but also degraded soil and environment (Plate 4) [8]. The quantity and the quality of agricultural products are determined by the capacity of the system not only to meet the required agricultural inputs, but also to control pests, diseases and adverse interactions between the applied inputs and environments [16]. In principle, precision agriculture tries to fine-tune land management practices with an objective of maximizing agricultural production and its quality, while minimizing the environmental side effects. This is to be achieved by satisfying the immediate plant needs such that, what is good for business is also good for environment. For instance, accurate measurements of the exact levels of water and nutrients required at a given stage of growth is need to determine the quantities of inputs that would interact optimally to give the right size, appearance and taste of the products required in the market. For example, too much water would result into production of okra which is too long and rough textured, while too little would give small and coiled products, which are thrown away during the sorting out processes, hence a lot waste to the farmers [3]. Accurate measurements would also result into improved environmental quality and sustainability and reduced land degradation [16]. Low irrigation efficiency, not only results into excess capillary rise of the ground, salty water to the root zone, but also causes increased volume of drainage water (agricultural waste water) that finds its way into Ewaso Ngiro River water (Plate 5) [8]. In 1993, the Independent Board of Consultants cautioned that irrigation in the vicinity of Ewaso Ngiro river should be accompanied by appropriate mechanisms that limit the excess irrigation water into the river. This is because the degrading buffer zones are continually losing their capacity to filter the water flowing from the scheme, with increased movement of unfiltered drainage water into the river water [17]. The consequence of this contamination would be the extinction of the flamingos in Lake Natron where the river ends, hence loss of tourist attraction. Therefore, appropriate instruments, irrigation methods and practices must carefully be considered at the preliminary stage to ensure enhanced environmental quality and sustainable agricultural production [17].

The design stage: data required and challenges

In Nguruman Irrigation Scheme 800 ha is targeted (in three phases) under the assumed cropping patterns indicated in Table 1 to be irrigated using annual water supply of 5,299,479 m³. The discharge of water from the headworks through mains, sub-mains and laterals into tertiary system and irrigated fields, are manually operated and

Phase I		Phase II		Phase III	
Crops	Area (ha)	Crops	Area (ha)	Crops	Area (ha)
Beans	40	Soybeans	65	Green grams	30
Onions	40	Ravaya	80	Okra	20
Water melon	40	Tomatoes	20	Karella	80
Pasture	40	Pasture	40	Pasture	40
Bananas	40	Mangoes	30	Lemons	40
Maize	65	Sorghum	35	Sweet potatoes	55
Total acreage	265		270		265

Table 1: Assumed cropping patterns, based on the farmers' preference.
Source: Muya et al. [14].

hardly any measurement is done [8]. According to Shanan [10], irrigation system is designed to facilitate regulation and measurement of water flows from the point of abstraction through the distribution system into the irrigated field units. To derive the data for design and operation of the irrigation distribution system, a detailed evaluation of the supply schedules is made [4]. This should start from the lowest and furthest irrigation field, through blocks of field, served by the tertiary, sub-areas, served by the laterals and the entire project area, served by the mains [6]. The detailed design should be followed by putting down the operation criteria of the system which are based on the irrigation supply schedules, i.e. size, duration and interval of the supply as well as the method of application (rotation, on demand or continuous supply) as is demonstrated by Withers and Vipond [18]. Rehabilitation, infrastructural improvement and completion of the construction of any irrigation scheme will have positive impacts in terms of enhanced water savings and irrigation efficiency if the system design is done in consideration of soil-water-plant-relationships and subsequent irrigation scheduling [7]. Determination of these relationships will result into appropriate system design that not only sustains the reasonable irrigation efficiency, but also the infrastructural stability including water distribution structures [19]. Hillel indicated that the soil aggregate stability can be sustained through appropriate soil water management strategies that must be put in place prior to any agricultural development [20].

The water supply requirements at field level are determined by the depth and interval of irrigation, where the supply requirements of the individual fields have to be expressed in flow rates or stream size, shown in the following equation that determines the quantitative aspects of soil-water-plant relationships:

Where: $qt = 10 \dfrac{P.Sa.D.A}{Ea}$

q=Stream size in m^3/s

t=Supply duration in seconds

P=Fraction of the available water permitting unrestricted evapotranspiration

Sa=Total available soil water in mm/m soil depth

A=Acreage of a specified crop in the assumed cropping patterns

The relative values of q and t in the in the soil-water-plant relationships have to be determined as a measure of water uptake and retention capacities and should be the basis of designing the stream size such that it may, not only be easily handled by the irrigators, but also causing no water wastage through high volume of run-off into the drainage system, deep percolation and surface ponding with eventual losses through evaporation and non-destructive to the soil.

The biggest challenge that requires redress is the fact that the actual water distribution to the groups of farmers in most irrigation scheme deviates from the irrigation supply schedules designed, based on the soil-water-plant relationships and efficiency considerations [10]. This means that the procedures on the regulation, measurement and monitoring of the water flows are either not provided as guidelines after the design and construction of the scheme or provided in the design report but ignored by the field extension staff and farmers. This discrepancy has a serious implication on the amount of water saved or lost measured by the quality of agricultural product and water use efficiency [21]. In Nguruman Irrigation Scheme, the irrigation methods currently practiced are not based on the design arrangement. The irrigation interval is once a week and water application duration ranges from 4 to 12 hours. These irrigation practices are based on the rule of thumb rather than soil and crop water demands [8]. Therefore, the main cause of low irrigation efficiency is uniform application of irrigation water, using the same irrigation methods and scheduling, throughout the scheme by most farmers, disregarding the differences in physical, hydraulic and chemical properties of soils that influence the choice of irrigation methods and scheduling [8]. Horst claims that one of the factors contributing to this problem is that the irrigation schedules, based on soil-water-plant relationships hardly feature in the design report, thereby, deliberately, assuming the farmers' practices, and how the irrigation manager, agencies or Water User Association will operate the irrigation system after the design and construction of the irrigation infrastructure [2,3]. It is normally taken for guaranteed the irrigation operators have the necessary skill to manage the scheme as per the design. This problem may be solved by understanding the relationships between these design assumptions, irrigation schedules and the operation reality. This can be achieved by bringing out clearly the picture of what happens to the four important parties engaged in irrigation design, development and practices as well as their role and challenges (Table 2).

Design assumption and their consequences from long-term perspectives

The systematic design procedures given by FAO were aim at translating the production objectives into adequate technical planning criteria [4]. This involves collection of the required information on water, soils, climate and crops. Based on the collected data, tentative plan is prepared, followed by the search for optimal plan through systematic analysis of the serial modifications of the tentative plan. In summary, FAO [4], describes the design procedures as comprising the following stages: (1) project identification and preliminary stage; (2) project design and (3) project operation. In practice, the design and construction of irrigation systems are normally contractual undertakings which are pressured by limited time, financial resources and timely availability of relevant technical expertise in diversified fields [2]. This often leads to deliberate assumption of important input-

Parties involved and challenges	Their role and problems
Research scientists	Delineate the extent and distribution of soil types together with their physical and chemical characteristics, particularly water-related properties including the fraction of the available water that permits unrestricted evapotranspiration; identify irrigable areas and clusters of production units in terms of their constraints/potentials/limitation to be superimposed on the detailed irrigation system design.
Planners, designers, or development agencies and projects (e.g. Small Scale Horticultural development Project)	Design the irrigation system on irrigation areas to facilitate regulation and measurement from the intake to the irrigated fields, based on soil-water-plant relationships and efficiency; and ensure that the supply system is at permissible level and satisfying the momentary irrigation requirements of each crop and area in terms of stream size, duration and interval of supply.
Operational staff (e.g. District Agricultural Officer or Irrigation Engineer).	Mainly concerned with water allocation and distribution scheduling.
Field staff, Irrigation Water User Association and farmers	Determine how the water is distributed in reality. The actual water distributed is the product of interactions between the field staff and farmers.
Challenges	Often there is no effective communication between the research scientists and the designers, who develop the detailed layout design without regards to the soil investigation and clustering results; designing irrigation schedules, based on soil-plant-water relationships requires enormous amount of data collection, processing and dissemination to arrive at stepwise-procedures and manual on the guidelines on the operation of the irrigation systems, based on the appropriate water supply scheduling.

Table 2: The important parties involved in irrigation development.
Source: Horst and Muya et al. [2,14].

variables required in each stage of design [3]. One of the most severely affected areas is soil resource inventory, hydraulic characterization, and clustering production/land units, required for preliminary, design and operation stage respectively. For example, lack of adequate soil resources inventory at the planning stage would lead to unclear boundary between irrigable and non-irrigable areas, and consequently, the design capacities of the water distribution and control structures would be based on the assumption that all the project area will be irrigated [21]. This is particularly the case if the preliminary irrigation layout was not superimposed (during design) on delineated clusters of soil units, with each cluster having well defined irrigation potentials, limitations, productivity index for envisaged crops and detailed hydraulic characterization. The detailed hydraulic characterization is required for the determination of the field irrigation schedules, field irrigation supply calendar, and field irrigation layout and water distribution plan [4]. According to Muchangi et al. lack of these important soil data would lead to the design of the capacity of water supply structure without considering the soil and soil-related variables that influence the operation of an irrigation system [9]. These include: the (net and gross) depth of water application, available soil moisture, allowable soil water depletion level, effective rootzone depth, irrigation cycles, number of irrigation shifts per day and irrigation time per shift. This may lead to over or under irrigation in the long-run.

Water saving through redress of the discrepancy between the designed irrigation schedules and operational realities

The problem of incompatibility between the designed irrigation schedules and operational reality can be addressed by sticking to the principle that irrigation practices imply optimization of production. By keeping the soil-water-plant relationships and water use efficiency as the main focus of the design, water saving can be achieved, based on measurement by newly developed techniques (Horst, 1996). The accurately measured data will facilitate the use of water stress indicators, water-yield functions and simulation models, aimed at achieving economic yield of high quality product [22]. The techniques include automation of the irrigation system. Automatically controlled irrigation system, either hydraulically, electronically or electro-magnetically by micro-processors or computers will be more efficient with less labour inputs [23]. For determination and monitoring soil water balance, web based controlled instruments remotely operate the irrigation system with chain interface on board, allowing connections

soil moisture sensors [24,25]. Application of these techniques can result into increased water saving and improved irrigation efficiency if they are developed through multi-level stakeholders' platforms including the irrigators. Stakeholders on the ground, especially farmers, would adopt the recommended practices if they are demonstrated as addressing practical problem issues. For example, in Nguruman Irrigation Scheme, one of the most critical problem is low product quality associated with non-optimal interactions between the applied irrigation water and available nutrients. Okra is one of the most commonly grown vegetable for direct export market, and has, since suffered massive rejection due to low quality (Box 1).

The design of water supply structures and schedules without considering soil and soil-related variables may have adverse consequences in the long-run, particularly in Nguruman Irrigation Scheme, if not addressed immediately after commissioning the project. This is a justified course of action because no irrigation system worked perfectly the day it was commissioned. According FAO, the function of an irrigation system is to satisfy as much as possible, the momentary irrigation requirements of each crop and each area in terms of stream size (Q), duration (T) and the interval of supply (I) and expressed in the following equation after [4,9]:

Where: $Q = 10\dfrac{A.dgross}{I.Ns.T}$

Q = System capacity (m³/hr)

A = Design area (Ha)

Dgross=Gross depth of water application (mm)

NS=Number of shifts per day

T=Irrigation time per shift

For the quality and quantity of the agricultural products to be maintained at the required level without huge losses, a combination of appropriate irrigation water application method, appropriate water measuring techniques and reasonable level of irrigation efficiency is required. Since field irrigation requirements will vary for each crop during the growing season, water supply should follow these changes over area and time, based on the analysis and evaluation of the field variables that include water supply requirement factor, irrigation system design factor, supply duration factor and supply factor [4].

Analysis of Water Savings through Different Technologies

Need for irrigators to have designed irrigation scheduling

Critical to any irrigation system is an accurate estimate of the amount of water to be applied from the water source into an irrigated field to meet the crop water requirements over the growing season [9]. Within the growing season, the amount of water required by a given crop varies with the growth stages, thereby requiring appropriate irrigation schedules to meet the changing water needs [12]. In many cases, farmers continue irrigating as long as water is available for irrigation without taking into consideration their system's efficiency [8]. Since irrigation has a major impact on both cost and yields, one must install an irrigation system that will deliver sufficient amount of water at minimum cost without compromising on the yields predicted [9]. Mike and Lynne emphasizes that a basic understanding of an irrigation system's capacity to deliver water is a very powerful piece of knowledge since it allows one to take a more scientific approach to irrigation process, greater control and conservation of water without compromising the crop yields. For this to be achieved, Muchangi et al. suggested an application of appropriate irrigation methods and schedules that indicate when to irrigate, how to irrigate, how much water to apply and how deep the water penetrates into the soil after irrigating [9]. In this regards, all the irrigators need to know their system's net water application rate and the methods applied in their determination as well as guidelines on measurement and monitoring soil moisture level. Against this background, detailed analysis of the performance of different technological options is carried out in terms of their efficiency, how much water is saved and additional acreage. The three technologies examined in relation to the farmers' methods are: drip plus digital instruments, drip plus analog instruments, sprinkler plus digital instruments, and sprinkler plus analog instruments.

Drip and sprinkler plus digital instruments: Overhead irrigation methods (drip and sprinkler) are away of applying irrigation water with reasonably high efficiencies (75-90%). This can be further improved when accompanied by advanced digital instruments [9]. When combined with digital measuring instruments, these technologies can relax constraints on irrigation rates and timing by permitting better adjustments of irrigation scheduling to varying crop water needs within the growing period [24]. The advanced brands of instruments are currently available and offering a range of wireless solar-powered measuring and monitoring systems with web based software for improved irrigation scheduling [25]. According to Babajimopoulus et al. optimal irrigation water management and scheduling take into account the effectiveness of the irrigation water use in crop production, which is attainable through the use of improved measuring instruments [22]. The web based instruments remotely operate the irrigation system through instruments chain interface on board that allows the wireless intelligent sensor mesh network to connect all the sensors for various soil and climatic attributes from the field to mobile that signals the irrigators to irrigate as per the pre-determined irrigation schedules. Because of limited and controlled loss of irrigation water through the use of these instruments, the efficiency of both drip and sprinkler will step up significantly.

Drip and sprinkler plus analog instrument: Drip irrigation, accompanied by analog instruments, involves measurement, laboratory determinations and synthesis of all the water-related properties of soils and climatic characteristics, using of analog instruments. Determination and monitoring of the hydraulic properties of soils

and climatic variables required for designing the irrigation schedules may result into huge water losses, hence relatively low water use efficiency because of the reasons demonstrated by Muya et al. [21]. Determination of the available soil water for a given crop in a given field and soil, involves laborious exercise of digging the soil profiles, wetting all the horizons to be sampled, taking the disturbed and undisturbed samples and transporting them to laboratory, where the they are subjected to stepwise incremental suctions and determination of the equilibrium moisture content for each suction. One of the biggest challenges is that the access roads from the field to the main roads may be rough with potholes such that the samples reach laboratories after undergoing considerable disturbance, so that they no longer represent the structure of undisturbed soils in-situ. Therefore, the soil moisture retention characteristics determined from the laboratory usually give wrong results in the system design [21]. The second challenge is that the samples have to be saturated before being closed into the pressure chamber, which may take several days, depending on the structure, degree of compaction, quantity and type of clay. The third challenge is the line between the air entry point that indicates equilibrium moisture content is so thin that values of soil moisture at each suction is a mere approximation, and the time taken to reach equilibrium is more than two weeks.

In Nguruman Irrigation Scheme, the planned sprinkler irrigation system design require accurate measurements of hydraulic properties of soil so as to give the farmers irrigation calendar that deviates from the traditional schedules, based on the rule of thumb. The types of data on the hydraulic properties of soil required for accurate design of irrigation schedules or cycles in this scheme are those that reflect the spatial and temporal variations in the rate and level of capillary rise of ground water, salinity and sodicity, available soil moisture holding capacity, the water uptake capacity, allowable soil water depletion level for various crops under different clusters of soil units identified [8]. Collecting these data for the determination of irrigation cycles, using analog instruments, is constrained by the magnitude of time and personnel required since their collection involves in-situ determinations and monitoring, which can only be achieved successfully using digital instruments [8,24]. This explains why the design of sprinkler irrigation systems in Nguruman Irrigation Scheme, was not based on these measurements. According to Muchangi et al. calculation of the irrigation cycle, based on these hydraulic properties is a must because it would lead to the determination of the parameters required for the design of the field layout for water application [9]. To obtain the highest yields, it is important to treat each field separately as one irrigation unit with specified dimensions, so each irrigation unit is irrigated according to its soil and crop water requirements [8].

To achieve the maximum yield from each irrigation unit Withers and vipond suggested booking methods to be applied in determining the permissible water depletion level by making observations of yields under various irrigation regimes, using instruments with reasonable level of accuracy [18].

For the determination of the net irrigation requirements, reference crop water requirements (ETo) are needed. For this purpose, prediction methods for ETo have been developed owing to the difficulty of obtaining accurate field measurements of the necessary climatic variables. These methods are Blaney-Criddle, Radiation, Penman and Pan Evaporation [8]. However, advances in research and more accurate assessment of crop water use have revealed weaknesses in these methods that include: need for local calibration to achieve satisfactory results with Penman method; erratic performance of the

Radiation method in arid conditions; and shortcomings in predicting crop evaporation [8]. The relatively accurate and consistent method is FAO Penman-Monteith which is recommended as the sole standard method with strong likelihood of predicting ETo with reasonable accuracy as was applied by Zoratelli.

Irrigation water supply requirements under different technologies

The monthly net irrigation requirements for different crops are the starting point for the determination of the water supply requirements. The water supply requirements of different crops depend on the efficiency of different irrigation techniques, assuming other factors are under control [4]. Therefore, determination of the irrigation water requirements for the selected cropping patterns in Nguruman includes the net irrigation requirements, irrigation efficiency of water distribution system that depends on the irrigation technology applied and other water needs such as leaching requirements (LR) of the excess salts that depend on the soil and crop types [13]. Table 3 shows different technologies and their efficiencies. Determination of irrigation water supply requirement in m³/ha/month is used to determine the acreage that can be irrigated from the available water supply, depending on the irrigation technique in question as is given by the following equation:

$$\text{Where:} \ Vc = 10 \frac{\sum A.In}{Ea(1-LR)}$$

Vc=Volume of water required for a specified crop

Ea=Project irrigation efficiency (55% for Nguruman after Knight Piesold

A=Area under a specified crop, ha

In=Net water requirement of a given crop, mm/month

LR=Leaching requirement, fraction (assumed to be 20% for Nguruman)

As is shown in the equation, the amount of water required to irrigate a given crop depends on the efficiency of the irrigation technique applied, acreage to be irrigated, net irrigation and leaching requirements [4]. Therefore, holding other factors constant, different irrigation techniques can be compared in terms of water saving, the additional area that can be irrigated with the amount of water saved [26]. The first step to determine the water savings from different techniques is to define the cropping patterns for a given irrigation scheme or system. Table 4 gives the cropping patterns selected for Nguruman Irrigation Scheme by Muya et al. [13]. For comparative analysis of the water savings from different irrigation techniques, pasture, beans, onions and maize were selected which fall in phase of the cropping systems.

The total water supply requirement to irrigate 800 ha with efficiency of 55% reported by Muya et al. is 5,333,809 m³/year, which is only 19% of the available water supply. This amount is required to irrigate all the crops in three crop phases given in Table 4 [13]. However, the amount of water required to irrigate pasture, beans, onions, and maize for which different water saving techniques are compared, using the farmer's method is shown in Table 5.

Based on the proposed cropping patterns for Nguruman Irrigation Scheme, the crop water requirements in different months can be used as guide for the scheme in planning the irrigation schedules. The total irrigation water supply requirements may be matched with the water supply from the source, and the result of matching is the basis of making adjustment in cropping patterns in both spatial and temporal terms [9].

As was shown by Muya et al. (Table 6), drip plus digital instruments had the highest water saving potential of 465,600 m³/year for irrigating additional area of 146.3 ha under the four crops (pasture, beans, onions and maize) [21]. This is followed by drip plus analog instruments which can save 433,132 m³ with the potential of irrigating additional area of 136.7 ha. Sprinkler plus digital and sprinkler plus analog can save 365,520 m³ and 323,637 m³ that can irrigate additional area of 115.4 and 104.1 ha respectively. Promotion of the precision agriculture in Nguruman scheme was proposed by the LOG ASSOCIATES to improve the production efficiencies through the design and application of the sprinkler irrigation system, using efficient water measuring

Description of technologies	Efficiency (%)	Remarks
Drip irrigation plus digital instruments	97	The instruments to monitor soil water balanced and weather conditions can either be analog or digital with the later being more efficient
Drip irrigation plus analog instruments	90	
Sprinkler plus analog instruments	85	
Sprinkler plus digital instruments	80	
Farmers irrigation methods and practices	55	Farmers' method is surface irrigation and water supply and distribution are manually operated with no measurements throughout the system.

Table 3: Different technologies and their efficiencies.
Source: Muya et al. [21].

Cropping systems		Cropping phases and proposed acreage		
Option No	Classification of crops	I	II	III
1	Legumes	Beans (40 ha)	Soybeans (65 ha)	G/grams (30 ha)
2	Vegetables	Onions (40 ha) Melon (40 ha)	Ravaya (80 ha) Tomatoes (20 ha)	Okra (20 ha) Karella (80 ha)
3	Pasture	Pasture (40 ha)	Pasture (40 ha)	Pasture (40 ha)
4	Fruit trees	Bananas (40 ha)	Mangoes (30 ha)	Lemons (40 ha)
5	Food security	Maize (65 ha)	Sorghum (35 ha)	Sweet potatoes (55 ha)
	Total acreage (800 ha for the three phases)	265 ha	270 ha	265 ha

Table 4: Proposed cropping patterns and cycles for Nguruman irrigation scheme.
Source: Muya et al. [13].

Crop	Net irrigation requirements of different crops and technological options	Crop water requirements in different months (m³) for different crops under different technological options				Total water requirements
		January	February	March	April	
Pasture	Net irrigation requirements for pasture in different months	13.7	139.4	85.6	12.0	
	Farmers' practices with 55% efficiency	9,963	101,381	62,254	87,27	182,325
	Drip irrigation plus digital instruments with 97% efficiency	5,649	57,484	35,298	4,948	103,379
	Drip irrigation plus analog instruments with Efficiency 90%	6,088	61,955	38,044	5,333	111,420
	Sprinkler irrigation plus digital instruments with efficiency 85%	6,447	65,600	40,282	5.647	117,976
	Sprinkler irrigation plus analog instruments with efficiency 80%	6,850	69,700	42,800	6,000	125,350
		May	June	July	August	
Beans	Net irrigation requirements for beans in different months	46.3	100.8	109	74.8	331
	Farmers' practices with 55% efficiency	33,672	73,309	79,272	54,400	240,653
	Drip irrigation plus digital instruments with 97% efficiency	19,092	41,567	44,948	30,845	119,280
	Drip irrigation plus analog instruments with Efficiency 90%	20,578	44,800	48,444	33,244	147,066
	Sprinkler irrigation plus digital instruments with efficiency 85%	21,788	47,435	51,294	35,200	15,5717
	Sprinkler irrigation plus analog instruments with efficiency 80%	23150	50,400	54,500	37,400	165,450
		September	October	November	December	
Onions	Net irrigation requirements for onions in different months	89.8	110.8	123.9	63.9	388.4
	Farmers' irrigation methods with 55% efficiency	65,309	80,582	90,109	46,473	282,473
	Drip irrigation plus digital instruments with 97% efficiency	37,031	45,691	51,092	26,351	160,165
	Drip irrigation plus analog instruments with Efficiency 90%	39,911	49,36	55,067	28,400	128,314
	Sprinkler irrigation plus digital instruments with efficiency 85%	42,259	52,142	58,306	30,071	182,778
	Sprinkler irrigation plus analog instruments with efficiency 80%	44,900	55,400	61,950	31,950	19,4200
		September	October	November	December	
Maize	Net irrigation requirements for maize in different months	21.8	84.8	148.9	23.9	279
	Farmers' practices with 55% efficiency	25,763	100,218	175,972	28,245	330,198
	Drip irrigation plus digital instruments with 97% efficiency	14,608	56,824	99,778	16,015	187,225
	Drip irrigation plus analog instruments with Efficiency 90%	15,744	61,244	107,538	17,261	201,787
	Sprinkler irrigation plus digital instruments with efficiency 85%	16,671	64,847	113,864	18,276	213,658
	Sprinkler irrigation plus analog instruments with efficiency 80%	17,713	68,900	120,981	19,419	227,012

Table 5: Crop water requirements.
Source: Muya et al. [21].

Irrigation techniques as compared to farmers method	The impacts of different techniques	Water supply requirements (m³), savings and additional area to be irrigated using different techniques for the following crops:				Totals
		Pasture	Beans	Onions	Maize	Four crops
Farmers' method	Water supply requirements	182,325	240,653	282,473	330,198	1,035,649
	Water saving	0	0	0	0	0
	Additional area to be irrigated	0	0	0	0	0
Drip plus digital instruments	Water supply requirements	103,379	119,280	160,165	187,225	570,049
	Water saving	78,946	121373	122,308	142,973	465,600
	Additional area (ha) to be irrigated	24.5	38.3	38.6	45.1	146.3
Drip plus analog	Water supply requirements	111,420	147,066	128,314	201,787	588,587
	Water saving	56,975	93,587	154,159	128,411	433,132
	Additional area (ha) to be irrigated	17.9	29.6	48.7	40.5	136.7
Sprinkler plus digital instruments	Water supply requirements	117,976	155,717	182,778	213,658	670,129
	Water saving	64,349	84,936	99,695	116,540	365,520
	Additional area to be irrigated (ha)	20.3	26.8	31.5	36.8	115.4
Sprinkler plus analog instruments	Water supply requirements	125,350	165,450	194,200	227,012	712,012
	Water saving	56,975	75,203	88,273	103,186	323,637
	Additional area to be irrigated (ha)	17.9	23.7	27.9	32.6	104.1

Table 6: Water savings from different techniques relative to famer's method.
Source: Muya et al. [13].

and monitoring instruments [1]. Therefore, since drip irrigation plus digital instruments is the most efficient technological option, it is recommended for the scheme. Regarding the farmers' irrigation methods and practices, no water saving is realized; hence no additional area may be irrigated.

Efficient irrigation technologies involving digital instruments in Nguruman

Although Nguruman scheme is graduating into a commercial farming enterprise where measurements of economic water use efficiency is extremely necessary, no single measurement has been done to determine and monitor soil water balance in relation to nutrient dynamics and solute transport within the soil profiles. This concern has been expressed by Muya et al., in response to the sentiments of many workers, that increased application of nutrient inputs into the irrigated soils without getting watering and irrigation scheduling right through accurate measurements of water and solute levels in the soils (using digital instruments) may lead not only to the decline of soil and environmental quality in the long-run, but also huge losses of resources [13]. Odongo indicated that improved irrigation scheduling based on measurement of soil-plant-water interactions, using accurate digital instruments, resulted into reduced production costs, controlled accumulation of unwanted salts, conservation of limited water resources and increased quality of agricultural products. Wilk et al. emphasized that a careful management of irrigation water through appropriate irrigation scheduling is a key factor in good yields and plant performance which can be attained through detailed assessment of hydraulic functions in relation to crop performance. Assessment of hydraulic functions for practical irrigation scheduling requires information on soil water/tension readings transferred from the irrigated fields to irrigation operator, farmer or irrigator who uses the information to decide on when, where and how much to apply irrigation water [9]. There are a number of digital instruments including wireless sensor networks now available to provide accurate information on when to irrigate and how much water to apply. Drip and sprinkler irrigation methods, accompanied by sensor and wireless transmission system can provide web-based irrigation scheduling through timely and accurately monitored field soil moisture balance in relation to crop performance and yields [21]. The wireless sensor network can be used to monitor various environmental parameters related to agriculture that are input variables into models developed to estimate the crop water requirements, field water supply requirements and irrigation scheduling. Monitoring of these parameters using these digital instruments results into significant minimization of time and costs; maximization of the quantity and quality of agricultural products; and increased economic water use efficiency [24].

Striking the balance between science and the reality on the ground

Many workers have demonstrated that rehabilitation of an irrigation scheme through infrastructural development and proper placement in agricultural production value chains without incorporating scientific principles in the implementation framework will not sustain the targeted production however efficient the planning, design and construction works was. Scientific principles required in the implementation of the designed irrigation scheme aim at achieving the enhanced soil-water-plant relationships for sustained irrigation efficiency. As reported by Muya et al. the main cause of low irrigation efficiency in Nguruman Scheme is uniform application of irrigation water, using the same irrigation methods and scheduling, throughout

the scheme by most farmers, disregarding the differences in physical, hydraulic and chemical properties of soils that influence the choice of irrigation methods and scheduling [8]. The cropping patterns, their acreage and the shape of the farms in relation to the designed layout are not yet defined by the irrigators. The proposed irrigation technology aim at maximizing the productivity on water through improved efficiency of its use by providing information, data, irrigation method and water measuring instruments that will assist farmers to decide how much acreage to irrigate, when to irrigate, how to irrigate and how much water to apply and the depth of water penetration in the soil after irrigation, and total production per unit of water consumed by the desired crops. For this to be achieved, farmers and the members of Water User Association (IWUA) should be equipped with data, instruments and capacity building that enable them to know how much water to supply individual farmers' fields, irrigation blocks and the entire scheme through the designed water distribution networks. Table 7 shows how to apply scientific principles in implementing the scheme. For Nguruman irrigation scheme, sprinkler irrigation method has been decided and designed. Therefore, a reasonable compromise must be arrived at between the recommended technologies (drip plus digital instruments) and those on the ground (sprinklers) to ensure sustainable use of limited water and nutrient resources. The guiding factors are the clusters arrived at, in which the use of sprinkler in certain area should be accompanied by the implementation of the recommended practices to reduce risks of land degradation associated with the use of the sprinkler technology. For example, one of the clusters identified in the scheme is relatively steep and extremely compact with potential of generating run-off under sprinkler irrigation which imitates the rain falling on steep and low-permeable soils. Increased land degradation under this method can be alleviated by selecting the sprinkler with discharge rates that are commensurate with water uptake characteristics and capacity of the soil. However, sprinkler method is ideally suitable for clusters with highly permeable soils which have been subjected to excessively nutrient leaching through basin irrigation. Therefore, the best approach would be to select and implement an irrigation practice for each cluster so that cluster-specific irrigation technologies are developed for the scheme. In this context, clusters which are highly degraded with compact surface and saline-sodic soils would certainly not benefit much from sprinkler because of the negative impacts this type of irrigation on soil structure stability [20]. In order to achieve the targeted crop production on long-term basis, the layout map of entire irrigation scheme, consisting production systems and Irrigation Water User Association, should be superimposed on the clusters of soils delineated to guide the irrigators on the management required and application of scientific inputs in order to sustain the irrigation and water use efficiency desired (Table 7).

Conclusion

Water saving for irrigating more land with less water can be achieved by planning, designing and operating an irrigation system by timely information and data inputs into the three stages of irrigation development. Planning stage requires data on seasonal and peak water supply requirements on the basis of which to determine the actual acreage that can be irrigated by the available water supply. For the design stage, water supply scheduling is designed, based on data timely collected on the hydraulic characteristics of soils.

In Nguruman Irrigation Scheme, the targeted 800 ha of land for irrigation, using farmers' method with efficiency of 55%, is only 40% of the total irrigable area, and the annual water supply requirement of 5,333,809 m^3 for irrigating the targeted area can be met by the water

Components	Management implications	Scientific inputs
Entire Irrigation Scheme	Physical infrastructure to deliver water to irrigated land.	• Determination of the irrigable area. • Calculation of seasonal and peak water supply required for the selected cropping pattern and intensity. • Calculation of the total irrigable area that can be irrigated by the available water from the source
Production system (plot use or farm enterprises)	Irrigation blocks with organized and input specific production systems with known water supply requirements, identified with irrigators, IWUA and other stakeholders.	• Organization of production systems into logical patterns in terms of acreage, temporal and spatial arrangements of the envisaged crops in each cluster identified. • Determination of the crop water requirements and water duties for the selected patterns. • Determination of the irrigation schedules/cycle for each cluster.
Organization (IWUA)	Group based irrigation systems imply an organization in charge of operation and management. Organizational performance is an important factor of productivity and sustainability of irrigation system.	• Development of the organizational structure that tallies with the layout of water distribution networks in relation to the irrigation blocks or clusters. • Development of irrigation calendar for efficient supply and management of irrigation water. • Installation of appropriate water control and monitoring structures/instruments.

Table 7: Application of scientific principles to various components of irrigation.
Source: Muchangi *et al.* with modification to suit the research area [9].

supply from the source, which is about, five times the current water requirements. However, with current increase in food demand, the remaining 60% of the total irrigable area will soon be brought under irrigated agriculture, hence potential water shortage.

The result of the review indicated that the highest potential water saving for Nguruman Scheme was reportedly 465,600 m³ with drip plus digital instrument, including wireless sensor networks, resulting in additional area to be irrigated of 146.3 ha. This was followed by 433,132 m³ saved by drip plus analog instrument. Sprinkler plus digital instrument and sprinkler plus analog instrument saved 365,520 and 323,637 m³ respectively. The additional area to be irrigated by sprinkler plus digital instrument and sprinkler plus analog instrument was reported to be 115.4 and 104.1 ha respectively.

Acknowledgement

Funding from African Development Bank for feasibility study for situational analysis of Nguruman Irrigation Scheme as well as timely facilitation by the Project Coordination Unit are highly appreciated as well as the logistic support provided by J. Gathoni, L. Wasilwa, V. Kirigua, V. Wasike and H. Goro.

References

1. LOG ASSOCIATES (2011) Mid-Term Review of Small Scale Horticultural Development Project.

2. Horst L (1996) Irrigation water divisional structures in Indonesia. Liquid Gold Special Report.

3. Muya EM, Jura MO, Mati BM (2009) Water management in irrigated agriculture: A training manual for technical staff.

4. FAO (1977) Crop water requirements. FAO Irrigation and Drainage paper, Rome.

5. FAO (1979) Yield response to water. FAO Irrigation and Drainage paper, Rome.

6. World Bank (1990) Discussion paper on operational issues at the scheme level.

7. Professional Training Consultants (2011) Training Curriculum for Extension Staff.

8. Muya EM, Obanyi S, Owenga PO (2012) Watershed characterization and analysis of water-related constraints of Mountains and Oasis areas of KASALS.

9. Muchangi P, Sijali IV, Wendot H, Lemperiere P (2005) Irrigation systems, their operation and maintenance.

10. Shanan L (1992) Planning and management of irrigation systems in developing countries. Agricultural Water Management Journal.

11. National Irrigation Board of Kenya (2013) Approaches for feasibility study for irrigation development.

12. FAO (1976) A framework for land evaluation.

13. Muya EM, Maingi PM, Kirigua V, Owenga PO (2014) Assessment of cluster-specific constraints and irrigation suitability of the degraded soils: A basis for formulating technological interventions for improved soil quality and productivity in Nguruman.

14. Muya EM, Owenga PO, Kahiga P, Kirigua V, Goro H, et al., (2013) A review of crop water requirements determination methods and their application in water supply scheduling: The blue print for Nguruman Irrigation Scheme.

15. Muya EM, Owenga PO, Kirigua V, Goro H, Atonya B, et al., (2013) Determination of crop water requirements for Nguruman Irrigation Scheme, Kenya Soil Survey.

16. Wanjogu SN, Gicheru PT, Maingi PM, Tanui K (2006) Land degradation and management in the river Yala catchments, Western Kenya.

17. Knight P (1999) Pre-feasibility study of Irrigating rangeland in Olkrimantian and Shompole Group Ranches for Ewaso Ngiro South Hydrao-Electic Power Development.

18. Withers B, Vipond S (1974) Irrigation design and practice.

19. Muya EM, Maingi PM, Chek AL, Gachini GN, Otipa M, et al., (2011) Analysis of soil quality, irrigation system and cluster-specific interventions for improved production of high value vegetables.

20. Hillel D (1982) Introduction to soil physics. Academic Press, California.

21. Muya EM, Owenga PO, Kirigua V, Goro H, Atonya B, et al., (2014) Determination of crop water requirements for Nguruman Irrigation Scheme. Kenya Soil Survey.

22. Babajimopoulus C, Panoras A, Geogoussis H, Arampatzis G, Hatzigiannakis E, et al., (2007) Contributing to irrigation from shallow water table under field conditions. Agricultural water management 92: 205-210.

23. Shasbank K (2014) Micro-controller-based automated irrigation system.

24. Okoth PFZ (2014) The use of mobile phones to access irrigation and weather information: Data collection and maintenance.

25. Pessl GJ (2013) Turning information into a profit through metos digital irrigation instruments.

26. Shani U, Tsur Y, Zemel A, Zilberman D (2007) Irrigation production functions with water capital substitution.

14

Exploring Estimation of Evaporation in Dry Climates Using a Class 'A' Evaporation Pan

Farai Malvern Simba*, Alois Matorevhu, David Chikodzi and Talent Murwendo

Department of Physics, Geography and Environmental Science, Great Zimbabwe University, Zimbabwe

Abstract

The rate of evaporation in dry climates is a concern and needs to be assessed and quantified for planning in water management activities. The main objective of the study was to investigate evaporation rate of a bare ground in Masvingo district in Zimbabwe using a class 'A' evaporation pan. Specific objectives include calibrating the pan using the FAO-Penman Monteith (P-M) method and obtaining typical evaporation rates for the area which could be extended to represent areas of relatively similar climates. To achieve these objectives an evaporation pan was installed on a wooden grid platform near a weather station that recorded wind speed, air temperature, humidity, maximum and minimum temperatures. Considering fetch dimensions the pan results were correlated against P-M method results to give pan calibration coefficient and the slope of the curve gave a Kpan of 0.91 and changes in water depths with respect to pan dimensions gave average evaporation rates of 5.1 mm/day at mean maximum temperature of 24.9°C. The evaporation rates obtained were not sustainable in the long term if water harnessing and conservation strategies are not employed.

Keywords: Evaporation; Water; Pan; Dry climate; Penman-monteith; Zimbabwe

Introduction

Water is a finite and important resource for agricultural production. In Sub-Saharan region there has been water scarcity due to climate change and an increasing demand by different competing various socio – economic users making water conservation practices important [1]. One critical way in which water is lost is through evaporation into the atmosphere due to high ambient temperatures associated with Global Warming. Evaporation is a parameter that is not easily estimated accurately yet it is of vital importance in the crop farming, water conservation, and in water inventory issues for surface and ground water sources. In this study the proposed evaporimeter, which is a Class 'A' pan will need calibration to give accurate readings, hence the need for the calibration exercise and ensuing evaporation measurements. These measurements will open future research doors in water related research at the Great Zimbabwe University and can be extrapolated to surrounding areas. A relationship established by Allen et al. relates that pan evaporation to evapotranspiration made is employed to achieve the calibration of the evaporimeter and ensuing evaporation measurements [2]. Evapotranspiration (ETo) estimated by the FAO-Penman-Monteith method is compared against the pan evaporation (Epan) method at Great Zimbabwe University in order to validate the use of the pan locally.

Energy enables water to change from liquid to gas (vapour). Direct solar radiation and to a reduced extent, the ambient temperature provide this energy. The driving force to remove water vapour from an evaporating surface is the difference between the vapour pressure at the evaporating surface and that of surrounding atmosphere. Increased saturation of air surrounding the evaporating surface slow down evaporation and may stop it if wet air is not transferred to the atmosphere. Therefore, solar radiation, air temperature, humidity, rainfall and wind speed are climatological parameters to consider when assessing the evaporation process. The rate of evaporation depends on these climatological parameters as well as soil water availability [3,4].

Many methods have been developed to derive reference evapotranspiration (ETo) and water evaporation (E) estimates [5]. Among these methods are FAO-Penman, FAO-corrected Penman, Blaney-Criddle, Hargreaves and FAO-Ration [2,6-10]. Many of these methods are subject to local calibration; thereby require local characterisation of topography, altitude and type of soils of the area under study to improve accuracy [11]. However, the high performance of the Penman Monteith method in estimating evapotranspiration in different parts of the world when compared with other methods has made it accepted as the sole method for computing evapotranspiration from meteorological data (temperature, relative humidity, solar radiation and wind speed) [12]. Roderick et al. observed that the universal application of the Penman Monteith approach enables it to be used as a standard to verify or calibrate other methods [13]. The pan evaporation method is also widely used because it is simple to use for estimating evapotranspiration. Before an evaporation pan can be used, it must be locally calibrated because its performance is site specific or depends on geographical location.

Water demand is controlled by the atmosphere depending on the vapour pressure deficit conditions and water supply is governed by the existing soil moisture and water bodies available. Some water is drained out of the soil due to gravitational forces exerted on the water in the soil. The water that remains in the soil via molecules and matric forces is known as the field capacity. Matric forces are generated by the soil particles adhesive and absorptive molecules attraction to water and cohesive attraction exerted by water molecules on other water molecules [14]. Capillary action enables water to move through the soil and plant, in the presence of potential difference .Water moves in the direction of decreasing energy that from areas of low to high matric potential. In the soil and plant system, this implies water moves from wet soil to dry and from soil to stomata [14]. This process continues until the soil reaches its maximum moisture deficit. Up to this point

---

*Corresponding author: Farai Malvern Simba, Department of Physics, Geography and Environmental Science, Great Zimbabwe University, Zimbabwe E-mail: faraims@yahoo.com

soil water extraction by the plant is not operating under stressed condition, thus water supply is virtually unlimited. Once the soil water content falls below the maximum soil moisture deficit, it becomes increasingly more difficult for the plant to extract water from the soil [15]. Therefore, it is vital that irrigation scheduling is efficiently done to meet plant water requirements.

Objectives of the study

The main objective of the study was to investigate evaporation rate of a bare ground in Masvingo district in Zimbabwe using a class 'A' evaporation pan. Specific objectives include calibrating the pan using the FAO-Penman Monteith (P-M) method and obtaining typical evaporation rates for the area which could be extended to represent areas of relatively similar climates.

Pan evaporation

An evaporation pan is used to hold water during observations for the determination of the quantity of evaporation at a given location. Such pans are of varying sizes and shapes, the most commonly used being circular or square [16,17]. The best known of the pans are the "Class 'A'" evaporation pan and the "Sunken Colorado Pan [18]. In Europe, India and South Africa, a Symon's Pan (or sometimes Symon's Tank) is used. Often the evaporation pans are automated with water level sensors and a small weather station is located nearby.

Penman-Monteith equation

The equation is stated as:

$$ETo = \frac{0.408\Delta(Rn - G) + \frac{Y900}{T} + 273U_2(e_s - e_a)}{\Delta + Y(1 + 0.34U_2)} \qquad \text{.....................(1)}$$

Where:

ETo=Reference evapotranspiration (mm/day)

Rn=Net radiation at crop surface (MJ/m² per day)

G=Soil heat flux density (MJ/m² per day)

T=Mean daily air temperature at 2 m height (°C)

U_2=Wind speed at 2m height (m/sec)

e_s=Saturated vapour pressure (kPa)

(e_s - e_a)=Saturation vapour pressure deficit (kPa)

Δ=Slope of saturation vapour pressure curve at temperature T (kPa/°C)

Y=Psychrometric constant (kPa/°C)

Penman Monteith uses standard climatological records of solar radiation (sunshine), air temperature, humidity and wind speed for daily, weekly, ten-day or monthly calculations [10]. The selection of the time step with which ETo is calculated depends on the purpose of the calculation required and the climate data available. Some data are measured directly in weather stations Parameters related to commonly measured data are derived using direct or indirect empirical equations. Integrity of computation is assured by making weather measurements at 2 m (or converted to that height) above an extensive surface of green grass shading the ground that is not short of water. Common units should be converted to standard units.

Parameters used in the Penman-Monteith equation are calculated using equations (2) to (8) given below.

$$Y = \frac{C_p P}{\varepsilon \lambda} = 0.665 \times 10^{-3} P \qquad \text{.................(2)}$$

Where:

C_p=Specific heat at constant pressure=1.013 × 10⁻³ MJ/kg/°C

P=Atmospheric pressure (kPa)

ε=Ratio molecular weight of water vapour/dry air=0.622

λ=Latent heat vaporization=2.45 MJ/kg (at 20°C)

$$T_{mean} = \frac{T_{max} + T_{min}}{2} \qquad \text{.................(3)}$$

Where:

T_{mean}=mean daily temperature (°C)

T_{max}=mean daily maximum temperature (°C)

T_{min}=mean daily minimum temperature (°C)

$$e_s = \frac{e^0(T_{max}) + e^0(T_{min})}{2} \qquad \text{.......................(4)}$$

Where:

e_s=mean saturation vapour pressure (kPa)

$e^0(T_{max})$=saturation vapour pressure at maximum air temperature (kPa)

$e^0(T_{min})$=saturation vapour pressure at minimum air temperature (kPa)

$$e^0(T) = 0.6108 \exp\frac{(17.27T)}{(T + 237.3)} \qquad \text{.....................(5)}$$

Where:

T=Mean air temperature (°C)

exp[...]=2.7183 (base of natural logarithm) raised to the power [...]

$$\Delta = \frac{4.098\left[0.6108\exp\left\{\frac{17.27T}{T + 237.3}\right\}\right]}{(T + 237.3)^2} \qquad \text{...................(6)}$$

Where: $0.6108\exp\frac{17.27T}{T + 237.3} = e^0$

Where:

T=mean daily temperature (Equation 5)

$$e_a = \frac{\left[e^0(T_{min}) \times \frac{RH_{max}}{100}\right] + \left[e^0(T_{max}) \times \frac{RH_{min}}{100}\right]}{2} \qquad \text{...............(7)}$$

Where:

e_a=Actual vapour pressure (kPa)

$e^0(T_{min})$=Saturation vapour pressure at daily minimum temperature (kPa)

$e^0(T_{max})$=Saturation vapour pressure at daily maximum temperature (kPa)

Net radiation estimation

The calculation of clear sky radiation (R_{so}), when n=N, is required to compute net longwave radiation. R_{so} is given by the following simplified expression:

$$R_{so} = \left[0.75 + \frac{2z}{100000} \right] \times R_a \qquad \text{................(8)}$$

Where:

R_{so}=clear sky solar radiation (MJ/m² per day)

Z=station elevation above sea level (m)

Net longwave radiation (Rnl): The rate of longwave radiation emission is proportional to the absolute temperature (Kelvin) of the surface raised to the fourth power. Rnl is calculated using the following expression:

$$R_{nl} = \left(\frac{\sigma(T_{max,K})^4 + \sigma(T_{min,K})^4}{2} \right) \times$$

$$(0.34 - 0.14\sqrt{e_a}) \times \left(1.35\frac{R_s}{R_{so}} - 0.35 \right) \qquad \text{...........(9)}$$

Where:

Rnl=net outgoing longwave radiation (MJ/m² per day)

σ is the Stefan Boltzmann constant (4.903 x 10⁻⁹ MJ/K⁴ per m² per day)

$T_{max,k}$=Maximum absolute temperature during the 24 hr period(K)

$T_{min,k}$=Minimum absolute temperature during the 24 hr period (K)

e_a=actual vapour pressure (kpa)

R_s/R_{so}= Relative shortwave radiation (limited ≤ 1)

R_s=Measured or calculated solar radiation (MJ/m² per day)

R_{so}=calculated clear sky radiation (MJ/m² per day)

Calibration equation

$$ETo = kp \times Epan \qquad \text{...(10)}$$

Where:

ETo is the evapotranspiration estimated by the FAO-Penman Monteith equation

kp is the pan coeffient

Epan is the pan evaporation

Pan siting

The accuracy of evaporation pans depend on the pan coefficient. Savva et al. defines pan coefficient as the ratio of the water body to pan evaporation [10]. Pan coefficients are pan specific and they depend on the colour, size, and position/site of the pan. Therefore when using a pan, consideration should be given to the pan type, the ground cover in the station where the pan is sited, its surrounding and the general wind and humidity conditions. This is particularly true if the pan is placed in fallow rather than cropped fields. Allen et al. identify two cases A and B of evaporation pan citing and their environments [2]. Case A is where the pan is sited on a fallow soil, and case B is where the pan is sited on fallow soil and surrounded by green crop as illustrated below in Figure 1.

Despite the apparent simplicity in use, evaporation pans need careful maintenance to provide accurate measurements. The water level must be kept to the prescribed level. Regular cleaning and periodic repainting are necessary, painting affects heat loss or absorption by the pan. The siting of the pan can have a major impact on the measurements. For instance, a pan sited on bare soil may record higher evaporation rates than one sited on the grass because the air moving over the pan tend to be drier. Heat storage in the pan can be appreciable and may cause significant evaporation during the night while plants transpire during the daytime. Differences also exist in the turbulence, temperature and humidity of air immediately above the respective surfaces. However, the evaporation pan has proved its practical value hence widely used to estimate ETo. Therefore, application of empirical coefficients to relate pan evaporation to ETo for periods of 10days or longer may be warranted [2].

Global warming and evaporation rates

Over the last 50 or so years, pan evaporation has been carefully monitored. For decades, nobody took much notice of the pan evaporation measurements. But in the 1990s scientists spotted something that at the time was considered very strange; the rate of evaporation was falling. This trend has been observed all over the world except in a few places where it has increased. As the global climate warms, all other things being equal, evaporation will increase and as a result, the hydrological cycle will accelerate. The downward trend of pan evaporation has been linked to a phenomenon called global dimming. In 2002 Roderick and Graham found that the "dimming" trend had reversed since about 1990 [19].

Location of the study site

The experiment was carried out at the Great Zimbabwe University (GZU) in Masvingo town in Zimbabwe. The location of the GZU is approximately on latitude 20, 10⁰ S and longitude 30, 86⁰ E (Figure 2). Like other places in Zimbabwe, the GZU experiences four seasons. The hot season starts in August up November. Temperatures rise up to 35°C. The rainy season begins in mid November up to March, with maximum temperatures up to 29°C. The post–rainy season from March to mid May is the transitional period to the cool season. The cool season called winter is from mid May to August. This study was at the end of April 2013 when temperatures had started failing.

Materials and Methodology

Instruments used

- Automatic weather station (AWS) called the RainWise MKIII Weather Transmitter and data logger

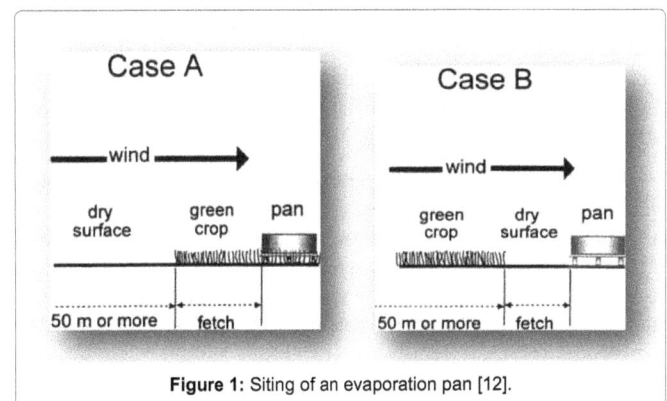

Figure 1: Siting of an evaporation pan [12].

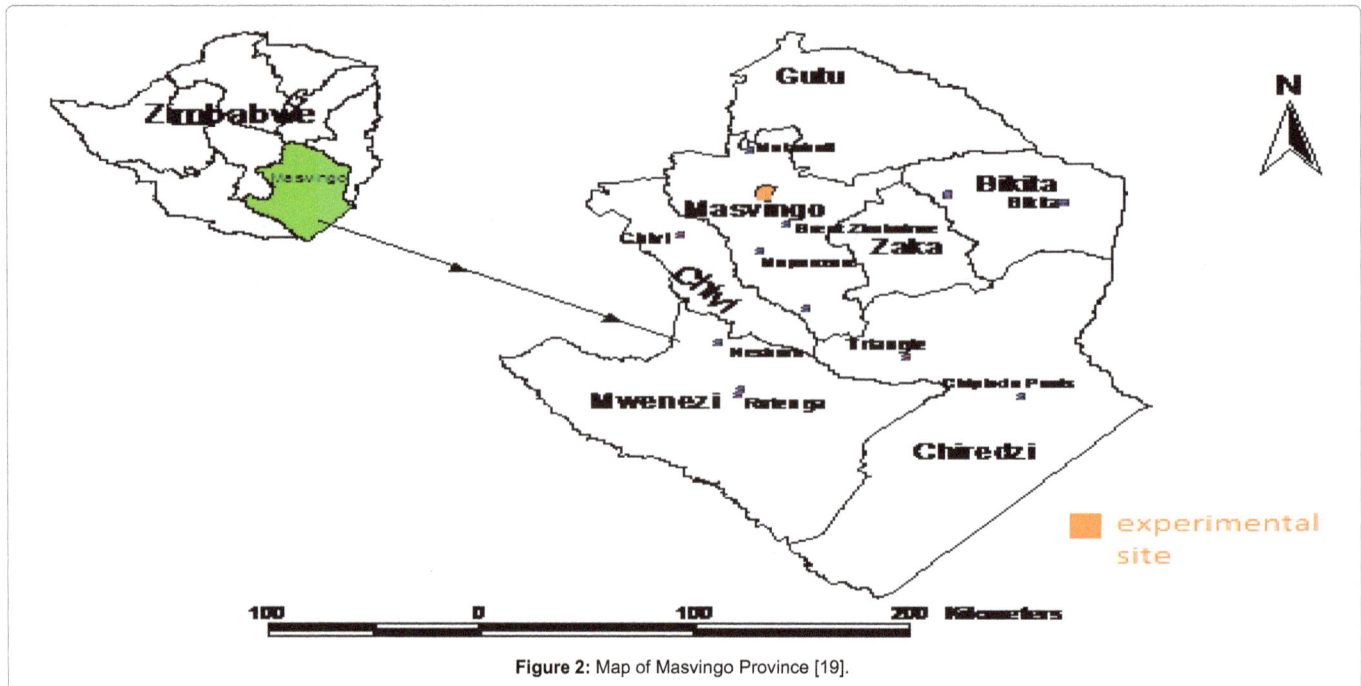

Figure 2: Map of Masvingo Province [19].

- Class 'A' pan
- Stevenson Screen (SS)
- Global Positioning System (GPS) 60 receiver
- Computer

Experimental setup

The experiment site comprised an automatic weather station, Class 'A' evaporation pan and a Stevenson screen with a hygrometer as illustrated in Figure 3. The experiment site was fenced to avoid unnecessary interference by animals and people.

Installation of instruments

All the instruments were levelled using the spirit level. Levelling is a pre-requirement for recording accurate readings. The automated weather station, pan and the Stevenson screen were all placed at least 1.5 m away from each other to avoid obstruction of measurements. The fetch of the study site was also of interest so the closest building was 15m away as this enabled correct measurements to be made.

RainWise MKIII Weather Transmitter: The MKIII automatic weather station was mounted at a height a standard height of 2 m as shown in Figure 3. The necked down end was placed into the MKIII assembly, until it bottomed the retaining screw in the slot and the screw was tightened. The solar panel was faced true north. The angle for optimum performance 60^0 on latitude 20^0 6′ was obtained from the MKIII user's manual.

RainWise MKIII Weather Transmitter mounted at experiment site: The RainWise MKIII Weather Transmitter shown in Figure 3 is designed to measure wind speed, wind direction, air temperature, humidity and rainfall. It is powered by a solar panel connected to rechargeable batteries. The station transmitted all the data via a wireless transmitter to a data logger. The received data was then downloaded to a computer, and analysed using MS Excel software.

Data logger: The RainWise CC-3000 MK-III Computer Interface

is a device that will record and store weather data received from MK-III sensor. The CC-3000 MK-III Computer Interface has a 2 MB flash memory which allows the device to log data independently of a computer. It provides both current and historical information.

Signal transmission: The RainWise MKIII Weather Transmitter generates and uses radio frequency waves in the frequency band 2.4 GHz for signal transmission. The transmitter was set to transmit the signal at 1 second interval. The data collect was later converted to daily averages. Transmission range is around 100m along the line of sight. However, obstruction between the sensor transmitter and the receiver may affect overall range.

Class 'A' pan installed at the experimental site: The United States Class 'A' pan (obtained from South Africa) which is cylindrical and is made of 0.8 mm thick galvanized iron. It is 25, 4 cm deep and 120, 7 cm in diameter. The pan was installed on bricks and levelled using a spirit level. Ideally a wooden grid platform could have been used but was not available. The bricks however allowed for the air circulation and detachment of the pan from the ground. The pan shown in Figure 3 was placed on a piece of bare land measuring 10 m by 10 m, fenced permitted free air circulation and prevent animals from drinking water from the pan . Water was filled to 5 cm below the rim. The level of the water was not allowed to drop to more than 7, 5 cm below the rim.

The Stevenson screen 100050 is a louvered cabinet with a hinged and chain – supported door, with door catch. It is constructed from hard wood and finished with white enamel paint. All the SS 100050 Stevenson screen was mounted on a stand 1.5 metres above the ground surface and it carried with it the wet and dry bulb thermometers.

Measurement of Weather Variables

Automatic weather station

The AWS sampled after every 30 minutes data on air temperature, humidity, pressure, rainfall, wind direction, and wind speed. Through wireless transmission the data logger received data from the automatic weather station and stored it. Daily averages were used in this study.

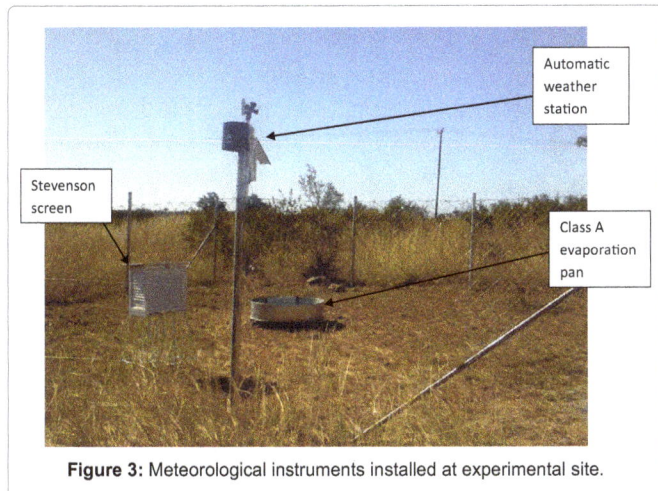

Figure 3: Meteorological instruments installed at experimental site.

Data was off loaded onto a computer via a USB cable and processed in Microsoft Excel spread sheets.

Measurement of evaporation

Decrease in height of water level was recorded at 30 minute intervals. Three readings were taken every 30minutes to minimise error and daily averages were computed. Microsoft Excel spreadsheets were used to process data.

Estimation of T_{max}, T_{min} and relative humidity

Readings were taken manually from the dry and wet bulb thermometers respectively, after every 30 minutes. The wet bulb thermometer gave minimum temperatures and the dry bulb thermometer gave maximum temperatures. The two temperatures were also used to estimate relative humidity using the tables in the manual. Daily averages were found and these were used during data processing with Excel spreadsheets

Estimation of net radiation

In the absence of a radiation sensor net radiation (R_n) was computed from solar and longwave radiation using equations (8) and (9). The computation is based mainly on T_{min} and T_{max} temperatures recorded by the Stevenson Screen. The magnitude of daily soil heat flux (G) beneath the reference grass surface relatively small, so it was ignored for 24 hour time steps used in this study.

Determination ETo by Penman–Monteith (P–M) method

Calculation of ETo with the Penman-Monteith equation on 24 hour time scales was done using equation 1. This generally provides accurate results. The daily ETo was computed from meteorological data consisted of the following parameters:

- Air temperature: maximum (T_{max}) and minimum (T_{min}) daily air temperatures.

- Air humidity: mean daily actual vapour pressure (e_a) derived from relative humidity data

- Wind speed: daily average for 24 hours of wind speed (u_2), measured at a standard height of 2 m.

- Net radiation was estimated using equations (8) and (9)

- Other parameters required for calculation of ETo by the Penman – Monteith method were calculated using equations (1) to (7).

Determination of Kp-the pan calibration constant

ETo values generated from the P–M method were plotted against E_p values to establish the correlation according to equation 10. A linear equation for correlation graph was generated using Microsoft Excel. The gradient of the curve gave the Kp value. The regression coefficient, R^2 indicated the strength of the relationship between variables calculated.

Corrective measures for evaporation measurements

Pan evaporation is used to estimate the evaporation from lakes [16]. There is a correlation between lake evaporation and pan evaporation. Evaporation from a natural body of water is usually at a lower rate because the body of water does not have metal sides that get hot with the sun, and while light penetration in a pan is essentially uniform, light penetration in natural bodies of water will decrease as depth increases. Most literature suggests multiplying the pan evaporation by 0.75 to correct for this [18]. Therefore in this study there are two sets of results, one for experimental pan measurements and for practical applications for dams and lakes which are multiplied by 0.75.

Class 'A' pan limitations

If precipitation occurs in the 24 hour period, it is taken into account in calculating the evaporation. Sometimes precipitation is greater than evaporation, and measured increments of water must be dipped from the pan. Evaporation cannot be measured in a Class 'A' pan when the pan's water surface is frozen [19].

The Class 'A' Evaporation Pan is of limited use on days with rainfall events of >30 mm (203 mm rain gauge) unless it is emptied more than once per 24 hours. Analysis of the daily rainfall and evaporation readings in areas with regular heavy rainfall events shows that almost without fail, on days with rainfall in excess of 30mm (203 mm Rain Gauge) the daily evaporation is spuriously higher than other days in the same month where conditions more receptive to evaporation prevailed. The most common and obvious error is in daily rainfall events of >55 mm (203 mm rain gauge) where the Class 'A' Evaporation pan will likely overflow.

Measurement of bare ground evaporation

Readings of water depth change in the Class 'A' pan were used to measure evaporation in units of mm day⁻¹. Graphs were plotted to show diurnal variation of evaporation with wind speed, net radiation.

The FAO-Penman-Monteith determined ETo, Class 'A' pan evaporation data and measured climatic data are presented. Data on variation of evaporation with relative humidity, air temperature, net radiation and wind speed various is analyzed and discussed.

Field measurements

Table 2 gives a summary of Ep values as determined by the Class 'A' pan ETo values estimated by the FAO Penman-Monteith method. The Penman-Monteith estimated values are higher than the Ep values because by definition ETo includes water lost both by evaporation from the soil and by transpiration. The average ETo is 5.6 mm per day, while the average Ep is 5.1 mm per day.

Table 1 and Table 2 shows that Ep has increased from 4.0 mm to 8.02 mm from day 1 to day 6. From day 6 to day 9, Ep decreased to 5.0 mm per day and then increased to 4.72 mm per day on day 10. The International Panel of Climate Change, IPCC (2007), Third Assessment Report [19], which assesses climate change research up to 2001, concludes that global average surface temperature has increased

Time (days)	Daily total net radiation (MJm⁻²day⁻¹)	Air temperature (°C)			Relative humidity (%)	Wind speed (m/s)	Pressure (kPa)	Class A pan evaporation Ep (mm/day)
		Max	Min	Mean				
1	23.59	26.3	18.9	22.6	50.0	8.4	89.9	3.59
2	17.48	28.9	17.9	18.9	71.3	13.1	90.5	3.80
3	20.21	26.6	18.4	18.5	65.2	9.6	90.6	4.91
4	20.60	24.3	19.9	20.1	52,4	12.1	90.4	5.33
5	21.71	25.5	18.1	19.8	54.7	13.9	90.6	7.12
6	17.91	24.3	13.7	19.0	50.2	15.7	90.7	8.02
7	23.77	25.7	18.5	20.1	40.8	10.7	90.3	5.65
8	21.37	23.5	19.1	19.3	38.4	8.4	90.4	5.02
9	22.22	21.1	13.5	17.3	50.3	10.7	90.7	2.90
10	19.51	23.0	14.4	17.8	53.6	11.4	90.6	5.72

Table 1: A summary of daytime climatic variables.

Time (days)	FAO Penman-Monteith Daily total evapotranspiration ETo (mm/day)	Class A pan evaporation Ep (mm/day)	Adjusted Ep (Lake Evaporation) (Ep x 0.75) (mm/day)
1	4.0	3.59	2.70
2	4.5	3.80	2.85
3	4.7	4.91	3.68
4	6.0	5.33	4.00
5	7.8	7.12	5.34
6	8.6	8.02	6.02
7	6.2	5.65	4.24
8	5.5	5.02	3.77
9	3.1	2.90	2.18
10	5.0	4.72	3.54
Average	5.6	5.1	3.83

Table 2: Evaporation rates.

by 0.6°C (±0.2°C) over the 20th century, and is predicted to increase by 1.4°C to 5.8°C between 1990 and 2100; average precipitation has increased over tropical latitudes by about 2% to 3% throughout the 20th century, and on average has decreased by about 3% in the sub-tropics . If water harnessing mechanisms are not put in place there will be water scarcity in Masvingo town due to evaporation.

Air temperature does not show a consistent pattern. For instance, when Ep increased from 4.0 mm to 4.91 mm per day, maximum temperature was almost constant from day1 to day 3 from 26.3°C to 26.6°C, then decreased to 23.0°C on another day to other day 10°C showing that Ep does not depend on air temperature alone.

Figure 4 shows the regression of the Class 'A' pan evaporation and evapotranspiration, ETo estimated by the Penman-Monteith method for a period of ten days. The linear regression coefficient R^2 with a value of 0.97 indicated a very strong correlation. Considering the equation (10) the pan coefficient Kp=0.91. The pan coefficient on an annual basis has an average value of 0.8 [20].

The experimentally obtained pan coefficient is within ± 1.5 of error which makes the reading acceptable.

Effects of weather variables on evaporation

Figure 5 shows that from day 1 to 2, RH increased from 50% to a peak value of 71%. In the same period Ep increased from 3.59 mm to 3.80 mm per day. After day 2 up to day 4, RH decreased to 52.4% while Ep continued increasing until it reached a peak value of 8.0 mm per day on day 6. RH increased from day 4 to 5 to 54.7% and up to day 8 to 38.4%. After day 8 RH increased up to day 10 to 53.6%. From day 6 to 9, Ep decreased to from 8.0 to 2.9 mm per day and then increased to 4.7 on day 10. It is clear that Ep does not always vary proportionally

directly with relative humidity. This shows that some whether variables other than relative humidity are also responsible for influencing the rate of evaporation, otherwise the rate of evaporation should have decrease whenever relative humidity increased.

Figure 6 shows that generally Ep increased when net radiation increased and decreased when net radiation decreased. This consistent with theory which identifies the main driver of evaporation is net radiation.

Figure 7 shows that Ep generally increases with wind speed. This because area around the pan remained unsaturated since water vapour was continuously swept away by wind. From day 2 to day 3 Ep increased although wind speed was decreasing. This clearly shows that Ep is dependent on other weather variables as well other than wind speed.

Conclusions

Evaporation rates seemed to be driven by net radiation and wind speed, and less by relative humidity and air temperature as shown by graphs. In most cases for the ten day period of the study, the rate of evaporation increased when net radiation and wind speed increased. This is because direct solar radiation, provided energy to change the state of water molecules from liquid to vapour. Net radiation also induced air motion which kept the vapour pressure deficit around the Class 'A' pan high hence increasing the rate of evaporation.

Generally the experiments gave close to literature readings for pan coefficient which in turn mean that the results from the pan can be accepted and used to estimate evaporation rates in the vicinity.

Typical evaporation rates estimated by the Class 'A' pan method

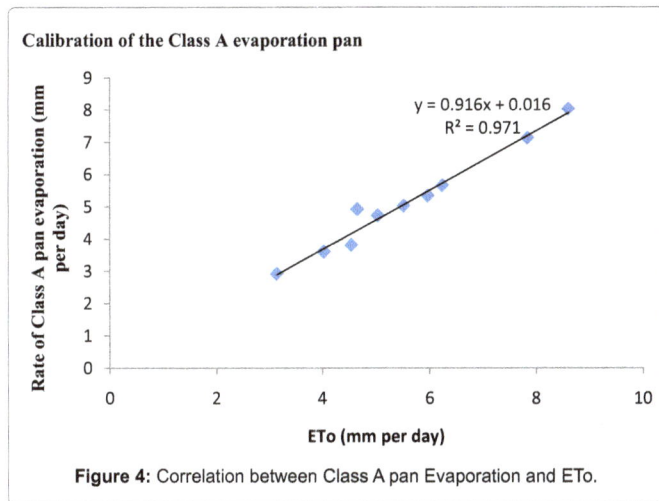

Figure 4: Correlation between Class A pan Evaporation and ETo.

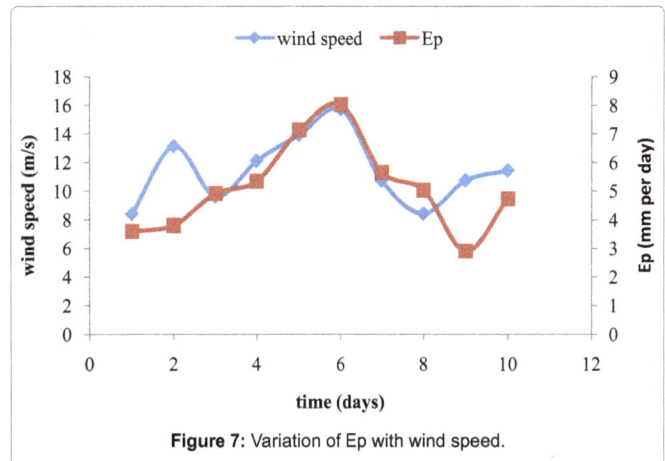

Figure 5: Variation of Ep with relative humidity.

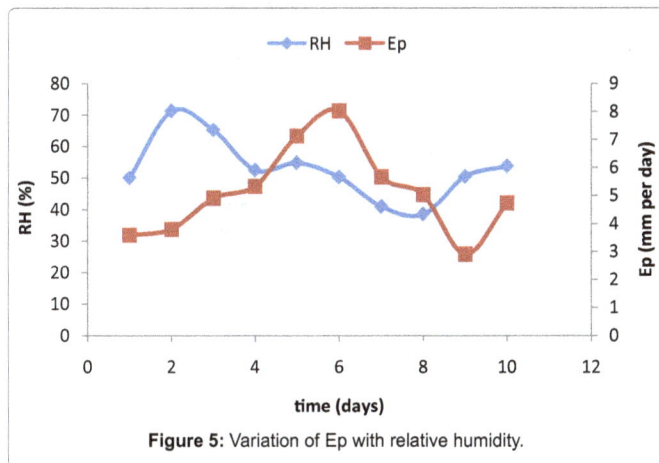

Figure 6: Variation of Ep with net radiation.

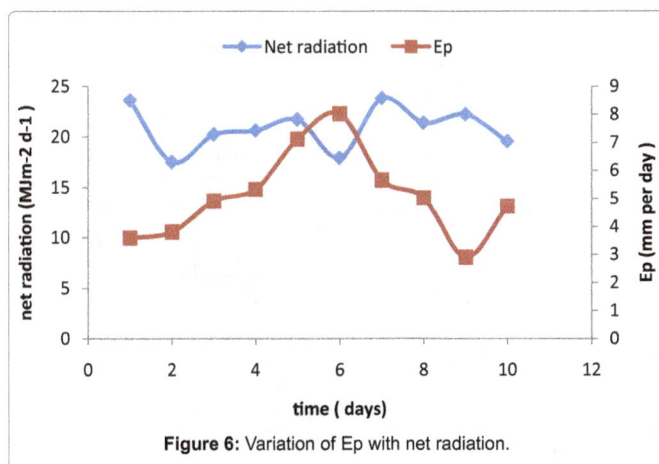

Figure 7: Variation of Ep with wind speed.

References

1. Butling F, Makadho J (1991) Estimating crop water requirements: comparative analysis between the Modified Penman method and the Pan Evaporation method used in Zimbabwe. Department of AGRITEX, Zimbabwe.

2. Allen RG, Pereira SL, Raes D, Smith M (1998) Crop evapotranspiration– Guidelines for computing crop water requirements – FAO Irrigation and Drainage Paper 56. Rome, United Nations.

3. Doorenbos J, Pruitt WO (1977) FAO Irrigation and Drainage Paper 24. Rome, Italy.

4. Feller MM (2011) Quantifying Evapotranspiration in Green Infrastructure: A Green Roof Case Study. A Thesis in Civil and Environmental Engineering submitted to the Villanva University in partial fulfillment of the requirements for the degree of Master of Science in Civil Engineering.

5. M Smith, D Kivumbi (1993) CLIMWAT for CROPWAT: A climatic database for irrigation planning and management. FAO Irrigation and Drainage Paper 49. Rome, Italy.

6. Kaboosi K (2012) The Investigation of Pan Evaporation Data, Estimation of Pan Evaporation Coefficient by Pan Data and Its Comparison with Empirical Equations. International Journal of Agriculture and Crop Science. 4: 1458 – 1465.

7. Taylor AS, Ashcroft GL (1972) Physical edaphology: The physics of irrigated and non-irrigated soils. WH Freeman & Co Ltd., New York.

8. Watermeyer JM (1980) Evaporation pans: their value to farmers. Cattle World.

9. Ambas VT, Baltas E (2012) Sensitivity analysis of evapotranspiration methods using a new sensitivity coefficient. Global NEST Journal 14: 335-345.

10. Savva P, Frenken K, Mudima K, Chitima M, Tirivamwe L (2002) Irrigation Manual Module 4: Crop Water Requirements and Irrigation Scheduling. Harare, FAO.

11. Che D, Gao G, Xu Chong-Yu, Guo J, Ren G (2005) Comparison of the Thornthwaite method and pan data with standard Penman-Monteith estimates of reference evaporation in China. Climate Research 28: 123-132.

12. Crop Water Needs, FAO Corporate Document Repository.

13. Roderick ML, Graham DF (2002) The cause of decreased pan evaporation over the past 50 years. Science 298: 1410–1411.

14. Simba FM, Mubvuma M, Murwendo T, Chikodzi D (2013) Prediction of yield and biomass productions: A remedy to climate change in semi-arid regions of Zimbabwe. International Journal of Advance Agricultural Research 1: 14-21.

15. Moore T (2007) Officials defend dam against attacks. Brisbane Times.

16. Linacre E (2002) Ratio of lake to pan evaporation rates.

17. NOAA Glossary: Evaporation Pan.

18. Bosman HH (1990) Methods to convert American Class A-Pan and Symon's Tank Evaporation to that of a representative environment. Water SA 16: 227-236.

ranged from 2.9 to 8.0 mm/day, while for the Penman-Monteith method they ranged from 3.1 to 8.6 mm/day. The average for the former method was less than that and for the latter. This is consistent with what was expected because the Penman – Monteith method estimated evaporation rates included water lost by evaporation and transpiration, while the Class 'A' pan method estimates involved only water lost by evaporation. The evaporation rates obtained were not sustainable in the long term if water harnessing strategies are not employed.

15

Mini Polders, as an Alternative of Flood Management in the Lower Bengawan Solo River

Wignyosukarto BS[1]*, Mawandha HG[2] and Jayadi R[3]

[1]Professor, Department of Civil and Environmental Engineering, Faculty of Engineering, Universitas Gadjah Mada, Yogyakarta, Indonesia
[2]Graduate Student, Department of Civil and Environmental Engineering, Faculty of Engineering, Universitas Gadjah Mada, Yogyakarta, Indonesia
[3]Associate Professor, Department of Civil and Environmental Engineering, Faculty of Engineering, Universitas Gadjah Mada, Yogyakarta, Indonesia

Abstract

Sumbangtimun Village and Kandangan Village, in the lower part of Bengawan Solo River, are regularly flooded almost every year. These flood prone area are situated in the river meander with low natural levee. Losses due to flood damage were estimated to reach 2.5 billion rupiah in the year 2007/2008 with the greatest losses from agricultural sector. These losses were caused by flooding that comes early before the harvest time. Surround dike to prevent flood inundation is recommended by the river authority. Regarding different land uses, i.e. rice fields, tree crops plantation, settlement, with different inundation tolerant, a set of mini polder is proposed to overcome the problem of inundation and its water management. The polder system is attributed by surrounded dike, local drainage system and pond retention with or without pumping system. The mini polder system that is equipped with lateral gate as flood control structure is proposed to solve the problem of inundation in the agriculture and settlement areas. Hydraulic mathematical models (HEC-RAS and Pond Pack) for simulating flow characteristics in the flood prone area are employed to determine the design of an appropriate mini polder system and its operational procedures. In addition of technical aspects, the design is also reviewed for compliance of the non-technical aspects including agricultural, economic and environmental in the local communities. According to the result of frequency analysis, flood discharge in the Bengawan Solo River was 1,500 m³/s for 2 year return period (Q2) and 2,525 m³/s for 10 year return period (Q10). Flood water elevation based on 2 year flood return period was obtained at +17.50 m or inundation depth average was around 1 m. The area that inundated at this condition was around 250.48 ha and the inundation duration was in 2 days. Flood water elevation based on 10 year flood return period was at +19.05 m or inundation depth average was around 2.5 m. The area that inundated at this condition was 382.58 ha with the inundation duration was in 3 days. The use of storage area and lateral structures models can describe flood phenomena in the modeled flood prone area more accurate.

Keywords: Flood mitigation; Polder; Mini polder; Bengawan solo river; Skotbalk

Introduction

Bengawan Solo River is the longest river in Java Island (± 600 km length) that flows started from Wonogiri, in the south part of Central Java Province, passing through the north part of East Java Province, and its mouth is in Java Sea at Ujung Pangkah. Lower Bengawan Solo River, forming wide and meandering river channel, with the river channel length ± 300 km flowing through the alluvial plains of ± 6,273 km², an area which is regularly flooded. Close to the river mouth, the channel passing through a broad marshy area called Jabung Swamp and Bengawan Jero. Flood inundation in the lower part of Bengawan Solo river is almost always occurred yearly, covering an area of about 11,000 ha, in the month of December to May [1]. Sumbangtimun and Kandangan villages, that are located in the inner bank of a meander of lower part of Bengawan Solo River, upstream of Bojonegoro city, were frequently inundated during high flood. Its location is depicting in Figure 1. Sumbangtimun and Kandangan villages covering 289 ha of rice field and 120 ha settlement. The irrigation system in these area are supplied by pumps, elevating water from Bengawan Solo river (90 ha) and local well (199 ha). Agricultural water resources in this area is highly dependent on rainfall and Begawan Solo River. In the dry season water level in the river is so low that the cost of pumping water to the land is too expensive. Land some distance from the river can only utilize groundwater pumped from farmer well. Considering this source of water, the cropping pattern is different between land near to the river and the land away from the river. Land close to the river, use the river water in the rainy season for rice cultivation, from the first planting season in October and harvesting in late January, and start the second planting season in March and harvested in June. In the dry season the land is planted dry crop such as corn. The land is located away from the river, because of limited irrigation water, pumping available soil

water during the dry season for dry crop (corn and cassava). Cropping pattern in this area only plant rice once during rainy season and twice

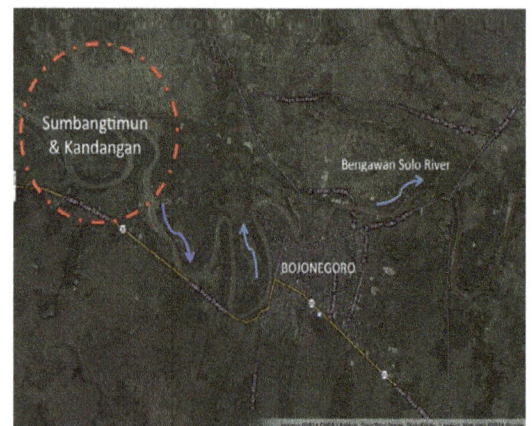

Figure 1: Location of Sumbangtimun and Kandangan villages.

*Corresponding author: Wignyosukarto BS, Professor, Department of Civil and Environmental Engineering, Faculty of Engineering, Universitas Gadjah Mada, Yogyakarta, Indonesia, E-mail: budiws@ugm.ac.id

dry crop. Figure 2 shows the land use pattern and the existing irrigation system.

Losses due to flood damage were estimated to reach 2.5 billion rupiah in the year 2007/2008 with the greatest losses from agricultural sector. These greatest losses were caused by flooding that comes early before the harvest time. Table 1 shows the cropping pattern and the occurrence of flood. Land elevation of this area varies from +14.00 m ~ +22.00 m above mean sea level (msl), with average elevation of rice field is +16.50 m and average elevation of the settlement is about +21.00 m. The extreme flood level could reach +17.00 m ~ +20.00 m above msl, while the average river water level during dry season is +8.00 m ~ +10.00 m above msl. Natural levee elevation is relatively low, between +17.00 m msl ~ +18.00 m msl, so it cannot be said as a flood control system. Pumping system is only used by farmers for irrigation and cannot be operated during floods. Adaptation action of society to flooding is to elevate ground elevation at several locations for the purposes of access roads and settlements.

Mini Polder

A polder is a piece of land protected from water by a system of dams and dikes and within which the water level can be regulated. Originally the name "polder" was used in The Netherlands, nine to ten centuries ago, and was related to the land protected with dams from sea floods. Later, the same system of land protection was used on the flood plains of rivers, lakes, and shallow reservoirs, and they were all called polders. The polders on the banks of the major rivers were protected against high water from the river and discharged into it. In several parts, polders are a special type of drained agricultural land typically found in low-lying coastal areas, river plains, shallow lakes, lagoons and upland depressions. Before impoldering, polder areas were either waterlogged or temporarily or permanently under water. An area becomes a polder when it is separated from the surrounding hydrological regime in such a way that its water level can be controlled independently of its surrounding regime. This condition is accomplished by various

combinations of drainage canals and dikes [2]. In other parts, polder are also dedicated to minimise the impact of flood water. The use of retention areas can be an efficient measure in flood risk management. By controlled flooding of sparsely or non-populated areas with relatively low damage potential, the risk of inundation for downstream areas with higher vulnerability can be reduced. The largest retention area along the Elbe River that is situated on the tributary Havel River, near its confluence with the Elbe is one of example. This retention area consists of six large polders and the floodplain of the Lower Havel River which together have a potential retention volume of up to 250 million m³. The six polders comprise an area of 100 km² with a volume of 110 million m³. The system was constructed in the 1950s, but was used operationally for the first time during the Elbe flood in 2002. By controlled flooding of the retention area the peak stages were attenuated by approximately 40 cm at the gauge of town of Wittenberge. Consequently, the risk of inundation for the town of Wittenberge and areas further downstream was significantly reduced. Before the extensive construction of dikes and water engineering works, the whole Lower Havel River floodplain was a natural retention area for the Elbe River and was characterised by frequent inundations. The dikes were constructed in order to protect settlements and agricultural areas from flooding [3]. Forster et al. [3] simulated the controlled flooding of the retention area by the use of a conceptual model and assessed economically for two flood scenarios. In a cost-benefit analysis, the damage to agriculture, the road network, buildings and fishery caused by the flooding of the polders was compared with the resulting reduction in potential damage in the town of Wittenberge, 30 km downstream. On the basis of a monetary assessment it was concluded that the use of the retention area for flood protection is highly cost-effective in economic terms.

Several methods have been applied to mathematically represent water flow in rivers with floodplain by using numerical model. Huang et al. [4] analysed the effectiveness of polders system along the middle reaches of the Elbe River, in capping flood discharges and how much this capping effect recedes as the flood wave propagates downstream of the polder, by using mathematical model. Along the floodplain of Lower Bengawan Solo River that forming wide and meandering river channel, there are some settlement and agriculture area that are inundated during high flood but in times of drought, they need irrigation water that can be pumped from the river. The same idea of flood protection in Elbe River, for agriculture area and settlement with a minimum risk and irrigated when is needed, a series of polder are introduced along the Lower Bengawan Solo River. The polders on the banks of the rivers with small area, called mini polder, are protected against high water from the river and discharged into it. Inside the dike the water will be controlled either by being run off naturally through a sluice or by being pumped out. Within the mini polder, the capillaries of the whole water control system, the ditches and drains between the

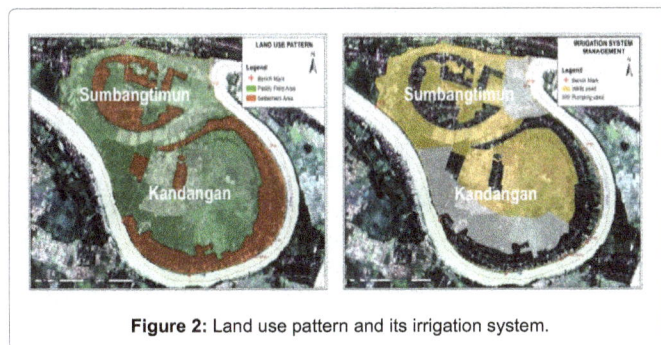

Figure 2: Land use pattern and its irrigation system.

Month	Sep		Oct		Nov		Dec		Jan		Feb		March		Apr		May		Jun		Jul		Aug		Losses
Half Monthly	I	II	I	II	I	II	I	II	I	II	I	II	I	II	I	II	I	II	I	II	I	II	I	II	(Billion Rp)
Pumping Irr	Land Prep		Rice Cultivation								Land Prep		Rice Cultivation						Land Prep		Maize				
		Land Prep		Rice Cultivation								Land Prep		Rice Cultivation						Land Prep		Maize			
Flood 2007																									2.5
Flood 2008																									2.5
Flood 2009																									1.2
Flood 2010																									0.75
Flood 2011																									0.35
Flood 2012																									no flood
Local Well Irr	Maize/cassava			Land Prep		Rice Cultivation						Land Prep		Maize/cassava											
	Maize/cassava				Land Prep		Rice Cultivation						Land Prep		Maize/Cassava										

Table 1: The occurrence of flood, 2007-2012.

strips of land, carry the water to the bigger watercourses through which it flows to the pumping station. In this way excess water is removed through the sluice or by a pump into the river directly.

Flood Mitigation

Sumbangtimun and Kandangan areas are located on the inner bank of meanders, surrounded by a natural levee that is not quite high, so that at the time of the great flood, always inundated by flood water of Bengawan Solo River. Flood water also come through mouth of Toweng Creek that has not equipped with flood gate. Drainage system in this area, is natural drainage system that generally serves to dispose of rain water into the Bengawan Solo River.

In the rainy season, the area has potential agricultural crop failure due to flooding. The main causes of crop failure is due to depth as well as duration of inundation, compare to the age of crop were flooded. The duration of inundation less than 24 hours is relatively safe, it will be less safe if the duration of inundation is more than 2×24 hours. Compare to the age of crop, inundation will give a high risk for crop with age of 0 ~ 30 days and 60 days (harvest time), and will give less risk for age of the crop of 30 ~ 60 days. In the dry season, as the river water drop too low, the farmers have a difficulty to lift up the irrigation water to the rice field. The farmers also face the problem of water scarcity during dry season, especially for domestic use and sanitation. Therefore flushing during rainy season using flood water is sometime needed (Figure 3). Regarding the water level measurements in Karangnongko and Bojonegoro during several years from 2009 up to 2012 [5], the land elevation of the area of Kandangan and Sumbangtimun generally always above the river water level, except during high flood that come during several days in the rainy season. Polder is an alternative of flood control measures to protect the area from being flooded. A polder is an area that is protected from outer water by a dike and that has a controlled water level on the inside of the diked area. Any water entering the polder (rain, seepage water) that is not used or stored, has to be pumped out [6]. Actually, this flood prone area is already surrounded by dikes along the river meander, but another dike in the north part along the Toweng Creek is still needed to form an artificial hydrological entity, and it should also be equipped with internal drainage system and pumping station. Considering hydro-topographical condition of the area, by comparing the land level to the river water fluctuation, the system of dikes is not really serve as polder system, because most of the time, more than 8 months/year, the land level is higher than the river water level, except 1 ~ 3 months/year during rainy season, the area is sometimes flooded. Due to the need of supply of water for irrigation purposes, several skotbalks (lateral gate) are proposed to be installed along the dikes that face to the Bengawan Solo River, as an intake of irrigation or flushing water (Figure 4).

Figure 3: (a) Existing drainage pattern**(b)** Existing Toweng Creek.

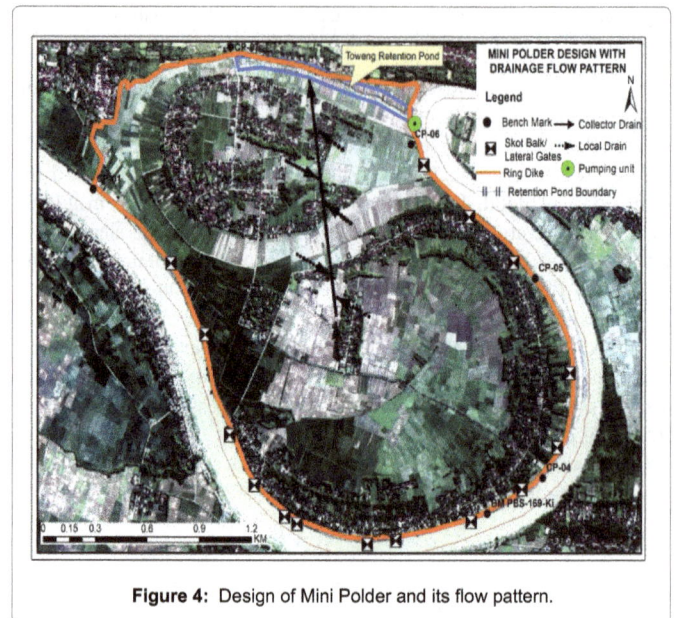

Figure 4: Design of Mini Polder and its flow pattern.

Flood simulation

Several methodologies have been applied to mathematically represent water flow inrivers with floodplain. The simplest method of river modeling consists of using a one-dimensional (1D) model and considering the floodplain as storage areas or adopting compound cross sections [7]. Horritt and Bates tested 1D and 2D models of flood hydraulics (HEC-RAS, LISFLOOD-FP and TELEMAC-2D) on a 60 km reach of the river Severn, UK. The three models are calibrated, using floodplain and channel friction as free parameters, against both the observed inundated area and records of downstream discharge. The predictive power of the models calibrated against inundation extent or discharge for one event can thus be measured using independent validation data for the second. The results show that for this reach both the HEC RAS and TELEMAC-2D models can be calibrated against discharge or inundated area data and give good predictions of inundated area. Huang et al. [4] simulated the effectiveness of polder system on peak discharge capping of flood along the middle reaches of the Elbe River in Germany, by using DYNHYD (1-D hydrodynamics) from WASP5 modelling package and extended to incorporate the quasi-2-D approach. It is sought in which a 1-D hydrodynamic model is used to allows the discretisation to be extended into polder system to give 2-D representation of the inundation area. In that simulation the inlet and outlet discharges of a polder are controlled by a virtual weir. To simulate the control of polders, the weir is opened gradually, by lowering the weir crest to the hinterland ground level, over a time period of up to 12 hours. The result show the quasi-2-D modeling is a powerful tool for provision and operational management of floods due to its robustness, ease of use and computational efficiency. The polder system can cap the flood peak effectively and the capping is relatively stable along the river down-stream.

In this research, a model for simulating flood propagation along the main river channel was combined with a reservoir model applied to the floodplain. The main flow along river channel was simulated with a 1-D hydrodynamic model (HEC-RAS), while the floodplain is considered a reservoir that is filled up gradually with a horizontal water surface, a catchment pond both function as flood control reservoirs pool (retention pond) as well as reservoirs and ponds reservoir (Pond Pack).

The hydrodynamic model HEC-RAS contains four one-dimensional river analysis components for: (1) steady flow water surface profile computations; (2) unsteady flow simulation; (3) movable boundary sediment transport computations; and (4) water quality analysis [8]. The equations that govern the unsteady flow of water in a stream are : (1) conservation of mass, and (2) conservation of momentum.

These laws are expressed mathematically in the form of partial differential equation as follows.

The continuity equation (conservation of mass)

$$\frac{\partial A}{\partial t} + \frac{\partial Q}{\partial x} - q_l = 0$$

The momentum equation (conservation of momentum)

$$\frac{\partial Q}{\partial t} + \frac{\partial QV}{\partial x} + gA\left(\frac{\partial z}{\partial x} + S_f\right) = 0$$

in which q_l is the lateral flow per unit length, A is wetted area, Q is flow discharge, V is flow velocity, z is channel bottom elevation, S_f is slope of energy.

Pond Pack allows users to model rainfall runoff urban and rural watershed to design detention and retention facilities, outlet structures and channels. The Pond Pack engine computes outlet rating curves with tail-water effects, account for pond infiltration, calculates pond detention times, analyses channels and performs interconnected pond routing with multiple outfalls. A pond may discharge to location influenced by downstream tidal effect or flooding [9]. Flood propagation of Lower Bengawan Solo River and its influence to the flood plain were simulated with HEC-RAS mathematical model combined with PondPack, for the reach of Karangnongko ~ Bojonegoro Barrage (Figure 5). Two schemes of simulation were done, started with simulation of existing condition and continued with simulation of system with mini polder scheme. The upstream boundary condition were 2- and 10-years return period flood hydrographs at Karangnongko, while downstream boundary condition was stage hydrograph at Bojonegoro Barrage [10]. According to results of the frequency analysis, the design flood discharge in the Bengawan Solo River at Karangnongko section for 2- and 10-year return period were 1,500 m³/s and 2,525 m³/s, respectively. By using stage gauge data in Karangnongko on December 31st 2007 and February 28th 2009, flood hydrograph of the two return periods could be determined.

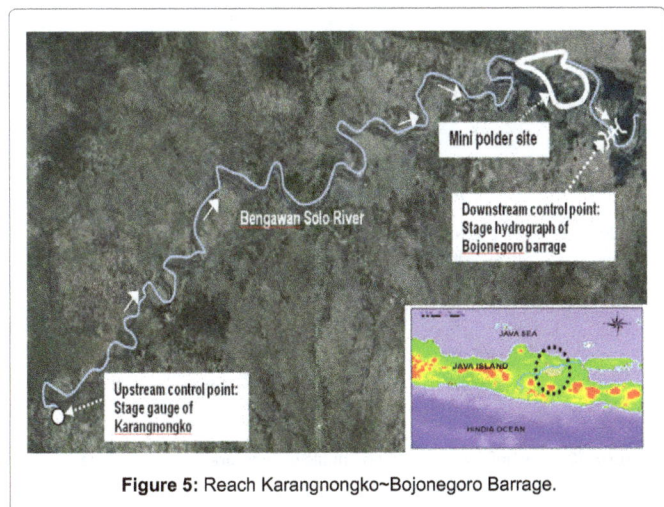

Figure 5: Reach Karangnongko~Bojonegoro Barrage.

Simulation of existing condition

Result of simulation of 2-year return period of flood, for existing condition, showed that the maximum water elevation in the project area was +17.50 m and average inundation depth was around 1 m. The inundated area was about 250.48 ha (Figures 6a and 6b). Stage hydrograph of 2-year return period flood with 2 days inundation time. In case of 10-year return period of flood, the simulation result show that the maximum water level in the project area was +19.05 m. This flood inundated the flood prone area up to 382.58 ha. Figure 7a shows the inundation area that covers almost the whole area of Sumbangtimun and Kandangan Villages. Results of hydraulic simulation explained that inundation due to 10-year return period of flood, occur for almost 3 days and the water level in Bengawan Solo River raised about 2.5 m above the ground elevation of storage area (Figure 7b).

Simulation with mini polder system

The proposed mini polder system should meet several objectives, i.e. this system should work as flood protection during high flood; it could give possibility to allow river water flows to the area for irrigation purposes during normal flow; the cost of construction and operation should not expensive, and it should be equipped with reasonable drainage system and pumping station.

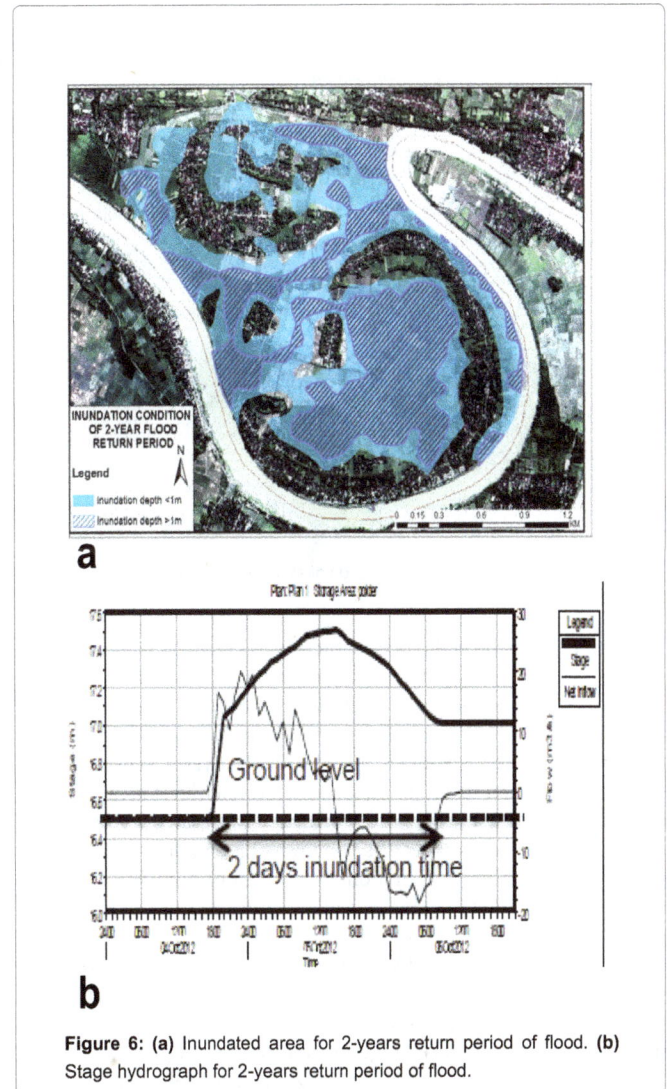

Figure 6: (a) Inundated area for 2-years return period of flood. **(b)** Stage hydrograph for 2-years return period of flood.

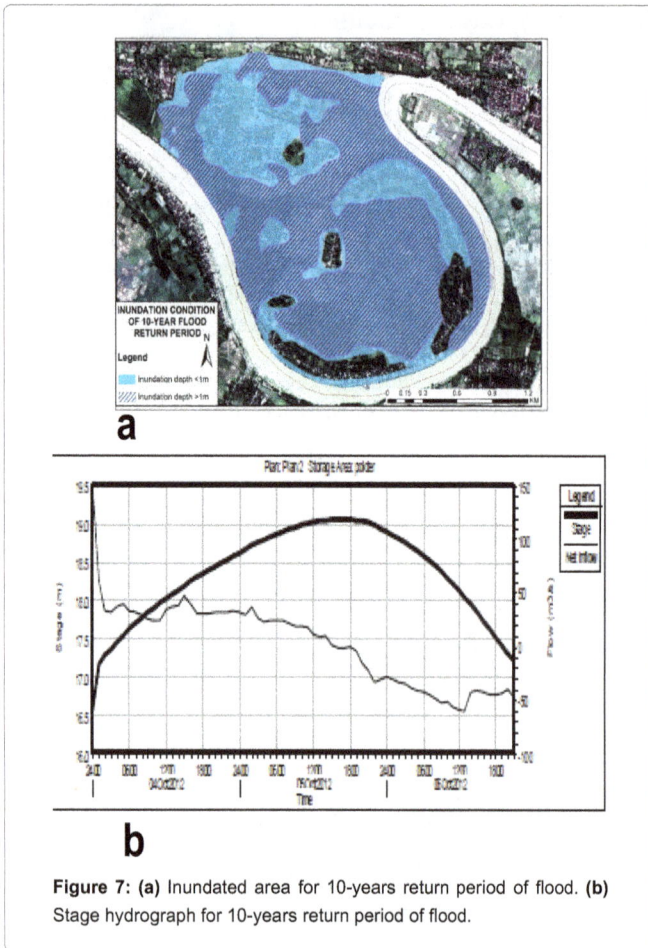

Figure 7: (a) Inundated area for 10-years return period of flood. **(b)** Stage hydrograph for 10-years return period of flood.

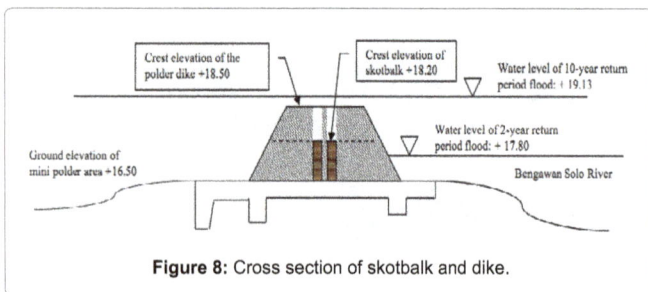

Figure 8: Cross section of skotbalk and dike.

Flow of water from the river to the area is regulated with skotbalks. In order to prevent the structural damages of dike due to overtopping risk of mean annual flood, skotbalks have to be fully opened when the water level in the river reach the crest of skotbalk, i.e. +18.20 m. Under this operation rule of skotbalk, the dike will be overflowed by 10-year return period flood in which the maximum water level is higher than 0.60 above the crest. Therefore when the river water level is rising close to the crest of skotbalk, the flood warning should be informed to people for evacuation preparation. The design of mini polder system was equipped with 15 skotbalks which has function as flood control gates. Situation of mini polder system can be seen in Figure 4. Figure 8 illustrates cross section of skotbalk and dike. According to result of simulation, for 2-year return period flood, the peak river water level was +17.80 m. Since the crest level of dike is +18.50 m, there is no water flows through the lateral weir (dike) and the lateral gates (skotbalk). Output of the simulation also the stage hydrograph in mini polder

area with constant value that equal to the ground elevation (+16.5 m) and zero discharges of the hydrograph. Result of simulation with 10-year return period of flood concluded that the dike of mini polder was overflowed. River Water level in the project area could raised up to + 19.13 m. Figure 9 describes fluctuation of water level and discharge of flood flow through dike and skotbalk.

Retention pond and pumping system

The computation of retention pond and pumping capacity was done by using Pond Pack software. The input design scenario of simulation are the overflow hydrograph for 2- and 5-years return period of rainfall. The existing capacity of Toweng Creek retention pond with the area of 7,000 m² and bottom elevation at +12.00 m, could not accommodate the runoff volume due to local rainfall of 2- and 5-year return periods. The capacity of retention pond without pumping system was 32,092 m² with the bottom elevation at +7.5 m. By using pumping system with the capacity of 6 m³/s, the storage area become 32,322 m² with the bottom elevation at +10 m. If the storage capacity is maintained in the existing condition (without dredging and widening) the required pump capacity becomes larger than 12 m³/s. The result of retention pond simulation, the inflow hydrograph, outflow hydrograph and surface water elevation are depicted in Figure 10.

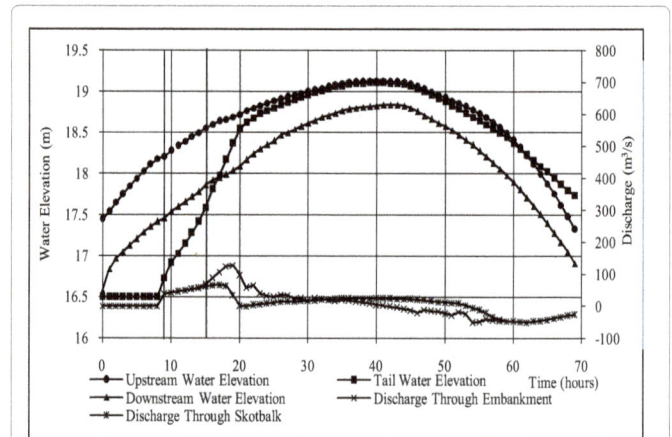

Figure 9: Fluctuation of water level and discharge of flood flow through dike and skotbalk.

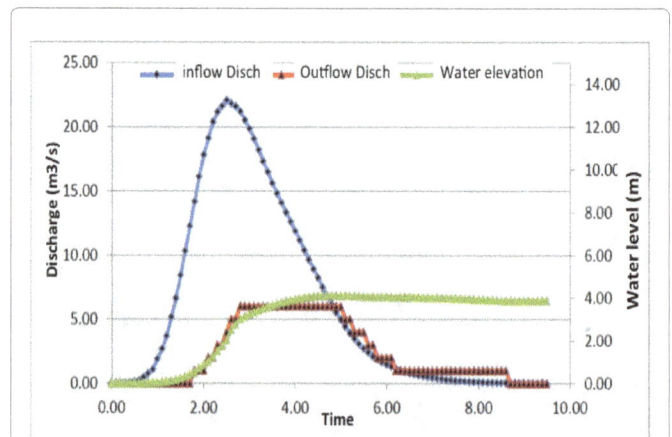

Figure 10: The inflow hydrograph, outflow hydrograph and surface water elevation in retention pond.

Conclusion

The use of 1-D hydrodynamic model (HEC-RAS) combined with Pond Pack for simulating the flood propagation along the main river channel and the floodplain storage area with lateral structures models can describe flood phenomena in the flood prone area more accurate. The mini polder system that is equipped with lateral gate as flood control structure could solve the problem of inundation in the agriculture and settlement areas. The crest of embankment is designed at elevation + 18.5 m. Mini polder area is still inundated by the flood that is greater than 2 year return period. When the greater flood occurred, people must follow the evacuation procedures. To anticipate a greater flood occurrence, the flood control gates are opened if the water level reaches elevation higher than +18.2 m (crest elevation of flood control gates) or opened for leaching the farming land. The analysis of retention pond concludes that by using pumping system with the capacity of 6 m^3/s, it showed reduction of storage capacity up to 30% becoming 205,309 m^3. If the storage capacity in the Toweng River is maintained in existing condition then the required capacity of pumping system become higher than 12 m^3/s.

Acknowledgement

This paper is part of the work of Hanggar G Mawandha in his thesis for his completion of Master Program in Universitas Gadjah Mada Yogyakarta Indonesia.

References

1. Sarwono (2000) Flood control and raw water supply in Lower Bengawan Solo River. Litbang Kimbangwil Sumber Daya Air, Bandung, Indonesia.

2. Luijendijk J, Schultz E, Segeren WA (1988) Polders, In: Developments in Hydraulic Engineering, Vol. 5, Elseviers Applied Science, London, Great Britain.

3. Forster S, Kneis D, Gocht M, Bronstert A (2005) Flood risk reduction by the use of retention areas at the Elbe River. International Journal of River Basin Management 3.

4. Huang S, Rauberg J, Apel H, Disse M (2007) The effectiveness of polder systems on peak discharge capping of floods along the middle reaches of the Elbe River in Germany. Journal of Hydrology Earth System Science 11: 1391-1401.

5. Bengawan Solo RBO (2013) Water Level Data Base.

6. Rijkswaterstaat (2011) Water Management in the Netherlands, The Ministry of Infrastructure and Environment, Directorate-General Water and Rijkswaterstaat, Centre for Water Management, The Netherland.

7. Cunge JA, Holly FM, Verwey A (1980) Practical aspects of computational river hydraulics, Pitman Advanced Publishing Program, Boston.

8. Brunner Gary W (2010) HEC-RAS River Analysis System. Hydraulic Reference Manual, ver. 4.1. U.S.Army Corps Engineers, USA.

9. Haestad Methods Inc. (2008) PondPack for windows User's Guide, Ver.8, Waterbury, USA.

10. Nippon Koei Co. Ltd. (2003) Detailed Design on Bojonegoro Barrage: Project Report. Bengawan Solo River Basin Organization. Surakarta, Indonesia.

16

Comparison of Reference Evapotranspiration Calculations for Southeastern North Dakota

Xinhua Jia[1]*, **Thomas Scherer[1]**, **Dongqing Lin[1]**, **Xiaodong Zhang[2]** and **Ishara Rijal[3]**

[1]Department of Agricultural and Biosystems Engineering, North Dakota State University, Fargo, North Dakota-58108, USA
[2]Department of Earth System Science and Policy, University of North Dakota, Grand Forks, North Dakota-58202, USA
[3]Department of Geography, MichiganStateUniversity, East Lansing, Michigan-48824, USA

Abstract

Potential water consumption for irrigation scheduling in North Dakota was typically calculated from a reference Evapotranspiration (ET_{ref}) using the Jensen-Haise method and its associated crop coefficient (K_c) curves developed in the 1970's and 1980's. The ET_{ref} method proposed by the American Society of Civil Engineers, Environmental and Water Research Institute (ASCE-EWRI) reference evapotranspiration task force has shown to be more accurate and therefore more widely used than any other methods. However, to apply the ASCE-EWRI method for irrigation scheduling requires a corresponding change of the K_c curves associated with the Jensen-Haise method. In this paper, a comparison of ET_{ref} estimates for 11 methods, including the ASCE-EWRI and the Jensen-Haise methods, was conducted using 18 years of data collected in southeastern North Dakota. The results show that the annual ET_{ref} by the Jensen-Haise method was nearly the same as the ASCE-EWRI grass ET_{ref}, but with a higher Root Mean Square Deviation (RMSD), 0.903 mm d^{-1}, and a lower coefficient of determination (R^2) 0.8659. The ET_{ref} comparison for the growing season only shows an RMSD of 1.007 mm d^{-1}, R^2 of 0.7996 and 8.13% overestimation. The ET_{ref} by the Jensen-Haisemethod has a higher monthly ET_{ref} than the ASCE-EWRI in June, July, and August, and a lower monthly ET_{ref} for all other months in an 18 year period. The ET_{ref} comparisons also show that the modified Penman method used by the High Plains Regional Climate Center (HPRCC Penman) has the best accuracy and correlation with the ASCE-EWRI ET_{ref} method. Indeed, all alfalfa based ET_{ref} methods, including Kimberly Penman and HPRCC Penman, show better performance than the grass based ET_{ref} methods, including FAO24 Penman, FAO24 Radiation, FAO24 Blaney-Criddle, Priestley-Taylor, Hargreaves, and the Jensen-Haise methods.

Keywords: Reference evapotranspiration; Jensen-Haise; ASCE standardized reference ET

Introduction

Evapotranspiration (ET) is defined as evaporation of water from land and water surfaces [1] and transpiration by vegetation [2]. Knowledge of ET is important for water resource planning, efficient water management, and water permitting application. Direct measurement of ET is time consuming and costly [3]. Therefore, ET is normally determined indirectlyby relating to a reference evapotranspiration (ET_{ref}) to a crop coefficient (K_c), namely, ET = $K_c \times ET_{ref}$ [3]. ET_{ref} is defined as the ET rate from a uniform surface of dense, actively growing vegetation having specified height and surface resistance, not short of soil water, and representing an expanse of at least 100 m of the same or similar vegetation [1]. It represents the evaporative power of the atmosphere at a specific location and time of the year, but does not consider the crop characteristics and soil factors [4]. ET_{ref} can be calculated from weather data collected by weather stations. The K_c curve represents crop growth characteristic for a growing season. Both ET and K_c are influenced by crop characteristics, such as crop variety and cultivar, growth stage, crop height, and surface roughness. ET can also be affected by soil characteristics, including soil salinity, fertility, impenetrable soil layers, and plant residue [4]. The K_c curve for a specific crop is normally developed from research data for a specific region.

Many methods have been developed to estimate ET_{ref}. These can be categorized into four basic groups: combination, radiation, temperature, and pan evaporation methods [3]. The combination method, accounting for radiation (energy balance) and aerodynamic (heat and mass transfer) terms [2], was first proposed in 1948 by Penman [5]. The Penman equation was subsequently modified as the FAO24 Penman method [3], the Kimberly Penman [6], the Penman-Monteith [7], the FAO Penman-Monteith [4] and the American Society of Civil Engineers, Environmental and Water Resources

Institute (ASCE-EWRI) Penman-Monteith [1] equation. Radiation based ET_{ref} equations include the Priestley-Taylor [8] and FAO24 radiation methods [3]. Temperature based ET_{ref} equations include the Thornthwaite [9], Jensen-Haise [10], FAO24 Blaney-Criddle [3], and Hargreaves [11]. The pan evaporation methods are termed FAO class-A Pan [3] and Christiansen Pan [12]. While the availability of reliable weather data is limited, temperature methods (e.g. Jensen-Haise method) have been shown to provide reasonable ET_{ref} estimates. Among all the methods, the one that was developed by the ASCE EWRI standardized reference evapotranspiration task committee [1] was recommended as the standardized reference ET method [13-15]. Application of this method requires solar radiation, air temperature, relative humidity and wind speed as the input parameters.

Weather data used for estimating ET_{ref} are normally collected from a reference crop surface, either a tall crop similar to a full-cover alfalfa or a short crop similar to a clipped, cool-season grass. While most ET_{ref} methods are only applicable for one reference surface, the ASCE-EWRI method [1] can be applied to both full cover crops of alfalfa and grass. The ET_{ref} on an alfalfa reference surface is abbreviated as ET_r, and the ET_{ref} on a grass reference surface as ET_o. Most methods, such

*Corresponding author: Xinhua Jia, Department of Agricultural and Biosystems Engineering, North Dakota State University, Fargo, North Dakota, USA
Email: xinhua.jia@ndsu.edu

as the FAO24 Penman [3] and the Penman-Monteith [7], are based on the grass reference surface, but some, such as the Kimberly Penman [6] and the modified Penman methods [16] used by the High Plains Regional Climate Center (HPRCC, http://www.HPRCC.unl.edu) are based on an alfalfa reference surface.

In North Dakota, the Jensen-Haise equation is used to calculate the ET_{ref} [17-19]. The Jensen-Haise method only requires temperature and solar radiation as the input parameters. It was originally developed from data collected in the western United States over 35 years using 15 field and orchard crops [10,20]. The North Dakota Agricultural Weather Network (NDAWN, http://ndawn.ndsu.nodak.edu/) calculates ET_{ref} values using the Jensen-Haise method and the modified Penman (or HPRCC Penman) method for each weather station on the network. North Dakota is part of the High Plains Regional Climate Center. As indicated by Irmak et al. [21], the HPRCC Penman method applies when vapor pressure deficit (VPD) and wind speed do not exceed 2.3kPa and 5.1 ms^{-1}, respectively. Weather records from the Oakes NDAWN weather station indicate that higher values for wind speeds and VPD are not rare. For the period of record from 1991 to 2008 (6575 days), there are 12 days with VPD over 2.3 kPa, and 724 days (or 40 days per year) with wind speed above 5.1 m s^{-1}. Irmak et al. [21] found that at the higher end of the ET_r values, the HPRCC Penman method provided consistently lower ET_r values than those using the ASCE-EWRI method, which was attributed to the upper limits of applicability by the HPRCC Penman method.

The standardized ET_{ref} method [1] has not been widely used in North Dakota. Most crop coefficient curves were developed using the Jensen-Haise method for this region [22-25]. As indicated by Snyder et al. [26], K_c values are developed specifically for a region, and are highly dependent on the methods used for actual ET measurement and reference ET calculations. This indicates that all K_c curves were bonded specifically to the ET and ET_{ref} methods used to develop them because K_c values were derived as ET/ET_{ref}. The variable ET would only need to be figured initially before ET_{ref} and K_c could be applied. Applications of the ASCE EWRI method will require sequential changes to the K_c curves developed using other methods, such as the Jensen-Haise method.

Most irrigation research studies in North Dakota were conducted near Oakes in the southeast area of the state [17, 27,28]. There hasn't been much research in the west part of the ND state where it's drier and research is needed. Irmak et al. [21] categorized the Jensen-Haise method as an alfalfa reference based method, but Jia et al. [29] found that the ET_{ref} by the Jensen-Haise method is closer to a grass reference based method. Jensen and Haise [10] stated they developed the method based on data collected during the growing season over 35 years from 15 field and orchard crops in different regions of the Western US. The Oakes area does not have the most typical climate to represent the whole state and may not be the best place for irrigation based on its above average precipitation [29], but the sandy soil conditions, available water resources, and financial assistance from Garrison Division Conservancy District made the Oakes area one of the most irrigated areas in ND [30,31].

In this study, using weather data collected at the Oakes NDAWN station from 1991 to 2008, the daily ET_{ref} was calculated using 11 methods. The differences between the ASCE-EWRI ET_o method [1] and the Jensen-Haise ET_o method [10] as well as other 9 methods were compared on a daily, monthly, and yearly basis for the entire year and the growing season for the period of May 1 to September 30.

Material and Methods

Study site

The study site is located in Oakes, North Dakota. The weather station, surrounded by agricultural land, is located south of Oakes at latitude 46.07°N, longitude 98.09°W, and an elevation of 392 m. The soil at the weather station is Embden fine sandy loam (coarse-loamy, mixed Pachic Udic Haploborolls), and Maddock fine sandy loam (sandy, mixed Udorthentic Haploborolls) [32].

Weather conditions

The weather conditions at Oakes are typical continental; cold in the winter and semi-humid in the summer. The weather data recorded during the past 18 years showed that the average annual temperature was about 6°C, with the minimum in January and the maximum in July and August. Rainfall amounts ranged from 346 mm to 637 mm from May to September, with the highest rainfall amounts generally in June. The average ET_r during the growing season was 842 mm, which was 471 mm higher than the average precipitation amount. Wind speed averaged 3.3 m s^{-1} at 2 meter above the ground, with the highest average monthly wind speed of 4.0 m s^{-1} in May, and the lowest monthly average wind speed of 2.4 m s^{-1} in August. The average annual maximal wind speed was 8.8 m s^{-1}. The longest day time at Oakes is 16 hours in June and the shortest day time of 9 hours is in December [3]. There are 137 frost free days at Oakes, with the last killing frost in May and the first killing frost in October [33]. Monthly average, maximum, and minimum daily values for temperature, relative humidity, rainfall, and solar radiation over the 18 years at Oakes are listed in Table 1.

Data quality

"Data quality has the highest priority in the operation of the North Dakota Agricultural Weather Network (NDAWN) because erroneous data are worse than no data" [34]. Two procedures are performed daily for ensuring data quality control: locate missing and erroneous values and provide estimates using data from nearby stations. The data retrieved from NDAWN are further checked following the weather data integrity assessment procedures recommended by Allen [35] and ASCE-EWRI [1] for solar radiation, humidity, temperature, and wind speed to ensure that all data used in the calculation and analysis are good quality.

Weather parameters

Daily weather data, including maximal temperature (T_{max}), minimal temperature (T_{min}), wind speed (U), maximal wind speed (U_{max}), dew point temperature (T_{dew}), and shortwave incoming radiation (R_s) were downloaded from the NDAWN website for the period of 01/01/1991 to 12/31/2008. All the other required information, such as latitude, elevation, height of wind speed measurement and grass height were obtained either from the NDAWN website or from personal communications [34].

NDAWN measures wind speed at a height of 3 m immediately adjacent to the weather station, the grass in an area of about 40 m^2 has been maintained at a height of about 8-10 cm. However, to accommodate the fully mature crop heights typically taller than 0.5 m [34], equation 47 in FAO56 [4] was used to convert the wind speed at 3 m height to 2 m height:

$$u_2 = u_z \frac{4.87}{\ln(67.8z - 5.42)} \qquad (1)$$

where u_2 is the wind speed at 2 m above the ground surface in m s^{-1},

Month	T_{max} (°C)	T_{min} (°C)	T_{avg} (°C)	U_{avg} (m d⁻¹)	RH (%)	R_s (MJ m⁻²)	Day time (h)	PET (mm)	Rain (mm)
Jan	-7	-18	-12	3.4	75	5.9	9	18	
Feb	-3	-14	-9	3.6	74	9.4	10	27	
Mar	3	-7	-2	3.7	71	13.4	12	60	
Apr	13	0	6	3.9	57	17.4	14	131	35
May	20	7	14	4.0	56	20.1	15	187	73
Jun	25	13	19	3.4	65	22.1	16	186	102
Jul	28	15	21	2.6	70	23.1	15	183	79
Aug	27	14	20	2.4	68	19.9	14	160	51
Sep	22	8	15	2.8	63	14.9	13	126	66
Oct	14	1	8	3.1	61	9.4	11	84	51
Nov	4	-7	-2	3.3	70	5.8	10	36	
Dec	-3	-14	-8	3.4	75	4.7	9	19	
Annual	**12**	**0**	**6**	**3.3**	**67**	**14**	**12**	**1217**	**457**

Table 1: Monthly average maximal temperature (T_{max}), minimal temperature (T_{min}), daily temperature (T_{avg}), wind speed (U_{avg}) at 2 m height, incoming solar radiation (R_s), day time length (hour), monthly total potential evapotranspiration (PET) by Hprcc Penman method (mm), and monthly total rainfall (Rain) during the study period from 1991 to 2008 at Oakes, North Dakota. All parameters were obtained from the NDAWN website, except day time hours were calculated from Doorenbos and Pruitt (1977) using Oakes' latitude.

u_z is the measured wind speed at z m above ground surface in m s⁻¹, and z is the height of measurement above the ground surface in m, which is 3 m for this study.

The relative humidity is calculated from equation 6 to 8 of ASCE-EWRI method [1] using measured T_{max}, T_{min}, and T_{dew} from NDAWN weather data via saturated (e_s) and actual vapor pressure (e_a):

$$e_s = \frac{e^o(T_{max}) + e^o(T_{min})}{2} \qquad (2)$$

$$e^o(T) = 0.6108 \exp\left(\frac{17.27T}{T + 237.3}\right) \qquad (3)$$

$$e_a = e^o(T_{dew}) = 0.6108 \exp\left(\frac{17.27T_{dew}}{T_{dew} + 237.3}\right) \qquad (4)$$

where the T in equation (3) can be either T_{max} in °C or T_{min} in °C to be used in equation (2) to calculate the e_s and e_a in kPa. The relative humidity (RH) is calculated as the ratio of e_a to e_s. Details of sensor types, layout, and data quality control are detailed on the NDAWN website.

Reference ET calculations

The daily Jensen-Haise and HPRCC Penman ET_{ref} values are available on the NDAWN website. The ET_{ref} by these two methods will be directly used in the comparison. The ET_{ref} by the ASCE-EWRI method for grass and alfalfa references were calculated [1] using:

$$ET_{ref} = \frac{0.408\Delta(R_n - G) + \gamma C_n(T + 273) \times u_2(e_s - e_a)}{\Delta + \gamma(1 + C_d \times u_2)} \qquad (5)$$

Where ET_{ref} is the reference crop evapotranspiration for short grass (ET_o) or tall alfalfa (ET_r) [mm day⁻¹], R_n is net radiation at the crop surface [MJ m⁻² day⁻¹], G is soil heat flux [MJ m⁻² day⁻¹], T is mean daily air temperature at 2 m height [°C], u_2 is wind speed at 2 m height [m s⁻¹], Δ is slope vapor pressure curve [kPa °C⁻¹], and γ is the psychrometric constant [kPa °C⁻¹]. For a 24 hour time step, soil heat flux, G, is presumed to be 0. The values of C_n and C_d vary depending on the reference crops, and are 900 and 0.34 for the grass reference and 1600 and 0.38 for the alfalfa reference, respectively.

The downloaded weather data were arranged in the correct format for the REFET software [36], so that daily ET_{ref} by FAO24 Penman, FAO24 Radiation, FAO24 Blaney-Criddle, Priestley-Taylor, Hargreaves, Kimberly Penman 1982 and Kimberly Penman 1972 methods could be calculated.

A total of eleven methods were used to calculate the ET_{ref}; four methods are alfalfa based methods (ASCE-EWRI ET_r, HPRCC Penman, Kimberly Penman 1982 and Kimberly Penman 1972) and seven methods are grass based reference methods (ASCE-EWRI ET_o, FAO24 Penman, FAO24 Radiation, FAO24 Blaney-Criddle, Priestley-Taylor, Hargreaves, and Jensen-Haise).

Statistics analysis

The daily ET_{ref} values calculated from each method were compared to the ASCE-EWRI ET_r or ET_o values, depending on whether it was grass or alfalfa reference surface method. The root mean square deviation (RMSD) between the ASCE-EWRI ET_{ref} (method x, in Eq. (6)) and the compared method (method y, in Eq. (6)) was used to determine the difference:

$$RMSD = \sqrt{\frac{\sum_{i=1}^{n}(x_i - y_i)^2}{n}} \qquad (6)$$

where x_i is the ET_{ref} calculated by method x on day i; y_i is the ET_{ref} calculated by method y on day i; and n is the total number of days used in the calculation.

Since the ASCE-EWRI ET_{ref} is considered a standard value for comparison, the RMSD values between ET_{ref} values using the ASCE-EWRI and the compared method are considered a quantitative measure of all other methods. A smaller RMSD means a better comparison between the other method and the ASCE-EWRI ET_{ref} method. The slope and coefficient of determination (R^2) values are used to assess the bias of each method. The intercept of the regression line between the ASCE-EWRI ET_{ref} and the compared ET_{ref} values were forced to zero for an equal comparison among all methods. However, when forcing the regression curve to zero, it also assumes that at zero ET values, there is no atmosphere demand for water for all methods and the resulting slope can be used to indicate the error regardless of the magnitude of the readings. It also biases the results by placing heavier weight on points farthest from the origin. The purpose of this paper is to determine how widely the ET_{ref} values were different from the standardized ASCE-EWRI ET_{ref} values and the RMSD and R^2 values should be reasonable sufficient.

Results and Discussions

Daily ET_o and ET_r comparison

Comparison of daily ET_o and ET_r values between the ASCE-EWRI ET_o or ET_r method and the targeted method are shown in Figure 1a-1j. The slope of the fitting and coefficient of determination for each pair are also shown in the graph and in Table 2. In addition, the RMSD and the rank of all methods are also shown in Table 2. The rank is made according to the average of the R^2 and the RMSD ranks. For example, the R^2 ranks 9 and the RMSD ranks 7 between the Prestley-Taylor and ASCE-EWRI ET_o methods, the overall rank is the average, 8.

From 1991 to 2008, the HPRCC Penman method results were most similar to the ASCE-EWRI ET_r values using R^2 and RMSD. Even with limitations on high wind speed and high VPD, the HPRCC Penman method performed the best among all methods. It overestimated the ASCE-EWRI ET_r by a mere 1%; much better compared to reports by Irmak et al. [21] with a 5% underestimation. The Jensen-Haise method provided very close ET_o values when compared to the ASCE-EWRI ET_o values with less than 0.2% difference. However, the R^2 was only 0.87 and the RMSD was 0.903 mm d^{-1}. If one argues that forcing the equation to zero has caused the problem, the R^2 was only 0.89 without forcing the equation to zero. This proves that the Jensen-Haise method is not strongly correlated to the ASCE-EWRI ET_o values.

Winter in North Dakota extends from late November to early April. During this time period, average air temperature is normally less than 0 °C, while the ground is frozen, plants are dead or dormant, and most of the state is covered with snow. Under these conditions, no water evaporates from the soil surface or transpired by plants. Thus, these conditions seem to violate the definition of ET. There may be some water loss through sublimation, a phase change from solid ice or snow to vapor [37,38]. The calculation of ET during this time period is for comparison purposed only, and does not represent any actual ET lost. Evaluation of ET values during the growing season in North Dakota is more important.

Growing season ET_o and ET_r

Because the Jensen-Haise method was originally developed using data during the growing season, the ET_{ref} comparisons are performed using weather data from May 1 to September 30 over an 18-year period (Figure 2a-2j).

After changing the comparison days from 6575 days for the 18 years to 2966 days for the growing season only, the relationship between the ET_{ref} by ASCE-EWRI method and other methods did not change significantly. The HPRCC Penman method still performed the best among all the methods with the higher R^2 and smallest RMSD value. The Priestley-Taylor method performed better for the growing season than for the entire year. The FAO24 Blaney-Criddle method had the highest correlation (R^2) with the ASCE-EWRI ET_o values, but with 20.76% overestimation, and therefore, a higher RMSD value than that in Figure 1. The Blaney-Criddle method required mean daily temperature, mean daily percentage of total annual daytime hours, and an adjustment factor depending on minimum relative humidity, sunshine hours, and daytime wind estimates as the input parameters, which are similar to the ASCE-EWRI method, but without considering the crop factors, and thus do not strongly correlated. The Jensen-Haise method remained about the same rank with the ASCE-EWRI ET_o either for the growing season or for the entire year. For the growing

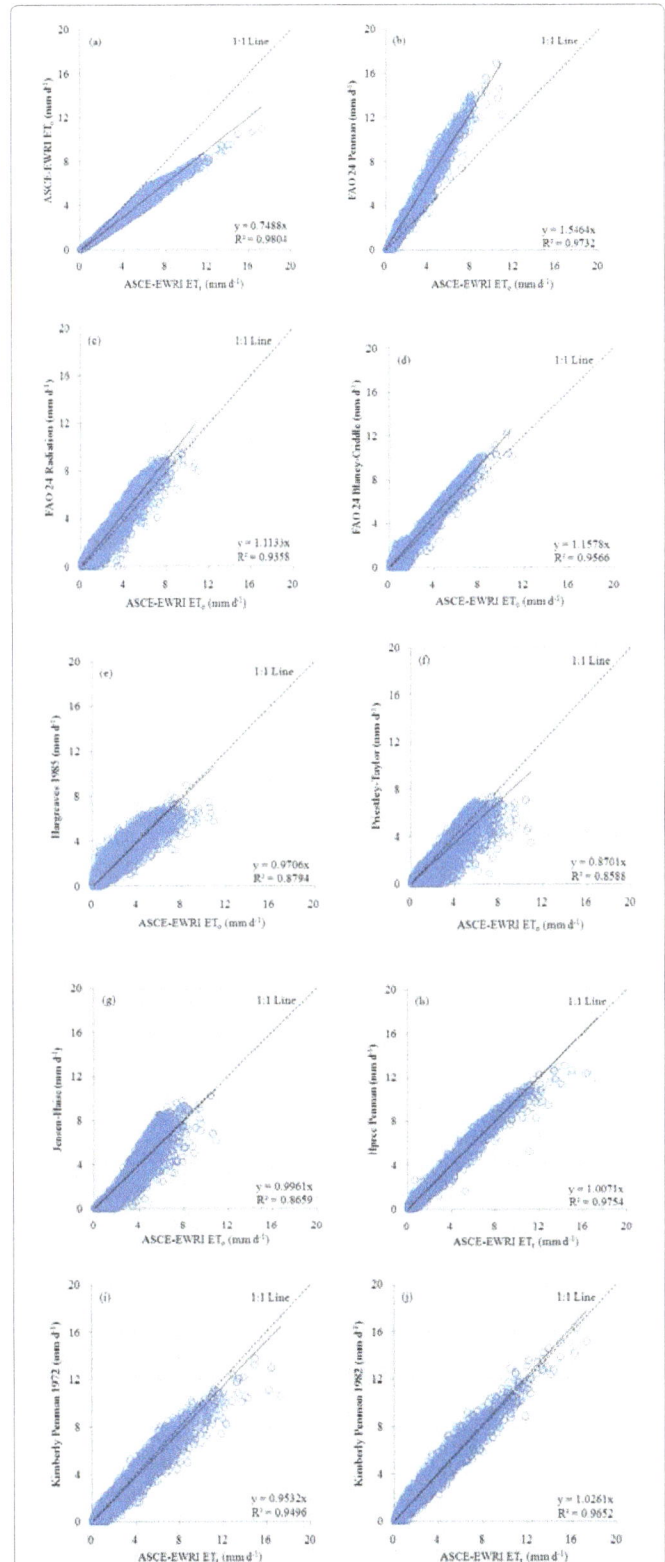

Figure 1: Daily reference evapotranspiration comparisons between (a) ASCE-EWRI ET_o and ASCE-EWRI ET_r; (b) FAO24 Penman and ASCE-EWRI ET_o; (c) FAO24 Radiation and ASCE-EWRI ET_o; (d) FAO24 Blaney-Criddle and ASCE-EWRI ET_o; (e) Hargreaves 1985 and ASCE-EWRI ET_o; (f) Prestley-Taylor and ASCE-EWRI ET_o; (g) Jensen-Haise and ASCE-EWRI ET_o; (h) Hprcc Penman and ASCE-EWRI ET_r; (i) Kimberly Penman 1972 and ASCE-EWRI ET_r; and (j) Kimberly Penman 1982 and ASCE-EWRI ET_r methods for Oakes, North Dakota in 1991-2008.

ID	Method y	Method x	Slope	R^2	Rank-R^2	RMSD	Rank-RMSD	Overall Rank
(a)	ASCE-EWRI ET_o	ASCE-EWRI ET_r	0.7488	0.9804		1.103		
(b)	FAO24 Penman	ASCE-EWRI ET_o	1.5464	0.9735	2	1.827	9	6
(c)	FAO24 Radiation	ASCE-EWRI ET_o	1.1133	0.9358	6	0.717	5	5
(d)	FAO24 Blaney-Criddle	ASCE-EWRI ET_o	1.1578	0.9566	4	0.746	6	4
(e)	Hargreaves 1985	ASCE-EWRI ET_o	0.9706	0.8794	7	0.707	4	5
(f)	Prestley-Taylor	ASCE-EWRI ET_o	0.8701	0.8588	9	0.854	7	7
(g)	Jensen-Haise	ASCE-EWRI ET_o	0.9961	0.8659	8	0.903	8	7
(h)	Hprcc Penman	ASCE-EWRI ET_r	1.0071	0.9754	1	0.429	1	1
(i)	Kimberly Penman 1972	ASCE-EWRI ET_r	0.9532	0.9496	5	0.624	3	3
(j)	Kimberly Penman 1982	ASCE-EWRI ET_r	1.0261	0.9652	3	0.522	2	2

Table 2: Comparison of daily reference evapotranspiration (ET_{ref}), Root Mean Square Deviation (RMSD), and coefficient of determination (R^2) from 1991 to 2008 at Oakes, North Dakota. ET_o is grass based reference surface and ET_r denotes alfalfa based reference surface. The overall rank is based on average ranks from RMSD and R^2 for annual ET_{ref}.

ID	Method y	Method x	Slope	R^2	Rank-R^2	RMSD	Rank-RMSD	Overall Rank
(a)	ASCE-EWRI ET_o	ASCE-EWRI ET_r	0.7671	0.9543		1.395		
(b)	FAO24 Penman	ASCE-EWRI ET_o	1.5697	0.9529	3	2.624	9	6
(c)	FAO24 Radiation	ASCE-EWRI ET_o	1.1387	0.8915	6	0.916	6	5
(d)	FAO24 Blaney-Criddle	ASCE-EWRI ET_o	1.2054	0.9625	1	1.001	8	3
(e)	Hargreaves 1985	ASCE-EWRI ET_o	1.0026	0.4373	9	0.901	5	7
(f)	Prestley-Taylor	ASCE-EWRI ET_o	0.9251	0.7272	8	0.858	4	5
(g)	Jensen-Haise	ASCE-EWRI ET_o	1.0813	0.7996	7	1.007	7	7
(h)	Hprcc Penman	ASCE-EWRI ET_r	1.0108	0.9541	2	0.483	1	1
(i)	Kimberly Penman 1972	ASCE-EWRI ET_r	0.9905	0.9148	5	0.588	2	2
(j)	Kimberly Penman 1982	ASCE-EWRI ET_r	1.0324	0.9316	4	0.586	3	2

Table 3: Comparison of daily reference evapotranspiration (ET_{ref}), Root Mean Square Deviation (RMSD), and coefficient of determination (R^2) from May to September in 1991-2008 at Oakes, North Dakota. ET_o is grass based reference surface and ET_r denotes alfalfa based reference surface. The overall rank is based on average ranks from RMSD and R^2 for seasonal ET_{ref}.

season, it overestimated the ET_o by 8.35% from the ASCE-EWRI ET_o method with a lower R^2 and a higher RMSD value. Considering the relationship between the ASCE-EWRI ET_o and ET_r, this might indicate more than 10% underestimation from the ET_r as others have reported [21,30]. Jensen [20] and Burman et al. [39] stated that the Jensen-Haise method is better suited for time intervals of five days to one month rather than for daily estimates. The daily estimated ET_{ref} by the Jensen-Haise method was used in the analysis for Figures 1 and 2. Therefore, a growing season comparison of ET_{ref} didn't improve the correlation between the Jensen-Haise method to the ASCE-EWRI method than for an entire year.

Monthly ET_{ref}

As shown in Table 3 and Figures 1 and 2, the total ET_{ref} by the Jensen-Haise method was very close to the ASCE-EWRI ET_o values both for annual or seasonal time scale, but with a poor correlation (R^2) and less accuracy (RMSD). Figure 3a-3c shows the monthly average ET_{ref} of the 11 methods over the 18 years. Most methods showed a similar trend as the ASCE-EWRI standardized equation; higher in the summer and lower in the winter. A higher difference was observed between winter and summer, but not between spring and fall. All combination methods showed similar trends for all seasons while comparing the ASCE-EWRI ET_{ref} methods. In Figure 4a, the FAO24 Penman method showed a comparable annual curve to the ASCE-EWRI ET_o method, while in Figure 4c, all the ET_r values were very similar to each other with less than 5% difference and followed the ASCE-EWRI ET_r curve. This is probably due to the fact that the ASCE-EWRI ET_r was developed using data at Kimberly, or originated from the Kimberly Penman methods [2]. The HPRCC Penman method also gave more similar results to the ASCE-EWRI ET_r method for all month. The local-adjusted HPRCC Penman method proved to be the best fit

for the Oakes area in southeastern North Dakota. The temperature and radiation based methods were quite different from the monthly ASCE-EWRI ET_o values. The FAO24 Radiation, FAO24 Blaney-Criddle and Prestley-Taylor methods showed underestimation in the winter and overestimation in the summer compared to the ASCE-EWRI ET_o values. The Jensen-Haise method had the greatest deviation from the ASCE-EWRI ET_o method with lower ET_o values from January to May and from September to December, and higher ET_o values from June to August. Though the annual ET_o values were close to the ASCE-EWRI ET_o values, the month to month difference was higher.

Figure 4 shows the average daily ET_{ref} for the ASCE-EWRI ET_r, ASCE-EWRI ET_o, and the Jensen-Haise ET_{ref}. The ASCE-EWRI ET_r peaked on May 21. Actually, the month of May has the highest ET_r, mainly due to the higher wind speed (Table 1). The Jensen-Haise method only accounts for temperature and solar radiation and does not include the effect of wind speed. This may be the reason that non-combination ET_{ref} methods do not have the same ET_{ref} pattern and peaked at different times than the combination methods. Also notice that the higher wind speed shifted the peak of alfalfa based ASCE-EWRI equation, but not the grass based equation. The grass based method peaked at the same time as the Jensen-Haise method. The difference between the grass and alfalfa based equation is the surface resistance, defined by Allen et al. [4] as "the resistance of vapor flow through stomata openings, total leaf area and soil surface". For the alfalfa reference surface, a constant surface resistance of 70 s m^{-1} was used, and for the grass reference surface, 45 s m^{-1} was used as the constant surface resistance for the standardized reference ET calculations [1].

A direct replacement of Jensen-Haise method by the ASCE-EWRI ET_o method may result in underestimation of ET_o during the growing season. Use of ASCE-EWRI ET_o values combined with the

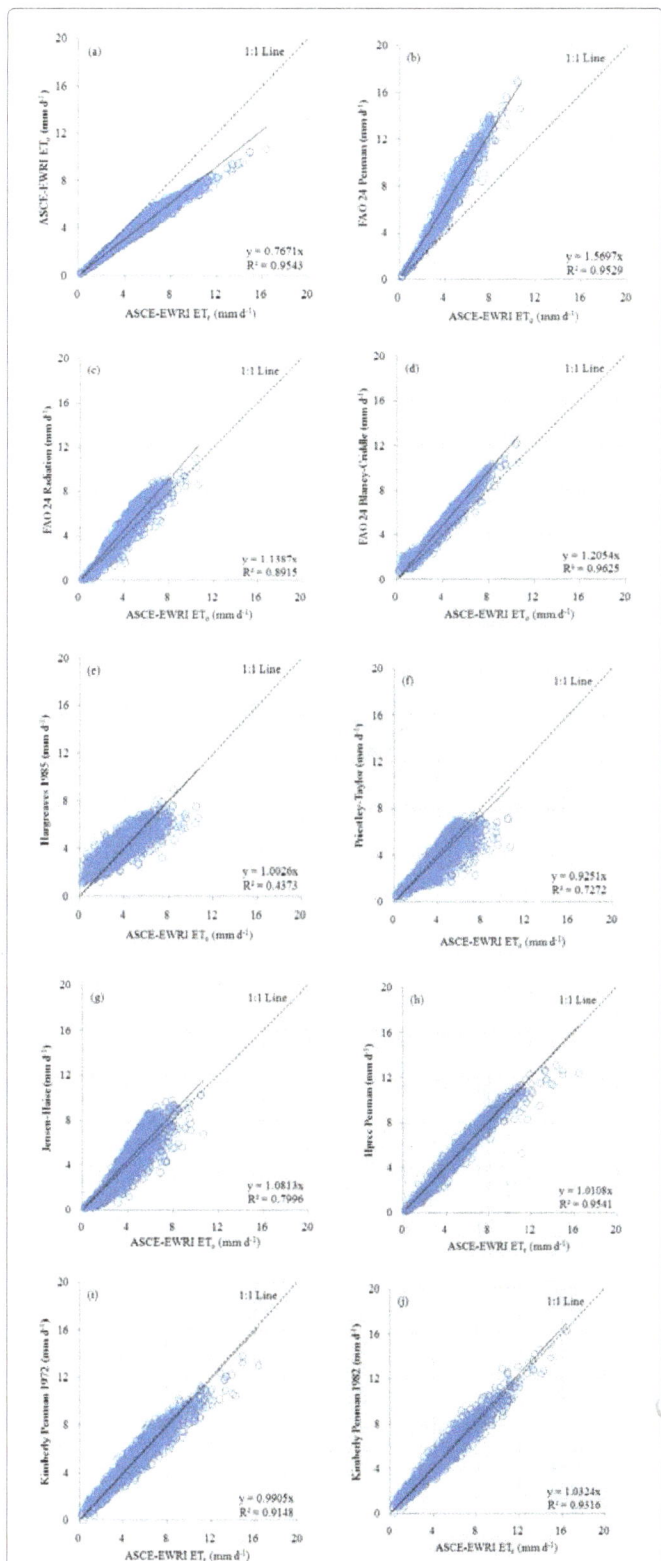

Figure 2: Daily reference evapotranspiration comparisons between (a) ASCE-EWRI ET_o and ASCE-EWRI ET_r; (b) FAO24 Penman and ASCE-EWRI ET_o; (c) FAO24 Radiation and ASCE-EWRI ET_o; (d) FAO24 Blaney-Criddle and ASCE-EWRI ET_o; (e) Hargreaves 1985 and ASCE-EWRI ET_o; (f) Prestley-Taylor and ASCE-EWRI ET_o; (g) Jensen-Haise and ASCE-EWRI ET_o; (h) Hprcc Penman and ASCE-EWRI ET_r; (i) Kimberly 1972 and ASCE-EWRI ET_r; and (j) Kimberly 1982 and ASCE-EWRI ET_r methods for Oakes, North Dakota for the growing season (May 1 – September 30) of 1991-2008.

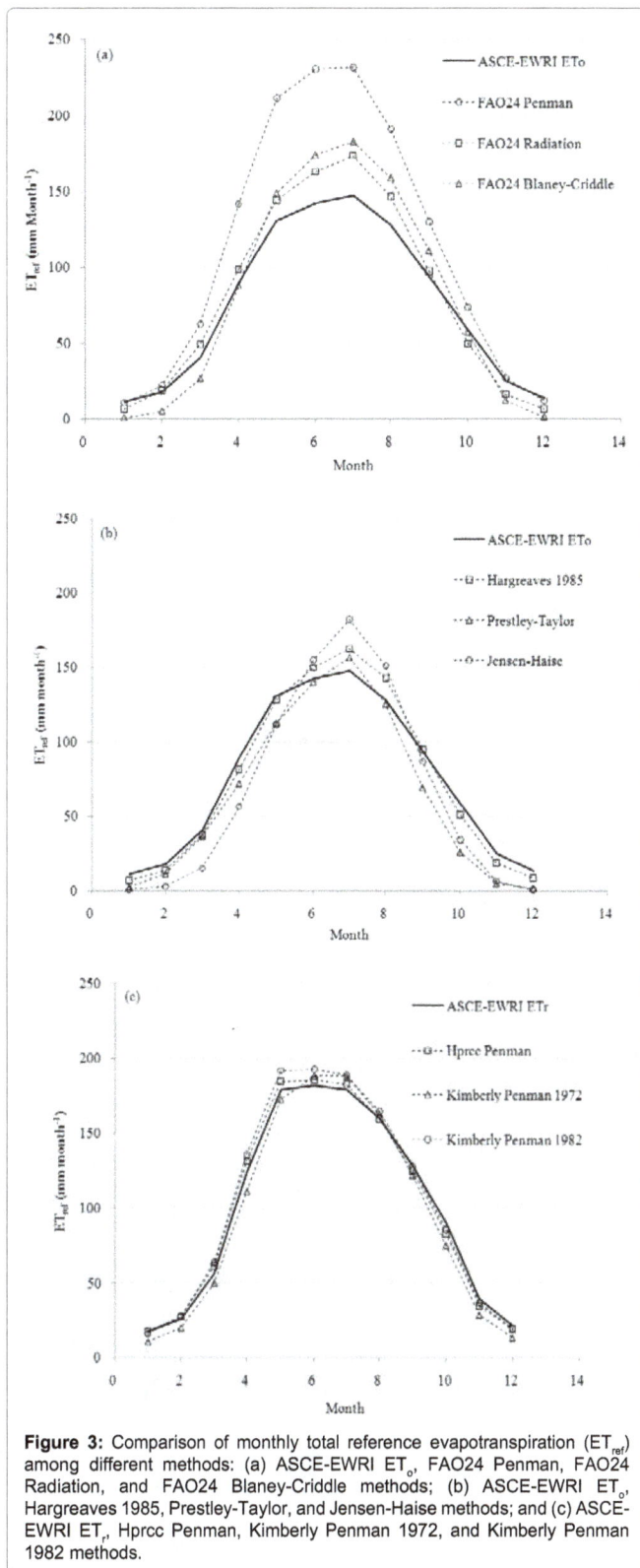

Figure 3: Comparison of monthly total reference evapotranspiration (ET_{ref}) among different methods: (a) ASCE-EWRI ET_o, FAO24 Penman, FAO24 Radiation, and FAO24 Blaney-Criddle methods; (b) ASCE-EWRI ET_o, Hargreaves 1985, Prestley-Taylor, and Jensen-Haise methods; and (c) ASCE-EWRI ET_r, Hprcc Penman, Kimberly Penman 1972, and Kimberly Penman 1982 methods.

K_c curve developed using the Jensen-Haise method would result in lower calculated crop ET, thus applying less irrigation than the crop actually needed. The K_c curve is tied to a particular ET_{ref} method and a replacement of the current ET_{ref} method used for irrigation scheduling will require changes to the K_c curves as well.

Annual ET$_{ref}$

Average annual ET$_{ref}$ values are shown in Figure 5, with error bars indicating the standard deviation across 18 years of data. Almost all grass based ET$_{ref}$ values showed lower annual ET$_{ref}$ than the alfalfa based methods. However, the FAO24 Penman method showed a similar total ET$_o$ as the alfalfa based method. The ET$_o$ showed lower standard deviation than the ET$_r$ values. Again, the HPRCC Penman method was the closest to the ASCE-EWRI ET$_r$ value, with only 12.4 mm or 1% annual difference. The Hargreaves method has a 0.5 mm, or 0.1% difference from the ASCE-EWRI ET$_r$ method.

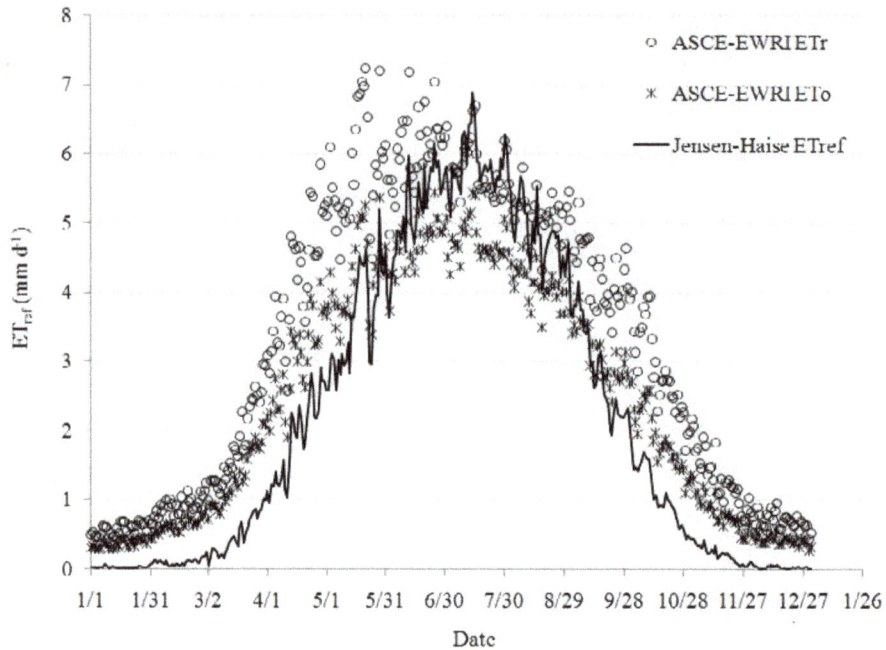

Figure 4: Comparison of average daily reference evapotranspiration (ET$_{ref}$) for 18 years using ASCE-EWRI ET$_r$, ASCE-EWRI ET$_o$ and Jensen-Haise ET$_{ref}$ methods at Oakes, North Dakota.

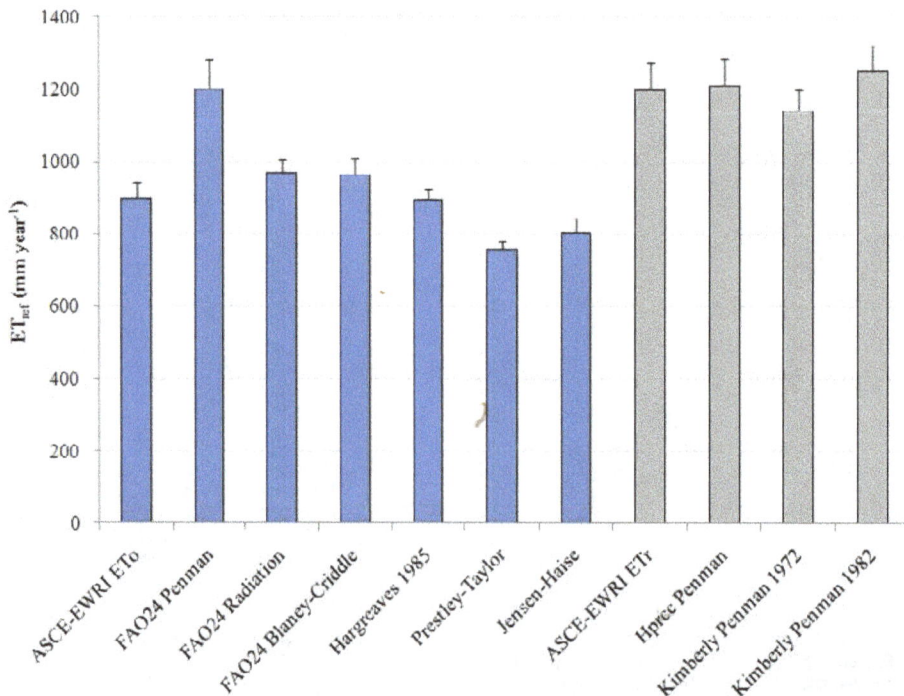

Figure 5: Comparison of annual reference evapotranspiration values (ET$_{ref}$) for all methods. The dark color bars indicate the grass reference based methods and the light color bars indicate the alfalfa reference based methods. The error bars indicate the standard deviation among the 18 years.

Conclusions

Crop water consumption use for irrigation scheduling in North Dakota is calculated from ET_{ref} by the Jensen-Haise method and the K_c curves developed in the 1970's and 1980's. The standardized ET_{ref} methods by the American Society of Civil Engineers, Environmental and Water Research Institute (ASCE-EWRI) reference evapotranspiration task force [1] has been widely accepted and applied across the world. However, application of the ASCE-EWRI method requires sequential changes to the K_c curves associated with the Jensen-Haise method. This paper compared ET_{ref} estimates for 11 methods, including the ASCE-EWRI and the Jensen-Haise methods using 18 years of data collected in southeast North Dakota. The results showed that the annual ET_o by the Jensen-Haise method was nearly the same (0.39% underestimation) as the ASCE-EWRI grass ET_o, but with a higher RMSD, 0.903 mm d^{-1}, and a lower R^2 0.8659, comparing to the ASCE-EWRI ET_o. Since the Jensen-Haise method was initially developed using growing season data collected from 15 crops, ET_o comparison for the growing season showed an RMSD of 1.007 mm d^{-1}, R^2 of 0.7996 and 8.13% overestimation. The ET_o by the Jensen-Haise method has a higher monthly ET_o than that y the ASCE-EWRI in June, July, and August, and lower monthly ET_o for all other months. The ET_o using the two methods does not show a strong agreement, so direct replacement of the Jensen-Haise method by the ASCE-EWRI method is not recommended. New K_c curves should be developed prior to the application of the ASCE-EWRI ET_{ref} method in southeastern North Dakota. In addition, interest in irrigating alternative crops, development of new crop cultivars of current irrigated crops and climate change will require the development of new K_c curves if ASCE-EWRI ET_{ref} values are used for irrigation scheduling. The ET_{ref} comparison also showed that the HPRCC Penman method has the best accuracy and correlation with the ASCE-EWRI ET_{ref} method overall. Indeed, all alfalfa based ET_{ref} methods, including Penman models, showed a better performance than grass based ET_{ref} methods.

Acknowledgements

This project is supported by the North Dakota Agricultural Experiment Station, CRIS project ND01464, and USDA CSREES project 2008-35102-19253. Mention of trade names is for information purposes only and does not imply endorsement by the authors or NDSU.

References

1. ASCE-EWRI (2005) The ASCE Standardized Reference Evapotranspiration Equation.

2. Jensen ME, Burman RD, AllenRG (1990) Evapotranspiration and irrigation water requirements. ASCE , New York, USA.

3. Doorenbos J, Pruitt WO (1975) Guidelines for predicting crop water requirements. Irrigation and Drainage paper 24.

4. Allen RG, Periera LS, Raes D, SmithM (1998) Crop evapotranspiration: Guidelines for computing crop requirements. Irrigation and Drainage Paper 56.

5. Penman HL (1948) Natural evaporation from open water, bare soil and grass. Proc Roy Soc 193: 120-145.

6. Wright JL (1982) New evapotranspiration crop coefficients. J Irrig Drain. Div 108: 57-74.

7. Allen RG, Smith M, Pereira A, Pereira LS (1994) An update for the definition of reference evapotranspiration. ICID Bull 43: 1-34.

8. Priestley CHB, Taylor RJ (1972) On the assessment of surface heat flux and evaporation using large-scale parameters. Mon Weather Rev 100: 81-92.

9. Thornthwaite CW (1948) An approach toward a rational classification of climate. Geogr Rev 38: 55-94.

10. Jensen ME, HaiseHR (1963) Estimating evapotranspiration from solar radiation. J Irrig Drain Div 89: 15-41.

11. Hargreaves GH, Samani ZA (1985) Reference crop evapotranspiration from temperature. Appl Eng Agric 1: 96-99.

12. Christiansen JE (1968) Pan evaporation and evapotranspiration from climatic data. J Irrig Drain Div 94: 243-265.

13. Itenfisu D, Elliott RL, Allen RG, WalterIA (2003) Comparison of reference evapotranspiration calculations as part of the ASCE standardization effort. J Irrig Drain Eng 129: 440-448.

14. Temesgen B, Eching S, Davidoff B, FrameK (2005) Comparison of some reference evapotranspiration equations for California. J Irrig Drain Eng 131: 73-84.

15. Farahani HJ, Howell TA, Shuttleworth WJ, Bausch WC (2007) Evapotranspiration: Progress in measurement and modeling in agriculture. T ASABE 50: 1627-1638.

16. Kincaid DC, Heermann DF (1974) Scheduling irrigations using a programmable calculator. Peoria, USDA, USA.

17. Steele DD, Stegman EC, GregorBL (1994) Field comparison of irrigation scheduling methods for corn. T ASAE 37: 1197-1203.

18. Stegman EC, Bauer A, Zubriski JC, Bauder J (1977) Crop curves for water balance irrigation scheduling in SE North Dakota.

19. Stegman EC, Soderlund M (1992) Irrigation scheduling of spring wheat using infrared thermometry. T ASAE 35: 143-152.

20. Jensen ME (1974) Consumptive use of water and irrigation water requirements. Irrig Drain Div : 215.

21. Irmak A, Irmak S, Martin DL (2008) Reference and crop evapotranspiration in South Central Nebraska. I: comparison and analysis of grass and alfalfa-reference evapotranspiration. J Irrig Drain Eng 134: 690-699.

22. Lundstrom DR, Stegman EC (1988) Irrigation scheduling by the checkbook method. NDSU Extension Service: 1-11.

23. Stegman EC (1988) Com crop curve comparisons for the central and northern plains of the U.S. Appl Eng Agric 4: 226-233.

24. Steele DD, Stegman EC, Knighton RE (2000) Irrigation management for corn in the northern Great Plains, USA. Irr Sci 19: 107-114.

25. Scherer TF, Morlock DJ (2008) A site-specific web-based irrigation scheduling program.

26. Snyder RL, O'ConnellN (2007) Citrus crop coefficients determined using a surface renewal method. J Irrig Drain Eng 133: 43-52.

27. Stegman EC (1983) Corn and sunflower yield vs. management of leaf xylem pressures. T ASAE 26: 1362-1368.

28. Steele DD, Scherer TF, Prunty LD, StegmanEC (1997) Water balance irrigation scheduling: Comparing crop curve accuracies and determining the frequency of corrections to soil moisture estimates. Appl Engr Agric 13: 593-599.

29. Jia X, Scherer TF, Steele DD (2007) Crop water requirement for major crops in North Dakota and its vicinity area.

30. Garrison Division (2006) Planning for the future. Garrison Diversion Annual Report.

31. Albus W, Besemann L, Eslinger H (2008) Oakes irrigation research site. North Dakota State University, North Dakota, USA.

32. Natural Resources Conservation Service (2009) Web Soil Survey.

33. Padbury G, Waltman S, Caprio J, Coen G, McGinn S, et al.(2002) Agroecosystems and land resources of the Northern Great Plains. Agron J 94: 251-261.

34. Akyuz FA (2013) North Dakota Agricultural Weather Network. Fargo, North Dakota, USA.

35. Allen RG (1996) Assessing integrity of weather data for reference evapotranspiration estimation. J Irrig Drain Eng 122: 97-106.

36. Allen RG (2001) REF-ET: Reference evapotranspiration calculation software for FAO and ASCE Standardized equations. University of Idaho, Kimberly, Idaho, USA.

37. BrookerDB (1967) Mathematical model of the psychrometric chart. T ASAE 10: 558-560.

17

Modeling Agricultural Drainage Hydraulic Nets

Luis Gurovich[1]* and Patricio Oyarce[2]

[1]*Faculty of Agronomy and Forestry, Pontificia Universidad Católica de Chile, Chile*
[2]*Agricultural Engineer, Departamento de Fruticultura y Enología, Pontificia Universidad Católica de Chile, Chile*

Abstract

A review on mathematical models available in the literature to design and evaluate agricultural drainage hydraulic nets is presented, including open ditch and buried pipe alternatives, for steady state and no steady state soil water flows, homogeneous and stratified soil profiles and smooth and corrugated drainage pipes.

Effective drainage pipe radius effects on drainage performance is considered, based on its perforation density and distribution, as it affects pipe weight bearing strength and pipe deformation intensity. Also, quantitative considerations on perforation density upon water flow resistance into the drain pipe are analyzed.

Keywords: Agriculture; Hydraulic nets; Pipe; Soil profile

Introduction

Agricultural land drainage consists of a set of technical strategies and hydraulic structures allowing the removal of excessive water and/or salts present in the soil volume occupied by root crops, to provide an adequately oxygenated environment, suitable for root normal development, keeping adequate water and air relative proportions according to crop physiological needs, to enable soil sustainability for crop productive conditions [1-4].

There are two drainage systems for controlling underground waters: open ditches (Figure 1a) and subsurface perforated piping (Figure 1b). Open ditches systems consist of excavations in the soil that collect the water stored at existing phreatic layers; it also can be used to remove surface run-off; it can account for significant land farming losses, smaller soil units for farm machinery operation and interference with irrigation systems, making agricultural tasks more expensive [5-7].

Subsurface pipe drainage systems consists on plastic tubes, either smooth or corrugated, with perforations, placed at specified distances and depths, buried within the soil; this system is used mainly to lower the water table in unconfined aquifers [8-10]. These drainage systems in most cases consist on a main drain pipe, a collector drain pipe and a network of field drains pipes (Figure 2); the position of the main drain depends on the field slope and the location of the lowest field level, through which the collected water is removed from the drained area. The collector drain and the network of field drains are usually located in parallel to each other; field drains are perforated pipes along their extension (Figure 3), and its function is the phreatic level control by receiving water excesses present in the soil profile and convey this effluent towards the collector drain. Secondary drains and the main drain main conduct water from the drain pipes to the site of water discharge. These conductive drains are either open ditch type or underground pipes, the selected option will depend on costs and dimensions of piping [8-13].

Subsurface drain design corresponds to a set of agronomic, hydraulic and engineering characteristics that a lateral drainage system must fulfill, to eliminate the excessive volume of soil water, enabling soil aeration values required to satisfy crop optimal growth and production [1-3,14,15]. In general terms, design features must define the proper criteria and parameters relevant to spacing among lateral drains, its depth placement inside the soil profile and the hydraulic characteristics of the hydraulic net, required to transport the volume of water to be collected and remove it from the cultivated area. In relation to construction aspects, drainage design must include definitions about drain hydraulic net layout, the materials to be used, the density and kind of perforations, as well as building techniques, network installation and maintenance.

Optimal distances between consecutive lateral drains are closely related to water flow towards the drains. The development of a mathematical model for quantitative description of the sub-surface

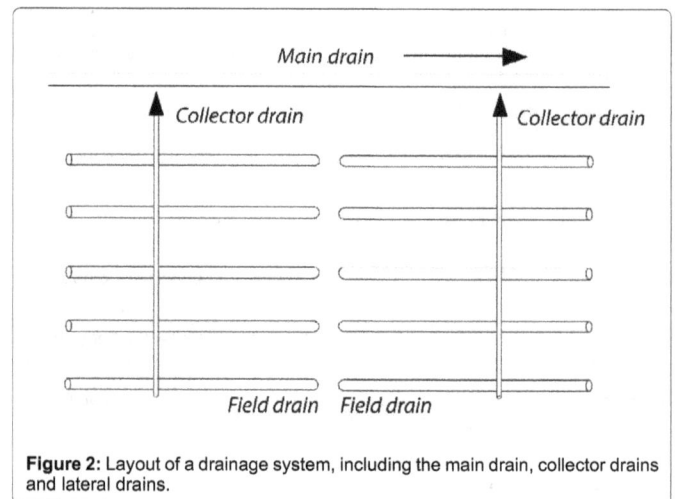

Figure 2: Layout of a drainage system, including the main drain, collector drains and lateral drains.

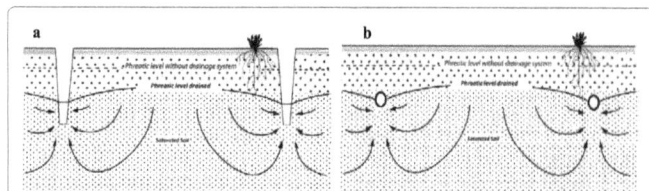

Figure 1: Subsurface drainage by open ditches (a) and underground pipes (b).

*Corresponding author: Luis Gurovich, Professor, Faculty of Agronomy and Forestry, Pontificia Universidad Católica de Chile, Chile
E-mail: lgurovic@puc.cl

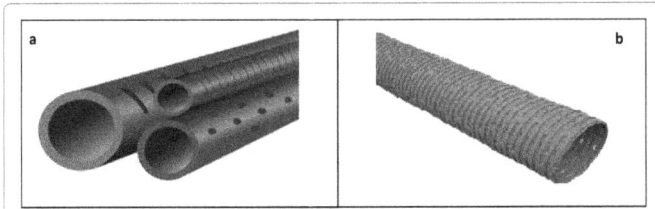

Figure 3: a. PVC drain pipes with circular perforations and discontinuous slots. b. Corrugated PVC drainage tiles with perforations.

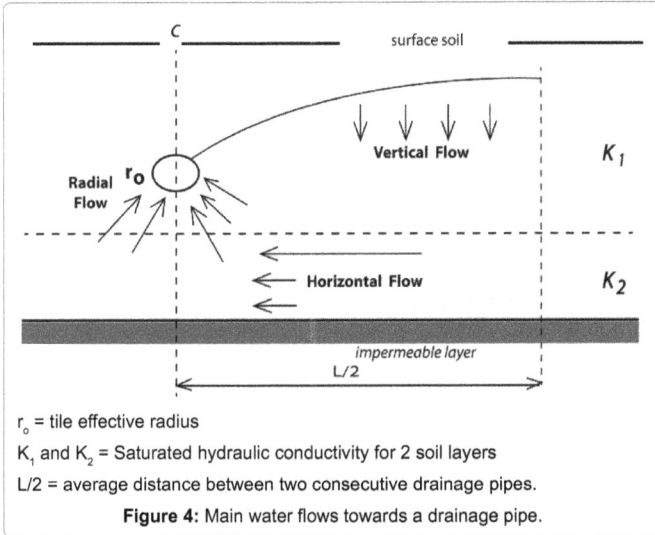

r_o = tile effective radius

K_1 and K_2 = Saturated hydraulic conductivity for 2 soil layers

$L/2$ = average distance between two consecutive drainage pipes.

Figure 4: Main water flows towards a drainage pipe.

flow towards lateral drains is possible only based on mathematics simplifications, deduced from the theory of underground water saturated flow, with pre-established initial and border conditions.

According to Ernst [16], water flow to a subsurface drain (Figure 4) consists of:

a) Descendent vertical flow from the phreatic level down to the drain level,

b) Horizontal flow towards the area nearby the drain,

c) Radial flow towards the drain and,

d) Input water flow into the drain.

Each flux magnitude and direction q occurring simultaneously, can be represented by means of vector components, using Darcy's law, where: q=the difference of the corresponding hydraulic potentials multiplied by specific resistances [10]. Differences in hydraulic potentials for saturated flow correspond to water hydrostatic pressure differences between the soil and the drain system (Figure 4).

In steady state regimes, water flow total resistance is the sum of the vertical, horizontal, radial and inflow (entrance) resistance. These resistances can be measured by means of piezometers strategically set (Figure 5). A piezometer consists on a small diameter tube without perforations, provided with a short filter in its lower end. Water level in the piezometer represents the hydraulic head in the soil around the filter.

Four head losses can be identified in Figure 5 [10]:

•   Vertical head loss (h_v) is the difference in water levels between

piezometers 1 and 2, located at the midpoint (half distance) between two consecutive drains, with its filters situated in the proximity of the phreatic level and at the depth where the drain is installed, respectively.

•   Horizontal head loss (h_h), mainly due to the horizontal flow towards the drain, corresponds to the difference in the water level between piezometers 2 and 3; with the filters situated at the level of the drain, one is at the midpoint (half distance) between two drains, and the other is in the close proximity to the drain.

•   The radial head loss (h_r) is given by the difference in the water levels between piezometers 3 and 4, with the filters located at the depth where the drain is installed: one besides the drain and the other at some specific distance.

•   The inflow head loss (h_e) is the difference in water levels between piezometer 4 and a piezometer situated over the drain.

Total head loss (h_t) is the sum of all those differences, as indicated in Figure 6. Head losses are measurements of the resistances to the corresponding flows; the relation between head loss and the corresponding resistance is:

$$h_i = q \, L \, W_i \tag{1}$$

where

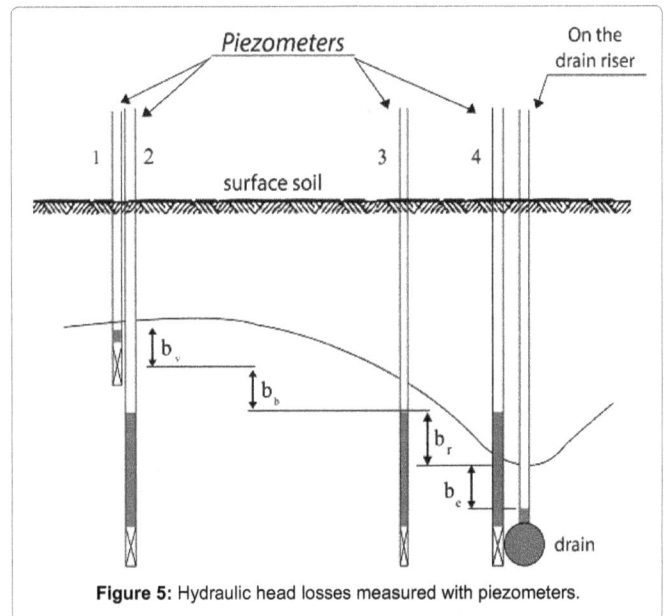

Figure 5: Hydraulic head losses measured with piezometers.

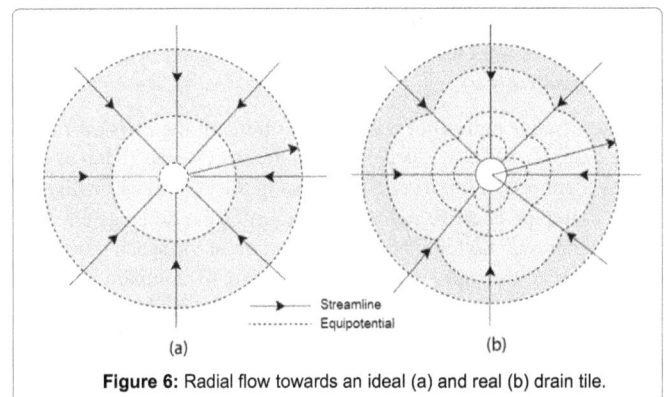

Figure 6: Radial flow towards an ideal (a) and real (b) drain tile.

h=head loss (m), L=distance between drains (m), q=specific flow (m d^{-1}), W= Resistance (d/m^{-1}), i=subscripts; v (vertical), r (radial), e (entrance), t (total)

Total head loss is:

$$h_t = h_v + h_r + h_h + h_e \qquad (2)$$

Sometimes, in mathematical models describing saturated water flow towards drains, resistances (W) are substituted by dimensionless coefficients α, which are independent of soil hydraulic conductivity:

$$\alpha = K*W \quad or \quad W = \alpha / K \qquad (3)$$

where:

K= saturated hydraulic conductivity (m d^{-1}), α=dimensionless geometric factor

Therefore, total head can be expressed as:

$$H_t = q\,L\left(W_v + W_h + W_r + W_e\right) = q\,L\left(\frac{\alpha_v}{K_v} + \frac{\alpha_h}{K_h} + \frac{\alpha_h}{K_h} + \frac{\alpha_r}{K_r} + \frac{\alpha_e}{K_e}\right) \qquad (4)$$

Mathematical models describing drainage systems are used for the calculation of drain spacing; these models are based in a set of assumptions related to drain hydraulic characteristics and to soil physical properties. One of these assumptions is that the drain is an ideal drain pipe, without inflow or entrance resistance, and flow is considered as an equipotential line (Figure 6). In these models it is assumed that the environment of the drain (surrounding materials and the soil altered by the trench excavated for perforated pipe installation) present a saturated hydraulic conductivity (K_s) much greater than the K_s of the natural unaltered soil, thus disregarding inflow resistance for the envelope material. However, practical experiences have shown that this condition is not always an adequate assumption [7,9].

Actual drains, being only permeable through its perforations, can be considered as continuously permeable drains, with an effective radius of drainage that is significantly lower than its physical radius; this fact is due to pipe mechanical resistance losses, due to a specific perforation density (distance between consecutive perforations). As the effective radius directly depends of the inflow resistance, it can be taken as an alternative to the entrance resistance; the smaller the inflow resistance, the larger will be the effective radius. Therefore, it is necessary to take into account the inflow (entrance) resistances in the equations allowing to define the optimal spacing between consecutive drain-pipes, and also to introduce the concept of effective radius in the outflow calculations, instead of using the physical radius of the lateral drain. If the physical radius is taken into account, calculated spacing between consecutive drains is larger than the spacing needed to optimize water outflow. Also, if the drain pipe physical radius is considered in these models, the actual phreatic layer depth after drainage is higher that the model results; under these conditions, optimal drain pipe layout criteria will not represent an optimal water extraction hydraulic drain net performance [8,9].

The effective radius not only depends both on the physical radius of the drain, as well as on its perforation density (Figure 7), but also, it is dependent on the inflow resistance; being this resistance smaller for larger drainpipe effective radius. Evaluation of the drainage effective radius is not only useful for the determination of the spacing between consecutive drain pipes, but it can be used to compare different materials' drainage efficiency [8,10,17]. Both from theoretical and experimental points of view, research on inflow resistances to drains are needed, since it can significantly affect calculations of the optimal distances between consecutive drains.

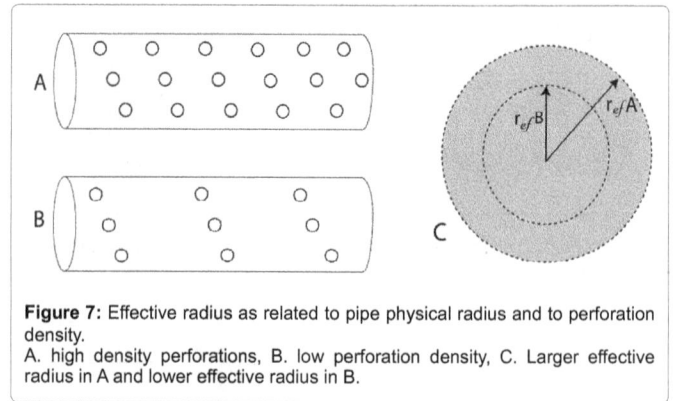

Figure 7: Effective radius as related to pipe physical radius and to perforation density.
A. high density perforations, B. low perforation density, C. Larger effective radius in A and lower effective radius in B.

Drain pipe external wall shape can be smooth or corrugated, affecting water inflow resistance. Similarly, soil particle sedimentation around the drain pipe perforations has highly significant effects on inflow resistance; if the corrugations are full with soil particles, the geometric limit between the soil and the perforation is important for the determination of the effective radius; also, if the corrugations are kept free of soil sediments, the limit of the interface soil-drain tile, when no filtering material is used as an envelope, significantly reduces inflow resistance. Corrugations shapes, (waves or blocks), have only a minor influence over water inflow resistance. For certain shapes and distribution of the perforations on a smooth wall drain pipe, inflow resistance may be determined for curved or flat borders. Dierickx [18] has made an extensive review on these analytic solutions and experimentally tested its accuracy.

Drains with circular perforations

a) Inflow resistance for a curved border (α_{ea}):

$$\alpha_{ea} = \frac{1}{\pi N}\left[2\sum_{n=1}^{\infty}K_0\left(\frac{n\pi\delta p}{\lambda r}\right) + 2\sum_{i=1}^{n-1}\left\{\sum_{n=1}^{\infty}K_0\left(\frac{4n\pi Ro}{\lambda r}\sin\frac{\theta i}{2}\right)\right\} + ln\frac{2Ro}{N\delta p}\right] \qquad (5)$$

b) Inflow resistance for a flat border (α_{ep}):

$$\alpha_{ep} = \frac{1}{\pi N}\left[2\sum_{n=1}^{\infty}K_0\left(\frac{n\pi\delta p}{\lambda r}\right) + 2\sum_{i=1}^{n-1}\left\{\sum_{n=1}^{\infty}Ko\left(\frac{4n\pi Ro}{\lambda r}\sin\frac{\theta i}{2}\right)\right\} + ln\frac{2Ro}{N\delta p} + \frac{\lambda r(\pi-2)}{2\delta p}\right] \qquad (6)$$

Where:

N=number of perforations rows, R_o=external radius of drainpipe, K_o=Bessel function of the second kind, of zero order, δ_p=perforation diameter, λ_r=perforation spacing between rows; **n** and **i** are integer numbers, θ_i=angle relative to the ith row, measured from the baseline (ordinate) explained in Figure 8.

The Bessel function is based on solutions to Laplace's and Helmholtz equations, by using the variable separation method in cylindrical or spherical coordinates [19].

Circular perforations

Engelund [20] considered water flow to drains with circular perforations distributed in a rectangular pattern, located over a flat surface (Figure 8b). Inflow resistance for a curved border (α_{ea}) in a cylindrical surface, having the same perforations pattern, can be described by:

$$\alpha_{ea} = \frac{1}{\pi m}\left\{\frac{1}{\delta_p} - \frac{1}{2\lambda_1}\left(3.91 - 2ln\frac{\lambda_1}{\lambda_2}\right)\right\} \qquad (7)$$

Inflow resistance for a flat border (α_{ep}) is:

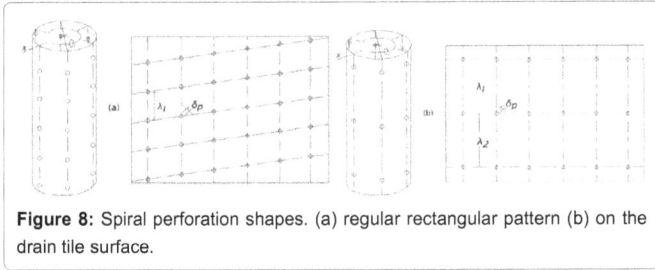

Figure 8: Spiral perforation shapes. (a) regular rectangular pattern (b) on the drain tile surface.

$$\alpha_{ep} = \frac{1}{\pi m}\left\{\frac{1}{2\delta_p} - \frac{1}{2\lambda_1}\left(3.91 - 2ln\frac{\lambda_1}{\lambda_2}\right)\right\} \qquad (8)$$

Where:

m=number of perforations per drain unit length, λ_1 and λ_2 correspond to the spacing between the smaller and larger perforations, respectively. These equations are valid when $\delta_p << 2\lambda_1$.

For square perforations, λ_1 and λ_2 in equations 7 and 8 are converted to

$$\alpha_{ea} = \frac{1}{\pi m}\left(\frac{1}{\delta_p} - \frac{3.91}{2\lambda_s}\right) = \frac{1}{\pi m\delta_p} - \frac{0.248}{\sqrt{mR_0}} \qquad (9)$$

and

$$\alpha_{ep} = \frac{1}{\pi m}\left(\frac{1}{2\delta_p} - \frac{3.91}{2\lambda_s}\right) = \frac{1}{\pi m\delta_p} - \frac{0.248}{\sqrt{mR_0}} \qquad (10)$$

Where α_{ea} and α_{ep} are the inflow resistances for a curved and a flat border, respectively.

For square orifices, perforation spacing λ_s can be described as

$$\lambda_s = \left(\frac{2\pi R_0}{m}\right)^{1/2} \qquad (11)$$

Drains with discontinuous grooving

Inflow resistances for a curved border (α_{ea}) and for a flat border (α_{ep}) are, respectively:

$$\alpha_{ea} = \frac{c}{2\pi^2 R_0}\left(ln\frac{c}{\pi\beta_s} - \frac{1}{Y}ln\frac{\pi\lambda_p}{2\beta_s} - \frac{4\lambda_c}{\beta_s} + 0.577\right) \qquad (12)$$

and

$$\alpha_{ep} = \frac{c}{2\pi^2 R_0}\left(ln\frac{2c}{\pi\beta_s} - \frac{1}{Y}ln\frac{\pi\lambda_p}{\beta_s} - \frac{8\lambda_c}{\beta_s} + 0.577\right) \qquad (13)$$

Where:

C= distance between rows, R_0=external radius of drainpipe, β_s=the slit width, λ_p=spacing between perforations over the drain circumference Explained in Figure 9 [18]

Corrugated drains

Equations developed to calculate inflow resistances for smooth drain pipes, can be applied only to corrugated drains having orifices on the top of the corrugations; however, corrugated drain pipes usually have its perforations on the corrugations valleys. For corrugated drains with a square wave profile, with an external major radius R_0 and with its external minor radius R'_0, provided with a circular opening β_s with the

same width that the valley β_v (Figure 10), the inflow resistance for the flat border (α_{ep}) condition is: [18]

$$\alpha_{ep} = \frac{c}{R_o 2\pi^2}\left(ln\frac{2c}{\pi\beta_v} - \frac{c}{4\pi R_0}\right) + \frac{c}{2\pi\beta_v}ln\frac{R_o}{R_o} - \frac{1}{2\pi}ln\frac{R_o}{R_o} \qquad (14)$$

For circular diameters smaller than the valley widths, inflow resistances result from the convergent flow lines towards drain perforations (Figure 11) [18].

The inflow resistance for a flat border (α_{ep}) is:

$$\alpha_{ep} = \frac{c}{R_o 2\pi^2}\left(ln\frac{2c}{\pi\beta_v} - \frac{c}{4\pi R_0}\right) + \frac{c}{4\pi^2 R_o}ln\frac{2 sinh\dfrac{2\pi\delta_v}{\beta_v}}{sin^2\dfrac{\pi\beta_v}{2\beta_v}} - \frac{1}{2\pi}ln\frac{R_o}{R_o} \qquad (15)$$

Research performed on saturated water flow towards drain pipes, based on mathematical models, have demonstrated that for circular orifices, the inflow resistance depends mainly on the distance between the orifices, as well as to the outer pipe diameter [18]. Efficient inflow resistance reduction into drainage pipes with circular orifices, can be achieved by increasing the number and diameter of perforations per drain pipe length unit, as compared to increments on discontinuous slot length.

Models for Agricultural Drainage

Darcy H and Dupuit J were the pioneer researchers formulating the basic equations for subsurface water flow across saturated porous

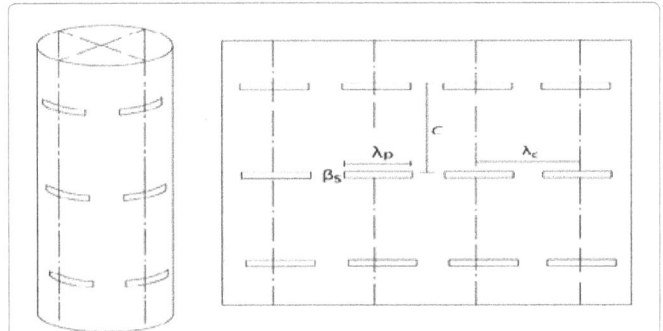

Figure 9: Discontinuous grooving rectangular perforations.

where:

C=distance between rows, R_o=external radius of drainpipe, β_s=the slit width, λ_p=spacing between perforations over the drain circumference.

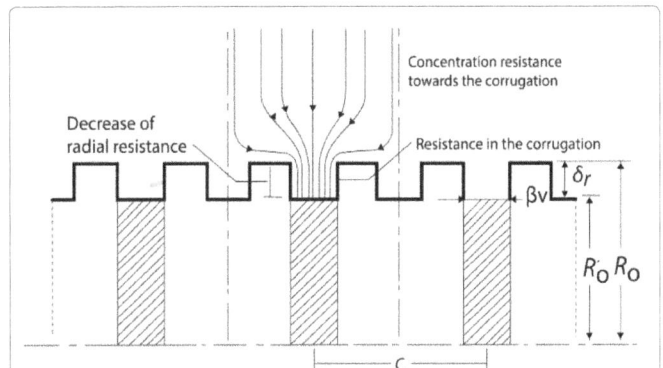

Figure 10: Flow patterns and inflow resistances for a corrugated drain, with a circular perforation diameter equal to the valley width.

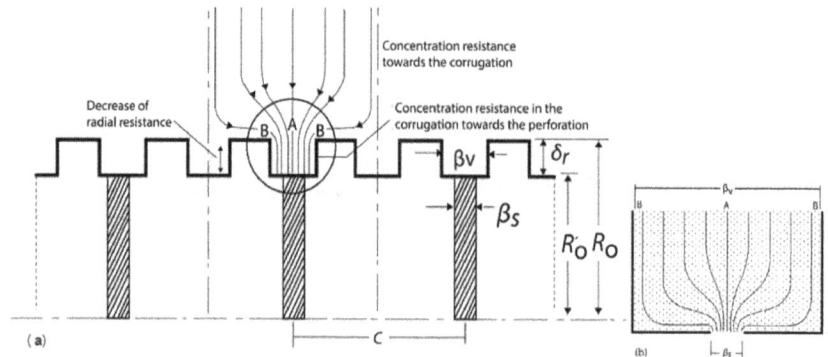

Figure 11: Corrugated drain (a) for perforations smaller than the valley width. (b) Inflow lines converging towards the circular perforations in the corrugated drain tile.

media and applying these models to describe water flow towards wells employed these same equations describing subsurface water flow to drains, thus describing the first drainage formulae reported [21-23]. Hooghoudt [24], was one of the first researchers on develop a rational analysis for the drainage problem, studying it in the context of the water - soil - plant system. Since then, scientists from all over the world, such as [25] from England, [26-28] in the United States, [29,30] in Holland, have contributed to the improvement of the rational analysis method first proposed by Hooghoudt [24], in 1940. These models are used for quantitative design of drainage systems, taking into consideration the correlation among some design characteristics (spacing and depth) with certain crops features, as well as to precipitation intensity and to soil saturated hydraulic conductivity (K_s) for each strata present in the soil profile within the area to be drained. Drainable pore space, the optimum depth of the phreatic layer respect to the effective depth of crop roots, phreatic layer rate of descent and the inflow resulting from either rainfall, irrigation or another water origin [1,31,32], are also relevant parameters to be considered in drainage network designs. Equations describing soil drainage can be defined into two major categories: steady state and non steady state water flow regimes.

Steady state regime

Equations describing situations of steady state flow regime assume that both the water recharge over an area, and the output of water through the drainage system are constants, meanwhile the level of phreatic layer stays in a steady state condition, thus it neither ascends nor descends. This condition properly describes the situation in wet zones, where rainfall is almost constant during a long period of time and its intensity fluctuations are not significant [32,33]. For the calculation of spacing between consecutive drains under a steady state condition, it is necessary to define (Figure 12):

*   The saturated hydraulic conductivity of the different soil profile strata (Ks) [m/d[;

*   The thickness of the flow region (over and underneath of drains);

*   The phreatic layer distance from the surface, at the midpoint (half distance) between two consecutive lateral drains (Pe) [m];

*   The depths of drains in the soil profile (Pd) [m];

*   The hydraulic head (Ah) [m];

*   The depth from drain basis down to the impermeable stratus

(D) [m]and

*   The recharge (R) [m].

Equations developed to validate drainage design for this type of steady state regime have been published [16,24,26,34,35].

Non-steady state regime

It assumes that the water recharge (R) over an area and the discharge of water (Q) by the drainage systems are not constants; for a condition characterized by a discharge smaller than the recharge, a phreatic level raise is produced during the recharge; afterwards, the phreatic level starts descending and subsequently it begins to increase again, when the following event of irrigation or rain starts. This non-steady state condition is found in zones with periodical irrigation or high intensities of rainfall followed by significant dry spells [12,13,35-37]. Equations describing this non-steady sate condition assume that soil hydraulic characteristics are homogeneous throughout the soil profile and that the depth from the surface down to the phreatic layer is such that the thickness of the flow region may be considered constant. Since these conditions are fulfilled on rare occasions in Nature and also, soil parameters like hydraulic conductivity, aquifer thickness and drainable porosity are difficult to measure with certainty, the drain spacing calculated with this kind of equations must be contrasted with spacing calculated using other procedures; such as the Hooghoudt equation [24] for a steady state regime, before making a definitive decision about the optimum burying depth and spacing between parallel drain pipes.

To calculate the optimal spacing between drains, under a condition of non-steady state regime, it is necessary to define the variables indicated in Figure 13 [35].

*   The unsaturated hydraulic conductivity {K=f(soil water content)} [m/d];

*   The drainable porosity (μ) [%];

*   The time (t) that water needs for descending from an initial position (h_o) down to a final   position   (h_t);

*   The instantaneous recharge rate (Ri) [m];

*   The drain depth (Pd) [m];

*   The effective depth (Pe) for crop root development [m]

Equations describing non-steady state regimes have been published [38-41].

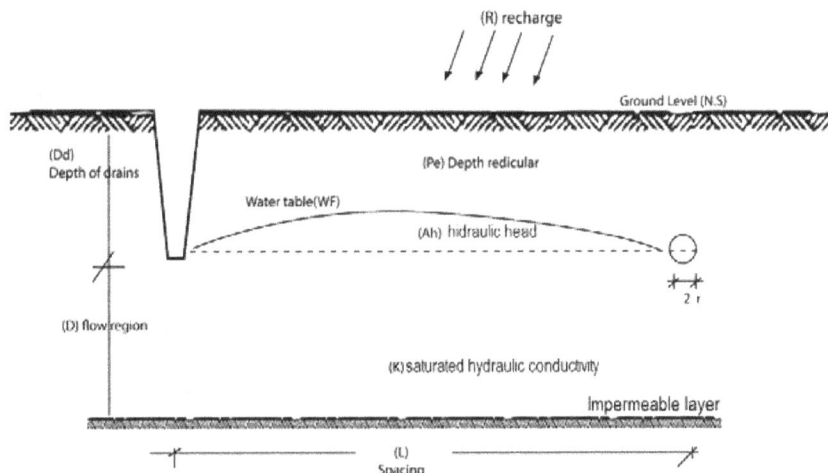

Figure 12: Generalized diagram for steady state drainage flow.

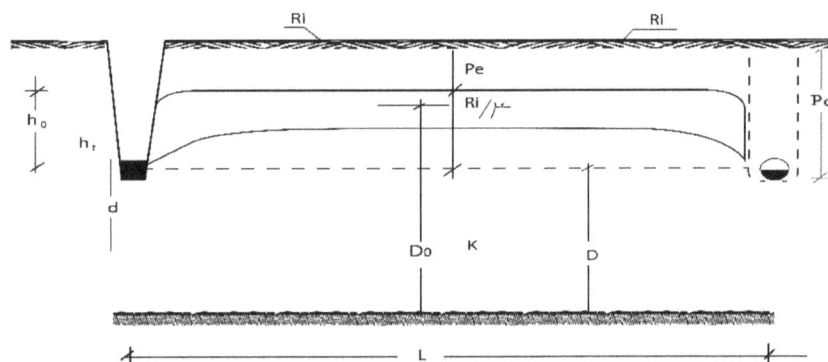

Figure 13: Generalized diagram for non-steady state water inflow to drain pipes.

Simulation Models in Agricultural Drainage

Since the beginning of digital computing technology, mathematical models have found wide application to different fields in applied natural sciences [42]; a mathematical model is defined as a set of equations and computing programs that can be used to quantify the performance of a natural system, in relation to specific functions [43]. For agricultural drainage, several mathematical models have been proposed, for different applications in basic research and applied engineering (Table 1). Most models are relatively simple and usually consider a single flow process or function, using finite element and finite differences analysis techniques [44], which enable calculating the main parameters needed for drainage design, (optimal depth and spacing between consecutive drains) (Table 1) [45].

Most simulation model results have been evaluated under field conditions, and applied to describe and characterize the phreatic layer behavior, and the agricultural drainage effects on agricultural production [46-48]. Other agricultural drainage models have been developed to predict soil salinity buildup, fertilizer leaching, and sediment and contaminant distributions in the soil drained profile [47-53]. Most commonly used drainage models are:

Swap

It is developed based on the agro-hydrological models SWATRE and SWACROP, and some of its later derivations, as SWASALT, for soil salt transportation and FLOCR, developed for the study of contraction and expansion in clay soils [54]. Swap integrates water flow, solutes transportation, and crop growth and development. It includes Richardson numerical solution methods [55], solutes incorporation and the heat transport, soil heterogeneity and the contraction and expansion of clay soils.

Drain mod

It is a field model based on poorly drained soil hydrology and the corresponding artificial drainage. The model accounts for water balances at the soil surface and within the soil profile, enabling to calculate drainage rates as a function of soil hydraulic properties, the phreatic layer water level and the design of the drainage network [56].

Sahys mod

It is a mathematical model used for simulating and predicting the increment in soil salinity, as related to soil moisture variations, underground water flow, phreatic layer depth and the drainage network discharge, under conditions of agricultural irrigation in soils with different hydro-geological conditions [57].

Espadren

It allows the calculation of drains spacing for a steady state regime,

Model/Code	Basic model of analysis	Purpose	Model application	Region
1D simulation Model	FDM	S	Gupta and Pandey (1983) and Gupta (1985)	INDIA
3D GWFM	FDM	-	Freyberg (1988)	USA
PLASM, MODFLOW, AQUIFEMM	FDM, FEM	C,V,S	Anderson and Woessner (1992)	-
-	FED,FEM	V	Konikow and Bredehoeft (1992)	USA
MODFLOW, GFLOW	FDM	C	Hunt et al. (1998)	USA
SWAP-SWATRE, SIMGRO	FEM	C,S	D'Urso et al. (1999)	ITALY
SWAP, SLURP	FEM	S	Kite and Droogers (2000)	TURKEY
MODFLOW, GFLOW, MODPATH	FDM	C,S	Pint et al. (2003), Budge and Sharp (2009)	USA
SWAP,WOFOST, SEBAL	FEM	C,S	van Dam and Malik (2003)	INDIA
GWFM	-	C,V	Hassan (2004)	-
GLUE		SA	Beven (2006)	-
UCODE, MMA		C	Poeter (2007)	-
MODFLOW	FDM	S	Michael and Voss (2008)	INDIA
WBM	FDM	C,V	Zhang et al. (2008)	AUTRALIA
HEM			Harou et al. (2009)	
UPFLOW		S	Raes (2009)	BELGIUM
MODFLOW, MT3D	FDM	V,S	Wondzell et al. (2009)	USA
ISOQUAD	FEM, FDM	S	Yang et al. (2009a)	TAIWAN
SEAWAT, UCODE	FDM	C	Sanford and Pope (2010)	USA
SVAT, MODFLOW, SIMGRO	FDM, FEM	V,S	vanWalsum and Veldhuizen (2011)	NETHERLANDS
MODFLOW	FDM	C, V, S	Sherif et al. (2012)	UAE
SGMP	FDM	C, S, SA, S	Singh (2013)	INDIA
HYDRUS-1D, SWMS2D	FEM, FDM	V, S	Zhu et al. (2013)	China

FDM- finite difference method
FEM- finite element method
GWFM-groundwater flow model
WBM-water balance model
HEM-hydro-economic model
S -simulation
C -calibration
V-validation/verification
SA -sensitivity analysis

Table 1: Validated and commonly used simulation models in drainage studies.

by using a specific set of equations [24,26,29,34]; it accounts also for a non-steady state regimes, by using [37,39] equations, either for open drains or for subsurface drainage pipes. It includes routines valid for homogeneous soil profiles as well as for soils having two layers (strata) with different Ks values [58].

The increasing development and applicability of these simulation models oriented to agricultural applications, has become an important tool for research on the quantification of crop productivity and the impacts of deficient drainage over the soil – root environment. However, most model results do not fully satisfy defining specific technical needs of agricultural drainage: also, accurate measurements of its input variables are difficult and costly and do not take into account soil hydraulic properties in terms of its geo – spatial variability. Therefore, for each specific drainage problem, it is necessary to consider specifics criteria to select and validate one or more of these simulation models.

Conclusion

The normative and protocols established for agricultural land drainage in countries having expertise in the subject, have not been validated for the specific conditions of soils and situations of deficient drainage existing in local agricultural conditions in different countries. Thus, no international standards for drainage network design are available. Specifications for drainage pipe resistances are also seldom available. For example, for plastic drains, standard norms specify only the use of limited proportions of recycled plastic raw materials

allowances. Additionally, physical and mechanical dimensions for drainage nets, like internal and external diameter, perforation size, location and density has not been universally defined.

The existing norms for drainage materials proceeding from countries with an extended drainage experience might be used as a reference, to define national standards, fitting specific local circumstances. Optimization of perforation density and shape for a PVC drainage pipe, allowing to increment water extraction efficiency and reducing pipe costs, is needed to define design and evaluation techniques of new components. A continuous, applied research program, carried on jointly by Universities, research institutes and industry, can provide technologies to develop efficient and low cost drainage systems, adapted to local conditions.

Acknowledgement

The authors thanks Conicyt (Chile National Research and Development Council) for the scholarship given to pursue Mr. Patricio Oyarce postgraduate studies and for the financial support received through Conicyt "Doctoral thesis in the industry" project N° 7813110015.

References

1. Ayars J, Christen E, Hornbuckle J (2006) Controlled drainage for improved water management in arid regions irrigated agriculture. Agricultural Water Management 86: 128-139.

2. Ritzema H, Nijland H, Croon (2006) Subsurface drainage practices: From manual installation to large-scale implementation. Agricultural Water Management 86: 60-71.

3. Ritzema H, Satyanarayana T, Raman S, Boonstra J (2008) Subsurface drainage to combat waterlogging and salinity in irrigated lands in India: Lessons learned in farmers' fields. Agricultural Water Management 95: 179-189.

4. Naz B, Ale S, Bowling L (2009) Detecting subsurface drainage systems and estimating drain spacing in intensively managed agricultural landscapes. Agricultural Water Management 96: 627-637.

5. Scholz M, Trepel M (2004) Hydraulic characteristics of groundwater-fed open ditches in a peatland. Ecological Engineering 23: 29-45.

6. Kröger R, Cooper CM, Moore M (2008) A preliminary study of an alternative controlled drainage strategy in surface drainage ditches: Low-grade weirs. Agricultural Water Management 95: 678-684.

7. Jia Z, Luo W, Xie J, Pan Y, Chen Y, et al. (2011) Salinity dynamics of wetland ditches receiving drainage from irrigated agricultural land in arid and semi-arid regions. Agricultural Water Management 100: 9-17.

8. Rimidis A, Dierickx W (2004) Field research on the performance of various drainage materials in Lithuania. Agricultural Water Management 68: 151-175.

9. Stuyt L, Dierickx W (2006) Design and performance of materials for subsurface drainage systems in agriculture. Agricultural Water Management 86: 50-59.

10. Stuyt, L Dierickx W, Martínez J (2005) Materials for subsurface land drainage systems. FAO, Irrigation and drainage 60.

11. Skaggs R, Van Shilfgaarde J (1999) Agricultural Drainage. American society of agronomy. Agronomy. USA.

12. Vander Molen W, Martínez B, Ochs W (2007) Guidelines and computer programs for the planning and design of land drainage systems. FAO, Rome.

13. Nijland B, Croon F, Ritzema H (2005) Subsurface Drainage Practices Guidelines for the implementation and operation. Wageningen, Alterra, ILRI Publication.

14. Welderufael W, Woyessa Y (2009) Evaluation of surface water drainage systems for cropping in the Central Highlands of Ethiopia. Agricultural Water Management 96: 1667-1672.

15. Rahman M, Lin Z, Jia X, Steele D, DeSutter T (2014) Impact of subsurface drainage on stream flows in the Red River of the North basin. Journal of Hydrology 511: 474-483.

16. Ernst L (1954) Het berekenen van stationaire ground water storming, welke in een vertikaal vlak afgebeeld kunnen worden. Rapport IV.

17. Rimidis A, Dierickx W (2003) Evaluation of subsurface drainage performance in Lithuania. Agricultural Water Management 59: 15-31.

18. Dierickx W (1999) Non-ideal drains: Agricultural Drainage. USA.

19. Abramowitz M, Stegun A (1965) Handbook of Mathematical Functions with Formulas, Graphs, and Mathematical Tables. New York.

20. Engelund F (1953) On the laminar and turbulent flows of ground water through homogeneous sand. Trans Techn Sci.

21. Darcy H (1856) Les Fontaines Publiques de la Ville de Dijon. Dalmont, Paris.

22. Dupuit J (1863) Introduction to theoretical and practical on the movement of water in open channels and through the permeable ground.

23. Rothe J (1924) Die Strangentfernung bei Dranungen Landw. Jahrb 59: 453-490.

24. Hooghoudt S (1940) Bjidrage to the knowledge of some physical quantities of the ground.

25. Childs E, O'Donnell T (1951) The water table, equipotential, and streamlines in drained land: VI The rising water table. Soil Sci 71: 33-237.

26. Donnan W (1947) Model tests of a tile-spacing formula. American Society of Agronomy.

27. Luthin J, Guitjens J (1967) Transient solutions for drainage of sloping land. ASCE 93: 43-51.

28. Kirkham D (1957) The theory of land drainage. Am Soc Agron.

29. Ernst L (1956) Calculation of steady flow of groundwater in vertical cross-sections. Neth J Agric Sci 4: 126-131.

30. Wesseling J (1964) A comparison of the steady state drain spacing formulas of Hooghoudt and Kirkham in connection with design practice. J Hydrol 2: 25-32.

31. Darzi-Naftchali A, Shahnazari A (2014) Influence of subsurface drainage on the productivity of poorly drained paddy fields. European Journal of Agronomy 56: 1-8.

32. Doležal F, Kvítek T (2004) The role of recharge zones, discharge zones, springs and tile drainage systems in peneplains of Central European highlands with regard to water quality generation processes. Physics and Chemistry of the Earth 29: 775-785.

33. Youngs E, Rushton K (2009) Steady-state ditch-drainage of two-layered soil regions overlying an inverted V-shaped impermeable bed with examples of the drainage of ballast beneath railway tracks. Journal of Hydrology 377: 367-376.

34. Dagan G (1965) The steady drainage of a two-layered soil. ASCE 91: 51-65.

35. Salgado L (1999) Manual of technical and economic standards for drainage. National Irrigation Commission and University of Concepción. Chile.

36. Wiskow E, Van Der Ploeg R (2003) Calculation of drain spacings for optimal rainstorm flood control. Journal of Hydrology 272: 163-174.

37. Vandersypen K, Keita A, Coulibaly B, Raes D, Jamin J (2007) Drainage problems in the rice schemes of the Office du Niger (Mali) in relation to water management. Agricultural Water Management 89: 153-160.

38. Mishra GC (1954) A Drain spacing formula. Agric Eng 35: 726-730.

39. Jenahb S (1967) Development of a drainage function for the transient case, and a two-dimensional ground-water mound study to evaluate aquifer parameters. Utah state university, Utah.

40. Kraijenhoff van de Leur D (1958) A study of non-steady groundwater flow with special reference to a reservoir-coefficient. DeIngenieur 70: 87-94.

41. Maasland M (1959) Water table fluctuations induced by intermittent recharge. Journal of Geophysics Research 64: 549-559.

42. Chau K (2006) A review on the integration of artificial intelligence into coastal modeling. Journal of Environmental Management 80: 47-57.

43. Imboden D, Pfenninger S (2013) Introduction to system analysis. Springer.

44. Singh A (2014) Groundwater resources management through the applications of simulation modeling: A review. Science of The Total Environment 499: 414-423.

45. Bennett S, Bishop T, Vervoort R (2013) Using SWAP to quantify space and time related uncertainty in deep drainage model estimates: A case study from northern NSW, Australia. Agricultural Water Management 130: 142-153.

46. Singh R, Helmers M, Qi Z (2006) Calibration and validation of DRAINMOD to design subsurface drainage systems for Iowa's tile landscapes. Agricultural Water Management 85: 221-232.

47. Xu X, Huang G, Sun C, Pereira L, Ramos T, et al. (2013) Assessing the effects of water table depth on water use, soil salinity and wheat yield: Searching for a target depth for irrigated areas in the upper Yellow River basin. Agricultural Water Management 125: 46-60.

48. Turunen M, Warsta L, Paasonen-Kivekäs M, Nurminen J, Myllys M, et al. (2013) Modeling water balance and effects of different subsurface drainage methods on water outflow components in a clayey agricultural field in boreal conditions. Agricultural Water Management 121: 135-148.

49. Negm L, Youssef M, Skaggs, R, Chescheir G, Jones J (2014) DRAINMOD–DSSAT model for simulating hydrology, soil carbon and nitrogen dynamics, and crop growth for drained crop land. Agricultural Water Management 137: 30-45.

50. Renaud F, Brown C (2008) Simulating pesticides in ditches to assess ecological risk (SPIDER): II. Benchmarking for the drainage model. The Science of the Total Environment 394: 124-133.

51. Singh A, Panda, S, Asce M (2012) Integrated Salt and Water Balance Modeling for the Management of Waterlogging and Salinization: Validation of SAHYSMOD. Journal of Irrigation and Drainage Engineering 138: 955-963.

52. Ale S, Gowda P, Mulla D, Moriasi D, Youssef M (2013) Comparison of the performances of DRAINMOD-NII and ADAPT models in simulating nitrate losses from subsurface drainage systems. Agricultural Water Management 129: 21-30.

53. Moriasi D, Gowda H, Arnold J, Mulla D, Ale S, et al. (2013) Modeling the impact of nitrogen fertilizer application and tile drain configuration on nitrate leaching using SWAT. Agricultural Water Management 130: 36-43.

54. Kroes J, Dam J, Van Huygen J, Vervoort R (1999) User's Guide of SWAP version 2.0.

55. Richardson L (1920) The supply of energy from and to atmospheric eddies. Proc. R. Soc. London 97: 354-373.

56. Skaggs R (1980) DRAINMOD reference report.USDA-SCS South National Technical Center.

57. Oosterbaan R (1995) SahysMod: Spatial Agro-Hydro-Salinity Model. Description of Principles, User Manual, and Case Studies. International Institute for Land Reclamation and Improvement, Wageningen, Netherlands.

58. Villon M (2006) Espadren, software for calculating spacing tertiary drains. Technology underway. Publisher technology of Costa Rica. Costa Rica.

18

Physical and Social Factors in Management of Community Based Water Storage Structures in Gujarat: An Institutional Analysis of Local Governance

Pande VC[1]*, Bagdi GL[1] and Sena DR[2]

[1]Central Soil and Water Conservation Research and Training Institute, Research Centre, Vasad (Anand), Gujarat, India
[2]Central Soil and Water Conservation Research and Training Institute, Dehradun, Uttarakhand, India

Abstract

Policy intervention in the management of community based water storage structures (CBWS) depends on identifying the factors governing collective action and institutions. Institutional factors not only have a direct bearing on the functioning of CBWS but also often interact with physical and technical factors to influence their sustainability. The present study has examined these issues taking sustainability of CBWS as a function of two components, financial viability and CBWS functionality to draw policy implications in Indian context. Data collected from field surveys revealed that Panchayati Raj Institution (PRI) functionality, perception about change in water collection time and number of households served by the water resource significantly affected financial viability of CBWS. The CBWS functionality was, similarly, found to be significantly affected by factors like accessibility and use restriction with respect to the CBWS. PRI functionality in respect of community resource management, therefore, need to be addressed through better representation of women and weaker section of the community in management of these resources as these sections of society are largely affected by their management. Factors such as use restriction of community water source which affected the physical status of the resource and catchment land use and storage to catchment ratio, which affected operational status of the source, are critical while designing location and size of the water resource such as pond.

Keywords: Community water storage structures; Management; Physical and social factors; Pond

Introduction

Water as common pool resource is indispensable for human life and development. It is efficient, effective, and sustainable use is paramount for ensuring sustainable development. The institutional arrangements for water management are diverse, varying in their structure, scope and style. As a common pool resource, the management of water can be organized under different types of regimes. In open access regime, rules regulating access to and allocation of benefits from the resource are absent. In public property regime, access rights for the public are held in trust by the state. In Private property regime, on the other hand, tradable rights are owned by an individual, household or company. Common property regime (CPR) entails a set of rules to govern access to, allocation of, and control over water [1]. In CPR regimes, some form of organized collective action between the individuals constituting the user community is contemplated; since a collective effort is required to manage access to the CPR and allocation of the benefits it produces [2].

Failures under public and private management have lead to community participation as an alternative mode to govern the resource [3]. In fact, participatory approaches to natural resource management are increasingly being advocated, world over, to promote local stakeholders' involvement in effective management of resources [4]. The literature on Common Property Resource management has also taken cognizance of this fact [5,6].

Interaction of various factors and, hence, design of policy instruments in respect of community based water storage structures (CBWS), however, is quite complex. This is more so because of poor understanding of the interaction and lack of sufficient empirical insight into identifying factors affecting the interplay of local governance forces [7]. Ineffective institutions and their overlapping mandates are, however, also frequently seen as bottlenecks for sustainable natural resources use, with institutional reforms and increased institutional coordination promoted as a solution [8-10].

Policy intervention in the management of community based water storage structures (CBWS) depends on identifying the factors governing collective action and institutions. Studies have shown that institutional factors not only have a direct bearing on the functioning of tank irrigation but also often interact with physical and technical factors to influence tank sustainability [11]. In the present study, sustainability of CBWS is hypothesized to be a function of two components, financial viability and CBWS functionality. Examination of these factors within the frame work of collective action will develop an understanding of the interplay of various physical, technical and social factors, which in turn, will help strengthen, preserve and enhance the collective action through policy intervention on financial and functional parameters of CBWS. Since local institutions are shaped by collective action, these policy interventions will strengthen the institutions for management of community based water storage structures.

Methodology

Study location and survey instruments

The study was conducted in Dhanduka taluka of Ahmedabad district in Gujarat (Figure 1). The selection of study area was based on number of structures. Total geographical area of the district is about

---

*Corresponding author: Pande VC, Central Soil and Water Conservation Research and Training Institute, Research Centre, Vasad-388306 (Anand), Gujarat, India, E-mail: vcpande_2000@yahoo.com

770,000 hectares, out of which 65.3% of the geographical is under cultivation. About 32% of the cultivated land is irrigated, half of which is irrigated by tube wells.

The empirical core of this study derives from extensive, primary surveys and focus group discussions at the household levels. Structured questionnaires were prepared and finalized through pre-testing for socio-economic data elicitation. Apart from the socio-economic surveys, relevant hydro geological and engineering enquiries are also envisaged as an integral component of the study. The hydro geological data gathered through field trips (and supplemented by secondary information) were useful in establishing the potential sustainability of the community water storage structure.

The entire survey exercise was conducted in two rounds. The first round involved (i) finalizing the sample sites and the systems; (ii) collecting basic village level information including sources of water; (iii) household survey focusing on socio-economic characteristics and pattern of water use; (iv) focus group discussions to obtain villagers views and perceptions about specific system related issues. The second round included (i) field surveys for geo hydrological and structural features of the structures.

Selection of systems and sites: Following discussions with different stakeholders, including concerned government and NGO officials, community talavs (pond) were identified for study. Twenty two ponds were randomly selected for extensive study.

Sampling of households: The major emphasis in the selection of households was placed on the fact of their using the selected CBWS. Depending on the number of households using the CBWS in a given village, the proportion of sample households selected from each village

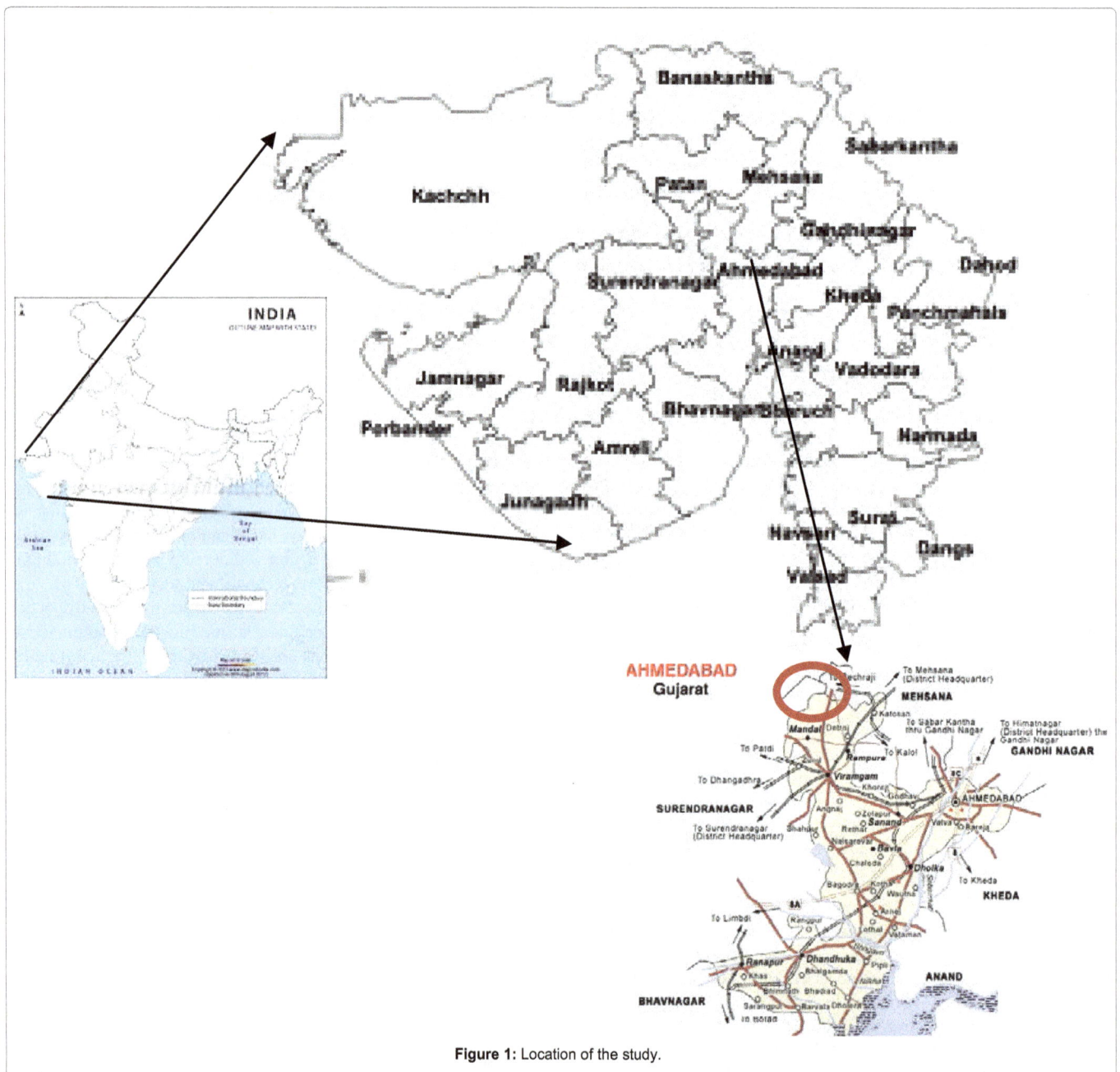

Figure 1: Location of the study.

varied. Factors such as topography, distance between the CBWS and the houses also influenced the sample size. An attempt was made to select beneficiaries staying at varying distances from structures. Ninety beneficiaries and two members of Panchayati Raj Institution managing each pond were identified for data collection.

Survey instruments: For the purpose of collection of both quantitative and qualitative data from the primary source, elaborate survey instruments were prepared. The survey was carried out in two distinct phases. In Phase I the village and household survey instruments were applied and in phase II detailed geo hydrological and engineering surveys were conducted.

Village level questionnaire: This was used to collect information on area, broad socio-economic characteristics of village population, access to ponds. In addition, information was elicited on existence of traditional and modern sources of water supply, crops grown, irrigation sources, and other relevant water related issues.

Household level questionnaire: This survey schedule was designed to canvass household level information on demographic profile of the family, social status, occupation, sources of income, housing details, land holding and also variety of information on domestic water collection and use.

Geo hydrological and engineering survey questionnaire: The schedule was used to collect information on location, design, hydro-climatic data and catchment characteristics of the structures.

The triangulation approach was followed to cross-examine responses to ensure similar result to a question with different methods [12]. This approach helped ascertain reliability of data collected even with low data base used in this study.

Conceptual framework

Sustainability of a community based water storage structure depends on its ability to reliably deliver services to the target community, through financial and physical maintenance support from the community, and with as little intervention from external sources as possible. This was hypothesized to be a factor of two components, viz., financial viability of the structures and functionality of the structure (Figure 2). The former would sustain the structure through regular maintenance, thereby, improving efficiency of the water delivery system, while the latter would ensure reliable service in perpetuity.

Financial viability index (FVI) was computed in terms of charges collected for domestic water use, charges collected for livestock water use, frequency of collection, utilization of collected saving (pond maintenance), mode of water charge collection. Factors that predict revenue generation for use of CBWS included household characteristics such as perception about change in water collection time, Panchayati Raj Institutions (PRI) functionality and number of household drawing water from resource and population below poverty line.

CBWS functionality was measured in terms of reliability (number of days the structure has water in a year). Factors affecting the functionality included the physical and technical factor associated with the structure, the quality of pond management, and the number of residents using the pond. Panchayati Raj Institutions functionality in water resource management was measured in terms of meeting and participation in decision making, amenability/ capability to resolve water management issues, social representation in the PRI executive body (resolving social conflict) and benefits perceived from community water source. The data collected pertained to the years 2009-10 and 2010-11.

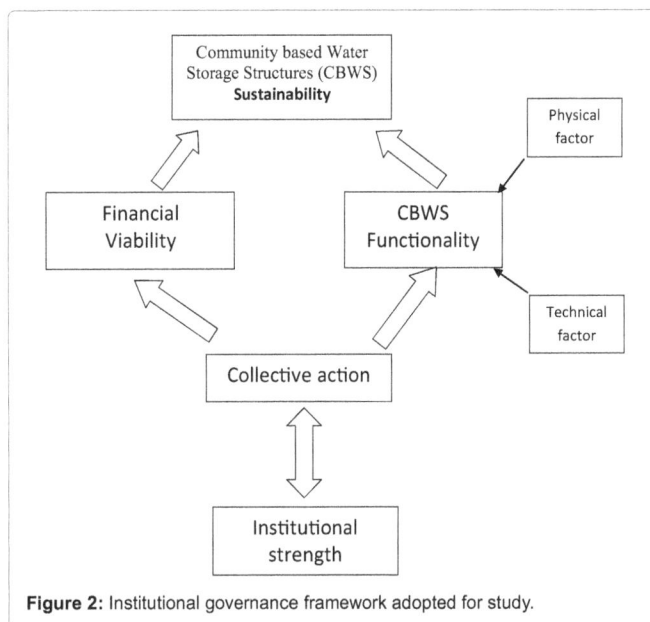

Figure 2: Institutional governance framework adopted for study.

Model used

Logit and regression models were fitted for establishing various relationships. The dependent variable in functional operationality and pond functionality perception are dichotomous in nature, logit model is best suited to examine the relationship. The financial viability model with dependent variable and index has been solved with multiple regression models, which is best to examine such relationships.

Logit model, also known as logistic regression model, is the functional relationship where the dependent variable is a dichotomous variable with probability of an event occurring or not occurring. Since the probability of an event must lie between 0 and 1, it is impractical to model probabilities with linear regression techniques, because the linear regression model allows the dependent variable to take values greater than 1 or less than 0. The logistic regression model is a type of generalized linear model that extends the linear regression model by linking the range of real numbers to the 0-1 range.

In the logistic regression model, the relationship between Z, an unobserved continuous model, and the probability of the event of interest is described by this link function.

$\pi_i = e^{z_i}/1 + e^{z_i} = 1/1 + e^{-z_i}$

This can be written as;

$z_i = \log(\pi_i/1 - \pi_i)$

Where,

π_i is the probability the ith case experiencing the event of interest

z_i is the value of the unobserved continuous variable for the ith case

The model also assumes that Z is linearly related to the predictors (X_{ip})

$Z_i = b_0 + b_1 X_{i1} + b_2 X_{i2} + ... + b_p X_{ip}$

Where,

X_{ij} is the jth predictor for the ith case, j=1, 2, ...p

b_j is the jth coefficient

p is the number of predictors

If Z were observable, we fit a linear regression to Z. However, since Z is unobserved, you must the predictors are related to the probability of interest by substituting for Z.

$$\pi i = 1/1 + e^{-(b_0 + b_1 X_{i1} + ... + b_p X_{ip})}$$

The regression coefficients are estimated through an iterative maximum likelihood method.

Pond operational functionality model:

$$Y = f(X_1, X_2, X_3, X_4) \ldots (1)$$

Dependent variable:

Y=Operational sustainability of pond (water stored during the year)

Dichotomous variable, more than six month=1, otherwise 0

Independent variables:

X_1=Catchment Land use (Non-arable land=1, Arable land=0)

X_2=Surplus arrangement (Separate inlet and outlet=1, otherwise=0)

X_3=Storage to catchment ratio (More than 0.1=1, otherwise=0)

X_4=Pond seepage behavior (No seepage=1, otherwise=0)

It is hypothesized that non-arable land, which in case of these structures is mostly open land with little scrubs here and there, would produce more run off into the ponds and would positively sustain the operationality of the pond. Pond with proper inlet and outlet systems were observed to retain water for longer time. Similarly, if rainfall runoff is to be used, and stored in a reservoir to supply the ponds, a ratio of 10 ha of catchment area to 1 ha of pond is required if the catchment area is pasture; a slightly higher ratio is needed for woodland, and less for land under cultivation [13]. It was, therefore, hypothesized that storage to catchment ratio of more than one would suitably keep the pond operational. Similarly, a pond with no seepage would retain water for longer time

Pond functionality perception model:

$$Y = f(X_1, X_2, X_3, X_4) \ldots (2)$$

Dependent variable:

Y=CBWS status (Perception of beneficiaries about present status, good=1, otherwise 0)

Independent variables:

X_1=Distance from village (Less than one kilometer=1, otherwise=0)

X_2=Accessibility to resource (Unrestricted to all=1, otherwise=0)

X_3=Use restriction (All uses (domestic, animal, irrigation)=1, otherwise=0)

X_4=Location (With village premises=1, otherwise=0)

Pond functionality perception affects beneficiaries' involvement with the management issues of the community owned water storage structures. A positive perception induces to participate in resource management. It was hypothesized that resource with less distance, unrestricted use and within village premises would receive better involvement of the beneficiaries. A pond outside the village premises but less than one kilometer was hypothesized to affect people's perception

positively. This draws from the concept of 'no source village' to identify villages with inadequate water supply.

Financial viability model;

$$Y = f(X_1, X_2, X_3, X_4, X_5, X_6, X_7) \ldots (3)$$

Dependent variable:

Y=Financial viability Index

Independent variable:

X_1=PRI functionality index (Panchayati Raj Institutions functionality in water resource management)

X_2=Perception about change in water collection time since constructing the CBWS (Positive change=1, no change=0)

X_3=Number of household dependent on resource (Nos.)

X_4=Number of BPL household (Nos.)

X_5=Total benefits accrued from the pond (Rs.)

X_6=Private water source owned by the members of PRI body (Yes =1, No=0)

X_7=Perception about change in water quality (Yes=1, No=0)

An index of CBWS's financial viability was computed from factors viz., fee collected for domestic, animal and irrigation uses, frequency of collection and mode of utilization. A community structure was hypothesized to be financially viable if more fees is collected on regular basis and is utilized with unanimous decisions of the members of the PRI. It was hypothesized that a functional PRI would positively contribute to the finances for the maintenance and up keep of the CBWS. PRI functionality was computed from factors, viz., meeting and participation in decision making, amenability to resolve water management issues, social and gender representation in PRI decision making body and benefits perceived by members and non-members of the body assigning equal weightage to each of them. A positive perception about change brought about by the CBWS would induce the beneficiaries to contribute to the finances. In the same manner, while higher number of beneficiary is positively related to financial viability of the community structure, the effect of a higher number of beneficiary household below poverty line would be contrary to that. Further, it was hypothesized with higher benefits accruing a community structure fee charged for water use would be higher as compared to those structures with lower benefits. A PRI with members owing their private water resources would not be much concerned about its maintenance and thereby, affecting the finances collected for the community structure. The perception about change in water quality available from the community structure would, similarly, play a role in beneficiaries' decision about contribution to finances for that structure.

Result

Village profile

The community based water storage structures selected for study were distributed over different villages varying in size from 50 ha to 7500 ha (Table 1). The share of agricultural land in total geographical area was quite high (varying between 70 to 90%) but irrigated land was very small. Most of the cultivation being rainfed, the water storage structures largely met the domestic and animal water requirements, though in some villages these also serve the supplementary irrigation requirements. The major crops irrigated through supplementary irrigation include wheat (*Triticum aestivum*) and cumin (*Cuminum cyminum*) in winter, fodder sorghum (*Sorghum bicolor*) in summer and cotton (*Gossypium hirsutum*) in rainy season (Table 2).

Technical and physical attributes of the structures

Though each villager was eligible to take water from village pond for any domestic use as per the requirement, the supply was limited by the pond's storage capacity and the quantity of water available to fill the tank depending upon catchment characteristics (Table 3). Some ponds retained water for the major part of the year during normal rainfall, while others became dry in five to six months. Similarly some ponds (60% of the sample surveyed) were filled more than once in a year while others were filled only once in a year. Some ponds (22%) also overflew during the season. Siltation and seepage problems (41%) had reduced

Village name	Geographical area (ha)	Agricultural land(ha)	Irrigated land (ha)
Rayka	1569	1382	114
Khadol	1204	1200	500
Khasta	1600	1584	16
Haripura	880	780	40
Fatepur	1120	1104	-
Jaska	2400	1600	83
Vagad	799	763	480
Pachcham	4238	3325	60
Gunjar	1000	800	280
Pipli	7500	6667	167
Bahadi	50	50	-
Tagadi	583	583	-
Morasiya	900	600	33
Zinkhar	1000	917	167

Table 1: Village profile of selected water storage structures.

Village	Method of irrigation	Crops and area irrigated		Irrigation	Irrigation depth	Crop yield
		Name	Area (ha)	Nos.	Depth (cm)	(kg/ha)
Khasta 1	Lift irrigation through pipe	Wheat	2	2	3-4	1440
		Cumin	1	3	2-3	840
Khasta 2	Lift irrigation through pipe	Wheat	4	3	3-4	1200
Pipli	Lift irrigation through pipe	Cotton	6	3	4-5	2400
		Wheat	4	3	2-3	1200
		Cumin	2	4	2-3	840
Panccham	Lift irrigation through pipe	Wheat	30	2	3-4	960
		Cumin	10	3	3-4	720
Zinkhar*	-	-	-	-	-	-
Tagadi*	-	-	-	-	-	-
Bahadi*	-	-	-	-	-	-
Jaska 1	Lift irrigation through pipe	Cotton	50	3	4-5	3000
		Cumin	33	3	2-3	960
Jaska 2				-		
Rayka 1	Lift irrigation through pipe	Wheat	10	3	2-3	1200
		Cumin	4	3	2-3	720
Khasta*	-	-	-	-	-	-
Paccham*	-	-	-	-	-	-
Fatepur*	-	-	-	-	-	-
Haripur*	-	-	-	-	-	-
Khadol	Lift irrigation through pipe	Wheat	300	3	3-4	1200
		Gram	100	1	2-3	960
		Jowar Fodder	100	2	4-5	6000
Rayka 2	Lift irrigation through pipe	Wheat	50	3	3-4	1200
		Gram	10	1	2-3	960
		Cumin	20	3	2-3	600
		Jowar Fodder	20	2	2-3	4800
Rayka 3*	-	-	-	-	-	-
Morasiya	Lift irrigation through pipe	Cotton	25	7	4-5	3000
		Cumin	4	3	2-3	600
		Jowar Fodder	4	1	3-4	6000
Vagad 1*	-	-	-	-	-	-
Vagad 2*	-	-	-	-	-	-
Vagad 3*	-	-	-	-	-	-
Gunjar*	-	-	-	-	-	-

*No supplementary irrigation provided from pond

Table 2: Details of supplementary irrigation, mode of supply and crops in selected ponds.

Pond Number	Pond name	Surface area (m²)	Depth at mid point (m)	Shape	Catchment area (ha)*	Major catchment Land use
1	Pipli	56121	2.0	Irregular	530.0	Non-arable
2	Zinkhar	360000	3.0	Irregular	400.0	Non-arable
3	Tagadi	450000	3.0	Irregular	600.0	Non-arable
4	Bahadi	78000	3.0	Irregular	200.0	Non-arable
5	Jaska talav 1	257300	6.0	Irregular	600.0	Non-arable
6	Jaska talav 2	50000	2.0	Irregular	40.0	Non-arable
7	Khasta talav 1	10000	3.0	Rectangular	15.0	Arable
8	Khasta talav 2	12500	2.0	Rectangular	24.0	Arable
9	Khasta talav 3	114100	4.0	Irregular	530.0	Non-arable
10	Panccham talav 1	233628	2.5	Rectangular	25.0	Arable
11	Panccham talav 2	200000	6.0	Irregular	600.0	Non-arable
12	Fatehpur	77700	3.0	Irregular	600.0	Non-arable
13	Haripur	41490	5.0	Irregular	300.0	Non-arable
14	Khadol	305100	4.0	Irregular	500.0	Non-arable
15	Rayaka talav 1	5625	3.0	Rectangular	7.0	Arable
16	Rayaka talav 2	8590	4.0	Irregular	150.0	Arable
17	Rayaka talav 3	30000	3.0	Irregular	200.0	Arable
18	Morasiya	14653	3.0	Irregular	200.0	Arable
19	Vagad talav 1	9000	2.5	Rectangular	100.0	Arable
20	Vagad talav 2	6375	2.5	Rectangular	50.0	Arable
21	Vagad talav 3	6715	2.0	Rectangular	17.0	Arable
22	Gunjar	24399	2.0	Irregular	150.0	Arable

*Approximation through observation and discussion with villagers

Table 3: Technical and physical attributes of village ponds.

the storage capacity of many ponds. The surplus arrangement (inlet and outlets) in the pond also affected the amount of water stored and thus, its availability to the beneficiaries. Though majority of the ponds (86%) had proper inlet and outlets, the remaining either had breached or were in defective condition. Absence of maintenance had reduced the water storage capacity of the ponds.

Sociology of community management

Only few ponds (less than 10% of the ponds) were managed by state department. The remaining ponds were managed by Panchayati Raj Institution (PRI), an elected body for local management. In majority of the cases (55% PRIs surveyed), however, the executive body did not hold meetings to discuss about water related issues. Women, who mostly bear the burden of arranging water for domestic and animal use, were not well represented in the panchayat executive body. Among the members of executive body, women were members in only few cases (45% PRIs). In these bodies, women as sarpanch, head of the executive body, was observed in only a few cases (15% PRIs). The other members did not bother to take up the issues related to water from pond. Similarly, in majority of the cases executive body members largely had own private sources. For drinking water, government source like Narmada canal pipe lines were laid in most of the villages. In a few villages, poorer farmers still depended on the village pond even for domestic uses.

Logit and regression analysis results

The general description of the variables used in the study is given in Table 4. Based on the technical and social attributes the variables for which consistent data could be procured from beneficiaries were used for analysis.

The pond operational functionality model had operational sustainability index as dependent variable. This variable was measured as dichotomous variable in terms of water storage. If water in pond remained for more than six months, the value of this variable was taken

as 1, if water remained stored for less than six months, the value was 0. The mean of the variable was 0.77 with a standard deviation (SD) of 0.43. The explanatory variables for the model included catchment land use, measured in terms of non-arable (1) and arable land (0), with mean 0.50 and SD 0.51; surplus arrangement, measured in terms of separate inlet and outlet (1) and no separate inlet and outlet (0), with mean 0.14 and SD; storage to catchment ratio, measured in terms of ratio of storage area to catchment area more than 0.1(1), otherwise (0), with mean 0.45 and SD 0.50 and pond seepage behavior, measured in terms of presence of seepage (0) and absence of seepage (1), with mean 0.72 and SD 0.45. The pond functionality perception from beneficiaries' view point was examined to understand the relationship and identify the factors that influence their perception about pond health. Such perceptions influence their involvement in pond management [14]. The dependent variable was measured as good status perception (1) and poor status perception (0) with mean 0.67 and SD 0.47. The explanatory variables viz., distance, accessibility, water use restriction and location had mean varying from 0.23 to 0.52 and standard variation, from 0.45 to 0.50. These variables were used as input in the software, SPSS 13 and analyzed using logistic regression module for best fit.

For the financial viability model, the dependent variable (financial viability index) was measured as an index with equal weightage given to each factor viz., fee collected for domestic, animal and irrigation uses, frequency of collection and mode of utilization. The variables were given values varying from 0 to 2. For fee collected a value was given '0', if no fee collected, '1' if less than Rs 100/- annum collected and 2, if more than Rs 100/- annum collected in case of a pond. The frequency of collection was valued as 0 (no collection), 1 (irregular collection) and 2 (regular collection). The fund utilization mode was valued as 0 (no utilization), 1 (decided by few) and 2 (decided with consensus). The index had mean 1.11 and standard deviation 0.17. The explanatory variables such as PRI functionality, perception about change in water collection time and water quality, private water source owned were dichotomous in nature with values 0 and 1. The remaining

Variable	Description	Mean	Standard deviation	Observations
Pond operational functionality model variables				
Dependent variable				
Operational sustainability Index	Water stored for more than six month	0.77	0.43	22
Explanatory variable				
Catchment Land use	Arable and non-arable land use	0.50	0.51	22
Surplus arrangement	Inlet and outlet system of the pond	0.14	0.35	22
Storage to catchment ratio	Ratio of storage area to catchment area	0.45	0.50	22
Pond seepage behavior	Presence or absence of seepage from pond	0.72	0.45	22
Pond functionality percpetion model variable				
Dependent variable				
CBWS status	Perception about present status of pond	0.67	0.47	22
Explanatory variable				
Distance from home	Distance of pond from home	0.44	0.50	22
Accessibility	Resource accessibility to users	0.23	0.49	22
Use restriction	Restriction in the use of water from pond	0.52	0.50	22
Location	Existence within village or outside the village	0.27	0.45	22
Financial viability model variables				
Dependent variable				
Financial viability Index	Revenue generation through collection of water charges	1.11	0.17	22
Explanatory variable				
PRI functionality index	Panchayati Raj Institutions functionality in water resource management	1.09	0.32	22
Collection time change perception	Perception about change in water collection time from water source	0.70	0.47	22
Household dependent on resource	No. of household dependent on water resource	463	575	22
BPL household	No. of household below poverty line dependent on resource	133	146	22
Gross benefits	Total benefits accrued from the pond	500498	6.57	22
Private water source	Private water source owned by the members of PRI body	0.70	0.47	22
Water quality change	Perception about change in water quality	0.30	0.47	22

Table 4: Model variables used in the study.

variables were measured as actually measured with relevant units. The dichotomous explanatory variables had mean varying from 0.70 to 1.11 and SD varying from 0.17 to 0.47. The number of households dependent on the resource varied from 15 to 1050, with a mean of 463 and SD 515. The BPL household, similarly, had mean 133 and SD 146. The gross benefits drawn from pond had mean Rs 5,00,498 and SD 6.57. This functional relationship was examined using the software, SPSS 13 with regression module.

The results of logit and regression model are given in Table 5. The variables were entered and model performance was checked. The final model with best fit was retained. For the pond functionality model, since the catchment land use was same in case of all the community ponds and the model fitted with this variable turned out to be poor, this was dropped in the final fit. The relationship of factors like surplus arrangement in the pond, storage to catchment ratio and pond seepage behaviour with operational status was examined with the response variable and the model slightly improved. Hence, these variables were retained for final analysis. Storage to catchment ratio turned out to be significantly affecting operation of ponds (significance level 11%). The other two factors turned out to be insignificant. The perception about current status of community pond was found to be affected by factors like accessibility to the resource, distance of community water resource from household and use restriction with respect to the resource in the final model. These factors significantly affected the current status of the resource (7%, 10% and 2% level of significance, respectively).

Examination of relationship of financial viability index with explanatory variables revealed that PRI functionality, gross benefit from pond and perception about water quality change were significantly related with dependent variable at 8%, 20% and 20% significance level, respectively. Perception about change in water collection time was closely related with location of the source from village. Resources closer to village periphery changed in water collection time and affected financial resource of the PRI positively.

Discussion

Pond with high demand for water for domestic, animal and irrigation uses against poor supply experienced water related conflicts (Table 6). The conflict management in some villages, though, was governed by the strength of the institution. While PRI an elected body entrusted with the task of pond management needs to be strengthened, factors such as design and location play important role in influencing beneficiaries' perception. The accessibility to the resource and use restriction with respect to the resource affected perception about present status of community based natural resources. Similarly, distance of resource also affected its current status in terms of maintenance. The ponds being located in the outskirt of village, only a few were observed to have easy access. Storage to catchment ratio affected operationality of the community based water storage structures. Similarly, catchment with arable land use was observed to have water storage for less than 6 months. In those structures with non-arable catchment use, storage was much higher than that. This catchment was devoid of vegetation except for some scrubs. Though this variable did not appear in the final model but the fact remains that non-arable catchment covered with little scrub contributed to more runoff in the pond in the study area. PRI functionality, perception about change in water collection time and number of households served by the water resource affected financial viability of the ponds. Perception about change in water collection time

S. No.	Variable	Coefficient	Significance level
Dependent variable : Operational sustainability of pond			
1	Surplus arrangement	-0.44	*
2	Storage to catchment ratio	2.08	11%
3	Pond seepage behaviour	-0.97	*
Number of observations 22			
-2 Log likelihood 9.85			
Pseudo R-Sq. (Cox & Snell R–Sq) 0.16			
Pseudo R-Sq. (nagelkerke R–Sq) 0.24			
Dependent variable : Pond status perception			
1	Distance from village	-2.20	10%
2	Accessibility	2.29	7%
3	Use restrictions	-3.13	2%
Number of observations 22			
-2 Log likelihood 47.60			
Pseudo R-Sq. (Cox & Snell R –Sq) 0.24			
Pseudo R-Sq. (nagelkerke R –Sq) 0.34			
Dependent variable : Financial viability			
1	PRI functionality index	6.63	8%
2	Collection time change perception	23.5	*
3	Household dependent on resource	0.001	*
4	BPL household	0.007	*
5	Gross benefit from pond	0.00002	20%
6	Private water source	-0.70	*
7	Water quality change	-2.58	20%
Number of observations 22			
-2 Log likelihood 19.82			
Pseudo R-Sq. (Cox & Snell R –Sq) 0.51			
Pseudo R-Sq. (nagelkerke R –Sq) 0.63			

*Insignificant

Table 5: Logit and regression model result for community based water storage structures.

Pond Number	Village name	Village population	Animal Population	Pond storage volume (m³)	Pond water usage	Pond Maintenance	Social conflict management
1	Pipli	760	750	100200	Domestic, animal, irrigation	Poor	Poor
2	Zinkhar	823	1520	240000	Domestic, animal	Good	Good
3	Iagadi	336	106	450000	Domestic, animal	Good	Good
4	Bahadi	45	23	52000	Domestic, animal	Good	Poor
5	Jaska	384	487	1029200	Domestic, animal irrigation	Poor	Good
6	Khasta	3885	382	55000`	Domestic, animal	Poor@	Poor
7	Paccham	2250	1270	349200	Domestic, animal	Poor	Good
8	Fatehpur	574	180	225000	Domestic, animal	Good	Good
9	Haripur	282	50	207460	Domestic, animal	Good	Good
10	Khadol	747	445	1220400	Domestic, animal, irrigation	Poor	No conflict
11	Rayaka	784	193	124360`	Domestic, animal irrigation	Poor@	No conflict
12	Morasiya	750	150	49590	Domestic, irrigation	Good	No conflict
13	Vagad	2100	1119	46015`	Domestic, animal	Good@	No conflict
14	Gunjar	12590	913	58000	Domestic, animal	Good	No conflict

`Sum of more than one pond volume
@ Includes all the structures
Domestic use includes cloth washing

Table 6: Maintenance and conflict management of selected ponds.

was closely related with location of the source from village. Resources closer to village periphery did perceive change in water collection time, quality and regularly paid for water charges. While the change in perception was governed by physical/technical factor in terms of pond size and location, PRI functionality turned out to be an important factor in managing finances for pond management. In fact, PRIs were observed to have poor gender sensitivity. The number of members in the executive body of panchayat varies from 7 to 10, women being member of the body in only few cases (45%). Similarly, women as sarpanch, head of the body, was observed in only a few cases (15%), and these bodies incidentally held executive body meeting at least once in a year. In other cases, the other executive body did not hold meetings (55%). Except for a couple of cases (10%), in other bodies the members were medium and large farmers, and having own private source of water such as tube wells. PRI functionality can, therefore, be strengthened by motivating and sensitizing PRI members to water governance issues by enhancing representation of women, who manage water uses at household level and weaker sections of farmers who did not have private water source and, primarily depended on these community resource. These observations find strength from similar observations elsewhere [15,16]. Both these groups were poorly represented in most of the panchayat body. The weak sensitivity of PRI towards these community based natural resources can also be partly explained in terms of network of Narmada Canal and pipeline to villages to meet largely domestic uses, animal uses like bathing, maintaining hygiene and in some villages drinking.

Conclusion and Policy Prescription

The active participation and local governance of community resources for more efficient, effective and equitable development need promotion of equitable participation of women and weaker section of rural community. The essential assumption here is that women and poor farmer represent a marginalized group in society whose lives are entrapped in an institutional framework characterized by gross inequalities of formal power and authority in the public sphere and denied equal access to and control over resources. The new institutional structures introduced under gender-equity based participatory models of local governance seek to balance out the inequalities by offering a platform or space where women can come together alongside men and be empowered to express their opinions as well as contribute effectively in decision-making processes. With respect to the water sector in general, women's participation seeks to correct imbalances perceived in terms of access to water resources and benefits from water development projects as well as exercise of decision-making powers with respect to the management of these resources [17,18]. Similarly, technical design and scientific planning in creating water resources would go a long way in not only serving the rural community but also efficiently as people's perception about resource utility was positively higher in case of ponds with right technical parameters. Storage to catchment ratio of more than 0.1 or more has been suggested appropriate [13] for pond utility such as aquaculture. Such ponds with water for sufficiently longer period of time would also serve other purposes of rural livelihood.

Further technical examination in terms of geo-hydrological factors contributing to efficient pond water delivery would further enhance the utility of such studies. There is debate on downstream water flow effect of watershed management programme being under taken in the country. Large scale implementation of these programmes and their impact on health of these ponds needs further exploration as pond health affects their functionality affecting people's perception for or against their involvement in the regular maintenance of these traditional sources of water. In the backdrop of poor perception about services delivered by pond, the financial resources generated are also adversely affected. Ponds, which traditionally have been the life line of a large section of Indian rural population, would be better managed if social factors are understood in the larger geo-hydrological context. The interplay of such technical and social factors can be better examined, understood and addressed if policy makers, local stakeholders and scientific community are brought to one platform and the cause-effect relationships amongst various region specific factors are established scientifically. The framework used in this study is one attempt. However, more efforts are required to test, modify and improve such models across the different socio-cultural and hydrological regions.

Notes

In the Fourth Plan, the concept of No Source Village (NSU) was introduced to identify problem villages with inadequate supply of water, and accordingly a village was an NSV if it did not have a reliable source of water. A village is a no source village if it has any of the following characteristics: (1) No public well, (2) has a public well that dries up in summer making villagers travel more than 1 km to fetch water, (3) a source of water supply more than 1 km away, (4) no possibility of a well, needed a tube well for drinking water, (5) there is a public well, but the supply is below 70 lpcd, (6) non potable water supply [19].

References

1. Edwards V M, et al. Developing an Analytical Framework for Multiple-Use Commons. Journal of Theoretical Politics. 1998; 10 (3): 347–383.

2. Taylor M. Governing natural resources. Society and Natural Resources. 1998; 11 (3): 251–258.

3. Agrawal A. Sustainable Governance of Common-Pool Resources: Context, Methods, and Politics. Annu Rev Anthropol. 2003; 32: 243-262.

4. Kiss A. Living with wildlife: wildlife resource management with local participation in Africa. World Bank Technical Paper, Washington, DC. 1990; WTP 130.

5. Wade R. Village Republics: Economic Conditions for Collective Action in South India. Cambridge: Cambridge University Press, 1988.

6. Ostrom E. Governing the Commons: The Evolution of Institutions for Collective Action. Cambridge: Cambridge University Press, 1990.

7. Heltberg R. Determinants and Impact of Local Institutions for Common Resource Management. Environment and Development Economics. 2001; 6: 183-208.

8. Mitchell B. Integrated Water Management: International Experiences and Perspectives. Belhaven Press, London, 1990.

9. Bandaragoda DJ. Status of institutional reforms for integrated water resources management in Asia: Indications from policy reviews in five countries. International Water Management Institute (IWMI), Colombo. 2006; Working Paper 108.

10. Rydin Y, et al. Networks and Institutions in Natural Resource Management. Edwar Elgar Publishing Limited, Cheltenham, UK, 2006.

11. Janakarajan S. In Search of Tanks Some Hidden Facts. Economic and Political Weekly, 1993; 28 (26): A53-A60.

12. Denzin NK. The Research Act: A Theoretical Introduction to Sociological Methods. McGraw-Hill, New York, 1978.

13. Kovari J. Considerations in the selection of sites for aquaculture. United Nations Development Programme, Food and Agriculture Organization of the United Nations, Rome, 1984.

14. Tyson B, et al. Facilitating Collaborative Efforts to Redesign Community Managed Water Systems. Applied Environmental Education and Communication. 2011; 10 (4): 211-218.

15. Aladuwaka S, et al. Sustainable development, water resources management and women's empowerment: the Wanaraniya Water Project in Sri Lanka. Gender & Development. 2010; 18 (1): 43-58.

16. Barnaud C, et al. Dealing with Power Games in a Companion Modelling Process: Lessons from Community Water Management in Thailand Highlands. The Journal of Agricultural Education and Extension. 2010; 16 (1): 55-74.

17. UNDP. Mainstreaming gender in water management: a practical journey to sustainability. UNDP/BDP Energy and Environment Group, New York, 2003.

18. GWA. Advocacy Manual for Gender and Water Ambassadors. WEDC (for GWA), Leicestershire, 2002.

19. Hirway I. Ensuring Drinking Water to all: A Study in Gujarat. Paper submitted to the 4th IWMI-TATA Annual Partners Research Meet, 2005.

19

Allocation of Canal Water Optimally Employing OPTALL Model

Upadhyaya A*

Division of Land and Water Management, ICAR Research Complex for Eastern Region, Patna, India

Introduction

Water is one of the most important inputs for agricultural production. But due to wide variation in its spatial and temporal availability, it is not always possible to apply it to the crop when it is essentially required. Sometimes, in spite of sufficient water availability in the reservoir, water is not released, distributed and allocated timely, adequately and equitably among the farmers due to many technical, hydraulic, socio-economic, institutional, financial and managerial problems as mentioned by Upadhyaya [1] and discussed in detail by Upadhyaya [2]. This leads to wide gap between water supply and water requirement in the canal command, resulting in either water logging or agricultural drought. Both the situations are detrimental and adversely affect the crop yield. This clearly indicates that there is sufficient scope to made water available to farmers in adequate quantity at right time, if canal managers properly plan, manage and release water as per crop water requirement. Upadhyaya [3] has mentioned number of water management technologies in agriculture to improve water productivity. In this paper, optimal and equitable releases of water in the distributaries of Patna Main Canal under various scenarios of water availability by employing OPTALL model have been presented and gap between actual and optimal releases has been assessed for conjunctive use planning.

OPTALL Model and its Capabilities

In order to equitably allocate water among farmers, total water availability, water requirement and a mathematical model capable of optimally and equitably allocating water are required. There are number of approaches to solve the optimization problem. These included dynamic programming, linear and non-linear programming. Of these, dynamic programming and non-linear programming (quadratic) have the potential to solve such problems in a better way because the optimization approach can offer significant benefits in terms of potential water saving, equity in water allocation subject to system constraints. OPTALL model developed at University of Edinburgh has the capability to allocate water optimally and equitably in various offtakes of Patna Main Canal keeping in view the requirement and system constraints (Figure 1). Development of objective function in problem formulation is very important. Since the problem addressed was related to irrigation water management and in particular to ensuring optimal and equitable allocation of irrigation water for farmers, the most appropriate objective function as defined by Wardlaw et al. was as follows [4]:

$$Minimize\ Z = \sum_{i=1}^{n} \frac{(d_i - x_i)^2}{d_i}$$

where n is the number of irrigation schemes, d_i the irrigation demand for scheme i, and x_i the irrigation supply to scheme i. The above equation is subject to canal capacity constraints, continuity constraints, and of course supply constraints defined mathematically as:

$$x_{i,w} \leq D_i$$

$$\frac{x_{i,w}}{d_{i,w}} \leq 1.0$$

$$\frac{x_{1,w}}{d_{1,w}} = \frac{x_{2,w}}{d_{2,w}} = \frac{x_{3,w}}{d_{3,w}} = \frac{x_{n,w}}{d_{n,w}}$$

Where D_i is design discharge capacity of off-take point 'i', $x_{i,w}$ is

irrigation supply in i^{th} off-take in 'w^{th}' week, and $d_{i,w}$ is demand in i^{th} off-take in 'w^{th}' week (Figure 1).

OPTALL was used by Wardlaw et al. [5] to model systems with complex distribution systems and has been demonstrated to be very robust. Wardlaw [6] presented the optimization approach and procedure to solve real time optimization of water for better water allocation. Wardlaw et al. [7] applied a genetic algorithm (GA) to the water allocation problem also, and while acceptable results were produced, they concluded that the GA offered no advantage over quadratic programming for this particular problem.

Data requirement of OPTALL model includes a schematic diagram of Patna Main Canal system network, which is first prepared manually and then with the help of software. Information about various reaches, irrigation schemes, nodes, branches, their capacity, seepage loss/gain and length is required to prepare the network input file. Other input files need weekly water demand in m³/s at all the nodes as defined in network file. Input file for weekly actual inflows diverted at the head of canal is also one of the input files.

With this input data OPTALL computes optimum and equitable release in the canal offtakes subject to system constraints. The optimal releases not only minimize the gap between water availability and water requirement but also give opportunity to equitably allocate and efficiently utilize the precious water in the command. This model is robust, relatively easy to apply, and has potential as a tool in decision support for real-time irrigation system operation. This approach can be used by scheme managers to improve the equity of their water distribution under various scenarios of water availability.

Actual and optimal canal water supply and comparison with irrigation requirement

Actual and optimal water releases in various distributaries of Patna Main Canal were studied and compared with irrigation requirement considering average, 75% dependable and actual rainfall of 2014 during Kharif and Rabi seasons in head, middle, tail and whole Patna Main Canal. Results are presented graphically for whole Patna Main canal in Figures 2-5.

Number of weeks having excess and deficit release in Patna Main Canal based on actual and optimal supply of canal water over irrigation requirement in head, middle and tail reaches as well as whole canal and range of variation have been summarized from graphs and are presented in Table 1 below. It may be observed from the figures that in case of actual water supply there is a wide gap between supply and irrigation requirement. In some weeks supply is excessively higher

*Corresponding author: Upadhyaya A, Principal Scientist, Division of Land and Water Management, ICAR Research Complex for Eastern Region, Patna, India
E-mail: aupadhyaya66@gmail.com

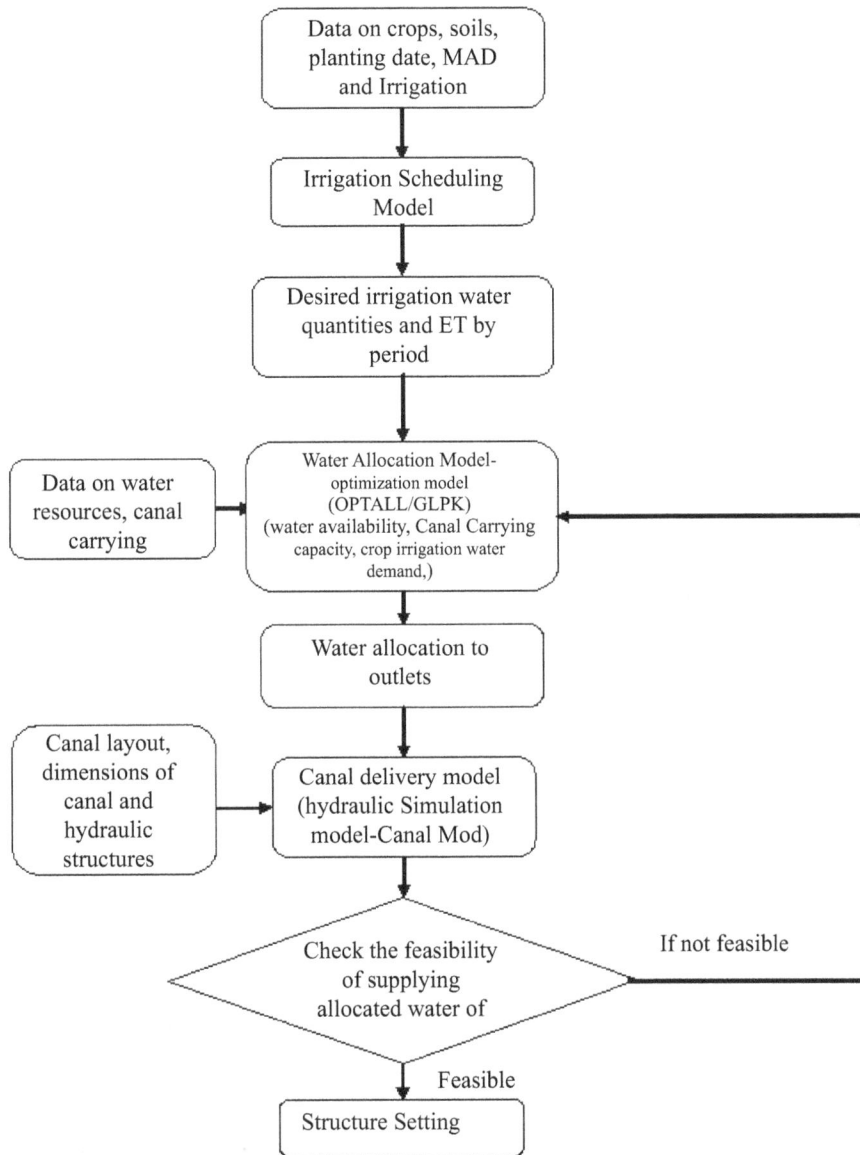

Figure 1: Flow chart of OPTALL.

than requirement, whereas in other weeks it is excessively lower than demand. Table 1 reveals that in case of actual release, during Kharif season number of weeks having excess supply of canal water than irrigation requirement are less than number of weeks having deficit supply of water, whereas during Rabi season trend is opposite and number of weeks having excess supply are more than the number of weeks having deficit supply. The range of deficit supply is less than the range of excess supply, which indicates that there is scope to increase area under irrigation during Rabi period. In case of optimal release of canal water, none of the weeks have excess water supply and number of weeks having deficit water supply are most of the time less than the case of actual release. The range of deficit water supply in case of optimal release is always less than the case of actual release. In Kharif season, deficit supply over irrigation requirement was highest for the case when irrigation requirement was computed considering 75% dependable rainfall followed by the cases of actual and average rainfall. There is significant variation in excess and deficit supplies in head, middle and

tail reaches also. Results clearly indicate that optimal releases obtained after employing OPTALL model never show excess supply and number of weeks having deficit supply as well as range of deficit supply reduce significantly and gap between supply and irrigation requirement minimizes in case of optimal release.

Actual and optimal supply-demand ratios in distributaries

Supply-demand ratios for all the distributaries of Patna Main Canal were computed for actual and optimal canal water supply as well as irrigation requirement considering average, 75% dependable and actual rainfall of 2014 during Kharif and Rabi seasons. Summary of distributaries with their supply-demand ratios in various weeks is presented in Table 2. It may be observed from this Table that in case of actual water supply, supply-demand ratios excessively exceeds 1.0 in many weeks in head, middle and tail reaches, whereas it is varying between 0.1 to less than 1.0 in other weeks. Only in very few weeks it is 1.0. On the other hand in case of optimal water supply, the supply-

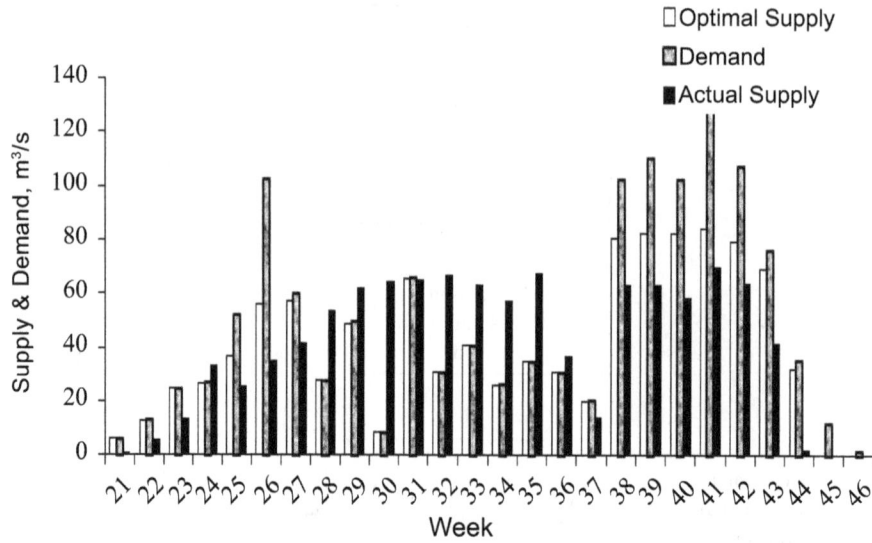

Figure 2: Weekly optimal and actual supply as well as demand considering average rainfall in PMC.

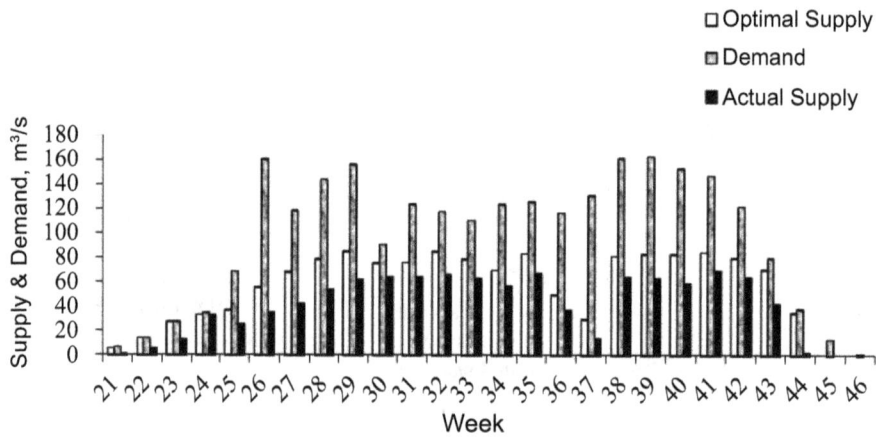

Figure 3: Weekly optimal and actual supply as well as demand considering 75% dependable rainfall in PMC.

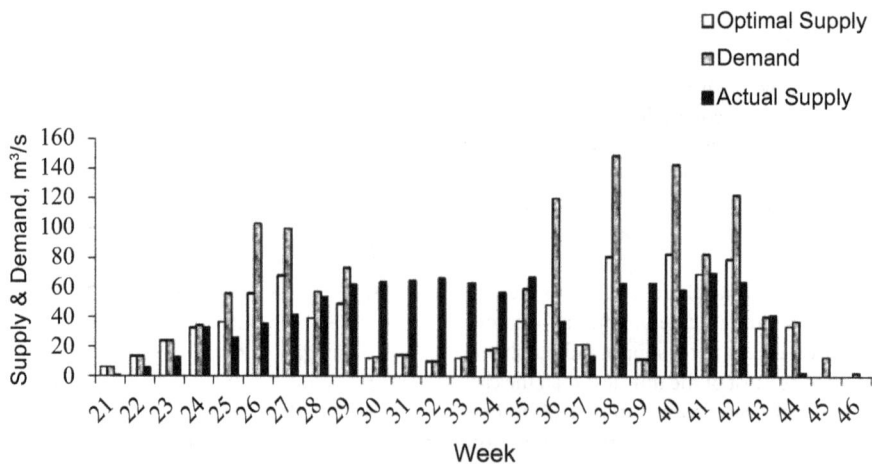

Figure 4: Weekly optimal and actual supply as well as demand considering actual rainfall of 2014 in PMC.

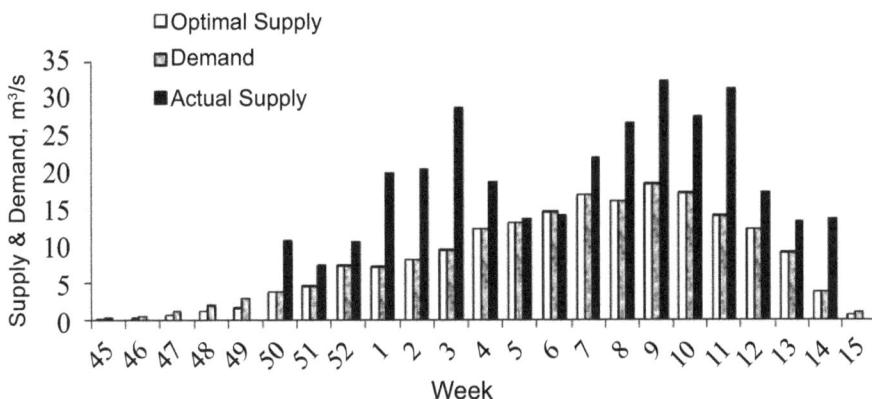

Figure 5: Weekly optimal and actual supply as well as demand during Rabi season in PMC.

	During Kharif considering average rainfall			During Kharif considering 75% dependable rainfall			During Kharif considering actual rainfall of 2014			During Rabi seson		
	Actual release		Optimal release	Actual release		Optimal release	Actual release		Optimal release	Actual release		Optimal release
	Excess over irrigation req. No. of weeks and Range (m³/s)	Deficit over irrigation req. No. of weeks and Range (m³/s)	Deficit over irrigation req. No. of weeks and Range (m³/s)	Excess over irrigation req. No. of weeks and Range (m³/s)	Deficit over irrigation req. No. of weeks and Range (m³/s)	Deficit over irrigation req. No. of weeks and Range (m³/s)	Excess over irrigation req. No. of weeks and Range (m³/s)	Deficit over irrigation req. No. of weeks and Range (m³/s)	Deficit over irrigation req. No. of weeks and Range (m³/s)	Excess over irrigation req. No. of weeks and Range (m³/s)	Deficit over irrigation req. No. of weeks and Range (m³/s)	Deficit over irrigation req. No. of weeks and Range (m³/s)
Head	11 (31.69-0.94)	15 (60.38-0.30)	10 (36.29-0.30)	1 (6.07)	25 (100.40-0.15)	22 (78.36-0.20)	13 (37.72-0.40)	13 (62.92-0.46)	10 (42.98-0.20)	15 (9.37-0.51)	5 (0.77-0.11)	6 (0.43-0.01)
Middle	9 (21.90-0.21)	17 (20.10-0.22)	13 (15.55-0.07)	-	26 (39.77-0.22)	23 (31.95-0.17)	7 (23.29-2.43)	19 (27.75-0.23)	18 (23.36-0.17)	15 (8.64-0.31)	7 (2.12-0.16)	6 (0.88-0.02)
Tail	11 (10.53-0.87)	13 (17.36-1.28)	14 (8.56-0.04)	5 (6.89-0.40)	19 (21.13-1.28)	23 (13.90-0.05)	7 (6.89-1.18)	17 25.07-0.67)	18 (16.32-0.05)	10 (4.24-0.38)	9 (0.80-0.02)	9 (0.04-0.01)
PMC	9 (55.70-5.86)	17 (67.33-1.13)	16 (46.25-0.15)	-	26 (125.80-1.13)	24 (104.73-0.37)	8 (56.37-1.26)	18 (86.07-1.13)	19 (68.92-0.37)	16 (19.24-0.48)	6 (2.98-0.16)	13 (1.35-0.01)

Table 1: Number of weeks and range of variation in excess as well as deficit in release over irrigation requirement in PMC.

demand ratio never exceeds 1.0. It is 1.0 in many weeks. It is also less than 1.0 in the weeks when canal water is available in limited quantity. It is most of the time greater than 0.5 and equitable in all the distributaries subject to the system constraints. It shows that OPTALL model not only allocates water optimally but equitably also, which may lead to efficient and effective utilization of canal water.

Gaps in optimal water supply and demand

Though, compared to actual water supply, optimal water supply could meet the demand equitably and optimally in many weeks during Kharif season and almost in all the weeks during Rabi season, yet during Kharif season in few weeks demand could not be met fully as shown in Figures 2-5. In order to meet this gap there is a need to explore the possibility of ground water use and promote the concept of conjunctive use of rain, surface and ground water in the canal command to improve yields.

Conclusions

In order to minimize the gap between water supply and irrigation requirement OPTALL model based on quadratic programming technique was employed and optimal as well as equitable water allocation plan to meet the irrigation requirement computed after considering average, 75% dependable and actual rainfall for various distributaries of Patna Main Canal under Sone Canal system was developed. The optimal water allocation schedule was found much better than actual release because in no case supply-demand ratio was more than 1.0, whereas in case of actual release it was excessively higher than 1.0 in many distributaries of Patna Main Canal showing inequitable water distribution. In order to utilize canal water equitably, efficiently and judiciously, canal operation schedules need to be developed and reviewed in consultation with water users under various situations of water availability. Through frequent meetings, interactions/dialogues between canal managers and water users and technical back stopping through such decision support tools, most of the water conflicts can be resolved. The gap between optimal water release and irrigation requirement indicates that there is a need to promote ground water utilization and explore the possibility of conjunctive use of rain, surface and ground water in canal command to minimize the gap and improve the yield.

Acknowledgements

This publication is an output of AP Cess funded ICAR Young Scientist Award Project. Author gratefully acknowledges the financial support provided by ICAR, New Delhi and DFID, U.K. in carrying out this study and its revision.

Reach	Supply-Demand Ratio	During Kharif considering average rainfall				During Kharif considering 75% dependable rainfall				During Kharif considering actual rainfall of 2014				During Rabi seson			
		Actual Release		Optimal Release		Actual Release		Optimal Release		Actual Release		Optimal Release		Actual Release		Optimal Release	
		Weeks	No. of Dist. (Range)	Weeks	No. of Dist. (Range)	Weeks	No. of Dist. (Range)	Weeks	No. of Dist. (Range)	Weeks	No. of Dist. (Range)	Weeks	No. of Dist. (Range)	Weeks	No. of Dist. (Range)	Weeks	No. of Dist. (Range)
Head	> 1.0	23	1-13	-	-	15	1-8	-	-	18	1-13	-	-	17	4-10	-	-
	1.0	9	1-2	16	13	6	1-2	5	13	4	1	8	13	7	1-2	20	3-13
	Partial	22	2-11	7	13	24	2-13	19	12-13	15	2-13	8	13	15	1-5	-	-
	0.0	5	1-9	1	13	8	1-9	-	-	6	1-9	8	13	21	1-13	8	1-13
Middle	> 1.0	22	1-10	-	-	11	1-9	-	-	19	1-10	-	-	17	1-10	-	-
	1.0	7	1-2	13	10	6	1-2	3	10	4	1	13	1-10	2	1	20	4-10
	Partial	19	1-8	7	1-10	22	2-10	21	1-10	15	2-10	14	1-10	8	1-5	-	-
	0.0	7	1-10	-	-	7	2-10	-	-	8	1-10	6	1-10	20	1-10	7	2-6
Tail	> 1.0	15	1-10	-	-	12	1-10	-	-	13	1-10	-	-	17	1-10	-	-
	1.0	4	1-2	7	10	4	1	2	2-3	4	1-2	9	2-10	4	1	19	3-10
	Partial	12	1-10	12	1-10	17	2-10	21	2-10	14	3-10	16	2-10	11	1-4	-	-
	0.0	7	1-10	8	1-10	6	1-10	6	1-10	5	7-10	6	1-10	16	1-10	15	1-10

Table 2: Summary of distributaries partially or fully meeting out demands in various weeks and supply-demand ratios.

References

1. Upadhyaya A (2002) Problems of water distribution in Patna canal command. Proceedings of International Conference on Hydrology and Watershed Management with a focal theme on water quality and conservation for sustainable development, Hyderabad, AP.

2. Upadhyaya A (2005) Development of operational plans/strategies for efficient water allocation and distribution from Patna Canal in Bihar. Final report of AP Cess Funded ICAR Young Scientist Award Project submitted to ICAR, New Delhi. pp: 76.

3. Upadhyaya A (2015) Water management technologies in agriculture: Challenges and opportunities. Journal of Agrisearch 2: 7-13.

4. Wardlaw RB, Barnes JM (1996) Real time operation and management of irrigation systems. International conference on new challenges for civil engineers of developing countries in the 21st century. Indian Society of Environmental Management, New Delhi.

5. Wardlaw RB (1999) Computer optimization for better water allocation. Agricultural Water Management 40: 65-70.

6. Wardlaw RB, Bhaktikul K (2001) Application of a genetic algorithm for water allocation in an irrigation system. Irrig and Drain 50: 159-170.

7. Wardlaw RB, Barnes JM (1999) Evaluating the potential of optimization in real time irrigation management. Proc Irrigation and Drainage Division, ASCE 125: 345-354.

20

Comparison of Different Methods to Estimate Mean Daily Evapotranspiration from Weekly Data at Patna, India

Upadhyaya A*

Division of Land and Water Management, ICAR Research Complex for Eastern Region, Patna, India

Abstract

Estimation of evapotranspiration is necessary for efficient water management and crop planning. Fifteen different methods of ET_0 estimation were employed to compute daily reference evapotranspiration for the period 2010 to 2014. In the absence of reliable open pan evaporimeter data, FAO-56 Penman Monteith method was considered as one of the reliable method of ET_0 estimation. The results showed that mean weekly evapotranspiration values obtained from Penman-Monteith method were very closer to FAO-56 Penman-Monteith method and values from all the other methods except FAO- 24 Pan, Christiansen Pan and Hargreaves methods generally predicted higher values of mean weekly daily ET_0 in comparison to FAO-56 Penman-Monteith method. The analysis shows that mean weekly daily ET_0 estimates of combination methods resulted better ET_0 estimates than radiation, and temperature and evaporation methods. Weekly ET_0 values estimated by FAO-56 Penman-Monteith method were found to vary in the range of 1.3 mm/day to 6.7 mm/day. Average annual reference evapotranspiration was found as 1517.1 mm.

Keywords: Evapotranspiration; Penman-Monteith method; Pan evaporation; Crop water requirement

Acronyms

Δ=slope of vapor pressure versus air temperature curve (KPa °C^{-1})

R_n=mean daily net radiation (MJm^{-2}d^{-1})

G=soil heat flux (MJm^{-2}d^{-1})

γ=psychometric constant (0.0671 Kpa °C^{-1})

T=mean daily air temperature at 2 m height

U_2=wind speed at 2 m height (ms^{-1})

e_s=saturation vapor pressure (Kpa)

e_a=actual vapor pressure (Kpa)

e_s-e_a=vapor pressure deficit VPD (KPa)

TD=difference between mean monthly maximum and mean monthly minimum temperatures in °C

RA=extraterrestrial solar radiation in MJ m^{-2} d^{-1}

T_{mean} is mean monthly air temperature in °C

$R_s^{'}$= solar radiation in cal cm^{-2} d^{-1}

$R_s^{'}$=R_s/0.041869

R_s=Total incoming solar radiation (MJ m^{-2} d^{-1})

RH=Mean relative humidity

Λ=latent heat of vaporization (MJ kg^{-1})

T_{mean}=mean air temperature in °C

Δ=slope of saturation vapour pressure-temperature curve (kPa °C^{-1})

e^0_{mean} (kPa)=saturation vapour pressure at mean temperature

γ=psychometric constant (kPa °C^{-1})

C_p=specific heat at constant pressure in kJ kg^{-1} °C^{-1}

P=atmospheric pressure at elevation in kPa

R_n=net radiation (MJ m^{-2}d^{-1})

α=albedo as 0.23

R_s=short wave or solar radiation received at earth surface

R_b=the net outgoing thermal radiation

E_v=class A pan evaporation (cm/d)

T_{mean}= mean air temperature in °C

W=wind speed in km/hr

RH_{mean}=the mean relative humidity in %

S=ratio of actual to maximum possible sunshine hours

E_{pan}=class A pan evaporation in mm/d

k_p=pan coefficient

$(e_z^0 - e_z)^{VPD\#1}$=vapour pressure deficit #1 (kPa) and equal to $(e^0_{mean} - e^o_d)$

e^0_{mean}=saturation vapour pressure at mean temperature

e^o_d=saturation vapour pressure at dew point temperature

W_f^{P1}=wind function for Penman (1963) VPD#1 method

u_2=wind speed measured at 2 m above the ground surface in ms^{-1}

k=von-Karman's constant and its value is equal to 0.41

u_z=horizontal wind speed at height z cm (m s^{-1})

d=zero plane displacement of wind profile in cm

Z_0=roughness length (cm) and it is set at arbitrary values of about 1/100 of crop height

---

*Corresponding author: Upadhyaya A, Principal Scientist, Division of Land and Water Management, ICAR Research Complex for Eastern Region, Patna, India
E-mail: aupadhyaya66@gmail.com

P=atmospheric pressure (kPa)

ρ=air density (kg m^{-3})

$W_f^{P3} = W_f^{P1}$

$(e_z^0 - e_z)^{VPD\#3}$=vapour pressure deficit #3 (kPa) and equal to 0.5 $(e_{min}^0 + e_{max}^0) - e_0^d$

u_d=day time wind speed

u_n=night time wind speed

Introduction

Crop water requirement mainly depends upon crop sowing date, crop growth stages, crop duration and climatic conditions during the growing season of an individual crop. A good estimate of evapotranspiration is essential for water balance irrigation scheduling and water resource planning and management. Since, direct measurement of ET_0 is difficult and time consuming, so most common practice is to estimate ET_0 from meteorological data available for the area employing empirical equations. There are numerous evapotranspiration equations developed and used by researchers but still there is confusion in selecting the reliable evapotranspiration estimation method for the area concerned. Some authors compared estimates of different ET_0 estimation methods for different regions of the world. Al-Ghobari compared five different methods: FAO-Penman, Jensen- Haise, Blaney and Criddle, Pan Evaporation and calibrated FAO-Penman for four areas under local conditions of Saudi Arabia [1]. The results showed that estimated ET values were highly correlated with measured evapotranspiration values. George et al. used DSS model for estimation of reference evapotranspiration for three climatic conditions: Davis, Jagdalpur, and Kharagpur by different applicable methods [2]. Comparison of the estimates of different methods with the Penman- Monteith ET_0 estimates showed that Hargreaves, FAO-24 Blaney Criddle and 1982 Kimberley- Penman methods ranked first, respectively for the Davis, Jagdalpur and Kharagpur stations. They recommended that DSS model is a user- friendly tool for estimating ET_0 under different data availability and climatic conditions. Itenfisu et al. made comparison between various methods of reference evapotranspiration with ASCE-Penman-Montieth equations in the United States [3]. Results showed that the ASCE standardized equation agreed best with full form of ASCE Penman-Montieth. The International Commission for Irrigation and Drainage and the Food and Agriculture Organization of the United Nations experts' consultation on revision of FAO methodologies for crop water requirements. Smith et al. recommended the use of FAO56-Penman-Monteith method as the standard method to estimate ET_0 [4]. Irmark et al. in north-central Florida, compared ET_0 values obtained from two equations derived from FAO56-PM (first equation was solar radiation (R_s) based and second net radiation (R_n) based) against FAO56-Penman-Monteith method and results were found correlated very well with the standard FAO56-Penman-Monteith method in all the locations [5]. Irmark et al. compared the estimates of 21 evapotranspiration methods with the estimate of FAO56- PM Method in humid climate of Florida [6]. The 1948 Penman method daily estimates were closest to the FAO56- PM method followed by 1963 Penman on the basis of standard error of estimate. Krishan et al. computed evapotranspiration rate for Pusa area in Bihar by various estimation methods [7]. They concluded that FAO-24 Penman(c=1) is the closest whereas Hargreaves method is the farthest. The objective of this study is to compare the estimates of various methods with the FAO56-Penman Montieth Method estimates and evaluate the performance of these methods for Patna canal command.

Methodology

Collection of meteorological parameters

In order to estimate daily reference evapotranspiration, available daily meteorological data for the period from 2010 to April 2015 were collected from automatic weather station located at 25° 27' North latitude and 85° 10' East Longitude in Indian Council of Agricultural Research- Research Complex for Eastern Region, (ICAR-RCER) Patna. These meteorological data include minimum and maximum temperature, minimum and maximum relative humidity, wind velocity, and solar radiation. Mean weekly meteorological parameters are presented below in Table 1.

It may be observed from the Table that mean maximum temperature varied from 18.9°C (1st week) to 37.4°C (17th week) and mean minimum temperature varied from 7.3°C (2nd week) to 26.8°C

Week No.	Max. Temp. (°C)	Min. Temp. (°C)	Relative Humidity Max. (%)	Relative Humidity Min. (%)	Wind Speed (m/s)	Solar Radiation, MJ/m²/ day
1	18.9	8.2	96.0	57.9	0.7	12.34
2	20.1	7.3	95.4	57.4	0.5	13.08
3	20.2	7.8	94.7	60.6	0.6	12.98
4	22.3	9.2	91.9	53.3	0.9	13.88
5	21.9	10.4	91.0	53.0	0.7	14.20
6	23.5	11.4	90.4	54.3	0.7	14.23
7	25.6	12.1	89.7	47.7	0.8	15.96
8	26.9	13.4	90.6	49.6	0.7	15.32
9	28.0	14.1	84.8	43.3	0.8	16.99
10	29.6	13.6	81.0	32.3	0.8	18.98
11	30.4	15.3	76.3	33.7	0.8	19.33
12	32.8	16.1	73.4	27.6	0.7	21.34
13	34.7	18.5	72.2	26.1	0.8	23.16
14	36.6	20.1	63.2	24.5	1.1	24.29
15	36.4	21.5	64.1	28.7	1.3	23.20
16	36.7	22.6	73.5	31.5	1.2	24.80
17	37.4	23.3	68.8	35.7	1.4	24.70
18	35.6	24.3	76.5	46.5	2.0	28.78
19	34.8	24.9	72.5	46.0	1.3	28.03
20	34.8	24.9	76.0	48.4	1.4	28.83
21	35.3	25.6	77.9	49.4	1.5	29.94
22	35.2	25.7	79.5	49.2	1.2	30.95
23	35.2	25.9	78.0	50.7	1.4	27.32
24	33.9	26.0	85.1	62.3	1.4	26.85
25	32.5	25.5	88.4	69.6	1.5	21.45
26	32.9	26.3	87.5	66.0	1.3	22.56
27	33.0	26.6	89.3	70.1	1.1	23.73
28	32.7	26.5	86.4	70.6	1.3	20.18
29	32.6	26.7	89.8	73.4	1.3	22.20
30	33.0	26.5	86.2	62.7	1.1	22.55
31	32.5	25.0	89.0	67.2	1.1	26.47
32	32.9	26.8	90.8	75.8	1.2	24.57
33	31.5	26.2	91.8	73.0	1.4	21.18
34	31.9	26.6	91.0	73.1	1.4	22.71
35	31.6	25.9	90.7	74.8	1.3	22.17
36	31.4	26.3	89.1	73.6	1.1	21.57
37	31.8	26.0	90.2	71.4	1.2	23.64
38	31.3	25.5	90.3	70.3	1.3	25.06
39	32.0	24.9	87.6	70.1	1.0	25.63
40	32.4	24.6	87.3	70.0	1.0	22.96
41	32.8	24.1	89.1	62.4	0.9	22.87
42	32.1	23.2	89.6	62.5	0.9	20.58
43	31.2	20.8	88.1	51.3	0.9	19.65
44	31.4	20.7	91.2	56.4	0.7	20.30
45	30.9	17.9	91.6	52.3	0.4	19.10
46	29.7	15.7	91.2	49.3	0.4	16.53
47	28.8	14.5	90.4	48.4	0.4	14.02
48	26.8	11.6	89.3	49.9	0.5	13.02
49	25.8	10.8	91.4	54.0	0.5	10.78
50	24.4	10.1	91.3	60.7	0.4	9.14
51	23.1	9.9	93.8	59.0	0.5	8.85
52	22.9	10.1	94.3	49.2	0.5	10.65

Table 1: Mean meteorological parameters as observed at ICAR-RCER, Patna.

(32nd week). Mean maximum relative humidity ranged from 63.2 percent (14th week) to 96.0 percent (1st week) whereas mean minimum relative humidity from 24.5 percent (14th week) to 75.8 percent (32nd week). Wind velocity varied in the range of 0.4 m/s to 2 m/s. Mean solar radiation was observed to be minimum of 8.85 MJm^{-2}day^{-1} in the 51st week and maximum of 30.95 MJm^{-2}day^{-1} in the 22nd week, respectively.

Computation of evapotranspiration

Computation of crop water requirement needs estimation of reference evapotranspiration (ET$_0$). Direct measurement of ET$_0$ is difficult and time consuming, so most common practice is to estimate ET$_0$ from meteorological data available for the area employing empirical equations. Daily reference evapotranspiration (ET$_0$) values for the period from 2010 to 2014 were estimated using thirteen different methods. All the methods used in this study are cited and described in DSS-ET model of ET$_0$ estimation. The methods used were grouped according to their classification as (1) combination methods such as FAO-56. Penman-Monteith Allen et al. [8] Penman-Monteith Monteith, [9] Kimberley Penman Wright, [10] FAO-PPP-17 Penman Frere and Popov, [11] Penman VPD#1, Penman VPD#3 Kimberley Penman Wright and Jensen, [12] FAO-24 Penman c=1 Doorenbos and Pruitt [13,14], Businger-van-Bavel, [15,16] and CIMIS Penman Snyder and Pruitt [17] (2) Radiation methods such as Turc [18] and Priestly-Taylor, [19] and (3) Temperature method such as Hargreaves [20]. The list of data requirement for each method is given below in Table 2.

The ICID and the FAO expert consultation on revision of FAO methodologies for crop water requirements have recommended that FAO-56 Penman-Monteith method be used as the standard method to estimate ET$_0$. In this study also FAO-56 Penman-Monteith method was used as an index to estimate daily ET$_0$. The FAO-56 Penman-Monteith is a grass reference equation that was derived from the ASCE equations by fixing grass height=0.12 m for clipped grass using latent heat of vaporization (λ)=2.45 MJ kg^{-1}; bulk surface resistance of 70 sm^{-1}, and albedo of 0.23. The equation to estimate daily ET$_0$ (mm/d) is given as:

$$ETo = \frac{0.408\,\Delta\,(R_n - G) + \gamma\,\dfrac{900}{T+273}U_2\,(e_s - e_a)}{\Delta + \gamma\,(1 + 0.34\,U_2)} \tag{1}$$

Methods to calculate evapotranspiration (ET$_0$)

ET methods recommended by ASCE are classified into four categories, namely, temperature, radiation, evaporation and combination methods. Description of each ET method is given below:

Hargreaves and Samani recommended estimation of solar radiation from extraterrestrial radiation and used temperature difference term [20]. Net solar radiation, relative humidity and wind parameters were not considered in the proposed equation mentioned below.

$$E_{to} = 0.0023\,R_A\,\sqrt{TD}\,(T_{mean} + 17.8) \tag{2}$$

Among the radiation methods, Turc proposed two equations of ET$_0$ estimation as a function of daily mean air temperature and total incoming solar radiation (R$_s$) for the mean relative humidity (RH) >50% and (RH) <50% as given below [18].

When RH$_{mean}$ > 50%

$$E_{to} = 0.013\,\frac{T_{mean}}{(T_{mean} + 15)}\,(R'_s + 50)\,\frac{1}{\lambda} \tag{3}$$

When RH$_{mean}$ ≤ 50%

$$E_{to} = 0.013\,\frac{T_{mean}}{(T_{mean} + 15)}\,(R'_s + 50)\,\frac{1}{\lambda}\left(1 + \frac{(50 - RH_{mean})}{70}\right) \tag{4}$$

Priestly and Taylor proposed an equation for surface area generally wet, which is a condition, required for potential evaporation [19]. The aerodynamic component was deleted and energy component was multiplied by a coefficient, α=1.26. The final equation is as below

$$E_p = \alpha\,\frac{1}{\lambda}\,\frac{\Delta}{(\Delta + \gamma)}\,(R_n - G) \tag{5}$$

Where Δ is expressed as

$$\Delta = \frac{4098\,e^0_{mean}}{(T_{mean} + 237.3)^2} \tag{6}$$

e$^0_{mean}$ (kPa) is expressed as

$$e^0_{mean} = 0.6108\,\exp\left[\frac{17.27\,T_{mean}}{T_{mean} + 237.3}\right] \tag{7}$$

Christiansen and Hargreaves developed equation for estimating reference evapotranspiration from USWB class 'A' pan evaporation and several weather parameters.

$$E_{to} = 0.755\,E_v\,C_{T2}\,C_{w2}\,C_{H2}\,C_{s2} \tag{8}$$

Methods	Temperature		Relative Humidity		Wind Velocity	Solar Radiation	Pan Evaporation
	Min.	Max.	Min.	Max.			
FAO-56Penman-Monteith	√	√	√	√	√	√	-
Penman-Monteith	√	√	√	√	√	√	-
1982Kimberley Penman	√	√	√	√	√	√	-
FAO-PPP	√	√	√	√	√	√	-
Penman 1963 VPD#1	√	√	√	√	√	√	-
Penman 1963 VPD#3	√	√	√	√	√	√	-
1972Kimberley Penman	√	√	√	√	√	√	-
FAO-24 Penman c=1	√	√	√	√	√	√	-
Hargreaves	√	√	-	-	-	-	-
Businger-van-Bavel	√	√	√	√	√	√	-
FAO-24 Pan	-	-	√	√	√	-	√
Christiansen Pan	√	√	√	√	√	-	√
Turc	√	√	√	√	-	√	-
Priestly-Taylor	√	√	√	√	√	√	-
CIMIS Penman	√	√	√	√	√	√	-

Table 2: Input parameters for different methods of ETo estimation.

C_{T2}, C_{w2}, C_{H2} and C_{s2} are estimated by following set of equations:

$$C_{T2}=0.862 + 0.179 \, (T_{mean}/20) - 0.041((T_{mean}/20)^2 \qquad (9)$$

$$C_{w2}=1.189 - 0.240 \, (W/6.7) + 0.051(W/6.7)^2 \qquad (10)$$

$$C_{H2}=0.499 + 0.620 \, (RH_{mean}/60) - 0.119 \, (RH_{mean}/60)^2 \qquad (11)$$

$$C_{s2}=0.904 - 0.0080 \, (S/0.80) + 0.088 \, (S/0.80)^2 \qquad (12)$$

Doorenbos and Pruitt suggested *FAO-24* pan method in which, ET_o depends on two different conditions for estimating the pan coefficient values, one is for vegetation while other is for bare soil [13].

$$E_{to}=k_p E_{pan} \qquad (13)$$

Frevert et al. has mentioned polynomial equations for estimating k_p.

Penman in Vapor pressure deficit *(VPD) #1,* combined the two components i.e., radiation and aerodynamic components and defined general form of resulting equation for a well-watered grass reference as [21]:

$$E_{to} = \frac{1}{\lambda}\frac{\Delta}{(\Delta+\gamma)}(R_n - G) + \frac{1}{\lambda}\frac{\gamma}{(\Delta+\gamma)} 6.43\, W_f^{P1} \, (e_z^0 - e_z)^{VPD\#1} \qquad (14)$$

Here,

$$W_f^{P1}=1.0 + 0.537 \, u_2$$

Businger, Monteith and van Bavel adopted a more theoretical vapour pressure function based on earlier work of resulting the following Businger Bavel equation for a wet surface with zero resistance to vapour transfer [9,15,16].

$$E_{to} = \frac{1}{\lambda}\frac{\Delta}{(\Delta+\gamma)}(R_n - G) + \frac{1}{\lambda}\frac{\gamma}{(\Delta+\gamma)} \frac{0.622\rho\lambda k^2}{P} \frac{u_z}{[\ln((z-d)/z_0)]^2} (e_z^0 - e_z)^{VPD\#1} \qquad (15)$$

Here 0.622 is the ratio of the molecular mass of water to the apparent molecular mass of dry air. Penman suggested to use VPD#3 in place of VPD#1 in his proposed equation, the resulting equation becomes [15]:

$$E_{to} = \frac{1}{\lambda}\frac{\Delta}{(\Delta+\gamma)}(R_n - G) + \frac{1}{\lambda}\frac{\gamma}{(\Delta+\gamma)} 6.43\, W_f^{P3} \, (e_z^0 - e_z)^{VPD\#3} \qquad (16)$$

$$(e_z^0 - e_z)^{VPD\#3} = 0.5\, (e_{min}^0 + e_{max}^0) - e_0^d$$

Penman-Monteith equation proposed by Monteith [9] was employed to estimate ET_o for clipped crop surface condition considering crop height as 0.12 for humidity and temperature measurement at the height of 2 m. The equation used variable daily value of latent heat of vaporization and bulk surface resistance and aerodynamic resistance. The equation is available in many text books.

Wright and Jensen recommended more common wind function for the Penman method. The revised method is referred as 1972 Kimberly Penman method and is given below [12]:

$$E_{to} = \frac{1}{\lambda}\frac{\Delta}{(\Delta+\gamma)}(R_n - G) + \frac{1}{\lambda}\frac{\gamma}{(\Delta+\gamma)} 6.43\, W_f^{72KP} \, (e_z^0 - e_z)^{VPD\#3} \qquad (17)$$

Here $W_f^{72KP}=0.75 + 0.993 \, u_2$.

Doorenbos and Pruitt [13,14] presented a modified Penman equation for estimating reference ET for grass which is known as FAO-24 corrected Penman [15]. If value of c is set to 1.0 in FAO 24 corrected Penman equation, then the resulting equation is called FAO-24 Penman (c=1) method.

$$E_{to} = c \left[\frac{\Delta}{(\Delta+\gamma)}(R_n - G) + \frac{\gamma}{(\Delta+\gamma)} 2.7\, W_f^{F24CP} \, (e_z^0 - e_z)^{VPD\#i} \right] \qquad (18)$$

Here W_f^{F24CP} is wind function for FAO Corrected Penman method and it can be calculated by $W_f^{F24CP}=1 + 0.864\, u_2$ and c is adjustment factor and expressed as:

$$c=0.68 + 0.0028\, RH_{max} + 0.018\, R_s - 0.068\, u_d + 0.013\, (u_d/u_n) + 0.0097\, u_d(u_d/u_n) + 0.430 \times 10^{-4}\, RH_{max} R_s\, u_d.$$ If c=1, it becomes FAO 24 Penman (c=1) method.

Frere and Popov suggested the different wind functions for various ranges of temperatures and used VPD#1 for estimating vapour pressure deficit. The equation can be written as below [11]:

$$E_{to} = \frac{1}{\lambda}\frac{\Delta}{(\Delta+\gamma)}(R_n - G) + \frac{1}{\lambda}\frac{\gamma}{(\Delta+\gamma)} 6.43\, W_f^{FP17P} \, (e_z^0 - e_z)^{VPD\#1} \qquad (19)$$

Here W_f^{FP17P} is wind function for FAO-PPP Penman method. The suggested wind functions depend upon temperature difference.

Wright presented variable wind function coefficients for E_{tr} using fifth order polynomials with calendar day D as independent variable at Kimberly, Idaho, Kimberly Penman equation used variable value of wind function coefficients using fifth order polynomials with calendar day as independent variable [10]. Simplified equation for wind function is expressed as:

$$W_f^{82KP} = a_w + b_w\, u_2 \qquad (20)$$

Where,

$a_w=0.4 + 1.4 \exp\{ - [(D-173)/58]^2\}$ and $b_w=0.605 + 0.345 \exp\{-[(D - 243)/80]^2\}$

D is calendar Day (i.e., day of the year). For southern latitude use D' in place of D and take D'=(D-182) for D≥ 182 and D'=(D+182) for D <182. Keeping all other parameters same, the 1982 Kimberly Penman may be written as follows:

$$E_{to} = \frac{1}{\lambda}\frac{\Delta}{(\Delta+\gamma)}(R_n - G) + \frac{1}{\lambda}\frac{\gamma}{(\Delta+\gamma)} 6.43\, W_f^{82KP} \, (e_z^0 - e_z)^{VPD\#3} \qquad (21)$$

Results and Discussion

Daily reference evapotranspiration were computed by the above mentioned methods and converted into mean weekly ET_0 (mm/day). The results are presented below in Figure 1.

The weekly reference evapotranspiration (ET_0) estimates in year 2010, 2011, 2012 and 2013, and 2014 computed using FAO 56 Penman Monteith method varied from 1.4 mm/day (3rd week) to 8.0 mm/day (22nd week), 1.4 mm/day (50th and 51st weeks) to 7.2 mm/day (22nd week), 1.1 mm/day (51st weeks) to 6.6 mm/day (18th week), and 1.3 mm/day (51st week) to 6.6 mm/day (16th week), respectively. The average of five years weekly ET_0 values ranged from 1.3 mm/day (51st week) to 6.7 mm/day (21st and 22nd weeks), respectively. Annual reference evapotranspiration in these years were computed as 1535.4 mm, 1523.5 mm, 1485.2 mm, 1501.6 and 1539.9 mm, respectively, and average annual reference evapotranspiration was found as 1517.1 mm.

It may be observed from Figure 1 that all the methods except Hargreaves, FAO 24 Pan and Christiansen Pan methods, follow the same trend of variation of ET_0 values in different weeks. ET_0 values computed by FAO 24 Pan and Christiansen Pan methods are always lower than ET_0 values computed by all other methods in all the weeks. ET_0 values computed by Hargreaves method are lower during 18th to 43rd week and higher during 44th week to 17th week compared to ET_0 values obtained from FAO 56 Penman-Monteith method. In rest of the methods the difference in ET_0 values obtained from Penman-Monteith and FAO 56 Penman-Monteith method is very less as compared to other methods.

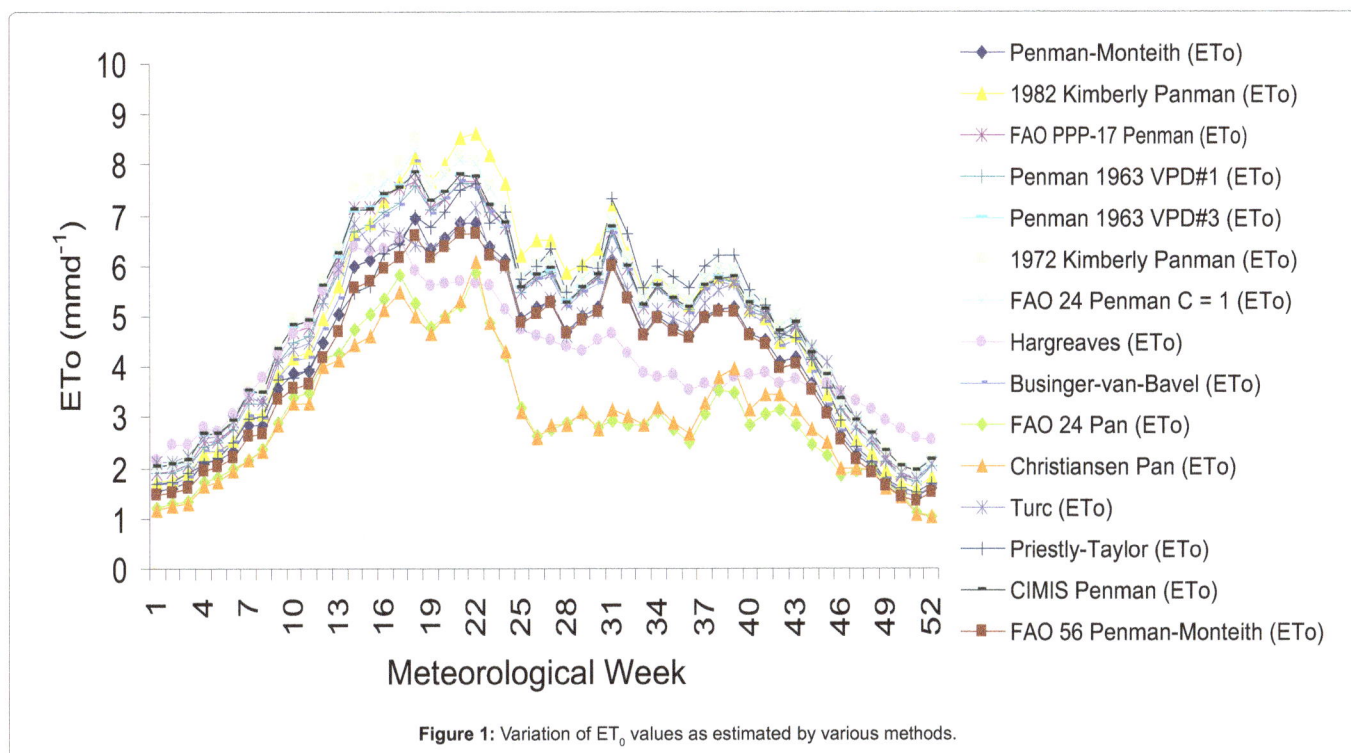

Figure 1: Variation of ET$_0$ values as estimated by various methods.

Methods	Average ratio of ETo(method)/ ETo(FAO56-PM)	Slope	Intercept	Coefficient of determination	Rank
Penman-Monteith	1.04	1.023	0.065	0.996	1
1982 Kimberley Penman	1.20	1.263	-0.251	0.986	10
FAO-PPP	1.19	1.107	0.344	0.977	4
Penman1963 VPD#1	1.17	1.093	0.329	0.990	2
Penman1963 VPD#3	1.21	1.089	0.513	0.978	3
1972 Kimberley Penman	1.26	1.176	0.325	0.971	9
FAO-24 Penman C=1	1.24	1.168	0.290	0.980	8
Hargreaves	1.03	0.591	1.795	0.657	14
Businger-van-Bavel	1.13	1.163	-0.126	0.990	7
FAO-24 Pan	0.76	0.681	0.305	0.719	13
Christiansen Pan	0.76	0.694	0.260	0.769	12
Turc	1.13	0.927	0.817	0.934	11
Priestly-Taylor	1.14	1.118	0.081	0.961	6
CIMIS Penman	1.22	1.090	0.511	0.976	5

Table 3: Various statistical parameters and rank of ET0 estimation methods.

The relationships between the ET$_0$ (mm/day) computed by fourteen various methods with ET$_0$ obtained from FAO56-Penman-Montetith method are given below in Figure 2a to 2n.

The ratio of Five years average of ET$_0$ (method) and ET$_0$ (FAO56-PM), slope (m), intercept (c) and coefficient of determination (r^2) were computed in order to know the over and under estimation of all the methods with respect to the FAO-56 Penman-Monteith method are given below in Table 3.

In the above Table, the average ratios above and below 1.0 indicate over and under estimation of ET$_0$ compared to ET$_0$ values computed by FAO-56 Penman-Monteith method. The average ratios lie in the range of 0.76 to 1.26, which indicates that Christiansen Pan and FAO-24 Pan under estimate, whereas other methods overestimate ET$_0$ values compared to ET$_0$ estimated by FAO-56 Penman-Monteith method [22]. Coefficient of determination represents the closeness of relationship of

other ET$_0$ estimation methods with FAO-56 Penman Monteith method and can be considered good indicator for deciding the rank. Penman Monteith with coefficient of determination of 0.996 was observed having 1st rank and Hargreaves with coefficient of determination 0.657 at 14th rank. The methods having rank between 1 and 5 were Penman Monteith, Penman1963 VPD#1, Penman1963 VPD#3, FAO-PPP, and CIMIS Penman, respectively and the methods having rank between 10 and 14 were 1982 Kimberley Penman, Turc, Christiansen Pan, FAO-24 Pan, and Hargreaves, respectively.

Conclusion

Fifteen different methods of ET$_0$ estimation were employed for climatic data of the period 2010 to 2014 and daily estimates of reference evapotranspiration were computed and converted into mean weekly daily values. The mean weekly daily ET$_0$ estimates from all the methods were compared against the ET$_0$ values obtained using FAO-56

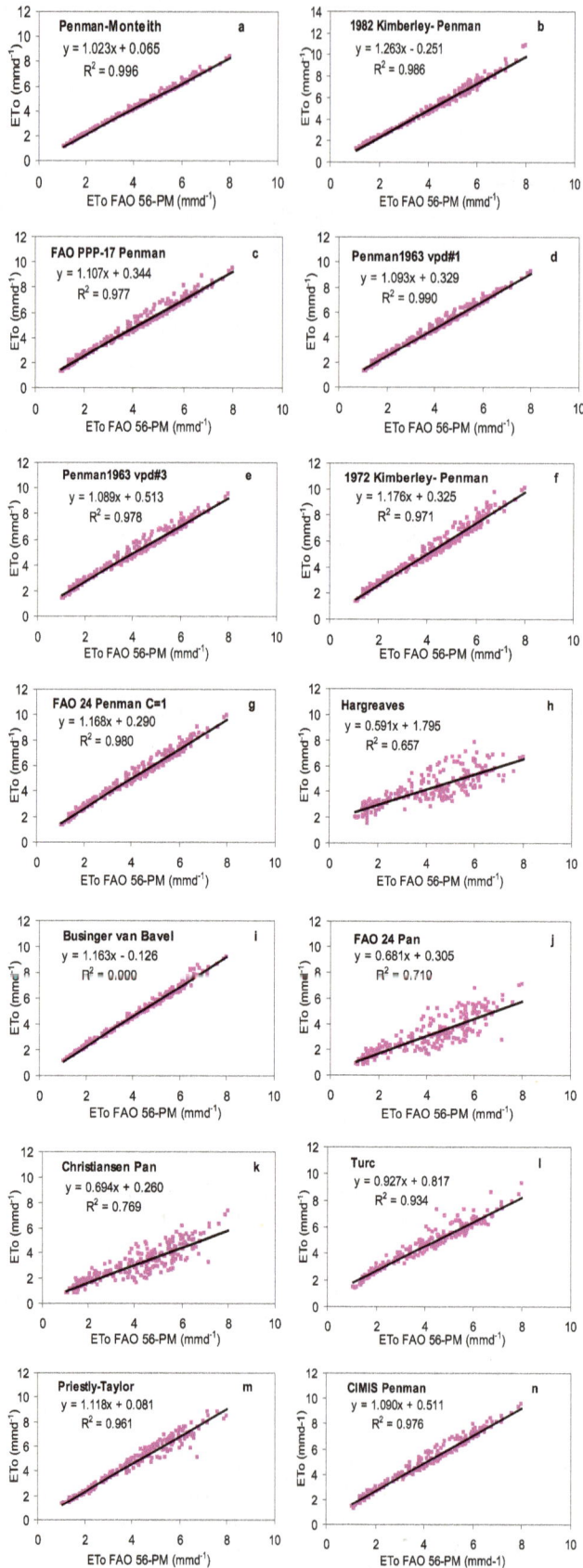

Figure 2: (a to n): Comparison of ET_0 (mm/day) estimated by various methods with FAO 56 Penman-Monteith method.

Penman-Monteith method. On the basis of results, it can be said that all the methods produced significantly different evapotranspiration estimates than the FAO-56 Penman-Monteith method. Mean weekly daily evapotranspiration rates obtained from Penman-Monteith method were found very closer to FAO-56 Penman-Monteith method and values from all the other methods except Christiansen Pan, FAO-24 Pan and Hargreaves methods, generally predicted higher values of mean weekly daily ET_0 in comparison to FAO-56 Penman-Monteith method. The performance of all the methods against FAO 56 Penman-Monteith was evaluated by assigning rank on the basis of coefficient of determination. The analysis shows that mean weekly daily ET_0 estimates of combination methods resulted better ET_0 estimates than radiation, temperature and evaporation methods. Penman-Monteith method produced best ET_0 estimate against FAO-56 Penman-Monteith estimated ET_0. Among the combination methods its performance followed by Penman VPD#1, PenmanVPD#3, FAO-PPP-17 Penman, CIMIS Penman, Businger-van-Bavel, FAO-24 Penman (c=1), 1972 Kimberley Penman method, 1982 Kimberley Penman. The performance of Turc radiation methods was less reliable than combination methods. Among the radiation methods Priestley-Taylor method performed much better than Turc method which may be due to consideration of wind parameter in its equation. The poorest estimates of temperature and evaporation methods were observed because these methods do not account for net radiation, vapor pressure deficit, solar radiation, or other important parameters, which have greater impact on ET_0 estimation.

References

1. Al-Ghobari HM (2000) Estimation of reference evapotranspiration for Southern region of Saudi Arabia. Irrigation Science 19: 81-86.

2. George BA, Reddy BRS, Raghubanshi NS, Wallender WW (2002) Decision sport system for estimating reference evapotranspiration. Journal of Irrigation and Drainage engineering 128: 1-10.

3. Itenfisu D, Elliott RL, Allen RG, Walter IA (2003) Comparison of reference evapotranspiration calculations as part of the ASCE standardization effort. Journal of Irrigation and Drainage engineering 129: 440-448.

4. Smith M, Allen RG, Montieth JL, Perrier A, Pereira L, et al. (1991) Rep of the expert consultation on procedures for revision of FAO guidelines for prediction of crop water requirements. Food and Agricultural Organization of the United Nations, Rome.

5. Irmak S, Irmak A, Allen RG, Jones JW (2003a) Solar and net radiation-based equations to estimate reference evapotranspiration in humid climates. Journal of Irrigation and Drainage engineering 129: 336-347.

6. Irmak S, Allen RG, Whitty EB (2003b) Daily grass and alfalfa-reference evapotranspiration estimates and alfalfa-to-grass evapotranspiration ratios in Florida. Journal of Irrigation and Drainage engineering 129: 360-370.

7. Krishan R, Upadhyaya A, Roy LB (2015) Comparison of evapotranspiration rate for Pusa area of Bihar, by employing different methods through DSS. J Indian Water Resour Soc 35: 14-22.

8. Allen RG, Pereira LS, Raes D, Smith M (1998) Crop evapotranspiration: Guidelines for computing crop water requirements. Food and Agricultural Organization of the United Nations.

9. Monteith JL (1964) Evaporation and the environment. Proceedings of 19th Symposium of the Society for Experimental Biology. Cambridge University Press. pp: 205-234.

10. Wright JL (1982) New Evapotranspiration crop coefficients. Journal of the Irrigation and Drainage Divison 108: 57-74.

11. Frere M, Popov GF (1979) Agrometeorological crop monitoring and forecasting. FAO Plant Production and Protection Paper 17. FAO. Rome. pp: 38-43.

12. Wright JL, Jensen ME (1972) Peak water requirements of crops in Southern Idaho. Journal of Irrigation and Drainage Divison ASCE 96: 193-201.

13. Doorenbos J, Pruitt WO (1977) Guidelines for predicting crop water requirements. Irrigation and Drainage Paper No. 24. FAO. Rome.

14. Doorenbos J, Pruitt WO (1975) Guidelines for predicting crop water requirements. Irrigation and Drainage Paper No. 24. FAO. Rome.

15. Businger JA (1956) Some remarks on Penman's equations for the evapotranspiration. J Agric Sci 4: 77.

16. Van Bavel CHM (1966) Potential evaporation: the combination concept and its experimental verification. Water resources Res 2: 455-467.

17. Snyder RL, Pruitt WO (1992) ET_0 User's Guide and Program Documentation. University of California, USA.

18. Turc L (1961) Estimation of irrigation water requirements, potential evapotranspiration: A simple climatic formula evolved up to date. Ann Agron 12: 13-14.

19. Priestley CHB, Taylor RJ (1972) On the assessment of surface heat flux and evaporation using large-scale parameters. Mon Weather Rev 100: 81-92.

20. Hargreaves GL, Samani ZA (1985) Reference crop evapotranspiration from temperature. Applied engineering in Agriculture 1: 96-99.

21. Penamn HL (1963) Vegetation and hydrology. Tech Comm No 53, Commonwealth Bureau of Soils. England. p. 125.

22. Penamn HL (1948) Natural evaporation from open water, bare soil and grass. Pro Roy Soc A 193: 120-146.

Physiology and Grain Yield of Common Beans under Evapotranspirated Water Reposition Levels

Magalhaes ID[1]*, Lyra GB[1], Souza JL[2], Teodora I[3], Cavalcante CA[3], Ferreira RA[4] and Souza RC[3]

[1]Department of Plant production, Federal University of Alagoas, Brazil
[2]Department of Agronomy, Federal University of Alagoas, Brazil
[3]Department of Agricultural engineering, Federal University of Alagoas, Brazil
[4]Department of Phytotechnology, Federal University of Alagoas, Brazil

Abstract

The aim of this work was to evaluate gas exchanges, photochemical efficiency and yield of common bean grains (crioula variety) grown under different irrigation levels in the state of Alagoas. The experimental design was a randomized block design with four replications. Treatments were composed of crop evapotranspiration fractions (25, 50, 75, 100, 125 and 150% of crop evapotranspiration). Gas exchanges were determined from measurements of internal CO_2 concentration, transpiration, stomata conductance, photosynthetic rate, instantaneous water use efficiency and instantaneous carboxylation efficiency. Chlorophyll a fluorescence evaluations were determined through the potential and effective quantum yield of photosystem II. Chlorophyll content was indirectly measured. The following production components were evaluated: number of pods per plant, number of grains per pods, mass of 1000 grains and grain yield. The water application variation promoted a significant difference for gas exchanges, causing a reduction in the potential and effective photochemical efficiency of common bean. The increase in the application of the irrigation levels directly influenced the SPAD index and the following production components: number of pods per plant and grain yield, obtaining significant values with irrigation level of 125% of ETc.

Keywords: *Phaseolus vulgaris* L.; Water deficit; Yield; Physiological variables

Introduction

Common bean (*Phaseolus vulgaris* L.) is an important crop for the world, since its grains are present in the diet of many countries due to its high nutritional value, being considered the main source of protein for low-income populations and a crop of great economic and social importance [1,2].

Brazil has been the world's largest bean producer in the last 10 years; however, as a result of successive atypical meteorological events and changes in the profile of producers, there were significant decreases in productivity and production, so that between 2012 and 2013, the country lost space in productive competitiveness for Asian countries, which have shown growth over the last few years [3]. The north-eastern region of Brazil is one of the main common bean producers in the country [4]. The state of Alagoas, which cultivation is performed in practically the entire territory, the state of Sergipe and the north eastern region of the state of Bahia compose a strong production belt, representing 22% of the winter production in the year of 2016 [3].

Although common bean is considered a versatile crop due to its easy adaptation to the most diverse environments, its yield is still considered low, with average of 800 kg ha⁻¹ [3]. One of the main causes of this small productivity is the spatial and temporal limitation of precipitation, characteristic of semiarid regions. The inadequate management of water resources in irrigation systems contributes to water scarcity in irrigated crops, contributing to waste, with undesirable consequences for the environment. In this sense, the rational use of irrigation water reduces losses through evaporation, runoff, and percolation among others [2]. Irrigation is an essential factor for a good performance of the crop in the field, especially when cultivated in a period of little rainfall. When well-managed, the plant can better express its productive potential, balancing the environmental issue, involving sustainability, when the subject is a shortage of water resources [5].

Like most crops, common bean is sensitive to water stresses, either due to water deficit or excess, being one of the environmental factors that most influence plant yield [6]. The reduction of water availability in the soil causes decrease in productivity for limiting the photosynthetic process, reducing the stomatal conductance and promoting stomatal closure, which in turn has direct implications for transpiration, photosynthesis and leaf temperature [7,8]. In this sense, it is of fundamental importance the knowledge of gas exchanges and photochemical efficiency for the management of plants cultivated under water deficit, aiming at a better crop development, being also determine the adaptation and stability of common bean to certain ecosystems [9].

Although there are studies on the cultivation of common bean (Mantovani et al. [10]; Souza et al. [2]; Brito et al. [4]) with irrigation levels and different management conditions, there is few available information regarding physiological parameters of irrigated common bean crop in north eastern Brazil. In this sense, the aim of this work was to evaluate gas exchanges, photochemical efficiency and grain yield of common beans cultivated under irrigation levels.

Materials and Methods

The experiment was conducted in the experimental area of the Center for Agricultural Sciences (CECA), Federal University of Alagoas (UFAL), Rio Largo, Alagoas (09 ° 28'02 "S, 35 ° 49'43" W, 127 m asl). According to climate classification, the climate of the region is humid and mega thermal, with moderate water deficiency in the

*Corresponding author: Magalhaes ID, Doctorate in Agronomy-Plant Production, Department of Plant Production, Federal University of Alagoas, Brazil
E-mail: ivomberg31@hotmail.com

summer and excess water in the winter, mean annual temperature of 25.4°C and total annual precipitation around 1,800 mm, with 70% of this occurring between April and August (Souza et al. [11]; Ferreira et al. [12]). The meteorological data of the experimental period were obtained from the agro meteorological station located 30 m away from the experimental area, and are presented in Table 1.

The local soil was classified as Cohesive Argisolic Yellow Latosol of medium clayey texture, according to analysis of the Department of Soil Physics at CECA/UFAL. The experimental area has topography with slope of less than 2%.

Fertilization was based on soil chemical analysis (Table 2), using 45 kg urea, 111 kg Simple Superphosphate and 78 kg of Potassium Chloride per hectare. At 20 days after planting (DAP), cover fertilization was applied using 89 kg ha^{-1} urea. When necessary, weed and pests were monitored by hand hoeing and chemical insecticides, respectively.

The seeds used in the experiment were *crioula rosinha* variety: undetermined growth habit (type II); upright; 78 days cycle; average of 34 days for flowering; white flowers; green pod, slightly pink in maturation and pigmented pod in harvest [13].

The study was conducted during the dry season, between November and January (2015/2016) using a randomized complete block design, with four replicates. Each experimental plot measured 8 m in length by 10 m in width (80 m^2). The total area of the experiment was 1.920 m^2. Treatments were composed of six irrigation levels established as a function of crop evapotranspiration fractions (ETc) according to Table 3. All treatments received the same irrigation levels in the initial period (15 DAP).

The study adopted Kc of 1.1 and 1.2 for the vegetative and reproductive phases, respectively. These Kc values are recommended by the FAO-56 bulletin for the intermediate stage, and then edaphoclimatic conditions and crop characteristics were adjusted during the experimental period, as recommended by Allen et al. [14]. The ETc values (mm day^{-1}) were calculated by means of equation (Eq. 1).

$$ETc = ET_0 * K_c \tag{1}$$

Where, ET0 is the reference evapotranspiration estimated by the Penman-Monteith method [14].

Seeding was manually performed with three seeds per pit at spacing of 0.50 m between rows and density of 13 to 15 plants per meter, totalling a final stand of 240,000 plants per hectare. At 15 DAP,

thinning was performed, leaving one plant per pit.

The irrigation method adopted was micro-sprinkler, with spacing of 2×2.5 m between sprinklers, service pressure of 14 mca, and average flow rate of 50 L h^{-1} per sprinkler and applied gross irrigation level of 5.06 mm h^{-1}.

Soil water content (SWC) was calculated using soil water balance suggested by Thornthweite and Matter [15] and adjusted by Lyra et al. [16] for agricultural crops. The total available water (TAW, mm) was generated for each stage of crop development as a function of the root system effective depth, according to equation:

$$TAW = 1.000 * \left(\theta_{CC} - \theta_{PMP}\right) * z \tag{2}$$

Where: θCC is the volumetric moisture in the field capacity (0.2445 m^3 m^{-3}) and θPMP is the moisture at the permanent wilting point (0.1475 m^3 m^{-3}) determined in laboratory by the retention curve of water in the soil Carvalho [17], and z (m) refers to the depth of the crop root system, ranging from 0.10 to 0.40 m.

Values for readily available water (RAW, mm) were calculated based on equation:

$$RAW = TAW * f \tag{3}$$

Where: f (0.45) refers to the water availability factor [0^{-1}] [14].

Gas exchange measurements were based on the internal CO_2 concentration rate (Ci), transpiration (E), stomatal conductance (gs) and photosynthesis rate (A). From these data, the instantaneous water use efficiency (A E^{-1}) and instantaneous carboxylation efficiency (EiC) (A/Ci) were also calculated. In these evaluations, an infrared gas analyser (IRGA, ADC model LCi, Hoddesdon, UK) with 300 ml min^{-1} air flow and coupled light source of 995 mmol m^{-2} s^{-1} was used. Measurements occurred at 35 DAP, between 8 am and 11 am at stage R5 on three useful plants between pre-flowering and pod formation, on the third fully expanded leaf, counted from the apex of the main branch of the plant (Figure 1).

Concomitant to gas exchanges, chlorophyll a fluorescence evaluations were performed using a portable light-modulated flora meter (Opti Sciences, model OS1-FL, Hudson, USA), from which the potential quantum yield of photosystem II (PSII) (Fv/Fm) was obtained after adaptation of leaves to the dark (H "30 min); yield was also quantified to obtain the effective quantum efficiency of photosystem II (ΦPSII). Evaluations occurred at different times (11 am to noon and noon to 1 pm).

Months	Tar (ºC)	RH (%)	Rainfall (mm)	WS$_2$ (m s^{-2})	ET$_0$ (mm)
November	25.39	69.38	10.42	2.21	140.17
December	25.39	72.88	120.39	1.93	160.40
January	25.75	77.35	170.43	1.65	136.81

Tar: Average air temperature; RH: Mean relative air humidity; WS2: Average wind speed; ET$_0$: Reference evapotranspiration.

Table 1: Agrometeorological data obtained during the experiment.

Lay	pH	*P	Ca	Mg	K	SB	Al	H+Al	T	m	v
Cm	mg dm^{-3}		..........................cmolc dm^{-3}............................							........%.......	
0-20	5.8	15.2	2.7	1.3	1.15	4.21	0.08	5.4	9.61	1.5	44.2

Lay: Soil layer; pH: Hydrogen potential; P: Phosphorus; Mehlich*; Ca: Calcium; Mg: Magnesium; K: Potassium, SB: Sum of bases; Al: Aluminum, H+Al: Hydrogen plus aluminum; T: Base exchange capacity; m: Aluminum saturation; v: Base saturation.

Table 2: Chemical soil attributes of the experimental area.

ETc %	25	50	75	100	125	150
Total irrigation (mm)	(462.5)	(504.9)	(543.6)	(577.9)	(614.5)	(654.2)

Table 3: Crop Evapotranspiration fractions (ETc %) and total applied water (irrigation+precipitation) in the research.

At stage R5, readings were taken for the indirect determination of the chlorophyll content using SPAD-502 chlorophyll meter (Soil Plant Analysis Development Section, Minolta Camera CO., Osaka, Japan) and expressed using the SPAD index.

At the end of the experiment (78 DAP), the following production components were evaluated: number of pods per plant (NPP) (und plant[-1]), number of grains per pod (NGP), mass of 1000 grains (M1000G) (g) and grain yield (GY) (kg ha[-1]). In order to determine the grain yield, the four central lines with length of 5 meters were collected, totalling 10 m² for each plot, the GY was calculated, correcting humidity to 13% (wet basis), determined by means of the greenhouse method at 105°C for 24 hours [18].

Data of variables were submitted to analysis of variance up to 5% probability. In case of significance, they were submitted to regressions.

Results and Discussion

The accumulated precipitation during the experiment was 291.6 mm, with a very uneven distribution, therefore justifying the use of irrigation to supply the water requirement of the bean crop. According to Cunha et al. [19], common bean crop requires 300 to 600 mm to obtain high productivity. Thus, average rainfall event of 3.8 mm d[-1] and frequency of approximately 1 event every 2.2 days were observed, corresponding to 45.4% of days with precipitation. However, there was a very marked intermittent rainfall with 98% concentration of precipitation occurring from the second fortnight of December, when the crop was beginning the reproductive phase, in which the greatest rainfall accumulation was observed in the interval between days 09 to 23 of January 2016, with 135.1 mm, accounting for 46.3% of the precipitation that occurred during the crop, corresponding to a mean per rainfall event of 9 mm d[-1]. During this period, the maximum rainfall event that was 46.2 mm day[-1] also occurred (63 DAP). ET0 totalled 382.6 mm, with daily mean of 5 mm and total ETc of 310.2 mm.

The influence of irrigation levels as a function of ETc was evidenced during stages R5 to R7. In these periods, plants presented maximum development of their vegetative canopy and, consequently, high transpiratory surface. As these stages coincided with the second fortnight of December, the period of greatest precipitation, it resulted in low evapotranspiration values.

With 25% of ETc (Figure 2A), the bean crop underwent 32 days in the vegetative stage, with SWC of 210.30 mm and RAW of 228.05 mm; during this period, rainfall was 10.4 mm, with irrigation of 153.75 mm. During the reproductive stage from December 22 to 31, 2015, rainfall was 96 mm, and irrigation was not necessary, because RAW was close to TAW. From January 2, 2016, irrigation was resumed until the first decennial of January. Taking into account the crop cycle, it was verified that 25% of ETc presented 29.9% of the cultivation period below RAW, characterizing water stress for this water level. For 50% of ETc (Figure 2B), water level was 189.39 mm (precipitation and irrigation) for the vegetative phase. In the reproductive phase, the crop remained for a short period with water deficit (6 days) from January 3 to 8, 2016. According to Cunha et al. [19], the most critical stages for bean cultivation, in terms of water deficiency, is from flowering to grain filling, stages R5 to R8 respectively, thus, in this study, the crop was penalized for presenting 7.8% of cultivation period below RAW, evidencing a period of water deficit.

For treatments with irrigation levels of 75%, 100%, 125% and 150% of ETc, no water deficit was verified, since the entire crop cycle was above RAW; however, it was observed that for irrigation levels of 75%

and 100% of ETc (Figures 2C and 2D, respectively) in the vegetative phase, SWC was lower than TAW, but never lower than RAW. This was favoured by the initial irrigation levels in treatments, which ranged from 1.5 to 9 mm per event and were applied during about 95% of the initial phase according to the need. Irrigation and precipitation in this period were 212.2 and 232.7 mm, respectively, keeping SWC very close to TAW until January 3, 2016.

Treatments with 125% and 150% of ETc (Figures 2E and 2F, respectively) always remained with SWC above RAW, being close to TAW up to stage R7, but a 32-day drought occurred (November 17 to 18 December of 2015), being necessary to perform irrigations of 245.71 and 273.29 mm, respectively. The total irrigation amount during the bean crop cycle for irrigation levels of 125% and 150% of ETc was 322.89 and 362.56 mm, respectively.

Figure 1: Photograph of the experiment under different irrigation levels at 35 DAP.

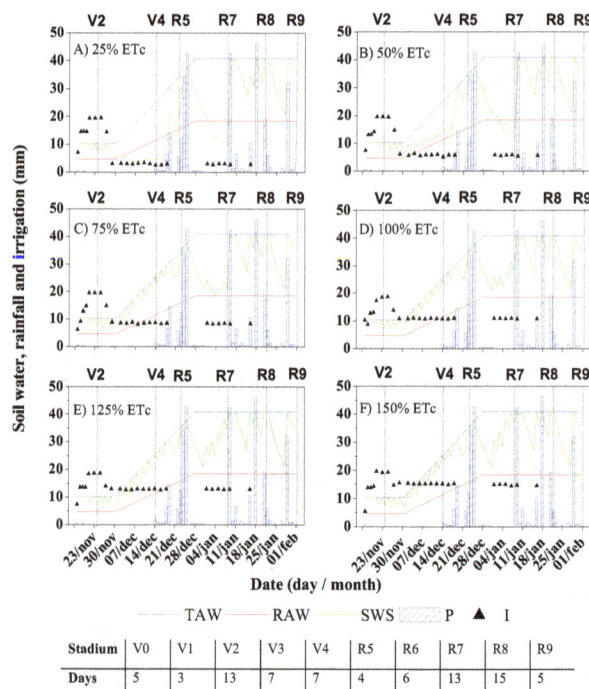

Stadium	V0	V1	V2	V3	V4	R5	R6	R7	R8	R9
Days	5	3	13	7	7	4	6	13	15	5

Figure 2: The water balance of the bean crop during the conduction of the experiment. Water balance of common bean crop, with emphasis on the total available water (TAW, mm), readily available water (RAW, mm), soil water content (SWC, mm), precipitation (P, mm) and irrigation (I, mm). Irrigated with 25% of ETc (2A), irrigated with 50% of ETc (2B), irrigated with 75% of ETc (2C), irrigated with 100% of ETc (2D), irrigated with 125% of ETc (2E), irrigated with 150% of ETc (2F). V0 Germination; V1 Emergency; V2 Primary leaves; V3 First trifoliate leaf; V4 third trifoliate leaf; R5 Pre-flowering; R6 Flowering; R7 Formation of pods; R8 Filling of pods; R9 Physiological maturation.

Significant difference among treatments with different irrigation levels was detected for the gas exchange variables: photosynthesis rate (A) ($p<0.05$), transpiration (E), stomatal conductance (gs), internal CO_2 concentration (Ci) ($P<0.01$), instantaneous carboxylation efficiency (EiC) ($p<0.05$) and no significant difference was observed for instantaneous water use efficiency (A/E). With respect to chlorophyll a fluorescence, significant effects were observed for the potential quantum yield of photosystem II (Fv/Fm) ($p<0.01$), effective quantum efficiency (ΦPSII) ($p<0.01$) and SPAD index ($P<0.05$). For production components, significant effects were observed for the following variables: number of pods per plant (NPP) ($p<0.05$) and grain yield (GY) ($p<0.01$) and no significant effects were observed for number of grains per pod (NGP) and mass of 1000 grains (M1000G) (Table 4).

The photosynthetic rate (A) of bean plants was adjusted to the quadratic model with lower value of 13.7 μmol m^{-2} s^{-1} and estimated irrigation levels of 25% of ETc, showing an increase of 13.3% in the A value compared to the highest rate (15.8 μmol m^{-2} s^{-1}) found with estimated 102.5% of ETc (Figure 3A), demonstrating that the net CO_2 assimilation rate was reduced by the water deficiency. Excess water in the soil caused by the application of water above ETc (102.5%) significantly reduced CO_2 assimilation by 5.7%, reaching a value of 14.9 μmol m^{-2} s^{-1} when soil was saturated with 150% of evapotranspirated water. These values corroborate those found by Dutra et al. [8] who verified 15.3 μmol m^{-2} s^{-1} with estimated irrigation level of 90% of ET0. These authors reported that in response to water deficit, plants reduce the opening of stomata, influencing other variables such as photosynthesis and transpiration rates, with negative consequences on crop productivity. Silva et al. [20] reported that under high soil water level, oxygen deficiency occurs, causing stomatal closure, photosystem II damage and photosynthesis reduction. The authors confirm this

information by explaining that under these conditions, there is an increase in the production of abscisic acid and reduction in stomatal conductance, in addition to the high CO_2 concentration found in the intercellular spaces of the mesophyll, suggesting that the stomatal closure is not the only cause of reduction in the photosynthetic rate.

For the transpiration rate (E), quadratic polynomial adjustment was also observed with increase of irrigation levels up to 125% of ETc, in which an increase of 15.3% was observed, calculated from the lowest value (6.09 mmol m^{-2} s^{-1}) obtained with irrigation level of 25% of ETc in relation to the highest value of 7.19 mmol m^{-2} s^{-1} transpired with irrigation levels of 125% of ETc (Figure 3B). Silva et al. [20] found similar results (7.78 and 7.40 mmol m^{-2} s^{-1}) in cowpea plants irrigated with 100 and 50% of water lost by evapotranspiration. E values lower than those described ones (2.96 and 3.41 mmol m^{-2} s^{-1} with irrigation levels estimated at 80 and 40% ET0, respectively) were presented by Dutra et al. [8] who evaluated the physiological parameters of common bean cultivated under water deficit. Lima [21], studying the physiological responses of common bean submitted to water deficit and found reductions up to 90% in the transpiration rate of non-irrigated plants compared to irrigated ones. For Silva et al. [20] the reduction in the transpiration rate is a response to water stress by plants. According to Shimazaki et al. [22], the loss of water by plants is regulated by the activity of guard cells. Pimentel and Perez [23] reported that during the day there is an increase in the transpiration rate of plants due to the inability of some plants to absorb enough water to replace that consumed in the transpiratory process.

The averages of the stomatal conductance variable (gs) had quadratic adjustment, with good predictive capacity ($R2=0.93$ **) (Figure 4A). The best result for gs (0.71 mol m^{-2} s^{-1}) was obtained with

Average Squares							
S.V	GL	35 days after application of treatments					
		A	E	gs	Ci	EiC	A/E
Blade	5	3,76*	0,83*	0.01*	11.13**	0.0074*	0.01ns
Block	3	3.03ns	0.15ns	0.00ns	0.33ns	0.0082ns	0.03ns
Quadratic	1	8.99**	1.92*	0.06**	11.64**	0.0119*	0.00ns
Linear	1	8.62*	0.94*	0.01*	13.03ns	0.0223*	0.02ns
Residue	15	0.68	0.27	0	0.97	0.002	0.03
V.C. (%)		5,45	8,55	9,04	4,41	13,7	0,74
		Fm/Fv		ΦPSII		SPAD	
Blade	5	0.0034**		0.008**		10.81*	
Block	3	0.0001ns		0.001 ns		2.96ns	
Quadratic	1	0.0077*		0.008*		11.40*	
Linear	1	0.0090*		0.031ns		40.43ns	
Residue	15	0.0003		0		2.37	
V.C. (%)		2,54		4,37		5,77	
		78 days after application of treatments					
		NPP	NGP	M1000G	GY		
Blade	5	49.51*	0.14ns	55.93ns	586442**		
Block	3	0.43ns	1.45ns	166.38ns	432097ns		
Quadratic	1	30.72ns	0.00ns	76.86ns	1708078**		
Linear	1	186.57*	0.34ns	3.47ns	952641*		
Residue	15	13.94	0.11	67.9	114062		
V.C. (%)		28,63	8,06	3,96	14,65		

S.V: Sources of variation; V.C: Variation coefficient; D.F: Degrees of freedom; **, *: Significant at 1 and 5% respectively; ns: Not significant by the F test at 5% probability; A: photosynthesis rate; E: Transpiration; gs: Stomatal conductance; Ci: Internal CO2 concentration; EiC: Instantaneous carboxylation efficiency; A/E: Instantaneous water use efficiency; Fv/Fm: potential quantum yield of photosystem II; ΦPSII: effective quantum efficiency; SPAD: Soil Plant Analysis Development Section; NPP: Number of pods per plant; NGP: Number of grains per pod; M1000G: Mass of 1000 grains; GY: Grain yield.

Table 4: Analysis of variance for physiological variables and production components of common bean under different irrigation levels.

the application of irrigation level of 100% of ETc, with an increase of 23% in relation to the lowest value (0.54 mol m^{-2} s^{-1}) obtained with irrigation level of 25% of ETc. These results are close to those found by Dutra et al. [8], who obtained gs of 0.51 mol m^{-2} s^{-1} in plants with the lowest irrigation level (40% ET0). Ferraz et al. [9], reported that under water scarcity, there is an increase in the resistance to water vapour diffusion by stomata, confirming the partial closure of stomata observed in this work, which can also be deduced by the less assimilation and accumulation of intracellular CO_2. According to Medrano et al. [24], gs are a factor linked to the global effect of water stress on the physiological parameters, responding to most of the internal and external factors of the plant. Paiva et al. [25], reported that, when the water deficit in

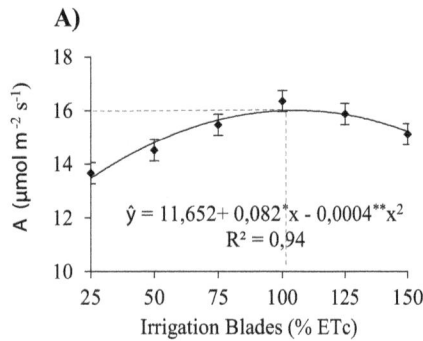

Figure 3A: Photosynthesis (A) of common bean submitted to different irrigation levels.

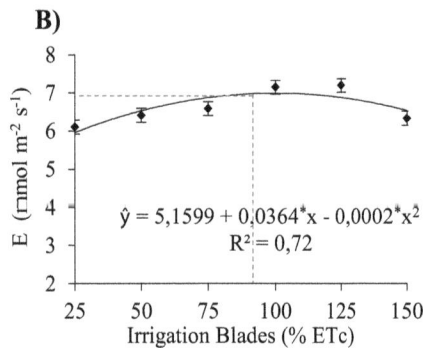

Figure 3B: Transpiration (E) of common bean submitted to different irrigation levels.

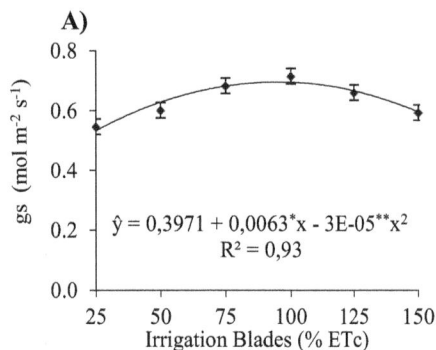

Figure 4A: Stomatal conductance (gs) of common bean submitted to different irrigation levels.

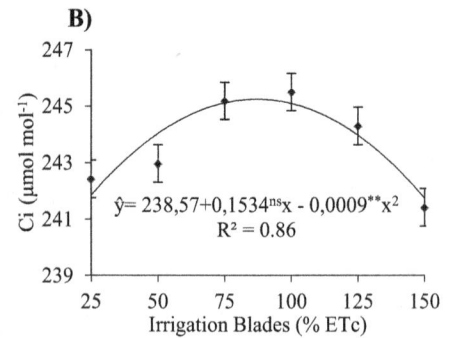

Figure 4B: Internal CO_2 concentration (Ci), of common bean submitted to different irrigation levels.

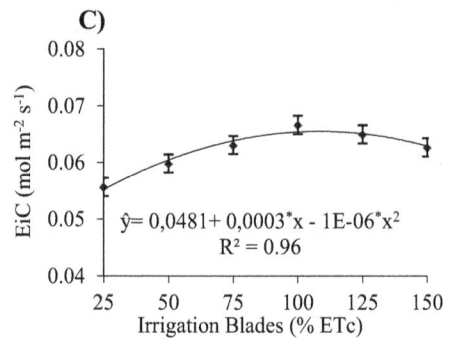

Figure 4C: Instantaneous carboxylation efficiency (EiC) of common bean submitted to different irrigation levels.

the soil is not relevant, gs variations follow the same trend of plants without water restriction, and ratify the importance of gs for directly participating in the growth and development of vegetables, playing a regulatory role in gas exchange activities.

The highest internal CO_2 concentration (Ci), average value of 245.5 μmol mol^{-1}, was observed in plants cultivated under 100% of ETc, being statistically above value observed with 25% of ETc (242 μmol mol^{-1}), evidencing an increment of the order of 4.75% (Figure 4B). This result, with small difference between plants conducted under soil water deficit and 100% of ETc, may be related to the fact that crioulo variety bean plants have greater tolerance to water deficit [26]. According to Bastos et al. [6], the increase in Ci values is indicative of restriction in CO_2 acquisition by the crop submitted to water deficit due to Ci accumulation in the leaf mesophyll, which is directly associated with stomata closure and reduction in CO_2 assimilation.

As for the instantaneous carboxylation efficiency (EiC), expressed by the relationship between net photosynthesis and internal CO_2 concentration (A/Ci), adjustments were similar to the other variables mentioned above (Figure 4C), in which an increase of 16% in relation to the lowest irrigation level was observed, with Ci equal to 0.056 mol m^{-2} s^{-1} and irrigation level of 100% of ETc providing EiC of 0.067 mol m^{-2} s^{-1}. Similar values were found by Dutra et al. [8] with 'BRS Marataoã' bean cultivar, finding EiC of 0.059 mol m^{-2} s^{-1} with estimated irrigation level of 84.5% of ET0. For Silva et al. [20], the A/Ci ratio is used to analyze the non-stomatal factors that hamper the photosynthetic rate and is related to the net photosynthesis rate and CO_2 concentration inside the substomal chamber.

In general, under water stress, plants adopt a conservative mechanism, reducing stomatal conductance and transpiration, thus increasing the water use efficiency. Under these conditions, the photosynthesis rate also ends up being influenced.

The potential photochemical efficiency of PSII measured by the Fv/Fm ratio of common bean under increasing irrigation levels obtained quadratic polynomial response (Figure 5A). A higher Fv/Fm ratio (0.78) was observed with the application of 125% of ETc, decreasing by 1.19% after this level. In the present study, it was observed that when irrigation level of 125% of ETc was applied, there was maximum efficiency in the use of radiation during the assimilation of carbon by plants [27]. The lowest value of this ratio (0.68) occurred in plants submitted to 25% of ETc. Silva et al. [28] reported that when a plant is not under water stress, the Fv/Fm ratio should be 0.75-0.85, while reductions in this ratio indicates some photo inhibitory damage in the reaction centres of PSII. Thus, the value observed in plants submitted to the lowest irrigation level is not within the previously mentioned range, so, it is suggested that the minimum efficiency of the Fv/Fm ratio occurred due to the low water replacement to which these plants were submitted.

According to Tsumanuma and Lunz [29], the Fv/Fm ratio consists of the maximum quantum yield of PSII and reflects the photochemical energy dissipation, indicating the energy capture efficiency through the open reaction centers of PSII. Based on the above, it should be pointed out that soil water deficit due to replacements of less than 50% of ETc results in damage to PSII, reducing the capacity of obtaining energy by the reaction centers, mainly due to the irregularity in the photochemical energy dissipation and reduction of the energy capture efficiency. This damage can also be related to the excess water in the soil, according to the significant reduction in the Fv/Fm ratio observed with irrigation level of 150% of ETc.

In relation to ΦPSII, an increase in its value was observed when the irrigation level was increased up to 15% of ETc (Figure 5B). By deriving the equation, the maximum efficiency value (0.7) was obtained with the aforementioned irrigation level. The lowest ΦPSII value (0.56) was obtained with 25% of ETc, showing a 21% increase when the irrigation level varied from 25 to 140% of ETc. Thus, it is understood that with irrigation (140% of ETc), greater efficiency in the electron transport was observed, reflecting in maximum photosynthetic efficiency.

As for the SPAD index, a quadratic increase as a function of the irrigation levels was also detected (Figure 5C). It was observed that the results of the SPAD index obtained with different irrigation levels increased from 40.93 to 45.75 SPAD units, resulting in a 10.5% increase when the irrigation level increased from 25 to 135% of ETc, with reduction of 0.3% after this level up to the application of 150% of ETc. This result indicates that the amount of chlorophyll in the bean leaf is related to the water status of the plant, which results in high chlorophyll synthesis, consequently increasing the photosynthetic activity and promoting an increase in the yield of this activity [30]. It should be emphasized that, under experimental conditions, values above references do not evidence the need for water replacement.

In relation to production components, Figure 6 shows the results obtained in the production evaluation performed at 78 DAP. In the number of pods per plant (NPP), there was a significant increase attributed to the application of irrigation level of 125% of ETc, with the lowest value observed, on average, 8 pods, with irrigation level of 25% of ETc and mean of 17 pods with application of 125% of ETc, achieving an expressive increase of 52.5% (Figure 6A). These results are higher

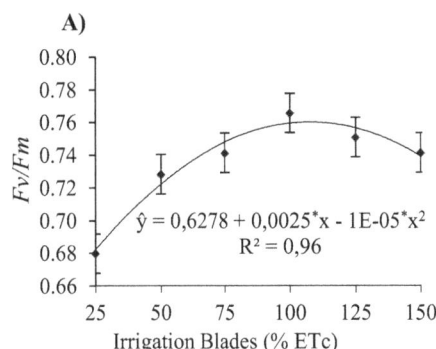

Figure 5A: Potential photochemical efficiency (Fm/Fv) of common bean submitted to different irrigation levels.

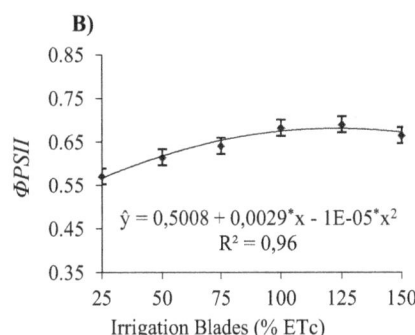

Figure 5B: Effective quantum efficiency (ΦPSII) of common bean submitted to different irrigation levels.

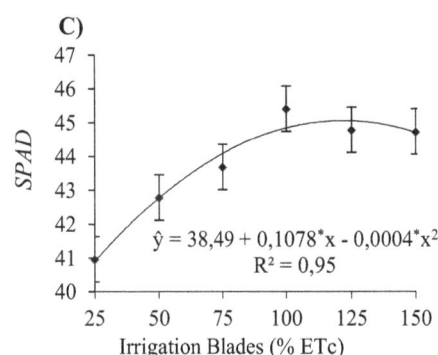

Figure 5C: SPAD index of common bean submitted to different irrigation levels.

than those found by Jadoski et al. [31], who found 8.8 pods per plant with 100% of ETc for Guapo Brilhante cultivar, working with irrigation management in Santa Maria, RS. Also working with irrigated bean cultivation, Silveira et al. [32] obtained 10.8 pods per plant. Guerra et al. [33] used Pérola cultivar under irrigation and obtained NPP equal to 14.

Grain yield (GY) and NPP showed quadratic behavior (Figure 6B). This shows that treatment with 125% of ETc obtained maximum productivity (2.230 kg ha⁻¹). Similar values were observed by Gomes et al. [34], who obtained productivity of 2.224 kg ha⁻¹ with the highest irrigation level (120% of ET0), studying the agronomic performance of common bean under irrigation in the north western region of

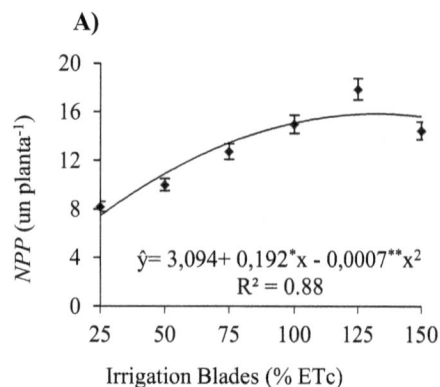

Figure 6A: Number of pods per plant (NPP) of common bean submitted to different irrigation levels.

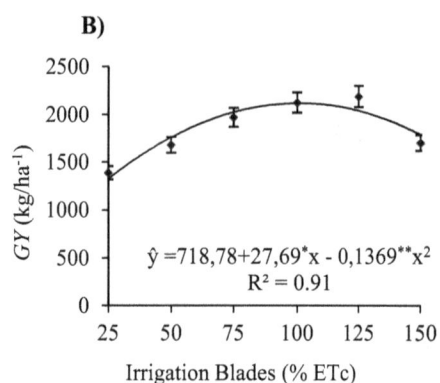

Figure 6B: Grain yield (GY) of common bean submitted to different irrigation levels.

The increase in the application of irrigation levels as a function of ETc significantly influenced the SPAD index.

Water application levels equal to 125% of ETc promoted a higher number of pods per plant and grain yield.

References

1. Zucareli C, Prando AM, Ramos EU, Nakagawa J (2011) Fósforo na produtividade e qualidade de sementes de feijão carioca precoce cultivado no período das águas. Revista Ciência Agronômica 42: 32-38.

2. Souza JVRS, Saad JCC, Sanchez RM, Rodriguez SL (2016) No till and direct seeding agriculture in irrigated bean: Effect of incorporating crop residues on soil water availability and retention, and yield. Agricultural Water Management, pp: 158-166.

3. Conab (2016) Companhia Nacional de Abastecimento. Disponível em: Acomp Safra bras Grãos Décimo primeiro Levantamento.

4. Brito JED, Almeida ACS, Lyra GB, Ferreira JRA, Teodoro I, et al. (2016) Produtividade e eficiência de uso da água em cultivo de feijão sob diferentes coberturas do solo submetido à restrição hídrica. Revista Brasileira de Agricultura Irrigada 10: 565-575.

5. Costa MS, Mantovani EC, Cunha FF, Aleman CC (2016) Avaliação dos níveis de lâmina de irrigação no desempenho do feijoeiro cultivado na região da zona da mata, MG. Revista Brasileira de Agricultura Irrigada 10: 799-808.

6. Bastos EA, Ramos HMM, Andrade JAS, Nascimento FN, Cardoso MJ (2012) Parâmetros fisiológicos e produtividade de grãos verdes do feijão-caupi sob déficit hídrico. Water Resources and Irrigation Management 1: 31-37.

7. Peixoto CP (2011) Curso de Fisiologia Vegetal. Cruz das Almas: Universidade Federal do Recôncavo da Bahia, p: 177.

8. Dutra AF, Melo AS, Filgueiras LMB, Silva ARF, Oliveira IM, et al. (2015) Parâmetros fisiológicos e componentes de produção de feijão-caupi cultivado sob deficiência hídrica. Revista Brasileira de Ciências Agrárias 10: 189-197.

9. Ferraz RLS, Melo AS, Suassuna JF, Brito MEB, Fernandes PD, et al. (2012) Trocas gasosas e eficiência fotossintética em ecótipos de feijoeiro cultivados no semiárido. Pesquisa Agropecuária Tropical 42: 181-188.

10. Mantovani EC, Montes DRP, Vieira GHS, Ramos MM, Soares AA (2012) Estimativa de produtividade da cultura do feijão irrigado em Cristalina-GO, para diferentes lâminas de irrigação como função da uniformidade de aplicação. Revista Engenharia Agrícola 32: 110-120.

11. Souza JL, Nicácio RM, Moura MAL (2005) Global solar radiation measurements in Maceió. Renewable Energy 30: 1203-1220.

12. Ferreira RA, Souza JL, Escobedo JF, Teodoro I, Lyra GB, et al. (2014) Cana de açúcar com irrigação por gotejamento em dois espaçamentos entrelinhas de plantio. Revista Brasileira de Engenharia Agrícola e Ambiental 18: 798-804.

13. Araújo AP, Teixeira MG (2012) Variabilidade dos Índices de Colheita de Nutrientes em Genótipos de Feijoeiro e Sua Relação com a Produção de Grãos. Revista Brasileira de Ciências do Solo 36: 137-146.

14. Allen RG, Pereira LS, Raes D, Smith M (1998) Crop evapotranspiration: guidelines for computing crop water requirements. Rome: Food Agriculture Organization of the United Nations.

15. Thornthwaite CW, Mather JR (1955) The water balance. New Jersey: Laboratory of climatology 8: 104.

16. Lyra GB, Souza JL, Teodoro I, Lyra GB, Filho MG, et al. (2010) Conteúdo de água no solo em cultivos de milho sem e com cobertura morta na entrelinha na região de Arapiraca-AL. Irriga 15: 173-183.

17. Carvalho OM (2003) Classificação e caracterização físico-hídrica de solos de Rio Largo, cultivados com cana-de-açúcar. Dissertação mestrado em agronomia – Rio Largo: Universidade Federal de Alagoas. P: 74.

18. Brasil (1992) Ministério da Agricultura e Reforma Agrária. Regras para análise de sementes, p: 365.

19. Cunha PCR, Silveira PM, Nascimento JL, Alves J (2013) Manejo da irrigação no feijoeiro cultivado em plantio direto. Revista Brasileira de Engenharia Agrícola e Ambiental 17: 735-742.

20. Silva CDS, Santos PAA, Lira JMS, Santana MC, Silva CD (2010) Curso diário das trocas gasosas em plantas de feijão-caupi submetidas a deficiência hídrica. Revista Caatinga 23: 7-13.

Paraná. Brito et al. [4] found 1.487 kg ha⁻¹ in cultivation without water restriction in the state of Alagoas. According to Silva et al. [28] bean crop can reach yields of more than 3.000 kg ha⁻¹ in irrigated crops with high technological level.

Thus, the potential of common beans in the production of NPP and PROD in situation of non-water restriction was demonstrated.

Water deficit in the vegetative period directly influences gas exchanges and photochemical efficiency, leading to reduced plant growth and consequently decreased productivity [35]. In study conducted by Ávila et al. [36] even though water deficit no longer occurred from the beginning of flowering, productivity was lower in relation to irrigated treatment due to the reduction in grain mass and lower number of pods per plant. Thus, the importance of irrigated bean cultivation in regions where there is a potential risk of the plant to go through periods of water stress should be highlighted, since there are soils with low water retention capacity especially in the sowing period.

Conclusions

The water application variation (% ETc) promoted a significant difference for gas exchanges, except for the instantaneous water use efficiency.

Water stress causes a reduction in the potential and effective photochemical efficiency of common bean (Crioula variety).

21. Lima EP, Silva EL (2008) Temperatura-base, coeficientes de cultura e graus-dia para cafeeiro arábica em fase de implantação. Revista Brasileira de Engenharia Agrícola e Ambiental 12: 266-273.

22. Shimazaki KI, Doi M, Assmann SM, Kinoshita T (2007) Light regulation of stomatal movement. Annual Review of Plant Biology 58: 219-247.

23. Pimentel C, Perez AJLC (2000) Estabelecimento de parâmetros para avaliação de tolerância à seca em genótipos de feijoeiro. Pesquisa Agropecuária Brasileira 35: 31-39.

24. Medrano H, Escalona JM, Bota J, Gulias J, Flexas J (2002) Regulation of photosynthesis of C3 plants in response to progressive drought: Stomatal conductance as a reference parameter. Annals of Botany 89: 895-905.

25. Paiva AS, Fernandes EJ, Rodrigues TJD, Turco JEP (2005) Condutância estomática em folhas de feijoeiro submetidos a diferentes regimes de irrigação. Engenharia Agrícola 25: 161-169.

26. Coelho CMM, Mota MR, Souza CA, (2010) Potencial fisiológico em sementes de cultivares de feijão crioulo (Phaseolus vulgaris L.). Revista Brasileira de Sementes 32: 97-105.

27. Verissimo V, Cruz SJS, Pereira LFM, Silva PBS, Teixeira JD, et al. (2010) Pigmentos e eficiência fotossintética de quatro variedades de mandioca. Revista Raízes e Amidos Tropicais 6: 222-231.

28. Silva JC, Heldwein AB, Martins FB, Maass GF (2006) Simulação para determinação das épocas de semeadura com menor risco de estresse hídrico para o feijão na região central do Rio Grande do Sul. IRRIGA 11: 188-197.

29. Tsumanuma GM, Lunz AMP, Feijoeiro (2008) In: Castro PRC, Kluge RA, Sestari I Manual de fisiologia vegetal: fisiologia de cultivos. Piracicaba: Ceres., pp: 77-91.

30. Pires AA, Araújo GAA, Miranda GV, Berger PG, Ferreira ACB, et al. (2004) Rendimento de grãos, componentes do rendimento e índice spad do feijoeiro (Phaseolus vulgaris L.) em função de época de aplicação e do parcelamento da aplicação foliar de molibdênio. Ciência agro técnica 28: 1092-1098.

31. Jadoski SO, Carlesso R, Melo GL, Rodrigues M, Frizzo Z (2003) Manejo da irrigação para maximização do rendimento de grãos do feijoeiro. IRRIGA 8: 1-9.

32. Silveira PM, Stone LF (2006) Irrigação. In: Vieira C, Paula TJ, Borém AF, (2ndedn) Viçosa, pp: 171-211.

33. Guerra AF, Silva DB, Rodrigues GC (2000) Manejo de irrigação e fertilização nitrogenada para o feijoeiro na Região dos Cerrados. Pesquisa Agropecuária Brasileira 35: 1229-1236.

34. Gomes EP, Biscaro GA, Ávila MR, Loosli FS, Vieira CV, et al. (2012) Desempenho agronômico do feijoeiro comum de terceira safra sob irrigação na região noroeste do Paraná. Semina Ciências Agrárias 33: 889-910.

35. Arf O, Rodrigues RAF, Sá ME, Buzetti S, Nascimento V (2004) Manejo do solo, água e nitrogênio no cultivo de feijão. Pesquisa Agropecuária Brasileira 39: 131-138.

36. Ávila MR, Barizão DAO, Gomes EP, Fedri G, Albrecht LP (2010) CULTIVO DE FEIJOEIRO NO OUTONO/INVERNO ASSOCIADO À APLICAÇÃO DE BIOESTIMULANTE E ADUBO FOLIAR NA PRESENÇA E AUSÊNCIA DE IRRIGAÇÃO. Scientia Agraria 11: 221-230.

22

Concept of Water, Land and Energy Productivity in Agriculture and Pathways for Improvement

Upadhyaya A* and Alok K Sikka

Division of Land and Water Management, Indian Council of Agricultural Research, India

Abstract

Land and water are finite natural resources, which are diminishing due to indiscriminate and unscrupulous exploitation. Ever increasing population is also posing a challenge to produce more from the available resources. Proper understanding of the concept of land, water and energy productivity has become quite relevant in recent days. The technologies/ strategies in which input application is more precise, efficient and cost effective and output is adequate in quantity, excellent in quantity, well in time and profitable, will lead to enhancement of land, water and energy productivity in agriculture. In this paper, concept of land, water and energy productivity, assessment procedure of crop and agricultural water productivity and case study of Pabnawa minor in Kurukshetra, Haryana (India), have been discussed and various means/ pathways to enhance productivity by employing suitable technologies/strategies have been highlighted.

Keywords: Natural resources; Energy productivity; Horticulture; Livestock

Introduction

Land and water are finite natural resources, which are diminishing due to indiscriminate and unscrupulous exploitation. Due to increasing population pressure, situation becomes more serious and calls for efficient and productive utilization of resources. Share of water diversion for agriculture is projected to reduce in future due to other competing demands. Growing water scarcity as a result of increasing demand is challenging to increase the productivity of land, water and energy to usher in the era of "evergreen revolution". Besides physical availability of water, it is economical accessibility of water with minimum energy input, which is causing concern to get maximum agricultural output or value for every drop of water used in agriculture. Understanding the concept of water, land and energy productivity and its enhancement in groundwater irrigated areas assume yet greater importance where water is either scarce due to faster depletion of water table as in the western IG basin or it is costlier to pump water in eastern part of the IG basin owing to diesel operated pumps.

Concept and definition of water productivity

Productivity is a ratio between a unit of output and a unit of input. Water productivity is used exclusively to denote the amount or value of product over volume or value of water used or depleted or diverted. Increasing the productivity of water means, in its real sense, getting more benefit from every unit of water used for various production systems. The definition of water productivity is scale dependent. From the farmer's viewpoint, it means getting more production per unit of irrigation water. But, at a river basin scale or at the country level, this means getting more value/benefit per unit of water resource used. Water productivity will depend on many factors other than quantity of water applied. Various factors, which influence water productivity, include cropping pattern, crop variety, level of other inputs applied, management factors etc. Though water is only one of the factors of agricultural production and cannot be meaningfully separated from the others, an estimate of its productivity and knowledge about the factors which influence it will help in making the future plans to improve water productivity in a particular area.

Raising crop water productivity means raising crop yields per unit of water consumed, though with declining crop yield growth globally, the attention has shifted to potential offered by improved management

of water resources [1]. It provides a means both to ease water scarcity and to leave more water for other competing demands.

The key to understanding the ways to enhance water productivity is to understand what it means [1]. Water productivity can be analyzed at the plant level, field level, farm level, system level and basin level, and its value would change with the changing scale of analysis. Many researchers have argued that the scope for improving water productivity through water management, or efficiency improvement, is often overestimated and re use of water is under-estimated [2].

Water productivity may be defined as the ratio of the net benefit from crop, forestry, fishery, livestock, and mixed agricultural systems to the amount of water required to produce those benefits. In its broadest sense it reflects the objectives of producing more food, income, livelihoods, and ecological benefits at less social and environmental cost per unit of water used, where water use means either water delivered to a use or depleted by a use. In simple words, it implies growing more food or gaining more benefits with less water. Within the broad definition of water productivity there are interrelated and cascading sets of definitions used for different purposes. Physical water productivity relates the mass of agricultural output to water use- "more crop per drop." Economic water productivity relates the economic benefits obtained per unit of water used and has also been applied to relate water use in agriculture to nutrition, jobs, welfare, and the environment.

Crop water productivity: Crop water productivity denotes the amount or value of product (i.e., crop) over volume or value of water used or depleted or diverted. It varies with location depending on the factors such as cropping pattern, crop genetic material, climatic conditions, irrigation technology, field water management,

*Corresponding author: Upadhyaya A, Division of Land and Water Management, Indian Council of Agricultural Research, India
E-mail: aupadhyaya66@gmail.com

infrastructure and on the labour, fertilizer, and machinery inputs as well as economic and policy incentives to produce. In general crop water productivity is a function of water applied, which depends on space scale and generally increases from small plots to large domains at basin scale because applied water is recycled and reused.

Agricultural water productivity: Agricultural water productivity takes into account multiple water uses like agriculture, horticulture, forest, livestock, fisheries, environment etc. It means that if all water users are taken into account and concept of recycling and reuse of water is considered in agricultural production system, then agricultural output per unit of total water input is referred as agricultural water productivity. Since agricultural water productivity assessment considers multiple uses of water, hence its value is higher than the crop water productivity.

Land and energy productivity: On the similar lines, concept of land productivity and energy productivity can also be realized. Basically these terms also indicate about agricultural production per unit of land and per unit of energy. The purpose of defining these terms is to measure the existing performance of these resources and suggest pathways to enhance productivity. Infact, there is a need to study tradeoff among water, land and energy productivity and attempts should be made to improve productivity by efficiently utilizing water, land and energy resources.

Water productivity and water use efficiency

The classical concept of irrigation efficiency used by water engineers to analyze the "productive use" of water omitted economic values and looked at the actual evapo-transpiration (ET) against the total water diverted for crop production [1]. Over and above, it does not factor in the "scale effect") [3]. Further, classical irrigation efficiency is defined as the crop water requirement (actual evapotranspiration minus effective precipitation) divided by the water withdrawn or diverted from a specific surface-water or groundwater source. 'Losses' in this approach include transpiration and evaporation (evapotranspiration), but also seepage, percolation and runoff, processes in which the water is not consumed. These latter so-called 'losses' may be captured or recycled for use elsewhere in the basin. Thus, classical measures of efficiency tend to underestimate the true efficiency and ignore the important role of surface irrigation systems in recharging groundwater and providing downstream sources of water for agriculture and other ecosystem services.

The notion of water productivity is evolved from two disciplines. Crop physiologists defined 'water use efficiency' as carbon assimilated and crop yield per unit of transpiration, and later as the amount of produce per unit of evapotranspiration. Irrigation specialists have used the term 'water use efficiency' to describe how effectively water is delivered to crops and to indicate the amount of water wasted. But this concept provides only a partial and sometimes misleading view because it does not indicate benefits produced, and water lost by irrigation is often gained by other users. Water productivity analysis can be seen as part of an ecosystem approach to managing water. Rain, natural flows, withdrawals, and evaporation support terrestrial and aquatic ecosystems, which produce numerous services for people. The primary service of agro-ecosystems is food and fiber production, but other important services are produced as well.

Units of expression of water productivity

Water productivity analysis can be applied to crops, livestock, tree plantation, fisheries, and mixed systems at selected scales-crop or animal, field or farm, irrigation system, and basin or landscape, with interacting ecosystems. The objectives of water productivity analysis range from assessing agricultural production (kilograms of grain per unit of water depleted by a crop on a field) to assessing incremental welfare per unit of water used in the agricultural sector. Because expressions for water productivity differ in each context, it is important to be clear about the agricultural output and input terms used.

As water moves down-stream, a drop may be transpired by a plant, be evaporated from the land, or continue to flow downstream to be used and reused by cities, agriculture, and fisheries to produce some output. Physical Water Productivity is generally expressed in terms of 'kg/m³' or 'kg/ha-cm' or any other unit of mass of agricultural output per unit volume of water used or depleted, whereas Economic Water Productivity is expressed in terms of 'Rs./m³' or 'Rs./ha-cm' or '$/m³' or '$/ha-cm' or any other unit of economic benefits obtained per unit of water used or depleted.

Water accounting diagram

The basic question in water productivity assessment is 'which crop/which drop' to be considered as numerator and denominator. This is aptly answered by identifying the scale at which we measure water productivity. As we move from one scale to another, the potential utility of some water changes. Basin level and/or irrigation system level water productivity takes into consideration beneficial depletion for multiple uses of water. Here, the problem lies in allocating the water among its multiple uses and users and also taking into account multiple uses of the same quantum of water. The same drop or quantum of water may serve hydropower, urban, fisheries and then agricultural needs before it is ultimately depleted. Water accounting employed to estimate the flows across the boundaries of the domain, provides a means to generalize water use across scales, and to understand the denominator of the water productivity [4]. Analytical framework of water accounting at field and basin scale is shown in Figure 1.

Assessment of water productivity

Water productivity may be computed during Crop period or whole year considering the production value first from crops only (crop water productivity) and then considering other water users including trees, livestock and fish (agricultural water productivity) in case of multiple uses per unit of water inflow (including rainfall, ground water and canal water) as well as water used (total inflow excluding runoff) in the field, farm, irrigation system, basin and landscape. The accuracy of water productivity assessment depends on water accounting procedures or estimation of water used by various users and water lost in various processes. Water accounting provides a means to generalize about water use across scales, and to better understand the denominator of the water productivity. Water accounting can be applied at all scales of interests, and requires the definition of a domain bounded in three-

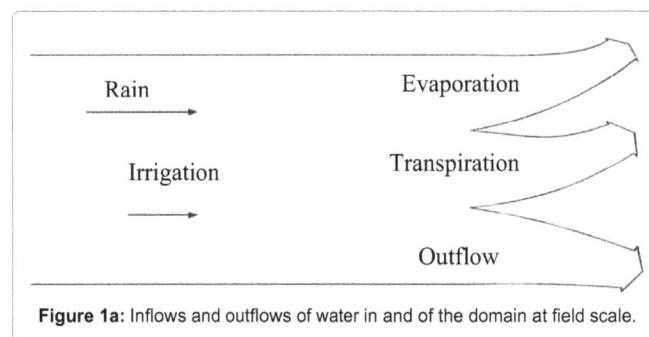

Figure 1a: Inflows and outflows of water in and of the domain at field scale.

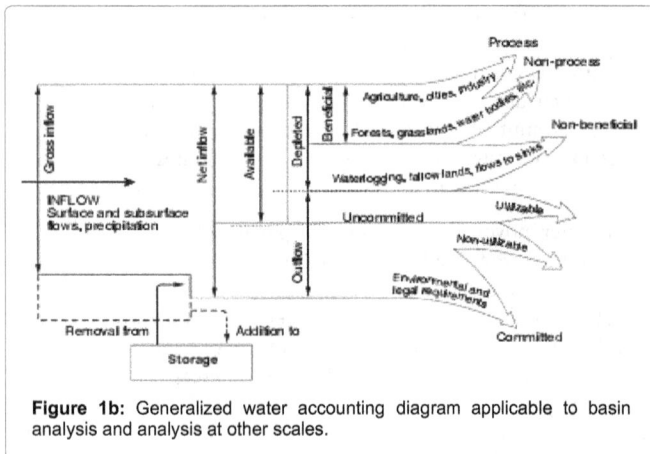

Figure 1b: Generalized water accounting diagram applicable to basin analysis and analysis at other scales.

dimensional space and time. For example, at the field scale, this could be from the top of the plant canopy to the bottom of the root zone, bounded by the edges of the field, over a growing season. The task in water accounting is to estimate the flows across the boundaries of the domain during the specified time period. At the field scale, water enters the domain by rain, by subsurface flows and, when irrigation is available, through irrigation supplies. Water is depleted by the process of growing plants: transpiration and evaporation. The remainder flows out of the domain as surface runoff or subsurface flows or is retained as soil moisture storage. In estimation of water productivity, we are interested in water inflows (rain plus irrigation, or just rainwater in rain-fed agriculture) and water depletion (evaporation and transpiration) as shown in Figure 1a.

Case Studies on WP Assessment

Case study 1

Most of the water productivity assessments deal with single crop. Sikka et al. discussed the methodology for assessment of water productivity in a multiple use scenario including crops, livestock, fisheries, and forests adopted in the command of RP Channel-V of Patna Main Canal in the Sone Command at Patna and tube well command in Vaishali, Bihar [5]. Crop water productivity was computed considering crops output per unit of irrigation water applied, inflow diverted (including rainfall, ground water and canal water) as well as water used (total inflow excluding runoff), whereas agricultural water productivity was computed considering output of various water users like cereal crops, horticulture, trees, fisheries and livestock. Water productivity was computed during Kharif (Monsoon) and Rabi (Winter) seasons considering the production value first from crops alone and then considering other water users including trees, livestock and fish.

The total water entering the domain (rainfall, canal and tubewell) was calculated for respective command area and the same was incorporated in the SWAP model The water diverted to the command areas of RP Channel–V and tubewell commands in Vaishali was found to be utilized by different crops. SWAP model was used to calculate interception, runoff, evaporation and transpiration separately. The water balance components for both Kharif and Rabi crops were calculated for all the three outlet commands and two tubewell commands.

Finally water productivity was computed considering value from crop production, trees, fish and livestock and water diverted or depleted by various users, as given in Table 1.

Crop water productivity (Rs/m³) considering applied water varied in the range of 4.79 to 8.39. Considering water inflow including rainfall, it ranged between 2.42 to 3.11 in the outlet commands. In tubewell commands the crop water productivity for applied water ranged from 14.03 to 29.61, whereas it was 2.39 to 2.81 in case of total water inflow including rainfall. Agricultural water productivity (Rs/m³) considering applied irrigation water varied in the range of 5.28 to 10.66. Considering total water inflow including rainfall, it ranged between 2.67 to 3.96 in the outlet commands. In tubewell commands the total water productivity ranged from 18.09 to 38.73 for applied water and 3.09 to 3.68 for total water inflow including rainfall. Lower water productivity considering total water inflow (irrigation + rainfall) in tubewell command may be attributed to higher proportion of rainfall in total water used. The analysis clearly indicates that in canal outlet commands, crop water productivity as well as agricultural water productivity are maximum in outlet 27 followed by outlet 17. Crop water productivity alone does not depict the actual use of water in the command. Since agricultural water productivity takes account of other water users like trees, fodder, livestock, fish etc., its value is higher and as such it gives the true picture of actual water use and productivity of water.

Case study 2

Chandra et al. carried out study in Pabnawa Minor of Bhakra Canal System with (Latitude 29°-31' and 30°-12' and longitude 76°-10' and 76°-43') in Kurukshetra (Figure 2 and Table 2) Irrigation Circle of North-west India in western IGP [6]. The study area is located in the semi-arid tropics in Haryana with average annual rainfall of 625 mm (80 percent during June to September). Minimum and maximum temperature varies between 5° to 25°C and 12° to 44°C, respectively. Soils are fine coarse in texture varying from sandy loam to clay loam. Many farmers have reclaimed sodic soils using gypsum as an amendment.

In each of the four selected watercourses, 15 farmers' fields were selected for detailed monitoring of water use and crop yields. In the selected fields in PH, PM1 and PT section of the water course, wheat was planted using zero tillage and bed planting techniques. However, wheat crop was planted using conventional tillage practices in the selected fields of PM2 command. Information related with different agricultural and water management practices adapted by farmers in the selected fields, was collected on a specially designed data collection form. Systematic observations were recorded for water use on a daily basis from the selected farmers' fields.

Water Productivity (Rs/m³)	Outlet 4 Head Reach	Outlet 17 Middle Reach	Outlet 27 Tail Reach	Tubewell 2 Land consolidation	Tubewell 11 Fragmented land holding
Area (ha)	30.61	43.68	4.65	18.74	13.21
Crop WP per unit of irrigation water applied	4.79	4.95	8.39	29.61	14.03
Crop WP per unit of water inflow including rainfall	2.42	2.73	3.11	2.81	2.39
Agricultural WP per unit of irrigation water applied	5.28	5.90	10.66	38.73	18.09
Agricultural WP per unit of water inflow including rainfall	2.67	3.25	3.96	3.68	3.09

Table 1: Crop and agricultural water productivity.

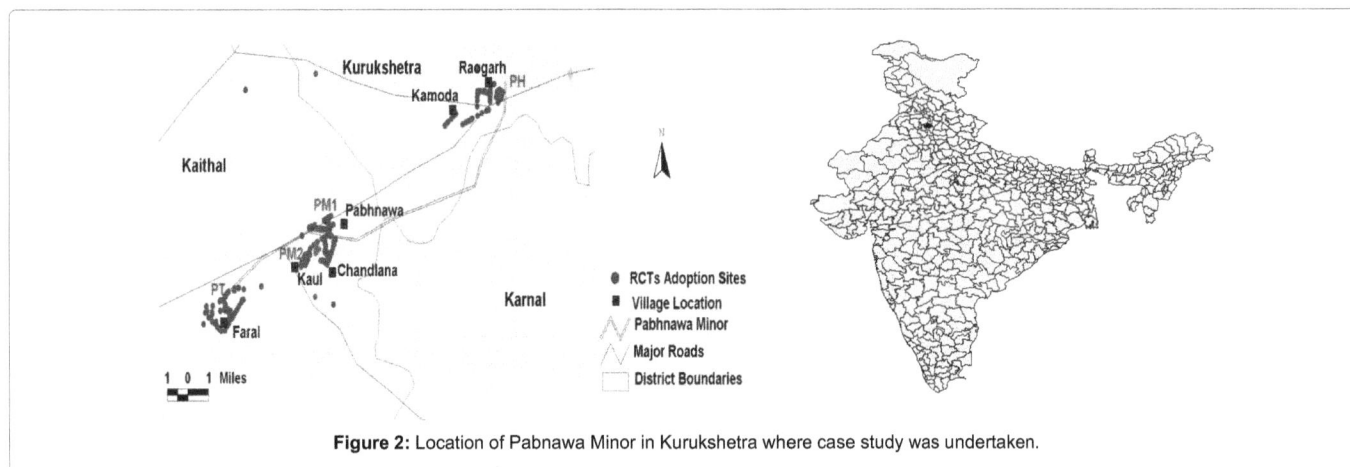

Figure 2: Location of Pabnawa Minor in Kurukshetra where case study was undertaken.

Irrigation Minor	Watercourse	Technology	Design discharge (m³/sec)	Gross command area (ha)	Cultivated command area (ha)
Pabnawa	Pabnawa Head-end (PH) 2820R'	Zero Tillage Bed Planting	0.028	231.6	208.9
Pabnawa	Pabnawa Middle (PM1) 53705L	Zero Tillage Bed Planting	0.041	320.2	300.0
Pabnawa	Pabnawa Middle (PM2) 53705 R	Conventional Tillage	0.025	341.3	169.6
Pabnawa	Pabnawa Tail-end (PT) 80000L	Zero Tillage Bed Planting	0.052	283.0	253.4

Letter L and R refer to left and right banks of the water course

Table 2: Details of Selected Watercourses.

The depth of flow at each outlet was measured on a daily basis in each rotation of canal flow. The relationship between measured discharge and depth of flow was worked out for each watercourse. The discharge-depth relationship was used to calculate canal water supplied to farmers' fields after taking into account seepage losses along the watercourse.

The discharge of tube wells owned by the farmers was measured by using the co-ordinate method. Relevant data on irrigation water supplies from the tube wells and canal water, cropping pattern, and yields were collected periodically in the selected fields at each of the four sites. Crop cutting trials were conducted in all the farmers' fields to determine the crop yields. To get realistic estimates of crop yields, a minimum of four crop-cutting trials per plot were conducted. Microplot yields were converted to crop productivity (Kg/ha).

The productivity of water is expressed in terms of Kg/m³ of water or Rs/m³ of water. The other terms used here are defined as under:

$$\text{WP gross inflow} = \frac{(\text{Yield, kg / ha})}{(\text{Gross inflow, m}^3 \text{ / ha})} = \frac{(\text{Gross income, Rs / ha})}{(\text{Gross inflow, m}^3 \text{ / ha})}$$

$$\text{WP irrigation inflow} = \frac{(\text{Yield, kg / ha})}{(\text{Irrigation inflow, m}^3 \text{ / ha})} = \frac{(\text{Gross income, Rs / ha})}{(\text{Irrigation inflow, m}^3 \text{ / ha})}$$

The term gross inflow represents the water from canal, groundwater pumped by the tube wells and rainfall received during crop season. Irrigation inflow includes only the canal water supplies and groundwater abstraction received from tube wells. Thus, the precipitation received in the area is not included in calculating irrigation inflow.

The following production indicators were used at the water course level:

$$\text{Output}_{\text{gross inflow}} (\text{Rs / m}^3) = \frac{\text{Gross value of produce (Rs)}}{\text{Gross inflow at watercourse inlet (m}^3)}$$

$$\text{Output}_{\text{gross inflow}} (\text{Rs / m}^3) = \frac{\text{Gross value of production (Rs)}}{\text{Gross inflow at watercourse inlet (m}^3)}$$

Water productivity in both Zero tillage and conventional tillage decreases (Figure 3) as one moves from plot level to watercourse level (i.e for the three level of analysis). This trend is same for all the three reaches of Pabnawa minor. Although absolute value of WP decreases from Plot to Watercourse level, higher percentage level of increase in ZT over CT at the farm and watercourse level (35 & 37% as against 20% at plot level) suggests benefit of ZT in water saving at watercourse level.

Wheat Water productivity in Bed planting method of crop establishment is generally higher than Zero tillage and conventional tillage at Plot level in different reaches. Water productivity of wheat in Bed planting is greater than zero tillage and wheat water productivity in zero tillage is greater than conventional tillage across Plot to Watercourse scales (Figure 4).

The comparison of benefit (gross margin) per unit of water used (consumed) under zero and conventional tillage practices at different locations of the watercourses studied are given in Table 3.

As can be seen from Table 3, the benefit for zero tillage over conventional tillage has increased by a margin of 35% to 66% per unit of irrigation water. Water and land Productivity of bed planted Rice and conventional tillage rice is given in Table 4.

Bed planted rice yielded 0.2 to 14% less yield over conventional tillage (Table 4). The irrigation water productivity for rice under BP is higher (22 to 28%) than that of CT but land productivity is lesser as compared to conventional tillage. There is a trade off between water productivity and land productivity in bed planted rice.

Water productivity of Wheat at Farmers' field computed using SWAP model is given in Table 5.

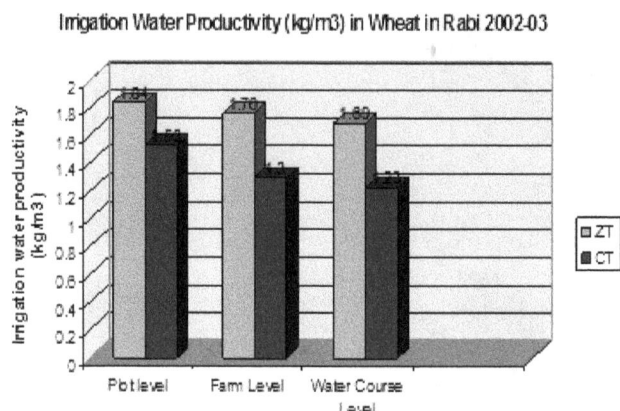

Figure 3: Irrigation water productivity at plot, farm and watercourse level.

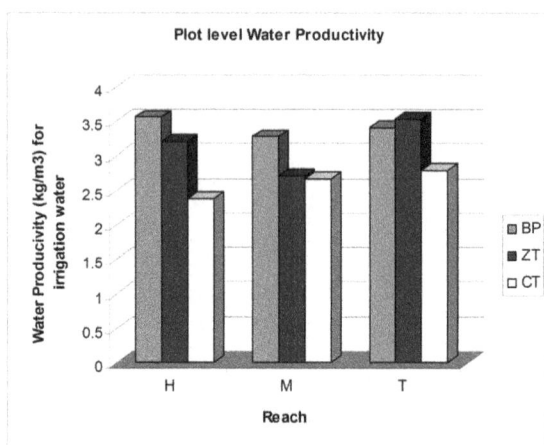

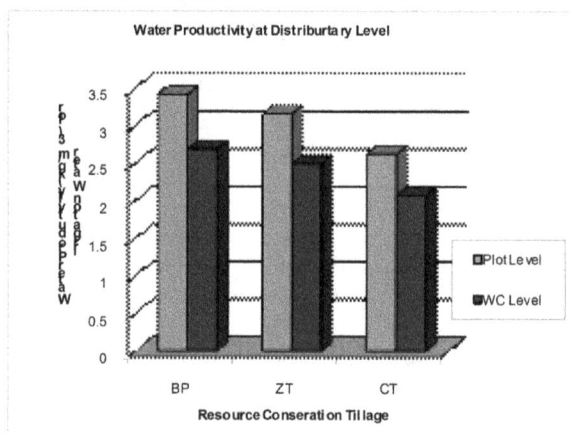

Figure 4: Wheat water productivity at different scales for rcts Pabnawa distributary, Haryana, India.

Location	Tillage Practice	Benefit (Rs /m³) of irrigation Water	Benefit (Rs/m³ of gross water)
Pabnawa Head-end 2820R*(PH)	Zero tillage	7.3	5.7
Pabnawa Middle 53705L (PM1)	Zero tillage	5.1	4.1
Pabnawa Middle 53705 R (PM2)	Conventional tillage	3.8	3.1
Pabnawa Tail-end, 80000L (PT)	Zero tillage	5.8	4.5

Table 3: Comparison of profitability per unit of water in different tillage practices among different watercourses.

Location	Method of Sowing	Average Depth of Irrigation. Water (cm)	Average depth of gross water (cm)	Average Yield (t/ ha)	Irrigation Water Productivity (kg/m³)	Gross water productivity (kg/ m³)	Average Yield (t/ha)
PH	BP	124.63	130.43	4.76	0.38	0.37	4.76
PM1	BP	137.95	143.75	5.43	0.39	0.38	5.43
PT	BP	100.56	106.36	4.93	0.49	0.46	4.93
PM2	CT	175.92	181.72	5.53	0.31	0.30	5.53

Table 4: Water and land Productivity of bed planted Rice and conventional tillage rice.

It may be observed from Table 5 that WPI under bed planting is higher than that of zero tillage by 25% and by 79% compared to conventional tillage. Zero tillage water productivity is higher by 42% from that of conventional tillage. These results are in conformity with the estimates from observed values.

Results of this analysis indicate the superiority of Zero tillage over the Conventional tillage both in terms of irrigation water productivity and land productivity in wheat besides profitability. Water productivity in both Zero tillage and Conventional tillage decreases as one moves from plot level to watercourse level (i.e. for the three levels of analysis). Higher level of increase in Zero tillage over Conventional tillage at the farm and watercourse level suggests benefits of Zero tillage in water saving at watercourse level. These results are based on limited but rarely

WP Water Productivity (Kg/m³)	BP	ZT	CT
WP$_I$ Water Productivity (Kg/m³ of Irrigation water)	2.83	2.25	1.58
WP$_T$ Water Productivity (Kg/m³ of Transpiration)	1.71	1.50	1.28
WP$_{ET}$ Water Productivity (Kg/m³ of Evapo-transpiration)	1.31	1.17	1.01
WP$_G$ Water Productivity (Kg/m³ of Rain and Irrigation water)	1.53	1.32	1.03

Table 5: Water productivity of Wheat at Rajender's Farm (15 ha) with SWAP model.

available field data. While analyzing the data some logical assumptions were made in view of data limitations. The results even if have some limitations but they definitely have suggestive indications of the benefits of adopting resource-conserving technologies at different scales.

Energy Requirement in Drip and Surface Irrigation System

Srivastava and Upadhyaya determined the size of prime mover and annual energy requirement to cover 5 ha area of sugarcane with drip and surface irrigation system for different water table depths and irrigation requirement [7]. Results are presented graphically in Figures 5 and 6 below.

It is evident from the Figure 5 that the size of prime mover is drastically reduced with adoption of drip irrigation. Similarly Figure 6 shows that contrary to common perception that drip irrigation is more energy intensive, drip irrigation consumes less energy.

Role of Micro-irrigation in Enhancement of Land, Water and Energy Productivity

Micro irrigation, which broadly includes drip and micro-sprinkler system, is one of the efficient methods of irrigation. Besides improving the quality and quantity of produce, this system helps in saving water even up to 50% as compared to surface irrigation system. But due to higher cost involvement, this system does not suit to small and fragmented landholders having small purchasing capability. Singh et al. and Singh et al. reported that ICAR-RCER, Patna has developed a Low Energy Water Application (LEWA) device, which can fit on risers

and can replace costly sprinklers [8,9]. It rotates at very low pressure and sprinkles water over crop like rain. LEWA system can be used to irrigate rice, wheat, vegetables and other close growing crops. The cost of LEWA system is less and it saves water, time and energy. So this is very efficient, effective and beneficial irrigation system for small and marginal farmers having fragmented land holdings (Figure 7). Low Energy Water Application (LEWA) device operates at 0.4 –0.6 kg/cm² operating pressure with throw diameter of 6-8 m, application rate 2.6-3.1 cm/h. The surface uniformity is observed greater than 70% when operated at an operating pressure of 0.5 Kg/Cm² or above and sub – surface uniformity is greater than 90%. Basically its discharge is higher than the infiltration rate unlike sprinklers just to keep soil surface wet. With flexible flat hose pipes a LEWA unit for 1000 m² costs approximately Rs. 15000/- (excluding prime mover). In other words considering 10 shifts its cost is Rs 15000/- per ha whereas cost of sprinkler system is 3-4 times of cost of LEWA system. Singh et al. also discussed the concept and applicability of LEWA and its role in improving water and energy productivity [9].

Water productivity and yield of wheat through LEWA was studied. It was found that there is 45% and 10% water saving as well as 50% and 55% energy saving over surface and sprinkler irrigation system. Yield values vary as 3.775, 3.581 and 3.525 and Water productivity values as 1.91, 1.62 and 0.95 kg/m³ in LEWA, sprinkler and surface irrigation system, respectively.

Pathways to Improve Water Productivity

Pathways to improving water productivity include improving the productivity of green and blue water; improving the water productivity of livestock and fisheries; applying an integrated approach to increase the value per unit of water; and adopting an integrated basin perspective to understand water productivity tradeoffs.

There are many well-known crop per drop improvements. These include more reliable and precise distribution and application of irrigation water, supplemental and deficit irrigation, improved soil fertility, and soil conservation practices. In smallholder livestock

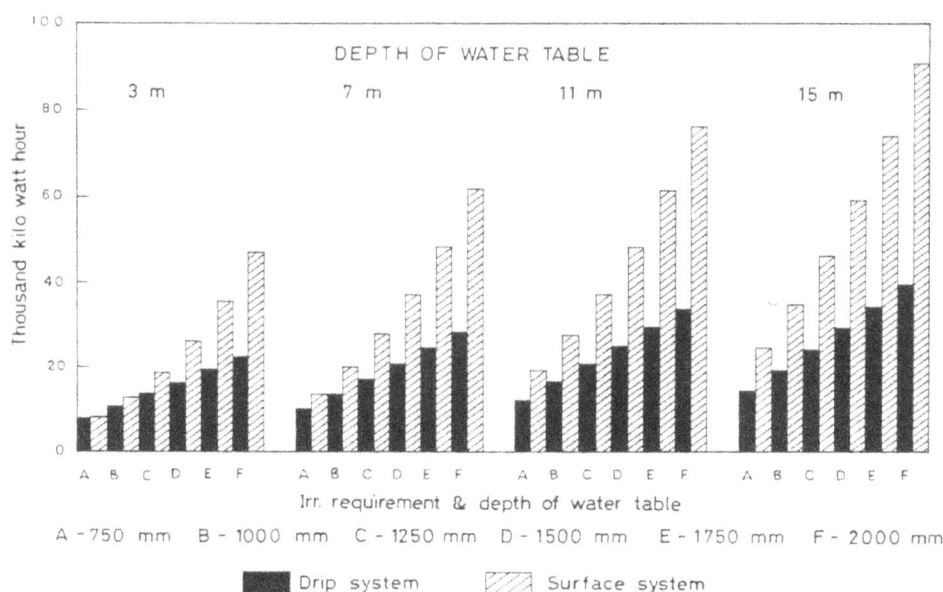

Figure 5: Size of prime mover required for drip and surface irrigation for different water table and irrigation requirement for command area of 5 ha.

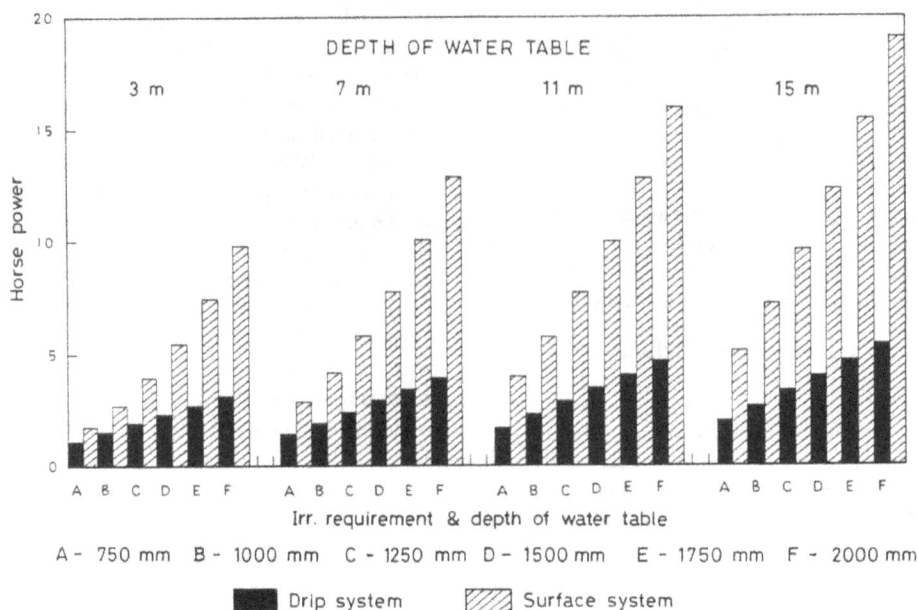

Figure 6: Annual energy consumption for drip and surface irrigation for different water table and irrigation requirement for command area of 5 ha.

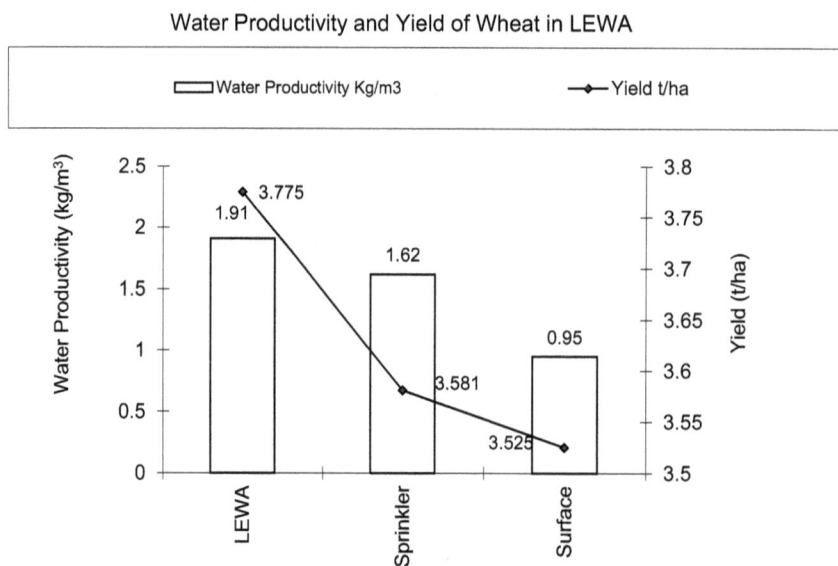

Figure 7: Water productivity and yield of wheat in LEWA sprinkler and surface irrigation methods.

systems, feeding animals crop residues can provide a several fold increase in water productivity. Integrated approaches are more effective than single technologies. Upadhyaya has reported many water saving and water use efficient technologies, which can improve yield, income and livelihood of farmers resulting in enhancement of land and water productivity [10]. Upadhyaya et al. studied spatial and temporal variation of soil moisture under different tillage practices in wheat crop and observed that soil moisture under zero tillage and in furrows of raised bed was found higher than that under conservation tillage and on the beds of raised bed method [11]. Contribution from various activities in enhancing WP is presented below in Table 6.

It shows that 50-60% improvement in WP is possible through land and water management. Only moderate impacts on crop water

productivity can be expected from genetic improvements to plants over the next 15-20 years.

Primary pathways to increase the productivity of water at different scales i.e. Plant, field and basin are as mentioned by Molden et al. are given in Table 7 [12].

Many known technologies and management practices promise considerable gains in water productivity. Achieving those gains requires a policy and institutional environment that aligns the incentives of various users at different scales- from field to basin or country- to encourage the uptake of new techniques and to deal with tradeoffs. It requires policies that: (i) overcome risks, (ii) provide incentives for gains in water productivity, (iii) adjust basin-level water allocation

Product	Current Water productivity (Kg/m3)	Potential contribution to increasing Water productivity		
		Genetic (%)	Water management (%)	Soil management (%)
Wheat	0.2-1.5	15	40	15
Cereals	0.2-2.4	20	40	15
Rice	0.15-0.6	10	45	15
Maize	0.3-2.0	20	35	10
Trees	1-20	25	10	20
Vegetables	5-20	50	20	10

Table 6: Contribution from various activities in enhancing WP.

Pathways	Plant level	Field level	System/ Basin level
1.Increase marketable yield per unit of water transpired	√	√	√
2. Reducing Outflows (drainage, seepage and percolation) and non-productive depletions (evaporation from soil and water, weeds)	√		√
3.Increasing non-irrigation inflows (Rainfall, stored water, marginal quality water, waterlogged/ drainage water)		√	√
4.Increasing the effective use of water from the storage		√	√
5. Using not yet committed flows			√
6.Reallocating and co-managing water (multiple use) among uses		√	√

Table 7: Pathways to improve water productivity at different levels.

policies, (iv) target the poor with sustainable, water productivity enhancing practices, and (v) look for the opportunities outside the water sector.

High priorities for water productivity improvement include:

- Areas where poverty is high and water productivity low, where the poor could benefit.

- Areas of physical water scarcity where there is intense competition for water.

- Areas with little water resources development.

- Areas of water-driven ecosystem degradation, such as falling groundwater tables and drying rivers.

Other Energy Saving Interventions

Considerable savings in energy requirement can be made if the level of efficiency of the pumping set is maintained at the minimum expected efficiency of 50 percent. In a survey of irrigation pump sets owned by the farmers in the Nainital Tarai region the average efficiency was found as 36%. The survey indicated that the possible reasons of low efficiency of installed pumping sets are: (i) The pumps are not selected according to the well conditions. (ii) The drive units are not matching the pump requirement. (iii) Improper sizing and excessive length of suction and delivery pipes and 90° bends are being used. (iv) Standard and good quality foot valve and pipe fittings are not being used. (v) Adequate technical service on the purchase, selection of a pump and drive unit matching with well conditions, installation, operation and maintenance of pumping sets is not available to the farmers.

Various factors influencing the energy requirement in irrigation indicate that by improvements in irrigation management practices irrigation saving can be done. The Potential energy saving in terms of percentage was given as below.

$$PES = 100\left[1 - \left(\frac{D_{n2}E_{i1}R_1H_2}{D_{n1}E_{i2}R_2H_1}\right)\right] \quad (1)$$

where the subscript 1 indicates initial values and the subscript 2 indicates values after modifications. D_n is the net amount of irrigation water applied (mm); E_i is irrigation efficiency, or fraction of pumped water that is stored in the crop root zone; R is performance rating of pumping plant; and H is total head required. The equation indicates that following modifications will reduce the pumping energy requirement: (i) reduction of net depth of irrigation ($D_{n2}/D_{n1}<1$); (ii) reduction in total head ($H_2/H_1<1$) or lower pressure requirements; (iii) improved performance rating of pumping plants by adjustment ($R_1/R_2<1$); and (iv) increased irrigation efficiencies ($E_{i1}/E_{i2}<1$).

In addition to the energy savings as calculated by above equation, three other possible cost and energy saving areas related to irrigation include (i) off-peak irrigation scheduling to reduce peak electrical demands, (ii) reduction of nitrogen losses through reduction of deep percolation losses, and (iii) reduced tillage practices.

Reduction in total water pumped can be achieved through either improvement in irrigation efficiency or reductions in net irrigation applications. Improvements in the application efficiency will reduce the gross water application and thereby reduce the total energy required to pump the water. The amount of water that must be pumped depends on the irrigation system type and the particular irrigation water management practices. It was found that the irrigation efficiency depended more on management than the type of system and he further noted that relatively "poor" systems under proper management were more efficient than "better systems" which were poorly managed.

The installation of runoff reuse systems can provide significant energy conservation to surface irrigators. Reuse systems require additional energy expenditures, however this additional requirement is usually small compared to the energy required to pump the water from the initial source.

Some irrigation systems may only need improved management (adjusting stream sizes, changing length of run, and/or timings) to obtain better efficiency. Land levelling may be needed on some fields before the desired efficiency can be obtained. Surge irrigation may be used to improve graded furrow irrigation. Sprinkler systems may require replacement of nozzles to match the site, reduced lateral spacings and changes in irrigation set times. Trickle irrigation systems may require cleaning or replacement of emitters to improve performance. Most of the irrigation systems can achieve improved efficiency through more intensive management and maintenance. This includes close supervision of the operation of the system and irrigation scheduling to prevent application of excess water.

Additional savings of water and energy are possible by inducing some moisture stress during part or all of the growing season by limiting irrigation. This procedure may result in reduced yield, thus economic constraints will govern the application of these practices in the field. It is usually possible to select an efficient scheduling process to produce the maximum yield for the attainable level of water use.

In future, farmers may shift to crops requiring less water because of reduced water supplies and/or increased energy costs. This change in cropping pattern will depend upon: the production function of the respective crops, the water supply policy, the type of irrigation equipment, the profitability of the respective crops, alternative energy sources and their costs, and governmental policies.

Reductions in total pumping head can be achieved by (i) reduced pressure sprinkler systems, (ii) substitution of surface water for ground water, (iii) pipeline modifications to reduce friction losses, and (iv) design changes in irrigation wells to reduce head losses.

A study indicated that well development procedures can reduce water table drawdown as much as 50% for wells constructed using bentonite drilling fluid. The amount of energy savings depends upon the initial well drilling method, construction materials and practices, and aquifer characteristics, all of which can affect the resulting drawdown.

Whenever long pipelines are needed to move water from the source to the field, pumping energy is required to overcome the frictional losses. The amount of friction loss is dependent on the type of pipe, its diameter and length of pipeline. Normally pipe length is fixed, and only the size of the pipe line can be increased or type of the pipeline changed to reduce pumping head, but it will increase the fixed cost. So selection of appropriate size pipeline distribution system depends on energy and cost considerations.

Reducing the pressure of sprinkler systems is a viable method to save energy. Conversion of conventional high pressure sprinkler systems to reduced pressure systems may require redesigning at additional expense. In all the cases, the energy savings resulting from reduced pressure will have to be greater than the cost of additional equipment and extra labour required for more frequent moves of the system.

Energy conservation through improved irrigation management is not limited to the energy used to pump water. Large quantities of energy are used to manufacture fertilizer, primarily nitrogen, which is used in irrigated agriculture. Over irrigation, especially on sandy soils, can leach nitrate nitrogen below the crop root zone, resulting in nitrate build up in the ground water, nitrogen deficiencies in the crop and decreased yields. Several investigators have found nitrate nitrogen loss ranging between 2.5 and 10.2 kg/ha per centimeter of deep percolation on sandy soils under surface irrigation systems. It was concluded that nitrate loss through percolation could be minimized by proper selection of the nitrogen amount, nitrogen source and irrigation management. Between 25 and 60 kg/ha of nitrate nitrogen can be saved with improved water-nitrogen management procedures. Approximately 1 kWh of electrical energy are required to produce 1 kg of nitrogen fertilizer in the anhydrous ammonia form. Thus, the annual energy requirements for nitrogen use in irrigation could be reduced by 25 to 60 kWh/ha of electricity.

Reduced tillage can be used to reduce the energy required in irrigation, but perhaps more important than the energy saved is the increased surface water storage capacity created by certain reduced tillage practices. Tillage systems that maintain plant residues on the soil surface protect soil from erosion, increase surface water storage, and improve infiltration by reducing the surface soil sealing caused by rain and irrigation drop impact. Such reduced tillage systems may allow the use of reduced-pressure irrigation systems on low-intake soils, thereby further reducing irrigation energy requirements. Reduced tillage systems will increase surface residues throughout the spring operations. Residues generally retard soil drying and temperature increases, thus delaying planting. Increased surface residue increases demand for proper management and the tillage system should not preclude, but rather complement, other good conservation practices.

Several irrigation management procedures for reducing energy use in irrigation have been presented. These energy saving techniques can be used to ensure continued high level production in spite of energy shortages, production cost increases, and limited water supplies. The improved management of existing irrigation systems help to conserve both energy and water. Pump irrigators in the future may have to accept soil water deficits and the corresponding yield reductions in some years because of limited energy supplies, scheduled electrical power interruptions, or water allocation. Improved management of the irrigation system will help minimize these losses.

Conclusion

Land and water are two finite natural resources, which are essentially required for agricultural production. To feed ever increasing population of our country, enhancement in production from diminishing resources is the need of hour. Agricultural productivity can be enhanced if land and water resources are utilized efficiently, timely and judiciously and energy is channelized properly in the positive direction. The case studies conducted in the head, middle and tail reaches of RP Channel V under Patna Main canal in the Sone Command and two tube well commands in Vaishali, Bihar show that agricultural water productivity (considering rain and irrigation water) varied in increasing order from 2.67 to 3.96 Rs/m^3 from Head to tail reach and 3.68 and 3.09 Rs/m^3 in tubewell 2 with land consolidation and tubewell 11 with fragmented land holdings. Similarly case study 2 in Pabnawa minor showed that wheat water productivity under bed planting was higher than that of zero tillage by 25% and by 79% compared to conventional tillage and zero tillage water productivity was higher by 42% from that of conventional tillage. It was also observed that contrary to common perception, drip irrigation was more energy efficient, since it required less energy than the surface irrigation. It is concluded that adoption of an integrated approach, which takes into account soil-water-crop-climate-human-livestock resources management and farm mechanization, planning and implementation of location specific, cost effective, energy efficient, simple, sound, sustainable, socially acceptable, and environmentally harmless technologies/ interventions/strategies are the pathways to enhance land water and energy productivity.

References

1. Kijne J, Barker R, Molden D (2003) Improving Water Productivity in Agriculture: Editors' Overview. International Water Management Institute.

2. Seckler D, Molden D, Sakthivadivel R (2003) The Concept of Efficiency in Water Resources Management. International Water Management Institute.

3. Keller A, Keller J, Seckler D (1996) Integrated Water Resources Systems: Theory and Policy Implications. International irrigation Management Institute.

4. Molden DJ, Sakthivadivel R, Christopher JP, Charlotte de F, Klozen WH (1998) Indicators for comparing performance of irrigated agricultural systems. International irrigation Management Institute.

5. Sikka AK, Abdul Haris A, Upadhyaya A, Thakur A, Reddy AR, et al. (2008) Water productivity assessment in the Canal and Tubewell Commands of Bihar-A Case Study.

6. Chandra R, Gupta RK, Sikka AK, Upadhyaya A, Shakthivadivel R (2007) Impact of conservation technologies on water use and water productivity in Pabnawa minor of Bhakra canal system. Rice-Wheat Consortium.

7. Srivastava RC, Upadhyaya A (1998) Study on feasibility of drip irrigation in shallow ground water zones of eastern India. Agricultural Water Management 36: 71-83.

8. Singh AK, Sharma SP, Upadhyaya A, Rahman A, Sikka AK (2010) Performance of low energy water application device. Water Resources Management 24: 1353-1362.

9. Singh AK, Islam A, Singh SR, Upadhyaya A, Rahman A, et al. (2015) Low Energy Water Application (LEWA) device: Concept and applications. Journal of Soil & Water Conservation 14: 344-351.

10. Upadhyaya A (2015) Water management technologies in agriculture: Challenges and opportunities. Journal of Agrisearch 2: 7-13.

11. Upadhyaya A, Singh SS, Prasad LK, Roy MK (2015) Spatial and Temporal Variation of Soil Moisture under different Tillage Practices in Wheat Crop. Journal of Agrisearch 2: 175-178.

12. Molden D, Oweis TY, Pasquale S, Kijne JW, Hanjra MA, et al. (2007) Pathways for increasing agricultural water productivity. International Water Management Institute.

23

Success Story and Factors Affecting Level of Income Earned from Improved Potato Farming in Damot Sore Woreda, Wolaita and Southern Ethiopia

Bassa Z*, Abera A, Zeleke B, Alemu M, Bashe A and Areka MS

Southern Agricultural Research Institute, Areka Agricultural Research Center, Areka, Ethiopia

Abstract

Increasing production and productivity of crop farming ,improving income of resource poor farmers and thereby enabling the producers to build assed in Southern Ethiopia in General and Damot Sore Woreda in particular require some form of transformation of the subsistence, low-input and low-productivity farming systems to full agricultural packages utilization and awareness creation. The study was employed in Irish AID Operational Research and Technology Dissemination Project (ORTDP) areas of Areka mandate. This study was undertaken to analyse factors affecting level of income earned from Potato and summarize benefits of utilizing improved potato variety and full agricultural packages in the district. A multi-stage sampling technique was used to select 80 sample households from two sample kebele. In the study, both primary and secondary data sources were used. Simple Linear Regression Model was employed to identify factors affecting level of income earned from improved potato production by resource poor farmers in the district. Results showed that using improved potato variety increase the production and productivity of the specific commodity and there help the resource poor farmers to build asset. From eight explanatory variables used the six determinant factors that affected significantly the level of income earned from improved potato adoption were comprised of Tropical livestock unit, Being beneficiary or not, family size and intervention period.

Keywords: ORTDP; Improved potato; Agricultural technologies

Introduction

In Southern Nations, Nationalities, and Peoples Region, (SNNPR) particularly the project target Woredas are characterized by persistent food insecurity with many farming household not producing enough food and income to meet household food requirements. Improved agricultural technologies largely focusing on increasing yield and market value have an important role in increasing productivity, income and building asset and improving household food security. The increased agricultural productivity also boosted by the availabilities and access of new and improved agricultural technologies. Improved agricultural technologies, management practices, and inclusion of resource poor household for enhanced technological access also have a proven track record on improving food security and decreasing susceptibility to individual stresses. Thus, investing in dissemination of improved agricultural technologies is key to improve the livelihood of low-income and food insecure households. By recognizing this, Irish Aid has launched technology dissemination initiative with aims to reducing poverty for poor and marginalized farmers, particularly women; driving agricultural growth by linking poor farmers into new and improved crop, livestock, and natural resource conservation technologies.

The Operational Research Technology Dissemination project (ORTDP) is addressing key agricultural development challenges prioritized by both the Ethiopian and Irish governments: improved food security, poverty reduction and greater gender equity, better nutrition outcomes and more climate resilient food and farming systems through supporting of rural poor household by accessing for improved agricultural technologies. Southern Agricultural Research Institute (SARI) in collaboration with Irish-Aid has been currently investing in agricultural research and dissemination of improved agricultural technologies focusing on crop, livestock, and natural resource management. The technologies being disseminated are tested and proven to have potential for up scaling to improve productivity, food and nutrition security, and climate resilience of resource poor farmers. For the past five years, the project has disseminated more

than 33 proven crop, livestock, and natural resource management technologies for more than 13266 resource poor household in seven food insecure Woredas of the region especially for whose landholding less than 0.25 hectare and women.

The project has reviewed its performance and status to lay out strategic directions and priorities for agriculture technologies dissemination and extension in the region. As one component to address the OR project goal particularly to improve income level and thereby build asset, reduce poverty or improve nutrition, potato is one of the crop technologies disseminated by the project and its performance has been evaluated giving an account in addressing issues related to productivity, income, nutrition and adaptability to ever changing environment. Of the technologies successfully disseminated by the project, the potato case studies presented as proven best-bet agricultural technologies and innovations that are available for uptake and up scaling. This case study was conducted in two Kebele of ORTDP in Wolaita Zone in Damot Sore Woreda. Farmers for the case study were selected using multistage sampling techniques. Thus, from the project Woreda 24 none beneficiaries (6F and 18M) and 56 beneficiary farmers (15F and 41M). The household survey included 50% of 2014/15 and 2015/16 beneficiary farmers. Data collection sheet was prepared to collect quantitative and qualitative data regarding the productivity, income, food, asset building and asset type, nutrition, trends on use of improved seed, adoption, and challenges. The data collected was subjected for simple descriptive statistical analysis.

---

*Corresponding author: Bassa Z, Southern Agricultural Research Institute, Areka Agricultural Research Center, Areka, Ethiopia
E-mail: bassazekarias@yahoo.com

Justification of the intervention

Potato is one of highland root and tuber crops produced on tow cropping seasons (Belg and Meher season) and mainly in Belg Season in the district. Potato is produced mainly as cash crop and in some extent as food crop by farmers in the area. Despite its role in the farming system and in supporting the national economy, yield has been low and stagnant for several years due to different reasons. There could be several reasons for this but the most important ones are lack of improved varieties with desirable agronomic practices, lack of awareness for farmers on hot to preserve improved seed, low yield potential of local varieties, and diseases. Especially low production and distribution of improved seed is limited among resource poor farmers. Besides this, the access of the improved varieties has been low for poorest farmers. Therefore, ORTDP project has proposed to promote various improved potato varieties (Belete and Gudene Variety) for five consecutive years and Belete for the past two years and Belete and Gudane for production period of 2015/16 that aimed to improve income, diversify diets, build asset and improves nutritional status of poorest households in six project intervention Woreda.

In Ethiopia, potato is produced on 66,361.67 hectares with an average national yield of 136.86 quintals in Meher cropping season. The major potato producing regions of Ethiopia and their Meher cropping land size in ha summarized as Oromiya 38256.15 ha , Amhara 17719.49 ha and SNNPR 10727.13 ha states in that order of production levels. From total Southern Nations Nationality People region area covered by potato, 1958. 12 ha potato area held by Wolaita zone (5.5%) in Meher cropping season. The average regional productivity of potato was 166.48 quintal per ha and that of Wolaita zone was 190 quintal per ha [1].

Methodology

Damot Sore is one of the Woreda in the Southern Nations, Nationalities, and Peoples' Region of Ethiopia. Part of the Wolaita Zone Damot Sore is bordered on the southeast by Sodo Zuria, on the west by Kindo Koysha, on the northwest by Boloso Bombe, and on the north by Boloso Sore. Damot Sore was separated from Boloso Sore woreda. Damot Sore has total of 20 Kebele and of which two are town administrative.

The Pre extension and Demonstration of Irish Potato trails were planted in Damotsore project Kebele namely Sheymba and Doge Hanchucho. The amount of improved potato disseminated per household amounted to 2 quintal in 2015 and 2014 intervention period per household. During provision of the improved potato farmers were trained on land preparation, ways and benefits of improved potato adoption. The beneficiary farmers were selected based on their participation in Safety Net program and food security level, land size owned in ha, gender base and considering other socioeconomic factors. The improved potato varieties introduced to the district were Bele potato Variety and Gudene potato variety.

Result/Achievement

In five project intervention years, ORTDP has addressed 310 direct beneficiaries and 930 indirect resource poor households with improved potato dissemination, especially for farmers who unable to access improved seeds and had smaller land (farmers their landholding less than 0.25 ha). The project had been provided 1020 quintals of four improved potato varieties (Jaleni, Belete and Gudane) and popularized on 51 hectare of land throughout the project intervention years. These varieties have been successfully promoted in all Woreda using cluster-based approaches accompanying with practical agronomic practice training and its related input as a package. Participatory technology dissemination method and cluster approach was a key element of the implementation of this project. Awareness creation of beneficiary farmers, development agents, Woreda agricultural experts and Office heads and demonstration of technology dissemination at Farmers Training centres were the basis of the Technology popularization and pre-scaling up in the district.

Improved productivity

Based on the suitability and agro-ecological adaptability of crops ORTDP has disseminated different crop varieties for beneficiaries. While the project proposes three potato varieties for demonstration, the primary criterion was its contribution to increase of productivity of potato. The advantages of theses potato varieties include their high yield potential, disease tolerance, drought tolerance, early maturity, high market value and nutritional values. The survey report confirmed that yields of improved potato have 100% higher than locally available potato variety in the districts and more than 60% beneficiary farmers built asset at housed level inform of cattle, sheep and poultry, able construct their house and purchase agricultural input without credit. In addition to these, the economic strata of certain farmers changed from very poor to be graduated from Safety net program and from zero livestock ownership to more than two units of livestock. The main reasons for increase in the productivity of improved potato listed as timely dissemination of inputs, full package application of the technologies, relatively suitable weather condition and practical capacity building of subject matter specialists. The survey result also revealed that farmers tendency to use adequate fertiliser especially UREA was improved in the district.

Food and nutrition security

Increased production has also led to significant improvement in food security and nutrition. As survey report, over 100% of project beneficiary household consume potato at least once in a week while only 23% of non-beneficiary household consume potato at least once in a week. Consumption of potato was much more prevalent in among beneficiaries, while lesser extent among non-beneficiaries. From potato production in 2014 fiscal year more than 90 per cent products sold for cash source that helped to build asset and this indicates that project beneficiaries tend to produce more for cash source and the remaining for food and seed source. Through their consumption household can have food nutrients like protein and iron that potato contains. While it reveal some interesting trends on contribution to nutrition, what is not mean that household access all required food items that meets the nutritional security standards (amount of kilo calories per day/week). The data to what extent reflects the availability and access of potato, which rich in carbohydrate through their own production. This also have a significant contribution to food security with many nutritional benefits, as it is rich in protein, iron, zinc, and dietary fibre. These constituent contributed to the improvement in food and nutrition and the potato acceptance by farmers.

Improved income

Farmers in two Woreda (Damot Sore and Mierab Badawacho) use potato both for cash generation and for home consumption although the majority use for cash. Potato is grown in both production seasons; Belg and Meher seasons with the main growing season being Belg. Farmers are in great need of cash for the Meher season especially wheat and teff producing Woreda to buy agricultural inputs like fertilizer and seed. The yield of the new varieties is about five times those of local variety and generated an additional income for household. Therefore, potato is strategic cash crop, which fulfils the immediate cash demand for input purchase thereby built asset. During interviewees farmers mentioned that most of the potato produced during Belg season is sold to generate income. As indicated in the baseline report, crop income for farmers in SNNPR ORTDP project Woreda comes from sale of cereals such as sweet potato, wheat, teff and common bean. After five years of project implementation, the annual mean cash income obtained only from sale of potato for both project beneficiaries and non-beneficiaries increased.

Adoption

The potato varieties disseminated by the project were evaluated against the local variety by using 13 criteria. More than 100% farmers in the survey districts evaluated Belete and Gudane Variety by seed size, taste, early maturity, yield, taste, and marketability. Over all rank calculated shows that Belete and Gudene Potato Variety have first preference by farmers in the all project location in comparison to local variety. Trends on planting of improved potato varieties in project Woreda during base lines ranged from 10% to 40% with mean value 27.8%. After four years of intervention, trends on use those potato varieties has become 75%. Three different successful aspects of the new Potato varieties were identified by farmers during the survey: improved taste, higher productivity, and market-preferred attributes reported among both beneficiaries and non-beneficiaries. As this was one of the outputs of the operational research programme, and 70% of households included in the survey were programme participants, it suggests more work has to be done through regular extension in disseminating and promoting of those improved potato varieties among non-beneficiary household.

Drivers to success

The higher productivity, marketability, seed size and taste of farmers were the main drivers for the successful dissemination of those potato varieties. Beside this, the project full package approach enables poorest to increase their productivity using fertilizer as a package. The use of adequate fertilizer for the potato especially NPS were uncommon in most projects Woreda and the project has tried to demonstrate the yield difference using NPS fertilizer. Moreover, the project's cluster based technology dissemination and transfer approach played a significant role to easily diffusion of knowledge and practice from one cluster to the other and created a critical mass on disseminated technologies.

Challenges

Despite the significant contribution of potato for food security, asset building, income and nutrition some challenges are also faced the small scale farmer (wilting disease, awareness problem of farmers on keeping and preservation up of seed). Farmers reported that the susceptibility to disease and pest and less tolerant to flood and heavy rainfall and management problem especially improper application of fertilizer. The significant number of farmers reported that they have not applied fertilizer mainly NPS as recommended rate and rarely used UREA. The other main challenges in dissemination of potato technology were erratic rain fall. Erratic rain fall in the district manifested as rain fall scarcity in 2014/15 and heavy rain during earthling up in the fiscal year of 2015/16. The other challenges faced in dissemination of the technology were high personal benefit expectation form Woreda experts and development agents and wrong belief of farmers over seed raising and preservation.

Opportunities for further scaling up

High productivity, attractive size of seed, early maturity, its seed size, its taste, high market value, and marketability played a significant contribution for the successful popularization and adoption of potato varieties disseminated in project Woreda as well as beyond project area.

Key lesson and recommendation

The interventions in agricultural research and dissemination have been strong components in strategies to promote sustainable agricultural development. Previously the technology dissemination approaches were focused on strengthening the productivity aspects of the technologies. Currently the project has disseminated the potato varieties focused demand of farmers (Figure 1). The project considers the multi-benefits and interactions of potato technologies disseminated with interest and demands of farmers for technologies. In all project Woreda, farmers give equal priority for income generating potential of technology as productivity potential. Therefore, potato technologies disseminated by the project has great demand by the community for food consumption as well as for income source and is a major crop in the area. A technology, which has great demand by beneficiaries, have multi-benefit and agro-ecologically suitable, ultimately leads to success. The Belete potato varieties distributed by the project fulfil most of farmers' interest and their production objectives and that is why it became successful. Therefore, the food and income security of poorest and marginalized people could be enhanced through accessing poorest household for demand driven better yielding and high valued agricultural technologies (Tables 1 and 2).

Moreover, the project's cluster based technology dissemination and transfer approach played a significant role to easily diffusion of knowledge and practice from one cluster to the other and created a critical mass on disseminated potato technologies. This approach can bridge the research with extension in more interlinked way and accelerate technology transfer between farmers. In addition, most of farmers prefers to plant potato as intercropping rather than mono cropping. Their preference varies with the primary objective of farmers. Farmers whose primary objective is for household consumption and have land shortage tends to cultivate as intercropping while farmers with their primary objective for cash tends to cultivate as mono-cropping (Table 3).

Interpretation of Linear Regression Result

The study used Simple linear regression model for sorting out factors that affect income earned from improved potato disseminated to small scale farmers in the district. Eight explanatory variables were identified and regressed over amount of income earned from selling improved

Figure 1: Potato yield collected at household level within 9 m².

Trt no	Trts NPS kg/ha	Plant height/cm				No of tubers per plant				Potato yield /3 m² (3 × 3 m) (kg/plot)				Potato yield/plot		Total yield qu/ha
		1	2	3	Avr.	1	2	3	Avr.	1	2	3	Avr.	Yield/plot	Yield kg/ha	
1	150	128	92	85	101.7	10	6	12	9.33	29.5	45.8	37.8	37.7	37.7	41888.9	418.88
2	175	72	76	94	80.66	13	13	1	9	22.2	31.5	35.6	29.766	29.766	33073.3	330.73
3	200	102	94	74	90	15	9	8	10.66	32.4	35.2	32.5	33.36	33.36	37066.7	370.66
4	225	102	94	92	96	9	14	9	10.66	41	26.2	24.5	30.56	30.56	33955.6	339.55
5	250	100	84	94	92.66	15	8	13	12	36.7	43	32.7	37.46	37.46	41622.2	416.22
1	150	112	110	110	110.7	16	10	9	11.66	35.2	36.6	48.4	40.06	40.06	44511.1	445.11
2	175	106	114	112	110.7	17	16	14	15.66	28.4	26	33.6	29.33	29.33	32588.9	325.88
3	200	108	114	100	107.3	13	13	20	15.33	27.8	34.2	33.6	31.86	31.86	35400	354
4	225	120	110	118	116	16	15	18	16.33	45.4	34.2	36.8	38.8	38.8	43111.1	431.11
5	250	114	108	114	112	12	15	15	14	30	34.2	39	34.4	34.4	38222.2	382.22
1	150	124	114	100	112.7	27	23	24	24.66	32.8	33.2	33.4	33.13	33.13	36811.1	368.11
2	175	110	112	136	119.3	24	18	26	22.66	31.9	30.5	28.6	30.33	30.33	33700	337
3	200	120	98	118	112	8	14	8	10	45.6	32	31.3	36.3	36.3	40333.3	403.33
4	225	116	114	126	118.7	37	14	9	20	40.8	32	31.3	34.7	34.7	38555.6	385.55
5	250	115	114	110	113	15	10	10	11.66	43.4	33.8	40	39.06	39.06	43400	434
1	150	98	102	100	100	10	9	13	10.66	29.4	27.8	27.2	28.13	28.13	31255.6	312.55
2	175	94	110	90	98	20	15	17	17.33	31	29.2	32	30.73	30.73	34144.4	341.44
3	200	98	110	104	104	18	21	25	21.33	28.8	28.8	21.2	26.26	26.26	29177.8	291.77
4	225	112	96	98	102	13	11	18	14	27.2	32	38	32.4	32.4	36000	360
5	250	114	108	100	107.3	9	22	17	16	22	26.7	25.2	24.63	24.63	27366.7	273.66
Total	4000	2165	2074	2075	2105	317	276	286	292.9	662	653	662.7	658.97	658.97	732185	7321.8
Mean	200	108	104	104	105.2	16	14	14	14.65	33.1	32.6	33.14	32.948	32.948	36609.2	366.09

Source: Field experimentation in Damot Sore Woreda (Abate Abera, 2015).

Table 1: Effects of different NPS fertilizer rate on potato crop yield and yield components at Damot sore Worenda 2015/16.

potato variety (Dependant variable) disseminated by ORTDP in the district. Explanatory variables used in the model regression comprised of household category (either beneficiary or not), tropical livestock unit, age of respondent, total family size, total land owned in timad (1/4 ha), quantity of potato used, project intervention period/year and quantity of seed reserved as seed. From eight explanatory variables used six were found determinant factors that affected significantly the level of income earned from improved potato adoption comprised of Tropical livestock unit, Being beneficiary or not, family size and intervention period. The study result was in line with Melesse [2] that confirmed access to improved seed affect proportion of the value of potato sold positively.

Household category

Household category was one of the explanatory variables that affect significantly the extent of income earned from improved potato adoption in the district. This variable defined as 1 for improved potato beneficiary farmers and 0 otherwise. The parameter estimates for the independent variable regressed was 14.01. This indicated that being beneficiary farmers of improved potato dissemination increases the level of income earned from the technology utilization by more than fourteen times. Since operational technology dissemination improves awareness of farmers, creates opportunity to avail and utilized improved agricultural packages at full level and supports through follow-up and training, beneficiary farmer's income from improved potato was significantly higher than the non-beneficiary farmers. Through participation in agricultural technology packages, awareness creation and linking the producers to improved agricultural inputs and full package utilization; it is possible to maximize the level of income earned from agriculture.

Tropical Livestock unit

The parameter estimate for the variable Tropical Livestock unit

owned is 6.04. This implies that for one unit increase in Tropical Livestock unit owned, the beneficiary farmers income increase by more than 6 Birr, holding all other explanatory variables constant. The size of Tropical Livestock unit reared by farmers is the variable that positively affects extent of income earned from crop farming. Tropical Livestock unit had significant effect on level of income earned from crop farming through its fertilizing that by its own helps crop productivity by decreasing cost of artificial fertilizer accessing. The findings is in inline Esmael et al. [3] that states livestock owned affect farmers extent of potato sales positively affects the extent of potato sales negatively.

Total family size

Family size of a respondent was one of independent variable (continuous variable) supposed to influence extent of income earned from improved potato by beneficiary farmers in the district. Its sign was positive that indicates household with large number of families' size earned grater income from potato that helped the household to weed, earth up, apply fertilizer and use improved agricultural technology packages in comparison to the household with low household size. The regression results confirmed that family size has significant effect in increasing income earned from crop farming for farmers in the area. It looked in to that family size was as such influential factors linked with crop farming that was their main income source and livelihood base. The result is non in line with the finding of Esmael et al. [3]; that justified family size of the household affect negatively the potato market participation and income earned and in line with Urgessa [4] that clarified Labour productivity significantly found as determinant factor of productivity of land and thereby affect income of farmers.

Land owned in timad (1Timad=1/4 ha)

Total land owned by farmers is one of explanatory variable significantly and positively affected the extent of income earned from improved potato adoption to the district. The parameter estimate of the

No.	Name of the farmers	Gender	Commodity	Quantity given	Yield collected	Sold	Income Earned	Consumed in qt	Saved as seed in qt	Kebele	Village	Asset Build
1	Aster Lea	F	Potato	2q	20 qt	15qt*220Birr	Y=3300Birr	3qt	2qt	Sheymaba	Takakacha	Cattle purchase, input purchase for belg(wheat, fertilizer)
2	Assefa Kussa	M	Potato	2qt	23qt	17 qt *235Birr	Y=3995	2qt	4qt	Sheymaba	Takakacha	Purchase of iron sheet for house construction
3	Gashau Abuche	M	Potato	2qt	22	18*250Birr	Y=4500Birr	1	3qt	Sheymaba	Takakacha	Heifer purchase
4	Ukumo Mana	M	Potato	3qt	30	20*240Birr	Y=4800Birr	8qt	2qt	Sheymaba	Takakacha	Ox purchase
5	Mulu Tantu	M	Potato	3	25	15*250Birr	Y=3150Birr	8	2	Sheymaba	Shoomolo	Bull purchase
6	Ayaanu Mamite/ Menta	F	Potato	3qt	22qt	15*210Birr	Y=3750Birr	6qt	1qt	Sheymaba	Shoomolo	Ox purchase
7	Mathwos Altaye	M	Potato	2qt	24	20qt*240Birr	Y=4800	2	2	Shemamba	Takatcha	Bull purchased
8	Gunushe Guffa	F	Potato	3qt	26	17qt*230Birr	Y=3910	4	2	Shemamba	Shomolo	Corrugated iron and window Purchase
9	Bancha Shito	M	Potato	3qt	20	11qt*230Birr	Y=2530	7	2	Shemamba	Shomolo	Purchase Heifer
10	Etagen Birhnu	F	Potato	3qt	25qt	15qt*230Birr	Y=3450	8	2	Shemamba	Takatcha	Purchased heifer
11	Churuko Borko	M	Potato	3qt	27qt	17qt*230Birr	Y=3910	7	3	Shemamba	Shomolo	Purchased ox
12	Abakao Anjulo	M	Potato	3qt	20qt	15qt*230Birr	Y=3450	3	2	Shemamba	Takatcha	
13	Adanech Tame	F	Potato	3qt	28qt	22qt*230Birr	Y=5060	4	2	Shemamba	Takatcha	
15	Mathwos Munea		Potato	3qt	22qt	17qt*230Birr	Y=3910	3	3	Shemamba	Gortchanco	Heifer purchase
16	Amanuel Anjulo		Potato	3qt	22qt	13qt*230Birr	Y=2990	6	3	Shemamba	Gortchancho	Bull purchase
17	Workinesh Nega		Potato	2qt	23qt	18qt *250Birr	Y=4500Birr	3	2	Shemamba	Takatacha	Bull purchase
18	Amanesh Mamo		Potato	2qt	23	20qt *230Birr	Y=4600Birr	3	3	Shemamba		Heifer purchase, Wheat seed and fertilizer purchase
19	Wondimu Shirko		Potato	2qt	16qt	12qt*230Bir	Y=2760Birr	2	2	Doge Anchucho		Calve Purhcase
20	Mengistu Utta		Potato	3qt	27qt	22qt*230Birr	Y=5060Birr			Doge Anchucho	Nazobo	Calve Purhcase and fettilzer and input purchase
21	Marta Godana	F	Potato	2qt	20qt	16qt*230	Y=3680Birr	2	2	Doge Anchucho		Bull Purhcase
22	Wogete Amona	F	Potato	2qt	21qt	18qt*230	Y=4140Birr	2	1	Doge Anchucho		Bull Purhcase and fettilzer and input purchase
23	Fekede Feleha	M	Potato	3qt	26qt	23qt*230	Y=5060Birr	2	1	Doge Anchucho		Heifer Purhcase and fettilzer and input purchase
24	Desta Ossa	M	Potato	3qt	26qt	24qt*230Birr	Y=5520Birr	1	1	Doge Anchucho	Nazibo	Heifer Purhcase and fettilzer and input purchase
25	Demisse Desta	M	Potato	3qt	27qt	25qt*230Birr	Y=5750Birr	1	1	Doge Anchucho	Nazibo	Heifer Purchase and fettilzer and input purchase
26	Asaye Anjulo	M	Potato	3qt	26qt	24qt*230Birr	Y=5520Birr	1	1	Doge Anchucho	Nazibo	Heifer Purchase and fettilzer and input purchase

Source: HH survey and Monitoring and evaluation report 2015/16.

Table 2: ORTDP potato technology success stories and in their project year of 2014/15 in Damot sore Woreda.

variable land resource owned was 3.11 that imply one unit increase in total land owned results in more than threefold increase in amount of income earned by small scale farmers, holding all the other explanatory variables constant. Since the farmers that acquainted with better land resource size in the district able to manage the resource in a better way, applies the agricultural package over the resource in better manner and consequently more yield and thereby better income than the farmers that owns lesser sized land. The result is not in line with the finding of Regassa [5] that stated income earned from potato negatively correlated with level total land owned.

Amount of potato produced

The parameter estimate for the variable termed quantity of potato produced at household level was 2.19. This indicates that the farming community that produced more potato earned more income in comparison to the household that produced lesser quantity of potato, holding other explanatory variables constant. When small scale farmers able to produce more of potato in quintals, their level of income earned increase parallel., in comparison to the household that produced less. Hence, through enabling resource poor farmers with provision of improved agricultural technology packages, it is possible to make the farmers produce more and thereby increase their earnings.

Project intervention period/year

The coefficients of parameter estimate regressed was -1.64. The regression result confirmed that potato technology intervention result decreased in the project period from 2014 to 2015 due to environmental factors (erratic rain fall). The extent of income earned

No.	Explanatory Variables	Coefficients				
		Unstandardized Coefficients		Standardized Coefficients	T	Sig.
		B	Std. Error	Beta		
1	(Constant)	32.81	9.45		3.473	.002*
2	Household category	14.04161	7.99.	0.195	1.757	.093***
3	TLU	6.04	1.78	0.445	3.389	.003*
4	Age	-4.936	12.345	-0.044	-0.4	0.693
5	family size	3.33	0.91	0.428	3.657	.001*
6	total land owned	3.11	0.81	0.462	3.855	.001*
7	quantity of potato consumed at home	2.19	0.81	0.334	2.709	.013**
8	Year of intervention	-1.64	0.47	-0.419	-3.476	0.002
9	amount of seed reserved	1.16	3.11	0.05	0.372	0.714
a. Dependent Variable: income earned from Selling of Improved potato						
Model Summary						
	Model	R	R Square	Adjusted R Square	Std. Error of the Estimate	
	1	.878a	0.771	0.688	1004.466	
a. Predictors: (Constant), amount of seed reserved , age, Household category, quantity of potato consumed at home, family size, total land owned, Year , TLU. The explanatory variables significantly affected income earned at 99%(*),95% (**) and 90% (***) significant level respectively.						

Table 3: Factors affecting level of income earned from improved potato selling in Damot sore Woreda, irish AID ORTDP project, Wolaita Zone.

by beneficiary farmers in 2014 project intervention year exceed 1.64 units in comparison 2015 period. For one unit increase in the year of intervention, the extent of income earned from improved potato dissemination decreased by 1.64 units. The study result indicated that favourable weather condition provokes the level of income earned from crop farming, while unfavourable weather condition worsens.

Conclusion and Recommendation

The Regression analysis result was used to identify the determinant factors of farm and crop income of small-scale farmers. The farm income regression result showed that independent variables such as landholding size, ownership of tropical livestock unit, being beneficiary farmer or not, family size, technology intervention period and amount of potato consumed were statistically significant variables that affected the farm income. This implies that a unit increase of total land owned, tropical livestock unit, and family size increases the farm income of the farmers.

The study results pointed out that through training and awareness creation, adoption improved agricultural technology packages and technical support, it is possible to increase the productivity of crop farming there by optimize the income of small scale farmers.

References

1. CSA-central statistical agency (2014) Report on area and production of major crops (private peasant holdings, meher season), p: 125.

2. Melesse KA (2016) Commercial behaviour of smallholder potato producers: the case of Kombolchaworeda, eastern part of Ethiopia. Economics of Agriculture 63: 159-173.

3. Esmael Y, Bekele A, Ketema M (2016) Determinants of level of smallholder farmers participation in potato sales in Kofele district, Oromia region, Ethiopia. Journal of Agricultural Sciences and Research 3: 23-30.

4. Urgessa T (2015) The Determinants of Agricultural Productivity and Rural Household income in Ethiopia. Ethiopian Journal of Economics, p: 123.

5. Regassa AE (2016) Income determinants of Irish potato (Solanum tuberosum L.) growers: The case of west Arsi Zone of Oromia Regional State. Net Journal of Agricultural Science 4: 1-8.

24

Estimation of Crop Coefficient and Water Requirement of Dutch Roses (Rosa hybrida) under Greenhouse and Open Field Conditions

Vikas Kumar Singh*, KN Tiwari and Santosh DT

Department of Agricultural and Food Engineering, Indian Institute of Technology, Kharagpur, India

Abstract

Precise estimation of reference evapotranspiration (ETo) and crop evapotranspiration (ETc) on a daily basis is important to apply water through drip system for crops grown in the greenhouse. Crop coefficients and crop water requirement were determined for the Dutch roses cultivated in the greenhouse and open field for the sub-humid climatic conditions of Kharagpur, India. Reference evapotranspiration (ETo) was estimated using the method suggested by FAO-56. The crop ET was determined using soil water balance approach. The soil moisture data was collected using TDR moisture meter at three depths. The maximum daily values of Crop ET were 4.99 and 5.28 mm day^{-1} for greenhouse and open field conditions respectively. Maximum values of crop coefficient were found during conjunction period of mid-season and late season stage of the crop growth period. During different growth stages of rose, crop coefficient values were found in the range from 0.48 to 0.96 and 0.59 to 1.01 for greenhouse and open field conditions. The Dutch rose planted in 200 μ diffused poly film cladded greenhouse and given 100% water requirement through drip irrigation resulted in maximum number of flowers/m² (212.3) annually. Total annual water requirement of rose plant was 999.51 mm and 1210.94 mm for greenhouse and open field condition respectively.

Keywords: Dutch rose; Crop coefficient; Crop evapotranspiration; Greenhouse

Introduction

Rose is a leading cut flower grown commercially all over the world. It ranks first in global cut flower trade. This flower has a worldwide consumption of more than 40 billion [1]. The various purposes for rose cultivation includes garden flowers, aesthetic value, decoration and preparation of various products such as rose oil, rose water, gulkhand rose attar, garland etc. The heavy demand of rose cut flowers in the European markets is mainly from November to March due to the shortage of local production because of severe winter. Fortunately, this is the most congenial condition for successful production of most of the cut flowers, including roses in India. It is pointed out that buyer at international market prefers a very high quality rose cut flowers. As it is difficult to obtain good quality cut flowers under open conditions throughout the year, the crops should be cultivated under the greenhouse to get good quality produce.

Greenhouse cultivation or protected cultivation is a kind of farming systems used to maintain a controlled or partially controlled environment suitable for maximum crop production. This includes creating an environment suitable for working efficiency as well as for better crop growth. The presence of a cover, characteristics of greenhouses, causes a change in the climatic conditions compares to those in open field. The property of cladding materials alters the magnitude and quality of solar radiation incidence in the greenhouse, thereby affecting the micro climate of greenhouse. UV stabilized diffused film does not allow the shadow formation of top canopy on the lower leaves. Diffused radiation penetrates deeper into plants canopy in comparison to direct radiation; thus, it is desirable to use diffused film. At high irradiation, diffused film greenhouse cover leads to better light distribution, lower plant temperature, decreased transpiration and increased photosynthesis and growth [2].

Drip irrigation has been proved as an effective water-saving irrigation method and it is important component of greenhouse cultivation system. Drip irrigation research studies carried out at Precision Farming Development Centre, IIT Kharagpur, India on several vegetable and fruit crops showed increase in yield, saving in water, higher water use efficiency and net increase in profit [3,4].

Crop evapotranspiration (ET) is required for determination of crop water requirements, irrigation scheduling and water productivity. The water balance method is a simple and easy to use as compared to several methods reported in the literature. ET is estimated using water balance method considering change in total soil water content between sampling dates plus rainfall minus any known drainage or surface runoff occurred during the period [5]. Time Domain Reflectometry (TDR) probes and capacitance-based sensors are popular devices for measuring inside soil water content directly. Advantages of these electronic sensors are in-situ measurements with continuous recording [6].

Crop coefficient is a significant parameter for estimation of crop evapotranspiration, because it accounts biological characteristics of crops, crop condition, soil texture, soil tillage conditions, crop growing environment etc. [7]. The crop coefficient Kc, is ratio of crop ET to reference ET, is needed to estimate crop evapotranspiration for irrigation planning at a regional scale. The crop coefficient value represents crop specific water use and is required for accurate estimation of irrigation requirement of single or more than one crop grown under different climatic conditions [8].

Several studies on estimation of water requirement and crop coefficient have been conducted for open field crops like Cotton (Abdelhadi et al. [9] Fisher [10]), Maize (Akinmutimi, [11] Chuanyan and Zhongren [12]), Onion and Spinach (Piccinni et al. [13]) Okra (Odofin et al. [14] Tiwari et al. [3]), Capsicum (Miranda et al. [15]), Cabbage Tiwari et al. [4] Paddy (Kuo et al. [16]), Wheat and Sorghum

*Corresponding author: VK Singh, Department of Agricultural and Food Engineering, Indian Institute of Technology, Kharagpur, India
E-mail: vikas@agfe.iitkgp.ernet.in

Tyagi et al. [17] Teff (Araya et al. [18]), Sapota (Tiwari et al. [4]), Jujube (Hu et al. [19]), Pomegranate (Parvizi et al. [20]) and greenhouse crops like Tomato (Maldonado et al. [21] Gómez et al. [22], Wahb-Allah et al. [23] Harmanto et al. [24]), Eggplant Senyigit et al. [25] Melon, Green beans, Watermelon and Pepper (Orgaz et al. [26]), Gladiolus (Bastug et al. [27]), Cucumber (Zhang et al. [28] and Blanco and Folegatti [29]) However, limited studies are found on cultivation of rose crop under greenhouse conditions (Ehret et al. [30] Katsoulas et al. [31]) and no literature is reported for the estimation of water requirement and crop coefficient of Dutch roses under greenhouse conditions. Therefore, the present study have been undertaken to determine the crop coefficient and water requirement of Dutch rose under greenhouse and open field conditions.

Materials and Methods

Description of the experimental area

Experiment was conducted at the Field Water Management Laboratory of Agricultural and Food Engineering Department, Indian Institute of Technology, Kharagpur, India. The experimental site is located on flat land at 22°18.5' N latitude, 87°19' E longitude and 48 m altitude above mean sea level. The local climate is sub-humid subtropical with an average annual rainfall of 1390 mm, of which about 80% is received during June to October. The mean monthly minimum temperature is 12°C in January, whereas the mean monthly maximum temperature is 41.5°C in May. The mean monthly relative humidity varies from 35% in February to 96% in July-August. The site was deep and well drained sandy loam soil. The average values of sand, silt and clay was 63%, 28% and 9% respectively in the soil depth of 1 m.

Experimental greenhouses

Two experimental sawtooth shape greenhouses (N-S oriented) one cladded with 200 μ diffused (PAR transmissivity as 90% and 42% diffusivity) film and another with 200 μ clear UV stabilized film installed at the Field Water Management Laboratory of Agricultural and Food Engineering Department, Indian Institute of Technology Kharagpur, India. The total ground area of each greenhouse was 84 m² with central height of 4.5 m and gutter height of 4 m (Figure 1). Greenhouses had provision to vary ventilation area maximum upto 60% (60% of the floor area) to allow hot air to escape during peak summer. Ventilation was provided at ridge and both sides of the greenhouses. These ventilated openings were covered with an insect-proof net of 60 mesh size to prevent the entry of insects. These greenhouses were also equipped with fogging system (16 L h⁻¹ discharge capacity) and shade net (75% shading intensity) beneath the greenhouse roof. Two exhaust fans with a capacity of 1100 m³/min with 2.2 kW power were installed to replace hot air with ambient air in the greenhouse. Natural ventilation was maintained in both greenhouses, however fans were operated only when the air temperature in the greenhouse was high during peak summer.

The diffused film used for cladding of one of the greenhouses has inbuilt property to scatter the radiation and does not form shadow on the top of the canopy of lower leaves.

Monitoring of greenhouse microclimate: Automatic weather station of M/S Campbell Scientific, Canada comprising a data-logger (model CR1000) and sensors were installed in the greenhouse to monitor soil temperatures (models 107 BL and CS616 L), air temperatures and relative humidity (model HMP 45 C), global radiation and Photosynthetically Active Radiation (SPLITE and PARLITE of Kipp and Zonen). Outside air and soil temperatures, relative humidity and solar radiation were measured manually at 8:30 AM, 12:30 PM and 4:00 PM in a day.

Reference evapotranspiration, ET$_O$

Reference evapotranspiration (ET$_O$) was estimated using the method suggested by FAO-56 [2]. The equation used to estimate ET$_O$ is described below [5]:

$$ET_o = \frac{0.408\Delta\left(R_n - G\right) + \gamma\dfrac{900}{T+273}u_2(e_s - e_a)}{\Delta + \gamma(1 + 0.34u_2)} \tag{1}$$

Where,

ET$_O$ - Reference evapotranspiration [mm day⁻¹], R$_n$ - Net radiation at the crop surface [MJ m⁻² day⁻¹], G - Soil heat flux density [MJ m⁻² day⁻¹], T - Mean daily air temperature at 2 m height [°C], u$_2$ - Wind speed at 2 m height [m s⁻¹], e$_s$ - Saturation vapour pressure [kPa], e$_a$ - Actual vapour pressure [kPa], e$_s$-e$_a$ - Saturation vapour pressure deficit [kPa], Δ - Slope vapour pressure curve [kPa °C⁻¹], γ - Psychrometric constant [kPa °C⁻¹].

In a greenhouse, wind speed approximately equals to zero; therefore, the Eq. (3) can be simplified written as [28]:

$$ET_o = \frac{0.408\Delta\left(R_n - G\right)}{\Delta + \gamma} \tag{2}$$

Crop details

Dutch rose plants (Variety First Red) were transplanted at a spacing of 0.5 m between rows and 0.3 m between plants in a row in both the greenhouses and also in open field on October, 2009. Rose nursery plants were planted on raised bed of 1 m wide with two rows per bed. A trench of 30 cm wide and 50 cm deep was dug in between two beds for removal of excess moisture from root zone and separation of one treatment from other. To improve the soil organic matter and nutrient (NPK) contents, manures and fertilizers were applied 1 month prior to the planting. The soils of beds were incorporated with 15 kg m⁻² of Farm Yard Manure, 5 kg m⁻² of vermi compost, 0.5 kg m⁻² Neem cake and 0.2 kg m⁻² of single superphosphate. The crop was drip-irrigated using one lateral per bed of two rows and one emitter of

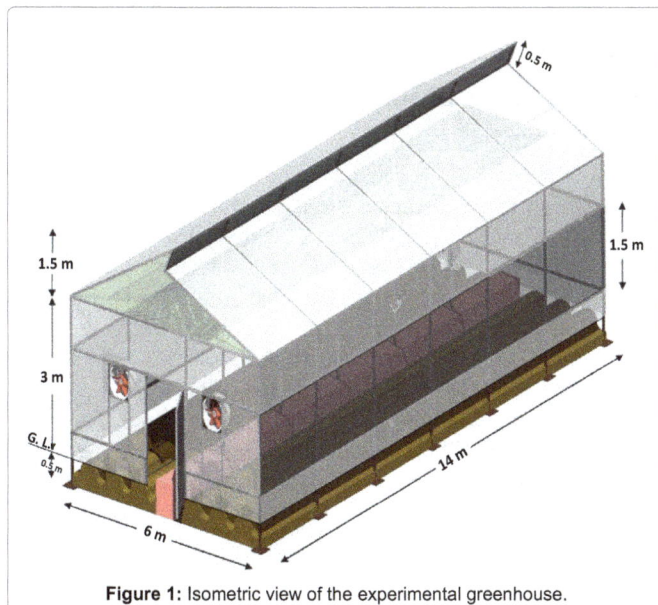

Figure 1: Isometric view of the experimental greenhouse.

2 L h^{-1} capacity, which met the water requirement of 4 plants of both rows. The drip irrigation system's emission uniformity was determined and maintained above 90%. Recommended dose of soluble fertilizers were applied twice in a week using a venturi injector. After 180 days of planting, plants were pruned to a height of 0.3 m above the ground. There after pruning was done once in a year to increase the number of branches for good flowering.

Measurement of crop evapotranspiration, ET$_C$

The actual crop evapotranspiration was estimated with the soil water balance method. The soil moisture changes in 0–90 cm soil layer during the study period were measured and used to estimate actual crop evapotranspiration with equation (1) [5]. Three circular shape drums having closed bottom, each having 0.5 m^2 cross sectional area, 1 m deep and 5 mm thick wall (Figure 2) were installed in the greenhouse. Three drums of the same size and specifications were installed outside in open field. Considering the rose plants spacing and the drum dimensions, one rose plant was planted in the each drum installed inside and outside the greenhouse. The packages of practices for the plant inside and outside the drum were maintained identical. Rose plant planted in the drum lysimeter installed outside the greenhouse is mulched with plastic film to prevent the entry of rainwater in the lysimeter during rainy season. Soil water content was measured daily using TDR moisture meter at 0-30, 30-60 and 60-90 cm depth. The root zone depth of rose plant was maximum up to 45-50 cm. The soil moisture measured at 60-90 cm was considered as vertical drainage and deducted from amount of irrigation water applied to estimate crop evapotranspiration. Appropriate amount of irrigation water was applied manually to the plant in lysimeters to replenish depleted soil moisture at field capacity. Soil-water depleted was determined by the TDR moisture meter readings and replenished by a factor of 1.1 on the same day, to meet the crop evapotranspiration in the next following day [15]. Amount of irrigation water varied according to crop water use in different growth stages. These measurements and irrigation water supply were done for the full rose crop season. Crop evapotranspiration was calculated using the change in soil moisture content and measurement of other water

balance parameters from following water balance equation

$$ET_C = P + I - R - D \pm \Delta W \qquad (3)$$

Where,

ET$_C$: Crop evapotranspiration (mm); P: Precipitation (mm); I: Irrigation water depth (mm); R is the surface runoff (mm); D: Amount of water drained from the root zone (mm); ΔW: change in soil water storage (mm). Contribution of water due to rain in the greenhouses was zero, hence, P = 0. The contribution of lateral flow and surface runoff during irrigation events was also zero, hence R = 0.

Thus Eq. (1) can be simplified as follows

$$ET_C = I - D \pm \Delta W \qquad (4)$$

The change in the amount of soil-water into the drum (ΔW) is computed by the difference between soil moisture observations of two consecutive days.

Establishment of crop coefficient (K$_C$) of Dutch roses for greenhouse and open field

The crop coefficient values for Dutch roses were determined for the crop grown both in greenhouse and open field conditions. The average crop coefficients for Dutch roses are estimated with an assumption that one annual cycle of crop is started after planting of crop in October and completed in next September month when pruning is done. The next annual cycle is started just after pruning and completed in next September month when pruning is done. Crop coefficient was computed on a daily basis from transplanting to pruning and pruning to pruning in next year using following equation:

$$Kc = \frac{ET_C}{ET_O} \qquad (5)$$

Average crop coefficients were calculated for the initial (Kc$_{initial}$), development (Kc$_{dev}$), mid-season (Kc$_{mid}$) and late-season (Kc$_{late}$) growth stages. The lengths of the growth stages were determined based on the crop phenological phase and the Kc curve. Kc$_{initial}$ begins from transplanting to plant covers 10% of the plant area. The Kc$_{Dev}$ begins after Kc$_{initial}$ (i.e., 10% canopy to 60% of canopy area). The Kc$_{mid}$ begins after Kc$_{Dev}$ ends (i.e., canopy attains 60% and ends in the later part of the season when the crop begins to enter in the end of the season) and the Kc$_{late}$ starts when crops has more than 60% canopy and end when crop attains dormant stage.

Water requirement of rose plant under greenhouse and open field conditions

To optimize the water requirement of rose plants grown under two saw tooth shape greenhouses cladded with UV stabilized clear and diffused film of 200 µ thickness, an experiment was conducted for three years (January 2012 to December 2014) with four irrigation levels of 100%, 80%, 60% and 40% of estimated crop evapotranspiration (ETc) application. The experiment was designed to assess the effect of irrigation levels and greenhouse cladding materials on biometric growth and yield (No. of flowers) of rose plant. Eight treatments replicated three times were planned for the greenhouses and one treatment for open field as a control to compare the results. Statistical analysis was done for analysis of variance (ANOVA) using the Least Significant Differences (LSD) test at the 5% probability level (P = 0.05).

Treatment details are stated below:

T1 : 100% water requirement supplied through drip system to rose

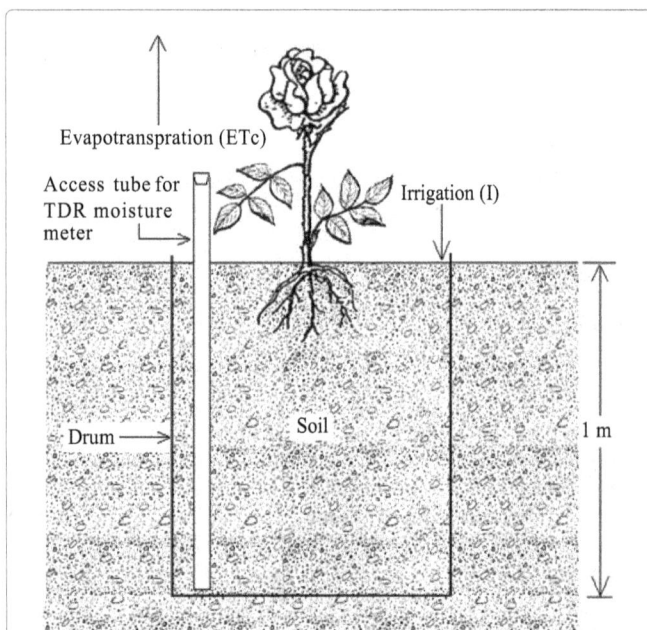

Figure 2: Cross sectional view of drum lysimeter used for determination of ET$_C$ of rose.

plants under greenhouse cladded with clear film

T2: 80% water requirement supplied through drip system to rose plants under greenhouse cladded with clear film

T3: 60% water requirement supplied through drip system to rose plants under greenhouse cladded with clear film

T4: 40% water requirement supplied through drip system to rose plants under greenhouse cladded with clear film

T5: 100% water requirement supplied through drip system to rose plants under greenhouse cladded with diffused film

T6: 80% water requirement supplied through drip system to rose plants under greenhouse cladded with diffused film

T7: 60% water requirement supplied through drip system to rose plants under greenhouse cladded with diffused film

T8: 40% water requirement supplied through drip system to rose plants under greenhouse cladded with diffused film

T9: 100% water requirement supplied through drip system to rose plants in open field conditions (Control)

Drip irrigation was installed to irrigate plants inside the greenhouse and outside (open field). The time of operation of drip system was computed using daily water requirement of four plants and emitter discharge for each irrigation. In drip system, irrigation was scheduled on alternate days and the quantity of total water applied was calculated on the basis of daily crop water requirement. The lateral lines of 16 mm diameter were laid parallel to plant rows. Online emitters of 2 L h^{-1} capacity were fitted with lateral and each emitter served the water requirement of four plants. Irrigation duration for delivery of water for different treatments was controlled with the help of gate valve provided at the inlet of each laterals. The operating pressure of about 1 kg/cm^2 was maintained to obtain design dripper discharge. The layout of the experiment and the division of beds along with the laterals are shown in Figure 3.

Bio-metric observations and water productivity: Bio-metric observations viz. plant height, number of shoots per plant, shoot length, number of flowers/m^2/year, flower diameter and number of petals per flower were recorded at 15 days interval to evaluate the effect of irrigation levels and cladding materials on yield and yield attributes of Dutch roses inside the greenhouse and in open field (outside).

The annual water productivity (number of flowers/m^2/mm) for greenhouse and open field treatments were calculated using the following equation [5]:

$$AWP = \frac{Y}{AWU} \tag{6}$$

Where,

Y is total flower yield (number of flowers/m^2) and AWU is annual water use (mm)

Results and Discussion

Estimation of ETo for inside the greenhouse and open field

The meteorological observations collected from automatic weather stations were substituted in FAO Penman-Monteith equation to calculate daily ETo for fifteen years (January 2000 to December 2014) for inside the greenhouse and outside (open field), and the results are shown in Table 1 and Figure 3. It can be seen from Figure 3 that average

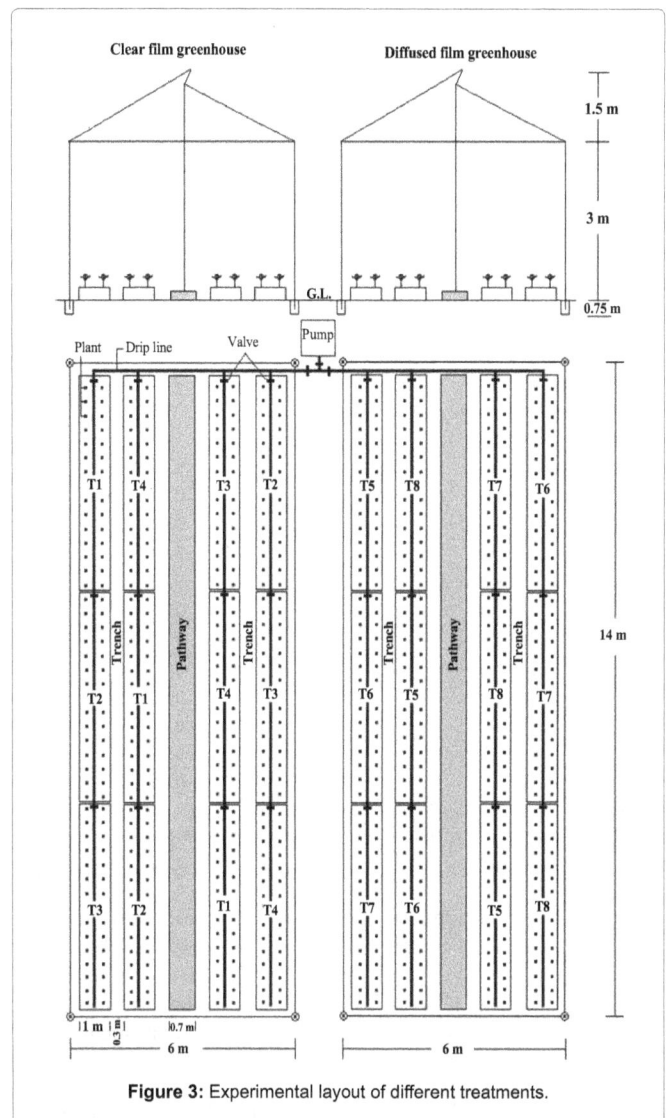

Figure 3: Experimental layout of different treatments.

ETo value inside the greenhouse and outside (open field) are relatively of greater amount from late March to October months as compared to January to early March and November – December. In April, the average daily ETo value of month is below 4 mm except a few high values occurred at the end of this month. With the increase of sunshine hours and the intensity of radiation, ETo value gradually increases. Peak value of the ETo (i.e., 5.39 mm) was observed in the month of May. June months onward average daily ETo values was reduced due to incidence of rainfall, low solar radiation, and high humidity especially during monsoon months (June to September). Further due to lowering of temperature from October months onward till March, daily ETo values reduced and reached to lowest in the month of January. ETo values get affected due to cloud cover and rains, hence fluctuating ETo values were found in the same months.

FAO Penman-Monteith equation contains two parts: one radiation and other aerodynamic components and they represent the impact of solar radiation and convection, turbulence and degree of drying above the evaporating surface respectively [32,33]. Both of these two components are dynamic processes vary with time with influenced by location and climatic conditions. Analysis of data shows that the part of solar radiation component is more than 60% of total evapotranspiration

Month	Inside/ outside condition	Reference evapo-transpiration, ETo (mm day^{-1})	Kc	Actual evapo-transpiration*, ETc (mm day^{-1})	Area occupied by four plants (4×0.5×0.3) (m²)	Required discharge through single emitter (L day^{-1})	Time of operation of drip system per day with 2 Lh^{-1} emitters (min)
		A	B	C=A×B	D	E=C×D	
Jan	Greenhouse	1.67	0.86	1.43	0.6	0.86	25.9
	Open field	2.40	0.89	2.14	0.6	1.28	38.5
Feb	Greenhouse	2.05	0.87	1.78	0.6	1.07	32.1
	Open field	3.12	0.90	2.81	0.6	1.69	50.6
Mar	Greenhouse	2.81	0.89	2.51	0.6	1.50	45.0
	Open field	4.18	0.93	3.89	0.6	2.33	70.0
Apr	Greenhouse	3.88	0.92	3.57	0.6	2.14	64.3
	Open field	4.99	0.96	4.79	0.6	2.87	86.2
May	Greenhouse	5.20	0.96	4.99	0.6	3.00	89.9
	Open field	5.39	0.98	5.28	0.6	3.11	93.4
Jun	Greenhouse	4.41	0.96	4.23	0.6	2.54	76.2
	Open field	4.59	1.01	4.63	0.6	2.78	83.4
Jul	Greenhouse	3.84	0.89	3.42	0.6	2.05	61.5
	Open field	4.05	0.97	3.93	0.6	2.36	70.7
Aug	Greenhouse	3.89	0.82	3.19	0.6	1.91	57.4
	Open field	4.11	0.84	3.45	0.6	2.07	62.1
Sept	Greenhouse	3.74	0.76	2.84	0.6	1.71	51.2
	Open field	4.02	0.78	3.14	0.6	1.88	56.4
Oct	Greenhouse	3.67	0.48	1.76	0.6	1.06	31.7
	Open field	3.96	0.52	2.06	0.6	1.24	37.1
Nov	Greenhouse	2.70	0.60	1.62	0.6	0.97	29.2
	Open field	2.96	0.63	1.87	0.6	1.12	33.6
Dec	Greenhouse	1.93	0.76	1.47	0.6	0.88	26.4
	Open field	2.34	0.81	1.90	0.6	1.14	34.1

* This is 100% of crop evapotranspiration estimated for Dutch rose crop with drum lysimeter in this study (treatments T1 & T5). ETc for treatment T2 & T6 is 80%, T3 & T7 is 60% and treatment T4 & T8 is 40% of this estimated value of ETc.

Table 1: Estimated water requirement and crop coefficient of Dutch rose crop and time of operation of drip system for different months.

during study period, and it is maximum in May. From May month to August month, the solar radiation decreases from the 23.5 MJ/m²/day in May to 18.5 MJ/m²/day in August, during which the value of each month are 23.5, 19.0, 17.4 and 18.5 MJ/m²/day during May, June, July and August respectively. This is due to decrease in solar intensity and day length during monsoon months.

Comparison of monthly ET_O values of greenhouse and open field conditions shows that ET_O of greenhouse is always lower and this is due to the reduced evaporative demand inside the greenhouse (Figure 3). The evaporative demand is lower inside the greenhouse than outside (open field) due to the decrease in solar radiation (20% on average) and the wind speed is nearly zero. The evaporative demand inside the greenhouse is 60 percent of that of outside and the same finding is reported by Möller and Assouline [34] Fernández et al. [35]. They used lime suspension on greenhouse cover surface to cut down the entry of solar radiation to reduce air temperature that caused ETo reduction. During May to August there is very less difference between ETo of inside and outside of greenhouse, is because of high humidity with minor difference in dry bulb and wet bulb temperature of prevailing micro climate.

Soil water content at different depth in drum lysimeter

Soil moisture content at three depths (0-30, 30-60 and 60-90 cm) of drum lysimeters kept inside and outside of greenhouse was measured daily with the help of TDR moisture meter. Analysis of the soil moisture data collected from TDR moisture meter shows that moisture content in the middle layer (30-60 cm) was found maximum (15.4-9.1%) in comparison to top layer (14.6-18%) and bottom layer (14-16.1%). The top layer has high evaporation due to greater soil temperature hence

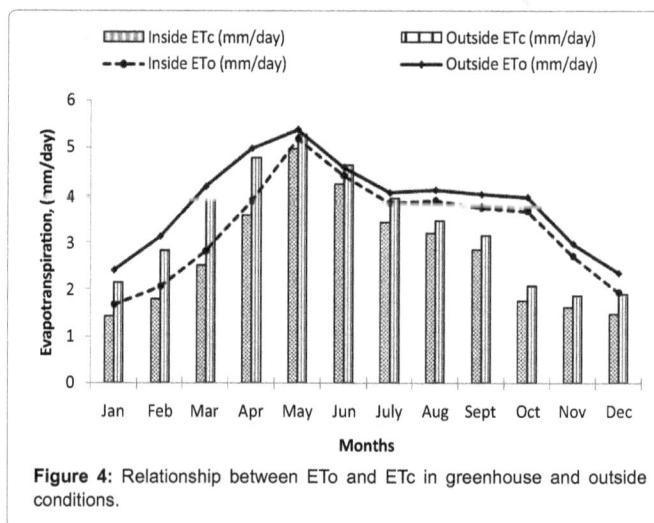

Figure 4: Relationship between ETo and ETc in greenhouse and outside conditions.

lesser soil moisture content than middle layer. The top soil (up to 50 cm) of lysimeters had rich organic matter which might have increased in water holding capacity of soil, hence moisture content at bottom layer was found to retain lesser than top and middle layers (Figure 4). Similar trend of soil moisture content was observed in all the three layers of soils of lysimeters kept outside the greenhouse (open field). Within the whole growth period, the change in values of soil water content at different layers reflects crop evapotranspiration demand. In the drip irrigation water supply, the evaporation of soil water is lesser than the surface irrigation and changes of soil water content indirectly reflect characteristics of the crop water consumption [19].

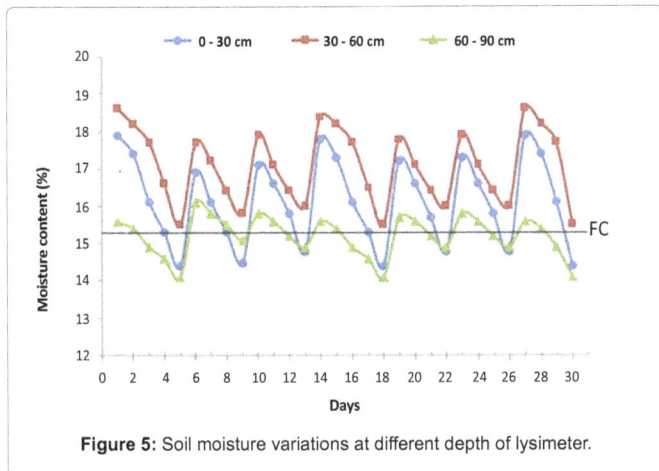

Figure 5: Soil moisture variations at different depth of lysimeter.

Crop evapotranspiration (ETc) and crop coefficient (Kc)

The evapotranspiration (ETc) of the rose crop was determined daily from the drum lysimeters kept inside and outside the greenhouse for the full crop season is presented in Table 1 and Figure 5. Daily Rose plant ET_C values varied from 1.43 to 4.99 mm day^{-1} and 1.86 to 5.28 mm day^{-1} for greenhouse and outside conditions respectively. Evapotranspiration (ETc) requirement of the plant in drum lysimeters in greenhouse is lower than the plant of drum lysimeters kept outside (open field) due to the reduced evaporative demand inside the greenhouse [34,35]. Maximum ET_C values occurred in the month of May when the crop is fully matured with greater level of ground cover [15]. ETc values reduced considerably and reached to 3.5 mm day^{-1} which coincided during the flower harvesting period. Then, during the second flowering peak, ETc values increased again rapidly, but did not reach the same level as observed during the first flowering peak.

Daily crop coefficient value was estimated for three years (October 2009 to October 2012) of experimental period and averaged on monthly basis. The daily average value of crop coefficient for one complete cycle under greenhouse and open field conditions are presented in Table 1 and Figure 6. The crop coefficient for greenhouse conditions is 0.48 to 0.6 during initial stage ($K_{C In}$), 0.6 to 0.86 during development stage ($K_{C Dev}$), 0.87 to 0.96 during middle stage ($K_{C Mid}$) and 0.96 to 0.76 during late season ($K_{C Late}$). In the same way the crop coefficient for open field conditions was estimated and found to vary from 0.50 to 0.63 during initial stage ($K_{C In}$), 0.63 to 0.89 during development stage ($K_{C Dev}$), 0.90 to 1.01 during middle stage ($K_{C Mid}$) and 0.97 to 0.78 during late season ($K_{C Late}$).

During different growth stages of rose, crop coefficient value varied from 0.48 to 0.96 and 0.59 to 1.01 for the plant under greenhouse and open field conditions. During the initial stage (30 days) of the plant crop coefficient is found lower value because of lesser ground cover and low crop water requirement whereas during the development stage (90 days) it is more than earlier [29]. During the mid-season (120 days) stage crop coefficient is found maximum for both greenhouse and open field conditions.

Crop coefficient values decreased steadily after the first harvest, reaching minimum values close to 0.75. After the pruning is done, a new cycle of vegetative growth and flowering began, that increased Kc values again. Finally, during the second harvest period Kc values decreased again. The similar trend of crop coefficient under both the conditions is found during all three years of experimental period. The behaviour of the crop coefficient curve throughout the Dutch rose growth period,

with a short period of low Kc values kept between two phases of high Kc values, matched with the crop phonological behaviour observed in the lysimeter. Although the Dutch rose produced incessantly after the first harvest, it presented basically two phases when flowering was more intense.

The rose plant present more than one harvest cycle, it is important to adjust the Kc value accordingly, increasing the Kc during periods of full ground cover and decreasing the Kc when pruning is done. Since there are no published Kc values specific for Dutch rose for both the conditions, it was assumed that for irrigation management purposes one would use Kc values recommended in this study.

Micro climate behavior of greenhouse cladded with UV stabilized clear and diffused film

During summer season (May to September), the average daily variation of temperatures was found to vary between 35°C to 44°C and 34°C to 42°C in the UV stabilized clear film and diffused film respectively and during winter season (November to February), average daily air temperature in the UV stabilized clear and diffused films at 12:30 PM varied from 25°C to 31°C and 23°C to 29°C respectively against open field condition (22 to 28°C). The maximum temperature reduction in winter at 12:30 PM in the diffused film was 3°C as compared to the clear film.

The study of temperature variation in the clear and diffused covers indicated that the UV stabilized clear film maintained 3°C higher temperature than diffused film during winter and 2°C higher temperature than diffused film during summer months. Temperature reduction in the UV stabilized diffused film is more prominent than in the UV stabilized clear film due to the diffusive property. UV stabilized clear film maintained 10-15% lesser relative humidity than the diffused film whereas outside relative humidity is even lesser than diffused film. Weekly variation of solar radiation levels in the greenhouse at 12:30 PM during winter season varied from 32 to 56 Wm^{-2} and 38 to 67 W m^{-2} in the clear and diffused film respectively against the ambient solar radiation of 74 to129 W m^{-2}. Whereas in summer months, it varied from 34 to 61 W m^{-2} and 41 to 82 W m^{-2} in the clear and diffused covers, respectively against the ambient solar radiation of 86 to 170 Wm^{-2}. UV stabilized diffused film passes 20-34% more solar radiation than clear film. This solar radiation is directly related to the Photo synthetically Active Radiation (PAR) which is used in the photosynthesis of crop, hence better crop growth is found under diffused film cladded

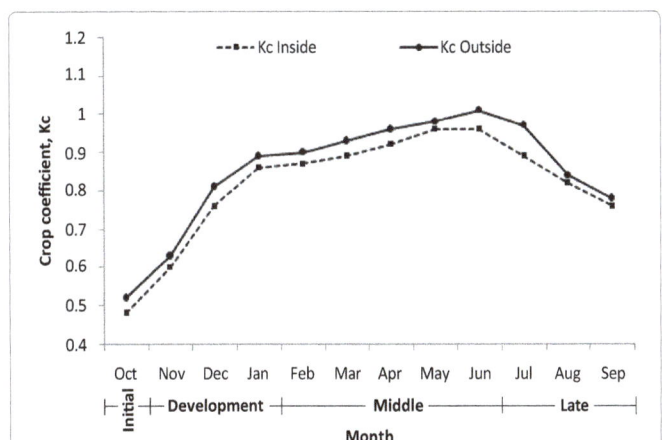

Figure 6: Crop coefficient for Dutch roses under greenhouse and open field conditions.

greenhouse. It is clear from other studies that Ultraviolet (UV), Photosynthetic Active Radiation (PAR) and Near-Infrared (NIR) is the part of solar radiation and which transmit through the greenhouse covering material. The PAR is essential for photosynthesis which is favourable to plants growth. Thus, greenhouse cladding films which diffuse the incoming solar radiation can offer several advantages and provide better micro climate (Chou and Lee [36], Kondratyev [37], Lamnatou and Chemisana [38,39], Espí et al. [40]). Soil temperature in the greenhouses cladded with clear and diffused film at 25 cm depth during winter season ranged from 19 to 21°C and 18 to 21°C respectively. During summer season, it varied from 25 to 29°C and 24 to 28°C in clear and diffused film cladded greenhouse. It was observed that no significant difference in the soil temperature was found in the greenhouses cladded with clear and diffused film.

Optimum water requirement for Dutch roses under two greenhouses of different cladding materials

To optimize the water requirement of the Rose crop under two different types of cladded film greenhouse (clear and diffused UV stabilized film of 200 µ thickness), the treatment wise biometric observations were recorded and analyzed for different irrigation levels.

The analysis of biometric data revealed that Dutch rose cultivation with 100% water requirement for the greenhouse cladded with diffused film (T_5) resulted in best performance than control plots in terms of plant height (68.8 cm), shoot length (46.2 cm), flower diameter (6.97 cm), number of petals per flower (25.1), number of shoots per plant (29.3) and number of flowers/m²/year (212.3) (Figure 7).

Time of operation of the drip irrigation system with 2 Lh^{-1} emitters for different months of growth period of rose crop planted inside and outside of greenhouse was also estimated and results are presented in Table 1. The maximum time of operation of drip system to irrigate rose plants was 89.9 minutes and 93.4 minutes required in May month under greenhouse cladded with diffused film and open field conditions respectively however, minimum time of operation is needed in January month i.e., 25.9 minutes and 38.5 minutes for inside and open field conditions respectively.

Effect of different levels of irrigation on biometric response of Dutch roses

The effect of different levels of irrigation on biometric parameters such as plant height, number of shoots per plant, shoot length, number of flowers/m²/year, flower diameter and number of petals per flower were analyzed statistically and compared with that of open field experimental treatment (T_9). The experimental results of these biometric observations for all the three years are presented in Figure 7. The analysis of variance results showed that variation among the replications for all the treatments was found to be statistically insignificant at 5% level of significance.

The plant height responded significantly with different irrigation levels of drip irrigation, however, plant height at 100% irrigation level (T_1 and T_5) was statistically at par with 80% irrigation level (T_2 and T_6). Similar trend was found for all the 3 years. The plant height was

found to be highest in the year 2012 as compared to other 2 years data. Irrigation levels had no significant influence on flower diameter and number of petals per flower.

Significant increase in the number of shoots per plant was obtained due to drip irrigation treatment at all the levels of irrigation. Among various levels of water supply 100% irrigation requirement met through drip resulted in best over 60% and 40% irrigation levels in terms of number of shoots per plant and shoot length. However, biometric response at 100% and 80% of water requirement supply through drip was statistically at par. The maximum shoot length and number of flowers/m²/year were recorded for the year 2012 in comparison to other two years (2011 and 2013). It can be seen that the 100% irrigation supply has significant influence on number of flowers/m²/year over two other irrigation levels that is 60% and 40% and 100% irrigation supply in open field treatment.

In all the three years the number of shoots per plant and number of flowers/m²/year at 100%, 80% and 60% irrigation levels was significantly greater than that of open field conditions. Based on the average yield (number of flowers/m²/year) of three years, all the irrigation levels at 100%, 80% and 60% (i.e., T_5, T_6 and T_7 respectively) resulted in 55.8%, 52.9% and 29.4% higher yield respectively as compared to open field treatment (T_9). Hence, it showed that even by 60% deficit water supply through drip irrigation resulted in 29.4% higher flower yield of Dutch roses as compared to open field cultivation (T_9). The flower yield (presented in Table 2) was found to decrease as the amount of irrigation water supply was reduced from 100 % to 40 % of irrigation water requirement.

Effect of cladding materials on biometric response of Dutch roses

The plant growth and flower yield were greater in the greenhouse cladded with diffused film as compared to clear film cladded greenhouse. Among the various treatments under both the greenhouses treatment T_5 i.e., 100% irrigation water level in greenhouse cladded with diffused film responded the highest plant height in all the years considered in this study. There was significant influence of irrigation levels and

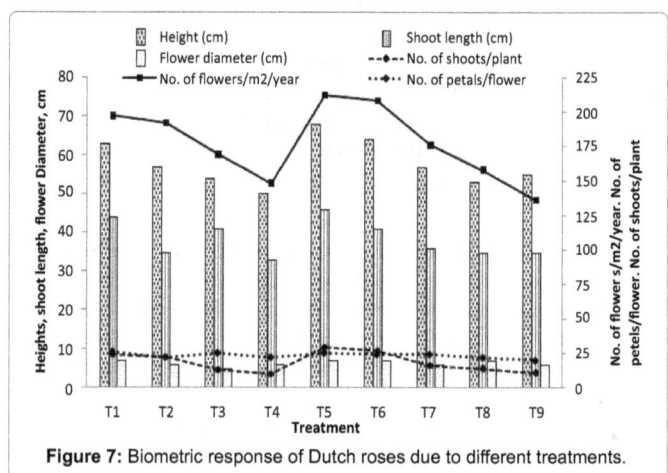

Figure 7: Biometric response of Dutch roses due to different treatments.

Treatment	T_1	T_2	T_3	T_4	T_5	T_6	T_7	T_8	T_9
Water applied (mm)	999.5	799.6	599.7	399.8	999.5	799.6	599.7	399.8	1210.9
Yield (No. of flowers/m²/year)	197	192	169	148	212	208	176	158	136
Annual water productivity (No. of flowers/m²/mm)	0.20	0.24	0.28	0.37	0.21	0.26	0.29	0.40	0.11

Table 2: Amount of water applied and yield response of Dutch roses for different amount of water applications under different treatments.

cladding material on plant height, number of shoots per plant, shoot length and number of flowers/m²/year. However, there is no significant influence of irrigation level and cladding material on flower diameter and number of petals per flower. Plant height and shoot length were found to be highest in case of greenhouse cladded with diffused film as compared to greenhouse cladded with clear film at the same level of irrigation water supply.

Analysis of biometric observations revealed that flower yield of Dutch roses (number of flowers/m²/year) was recorded greater in all the 3 years for the crop grown in the greenhouse cladded with diffused film at all the levels of irrigation. Maximum number of flowers/m²/year was found as 212 in treatment T_5 (i.e., 100% irrigation water level under greenhouse cladded with diffused film). With the same level of irrigation water application in both the cladded materials, the flower yield was always greater for greenhouse cladded with diffused film. This may be due to fact that the diffused film has inbuilt property to disperse solar radiation which penetrate into the crop canopy that increases photosynthesis process of the crop. Diffused film does not allow the radiations to interact directly to the plant leaves hence protect plats from scorching. Moreover, the plants in the greenhouse covered with diffused film were healthy compared to the clear film greenhouse during summer season. Lamnatou and Chemisana also observed the same results that diffused light have the ability to penetrate deeper into a plant canopy in comparison to the direct light [38,39]. The seasonal variation of solar radiation, affect variations in the crop response. The temperature in diffused film greenhouse was 2-3°C lower than clear film cladded greenhouse. Higher temperature prevailing in Kharagpur sub-humid climatic condition under greenhouse cladded with clear film resulted in inferior biometric response and reduced flower yield.

Water productivity

The annual water productivity (Number of flowers/m²/mm) values (AWP) ranged from 0.2 to 0.37 and 0.21 to 0.40 for greenhouse cladded with clear and diffused film, however it is 0.11 for open field conditions (Table 2). The AWP values are always greater for the same amount of water application under diffused film cladded greenhouse. The AWP value was found to increase as the amount of irrigation water supply was reduced from 100% to 40% of crop evapotranspiration [40-42]. From the analysis of AWP values, it is clear that for water scarce region Dutch roses can be grown under diffused film cladded greenhouse at 40% of crop evapotranspiration with the compromise in 25.5% reduction in yield [43-44].

Conclusion

The evapotranspiration requirement (ETc) of Dutch roses established under greenhouse and open field conditions is useful for the determination of the irrigation water requirement. The crop coefficient (Kc) values established on a monthly basis for both the conditions from the beginning to the end of each annual cropping period are the basic information for estimation of ETc. Four Kc values are the irreducible minimum for describing and constructing the Kc curve, namely Kc values at the initial stage ($K_{c\,Initial}$), development stage ($K_{c\,Dev}$), mid-season stage ($K_{c\,mid}$) and the end of the late season stage ($K_{c\,end}$). The optimum water requirement of Dutch rose crop estimated under both the greenhouses can be successfully used by the rose growers and irrigation planners in other parts of the sub-humid climatic conditions.

There is significant influence of cladding materials on the Dutch rose plant height, number of shoots per plant, shoot length and flower yield during winter and summer seasons. The number of shoots per plant in the clear and diffused film is almost same up to 30 days after

transplanting of the crop, and thereafter plant response gradually differs up to the flowering stage. The plant height is better under the diffused film greenhouse than that in the clear film greenhouse from vegetative stage to the flowering stage in winter and summer seasons due to inbuilt property of diffused film. The values of Crop ET are more for open field conditions than greenhouse. Total annual water requirement of rose plant is 999.51 mm and 1210.94 mm for greenhouse and open field condition.

Acknowledgement

Authors are thankful to the National Committee on Plasticulture Applications in Horticulture (NCPAH), Department of Agriculture and Cooperation, Ministry of Agriculture, Government of India for providing necessary funds and Department of Agricultural and Food Engineering, IIT Kharagpur, India for making available necessary facilities to conduct this research studies.

References

1. Singh AK (2009) Greenhouse technology for rose production. Indian Horticulture. 54: 10-14.

2. Hemming S, Mohammadkhani V, Dueck T (2008) Diffuse greenhouse covering materials-material technology, measurements and evaluation of optical properties. Acta Horticulturae 797: 469-475.

3. Tiwari KN, Singh A, Mal PK (2003) Effect of drip irrigation on yield of cabbage (Brassica oleracea L. var. capitata) under mulch and non-mulch conditions. Agricultural Water Management. 58: 19-28.

4. Tiwari KN, Kumar M, Santosh DT, Singh VK, Maji MK (2014) Influence of drip irrigation and plastic mulch on yield of Sapota (Achras zapota) and Soil Nutrients. Irrigation and Drainage Sys Eng. 3: 116.

5. Allen RG, Pereira LS, Howell TA, Jensen ME (2011) Evapotranspiration information reporting: I.Factors governing measurement accuracy. Agricultural Water Management 98: 899-920.

6. Evett SR, Tolk JA, Howell TA (2006) Soil profile water content determination: sensor accuracy, axial response, calibration, temperature dependence, and precision. Vadose Zone Journal 5: 894-907.

7. Ma HY, Jiao XY (2006) Research progress of the crop water demand calculation. Water Science Engineering Technology 16: 5-7.

8. Doorenbos J, Pruitt WO (1975) Guidelines for predicting crop water requirements. Irrigation and Drainage. Italy.

9. Abdelhadi AW, Hata T, Tanakamaru H, Tada A, Tariq MA (2000) Estimation of crop water requirements in arid region using Penman-Monteith equation with derived crop coefficients: a case study on Acala cotton in Sudan Gezira irrigated scheme. Agricultural Water Management 45: 203-214.

10. Fisher DK (2012) Simple weighing lysimeters for measuring evapotranspiration and developing crop coefficients. Int J Agric & Biol Eng 5: 35-43.

11. Akinmutimi AL (2015) Estimation of water requirements of early and late season maize in Umudike southeastern Nigeria, using Penman's equation of soil science and environmental management. J Soil Sci Environ Manage 6: 24-28.

12. Chuanyan Z, Zhongren N (2007) Estimating water needs of maize (Zea mays L.) using the dual crop coefficient method in the arid region of northwestern China. Afr J Agric Res 2: 325-333.

13. Piccinni G, Ko J, Thomas M, Daniel IL (2009) Crop Coefficients Specific to Multiple Phenological Stages for Evapotranspiration-based Irrigation Management of Onion and Spinach. Hort Science. 44: 421-425.

14. Odofin AJ, Oladiran JA, Oladipo JA, Wuya EP (2011) Determination of evapotranspiration and crop coefficients for bush okra (Corchorus olitorius) in a sub-humid area of Nigeria. Afr J Agric Res 6: 3949-3953.

15. Miranda FR, Gondim RS, Costa CAG (2006) Evapotranspiration and crop coefficients for tabasco pepper (Capsicum frutescens L.). Agricultural Water Management 82: 237-246.

16. Kuo SF, Ho SS, Liu CW (2009) Estimation of irrigation water requirements with derived crop coefficients for upland paddy crops in ChiaNan Irrigation Association, Taiwan. Agricultural Water Management 82: 433-451.

17. Tyagi NK, Sharma DK, Luthra SK (2000) Evapotranspiration and crop

coefficients of wheat and sorghum. Journal of Irrigation and Drainage Engineering. 126: 215-222.

18. Arayaa A, Stroosnijderb L, Girmayc G, Keesstrab SD (2011) Crop coefficient, yield response to water stress and water productivity of teff (Eragrostis tef Zucc.). Agricultural Water Management 98: 775-783.

19. Hu Y, Li Y, Zhang Y (2012) A study on crop coefficients of jujube under drip-irrigation in Loess Plateau of China. Afr J Agric Res 7: 2971-2977.

20. Parvizi H, Sepaskhah AR, Ahmadi SH (2014) Effect of drip irrigation and fertilizer regimes on fruit yields and water productivity of a pomegranate (Punica granatum (L.) cv. Rabab) orchard. Agricultural Water Management. 146: 45-56.

21. Maldonado AJ, Mendoza AB, Alba-Romenus Kd, Morales-Díaz AB (2014) Estimation of the water requirements of greenhouse tomato crop using multiple regression models. Emir J Food Agriculture 26: 885-897.

22. Gómez HV, Farías SO, Argote M (2009) Evaluation of water requirements for a greenhouse Tomato cropusing the priestley-taylor method. Chilean Journal of Agricultural Research 69: 3-11.

23. Wahb-Allah MA, Alsadon AA, Ibrahim AA (2011) Drought tolerance of several tomato genotypes under greenhouse conditions. world applied sciences journal. 15: 933-940.

24. Harmantoa, Salokhea VM, Babelb MS, Tantau HJ (2005) Water requirement of drip irrigated tomatoes grown in greenhouse in tropical environment. Agricultural Water Management 71: 225-242.

25. Senyigit U, Kadayifci A, Ozdemir FO, Hasan O, Atilgan A (2011) Effects of different irrigation programs on yield and quality parameters of eggplant (Solanum melongena L.) under greenhouse conditions. African Journal of Biotechnology. 10: 6497-6503.

26. Orgaz F, Fernańdezb MD, Bonachelac S, Gallardoc M, Fereresa E (2005) Evapotranspiration of horticultural crops in an unheated plastic greenhouse. Agricultural Water Management 72: 81-96.

27. Bastug R, Karaguzel O, Aydinsakir K, Buyuktas D (2006) The effects of drip irrigation on flowering and flower quality of glasshouse gladiolus plant. Agricultural Water Management 81: 132-144.

28. Zhang, Zi-kun, Liu, Shi-q, Su-hui L, Zhi-jun H (2010) Estimation of Cucumber Evapotranspiration in Solar Greenhouse in Northeast China. Agricultural Sciences in China. 9: 512-518.

29. Blanco FF, Folegatti MV (2003) Evapotranspiration and crop coefficient of cucumber in greenhouse. Revista Brasileira de Engenharia Agrícola e Ambiental 7: 285-291.

30. Ehret DL, Menzies JG, Helmer T (2005) Production and quality of greenhouse roses in recirculating nutrient systems. Scientia Horticulturae 106: 103-113.

31. Katsoulas N, Kittas C, Dimokas G, Lykas Ch (2006) Effect of irrigation frequency on rose flower production and quality. Biosystems Engineering 93: 237-244.

32. Qi SH, Li ZZ, Gong YS (2002) Evaluating Crop water requirements and Crop coefficients for three vegetables based on Field water budget. J China Agric Univ. 7: 71-76.

33. Zhao NN (2010) Calculation of crop coefficient and water consumption of summer maize. J Hydraulic Engine. 41: 953-959.

34. Möller M, Assouline S (2007) Effects of a shading screen on microclimate and crop water requirements. Irrigation Science 25: 171-181.

35. Fernández MD, Bonachela S, Orgaz F, Thompson RB, López JC, et al. (2010) Measurement and Estimation of plastic greenhouse reference evapotranspiration in a Mediterranean climate. Irrig Sci 28: 497-509.

36. Chou MD, Lee KT (1996) Parameterizations for the absorption of solar radiation by water vapor and ozone. American Meteorolological Society 53: 1203-1208.

37. Kondratyev KY (1969) Radiation in the atmosphere. New York.

38. Lamnatou C, Chemisana D (2013) Solar radiation manipulations and their role in greenhouse claddings: Fresnel lenses, NIR and UV-blocking materials. Renewable and Sustainable Energy Reviews 18: 271-287.

39. Lamnatou C, Chemisana D (2013) Solar radiation manipulations and their role in greenhouse claddings: Fluorescent solar concentrators, photoselective and other materials. Renewable and Sustainable Energy Reviews 27: 175-190.

40. Espí E, Salmerón A, Fontecha A, García Y, Real AI (2006) Plastic films for agricultural applications. J Plast Film Sheeting 22: 85-102.

41. Zheng J, Huang G, Jia D, Wang J, Mota M et al., (2013) Responses of drip irrigated tomato (Solanum lycopersicum L.) yield, quality and water productivity to various soil matric potential thresholds in an arid region of Northwest China. Agricultural Water Management. 129: 181002D193.

42. Pereira Ls, Cordery I, Iacovides I (2012) Improved indicators of water use performance and productivity for sustainable water conservation and saving. Agricultural Water Management. 108: 39-51.

43. Manohar RK, Khan MM, Shyamalamma, Kariyanna, Biradar MS (1996) Gerbera under low cost greenhouse. Published by Plasticulture Development Centre. Bangalore.

44. Sengar SH, Kothari S (2008) Economic evaluation of greenhouse for cultivation of rose nursery. African Journal of Agricultural Research. 3: 435-439.

25

Evapotranspiration Estimation using Six Different Multi-layer Perceptron Algorithms

Ozgur Kisi* and Vahdettin Demir

Canik Basari University, Civil Engineering Department, Samsun, Turkey

Abstract

Evapotranspiration has a vital importance in water resources planning and management. In this study, the applicability of six different multi-layer perceptron (MLP) algorithms, Quasi-Newton, Conjugate Gradient, Levenberg-Marquardt, One Step Secant, Resilient Back propagation and Scaled Conjugate Gradient algorithms, in modeling reference evapotranspiration (ET_0) is investigated. Daily climatic data of solar radiation, air temperature, relative humidity and wind speed from Antalya City are used as inputs to the MLP models to estimate daily ET_0 values obtained using FAO 56 Penman Monteith empirical method. The results of the MLP algorithms are compared with those of the multiple linear regression models with respect to root mean square error (RMSE), mean absolute error (MAE), Willmott index of agreement (d) and determination coefficient (R^2). The comparison results indicate that the Levenberg-Marquardt is faster and has a better accuracy than the other five training algorithms in modeling ET_0. The Levenberg-Marquardt with RMSE = 0.083 mm, MAE = 0.006 mm, d = 0.999 and R^2 = 0.999 in test period was found to be superior in modeling daily ET_0 than the other algorithms, respectively.

Keywords: Estimation; Reference evapotranspiration; Multi-layer perceptron; Multiple linear regression; Training algorithms

Introduction

Accurate estimation of reference evapotranspiration (ET_0) has a vital importance for many studies such as hydrologic water balance, the design and management of irrigation system and water resources planning and management. The Penman-Monteith FAO 56 (PM FAO-56) model is recommended as the sole method for calculation of ET_0 and it has been reported to be able to provide consistent ET_0 values in many regions and climates [1-2]. The main shortcoming of the PM FAO-56 method is, however, that it needs large number of climatic data and variables which are unavailable in many regions (especially in developing countries like Turkey).

Recently, the multi-layer perceptron (MLP) neural networks successfully applied in ET_0 estimation. Kumar et al. used MLP models for the estimation of evapotranspiration and they found that the MLP performed better than the PM FAO-56 method [3]. Trajkovic et al. applied radial basis function neural networks in ET_0 estimation [4]. Kisi investigated the accuracy of the MLP with Levenberg-Marquardt training algorithm and reported that MLP can be successfully employed in modeling ET_0 from available climate data. MLP models were compared with some empirical models and found to have better accuracy in estimating ET_0 [5]. Gorka et al. Rahimikhoob investigated the use of MLP for estimating ET_0 based on air temperature data under humid subtropical conditions and found that MLP performed better than the Hargreaves method [6]. Marti et al. estimated ET_0 by MLP without local climatic data [7]. Marti et al. examined the 4-input MLP models for ET_0 estimation through data set scanning procedures [8]. Several contributions on MLP modeling in ET_0 estimation were reviewed by Kumar et al. [3]. Shrestha and Shukla used support vector machine for modeling of ET_0 using hydro-climatic variables in a sub-tropical environment [9]. Gocic et al. applied extreme learning machine for estimation of reference evapotranspiration and compared with empirical equations. It is evident from the literature; there is not any published work that compares the accuracy of, in modeling daily ET_0 [10].

The aim of this study is to investigate the accuracy of six different MLP algorithms, Quasi-Newton, Conjugate Gradient, Levenberg-Marquardt, One Step Secant, Resilient Backpropagation and Scaled Conjugate Gradient algorithms, in daily ET_0 estimation.

Materials and Methods

Materials

Daily weather data from Antalya Station (latitude 36° 42' N, longitude 30° 44' E) operated by the Turkish Meteorological Organization (TMO) in Turkey were used in the study. The station is located in Mediterranean Region (Figure 1) of Turkey and 47 m below the sea level. It has a Mediterranean climate (dry summers and wet winters). The maximum temperatures are 24°C for winter and 40°C for summer.

The data sample is composed of 7743 daily (1973-2002) records of solar adiation (SR), air temperature (T), relative humidity (RH) and wind speed (U_2). First 4645 data (60% of the whole data) were used to train the MLP models, second 1549 data (20% of the whole data) data were used for validation and the remaining 1549 data (20% of the whole data) were used for testing. Statistical parameters of the used weather data are reported in Table 1. In this table, the x_{mean}, S_x, C_{sx}, C_v, x_{min}, and x_{max} denote the mean, standard deviation, skewness, coefficient of deviation, minimum, and maximum, respectively. It is clear from the table that the relative humidity shows a skewed distribution. SR seems to be most effective parameter on ET_0 according to the correlation analysis. T_{mean} and RH are the second and third most effective parameters on the ET_0.

Multi-layer perceptron

Multi-layer perceptron is inspired from biological nervous system, though much of the biological detail is neglected. MLP networks are massively parallel systems composed of many processing elements. The MLP structure used in the present study is shown in Figure 2.

---

*Corresponding author: Ozgur Kisi, Canik Basari University, Civil Engineering Department, Samsun, Turkey, E-mail: okisi@basari.edu.tr

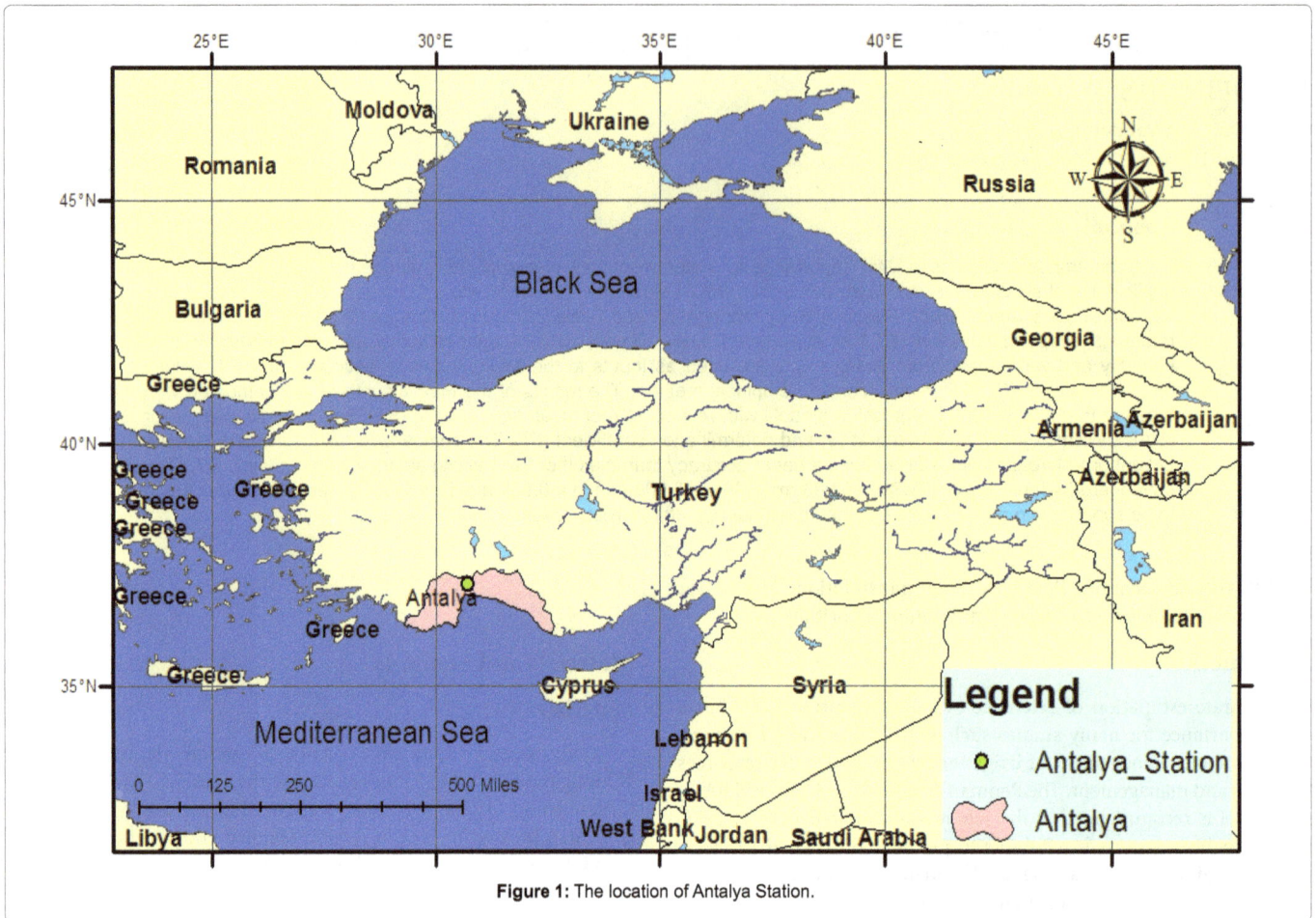

Figure 1: The location of Antalya Station.

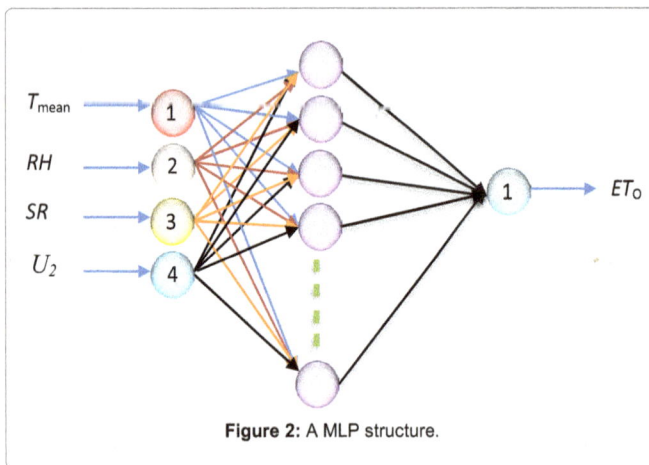

Figure 2: A MLP structure.

Data set	Unit	X_{max}	X_{min}	X_{mean}	S_x	C_v	C_{sx}	Correlation with ET_0
T_{mean}	°C	37.5	1.7	18.5	7.1	0.38	0.17	0.57
SR	Mj/m²/ day	33.4	0.15	17.2	7.5	0.43	-0.23	0.68
U_2	m/s	98	14	63.7	16.8	0.26	-0.45	0.07
RH	%	13	0	2.8	1.55	0.55	1.9	0.46
ET_0	mm	15	0	4.08	2.39	0.59	0.7	1

Table 1: Basic statistics of the weather parameters for the Antalya Station.

The network consists of layers of parallel processing elements, called neurons. Each layer in MLP is connected to the proceeding layer by interconnection weights. During the training/calibration process, randomly assigned initial weight values are progressively corrected. In this process, calculated outputs are compared with the known outputs and the errors are back propagated to determine the appropriate weight adjustments necessary to minimize the errors.

In the present study, six different training algorithms, Quasi-Newton (QN), Conjugate Gradient (CG), Levenberg-Marquardt (LM), One Step Secant (OSS), Resilient Back propagation (RB) and Scaled Conjugate Gradient (SCG), were used for adjusting the MLP networks. The detailed theoretical information about MLP can be found in Haykin [11].

Choosing optimal hidden nodes' number is a difficult task in developing MLP models. In this study, the MLP with one hidden layer was used and the optimal hidden nodes were determined by trial-error method. The sigmoid and linear activation functions were used for the hidden and output nodes, respectively. Two different iteration numbers, 1000 and 5000 were used for the MLP training because the variation of error was too small after 5000 epochs. A MATLAB code including neural networks toolbox was used for the MLP simulations. Four weather parameters were used as inputs to the MLP models to estimate ET_0. Root mean square errrors (RMSE), mean absolute error (MAE), Willmott index of agreement (d) and determination coefficient (R²) statistics were used for evaluation of the applied models. The RMSE, MAE and d can be defined as:

$$RMSE = \sqrt{\frac{1}{N}\sum_{i=1}^{N}\left(ETi_{PM\,FAO-56} - ETi_{predicted}\right)^2} \qquad (1)$$

$$MAE = \frac{1}{N}\sum_{i=1}^{N}\left|ETi_{PM\,FAO-56} - ETi_{predicted}\right| \qquad (2)$$

$$d = 1 - \frac{\sum_{i=1}^{n}\left(ETi_{predicted} - ETi_{PM\,FAO-56}\right)^2}{\sum_{i=1}^{n}\left(\left|ETi_{predicted} - \overline{ETi_{PM\,FAO-56}}\right| + \left|ETi_{PM\,FAO-56} - \overline{ETi_{PM\,FAO-56}}\right|\right)^2} \qquad (3)$$

In which the N and ET show the number of data sets and reference evapotranspiration, respectively.

Application and results

Training, validation and test results of the MLP algorithms are given in Table 2. Training duration is also provided in this table for each algorithm. It should be noted that the properties of the computer used in the applications are Intel(R) Core(TM) i5-3230M CPU@2.60GHz.

Optimal hidden node number that gave the minimum RMSE errors in the validation period was selected for each MLP model. In Table 2, (4, 10, 1) indicates a MLP model comprising 4 input, 10 hidden and 1 output nodes. The QN, CG, LM and RB algorithms has the same optimal hidden node numbers for the 1000 and 5000 epochs. The hidden node numbers of the OSS and SCG algorithms decrease by increasing epoch numbers. Actually, the runs of the LM, QN and CG algorithms were automatically stopped after 24, 830 and 354 epochs, respectively. It can be said that these epochs are enough for the training of QN, CG and LM algorithms because the error gradients are too small after these epochs. For this reason, the structure, training duration and accuracies of these three algorithms are same for the 1000 and 5000 epochs. It is clearly seen from Table 1 that the LM algorithm has the lowest RMSE and MAE and the highest R^2 values than the other algorithms for both 1000 and 5000 epochs. In the case of 1000 epochs, the accuracy ranks of the algorithms in training period are; LM, QN, SCG, CG, OSS and RB from the RMSE viewpoint. In the case of 5000 epochs, however, the ranks are; LM, QN, SCG, OSS, RB and CG. The algorithms are also

Phase	Epochs	Comparison criteria	Algorithm						
			MLR	QN	CG	LM	OSS	RB	SCG
Training	1000	RMSE(mm)	-	0.076	0.265	0.073	0.31	0.461	0.18
		MAE (mm)	-	0.005	0.07	0.005	0.096	0.213	0.032
		R^2	-	0.998	0.987	0.999	0.982	0.962	0.994
		Duration (sn)	-	31.7	3.92	2	39.6	20.6	37.7
		Structure	-	(4,10,1)	(4,9,1)	(4,5,1)	(4,7,1)	(4,7,1)	(4,6,1)
	5000	RMSE(mm)	-	0.076	0.265	0.073	0.128	0.131	0.101
		MAE (mm)	-	0.005	0.07	0.005	0.093	0.017	0.01
		R^2	-	0.998	0.987	0.999	0.997	0.996	0.998
		Duration (sn)	-	31.3	3.91	1.97	200	102	179
		Structure	-	(4,10,1)	(4,9,1)	(4,5,1)	(4,4,1)	(4,7,1)	(4,4,1)
Validation	1000	RMSE(mm)	-	0.077	0.281	0.071	0.273	0.512	0.18
		MAE (mm)	-	0.006	0.079	0.005	0.074	0.263	0.032
		R^2	-	0.999	0.986	0.999	0.987	0.955	0.994
	5000	RMSE(mm)	-	0.077	0.281	0.071	0.137	0.124	0.099
		MAE (mm)	-	0.006	0.079	0.005	0.018	0.015	0.009
		R^2	-	0.999	0.986	0.999	0.996	0.997	0.998
Test	1000	RMSE(mm)	-	0.089	0.327	0.083	0.334	0.524	0.205
		MAE (mm)	-	0.007	0.107	0.006	0.112	0.274	0.042
		R^2	-	0.999	0.983	0.999	0.982	0.996	0.994
		d	-	0.995	0.988	0.999	0.980	0.996	0.993
	5000	RMSE(mm)	-	0.089	0.327	0.083	0.147	0.161	0.125
		MAE (mm)	-	0.007	0.107	0.006	0.021	0.026	0.015
		R^2	-	0.999	0.983	0.999	0.996	0.995	0.997
		d	-	0.996	0.990	0.999	0.999	0.997	0.995
	MLR	RMSE(mm)	0.500	-	-	-	-	-	-
		MAE (mm)	0.250	-	-	-	-	-	-
		R^2	0.960	-	-	-	-	-	-
		d	**0.983**	-	-	-	-	-	-

Table 2: Training, validation and test results of the MLP algorithms in estimating PM FAO-56 ET_0.

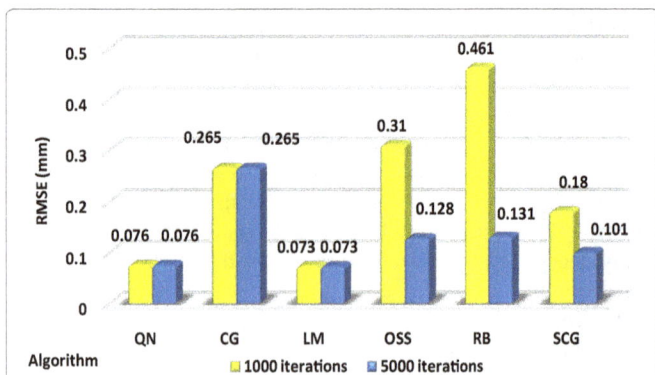

Figure 3: The RMSE accuracies of each MLP algorithm in simulation PM FAO-56 ET_0 in training phase.

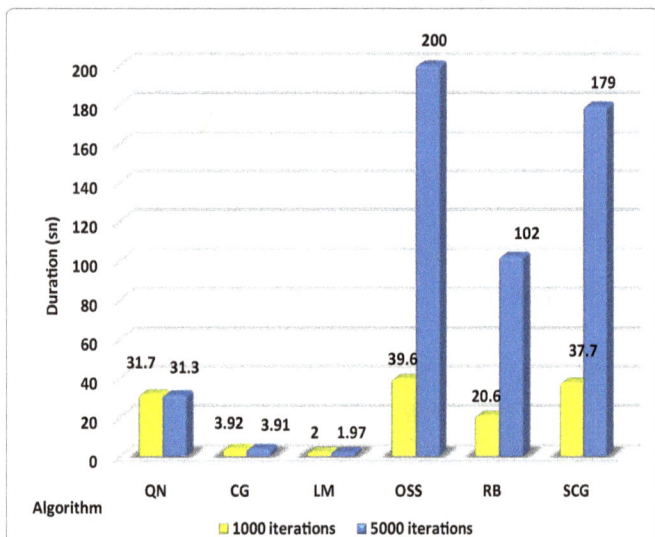

Figure 4: Training durations of each MLP algorithm in simulation of PM FAO-56 ET_0 in case of 1000 iterations (yellow color) and 5000 iterations (blue color).

compared in Figures 3-4 with respect to RMSE accuracy and training duration. Training speed of each algorithm can be obviously seen from this figure. Comparison of training times of the algorithms indicates that the LM is faster than the other algorithms. The training duration ranks are; LM, CG, RB, QN, SCG and OSS.

From Table 2, it is clear that the LM algorithm performs better than the other algorithms in daily ET_0 estimation in validation stage. There is a slight difference between the QN and LM algorithms. The accuracy ranks of the algorithms for the 1000 epochs are; LM, QN, SCG, OSS, CG and RB. In the case of 5000 epochs, the ranks are; LM, QN, SCG, RB, OSS and CG. The multiple linear regression (MLR) model results are also included in Table 2 for the test stage. It is obviously seen from the table that the LM algorithm has almost same accuracy with the QN and they perform better than the other four algorithms in test stage. In the case of 1000 epochs, the accuracy ranks of the algorithms in the test period are; LM, QN, SCG, CG, OSS and RB. In the case of 5000 epochs, however, the ranks are; LM, QN, SCG, OSS, RB and CG as found in the training period. All the algorithms are found to be better than the MLR in estimating daily ET_0.

The scatterplots of the ET_0 estimates for the 1000 epochs are illustrated in Figure 5. It is clear from the fit line equations and R^2

values in the figure that all the algorithms gave better estimates than the MLR model. It is evident form the scatterplots that the slope of the LM algorithm (0.9962) is closer to the 1 than those of the other algorithms. The CG and OSS algorithms have much more scattered estimates than the QN, LM, RB and SCG. Figure 6 demonstrates the ET_0 estimates of the six MLP algorithms for the 5000 epochs. Here also the estimates of the LM algorithm are closer to the corresponding FAO-56 ET_0 values than the other five algorithms. The CG algorithm gave the worst estimates.

Ladlani et al. modeled daily FAO 56 PM ET_0 in the north of Algeria using two different ANN methods, radial basis neural networks (RBNN) and generalized regression neural networks (GRNN) [12]. Climatic data of daily mean relative humidity, sunshine duration, maximum, minimum and mean air temperature and wind speed were used as inputs to the applied models. The optimal RBNN and GRNN models provided the R^2 of 0.934 and 0.945, respectively. Adamala et al. applied second order neural networks (SONN) and compared with MLP method in estimating daily FAO 56 PM ET_0 in India [13]. They used inputs of daily climate data of minimum and maximum air temperatures, minimum and maximum relative humidity, wind speed and solar radiation in the models and they found that the best SONN and MLP models gave R^2 of 0.998 and 0.995, respectively. Yassin et al. used MLP and gene expression programming (GEP) in estimating FAO 56 PM ET_0 in Saudi Arabia [14]. They used daily data of maximum, minimum and mean air temperatures, maximum, minimum and mean relative humidity, wind speed at a 2 m height and solar radiation as input s to the models [15-19]. They found R^2 of 0.998 and 0.954 for the best MLP and GEP models in in estimating ET_0. It is clear from Table 2 that the MLP models (R^2 values range 0.995-0.999) accurately estimate daily FAO 56 PM ET_0 of Antalya station from the R^2 viewpoint [20-24].

In overall, the LM and QN generally performed superior to the other algorithms in estimating daily FAO 56 PM ET_0. Like QN method, the LM algorithm was designed to approach second order training speed [25-28]. They can converge much faster than first order algorithms such as CG, OSS, RB and SCG. However, the main disadvantage of these approaches is that they require large memory space for approximation when training has large-sized patterns. LM algorithm is viewed as a standout amongst the most efficient algorithms for training small and medium sized patterns [29-30].

Conclusion

This study investigated the accuracy and training speed of six different MLP algorithms, Quasi-Newton, Conjugate Gradient, Levenberg-Marquardt, One Step Secant, Resilient Backpropagation and Scaled Conjugate Gradient algorithms, in estimating daily reference evapotranspiration. The results of the MLP algorithms are compared with those of the multiple linear regression models with respect to root mean square error mean absolute error and determination coefficient. The LM was found to be faster and had a better accuracy than the other five training algorithms in estimating daily ET_0. A slight difference exists between the QN and LM algorithms. The worst estimates were obtained from the CG algorithm. Comparison with multiple linear regression indicated that all the considered algorithms performed better than the MLR in estimating daily ET_0.

Acknowledgements

This study was supported by The Turkish Academy of Sciences (TUBA). The first author would like to thank TUBA for their support of this study.

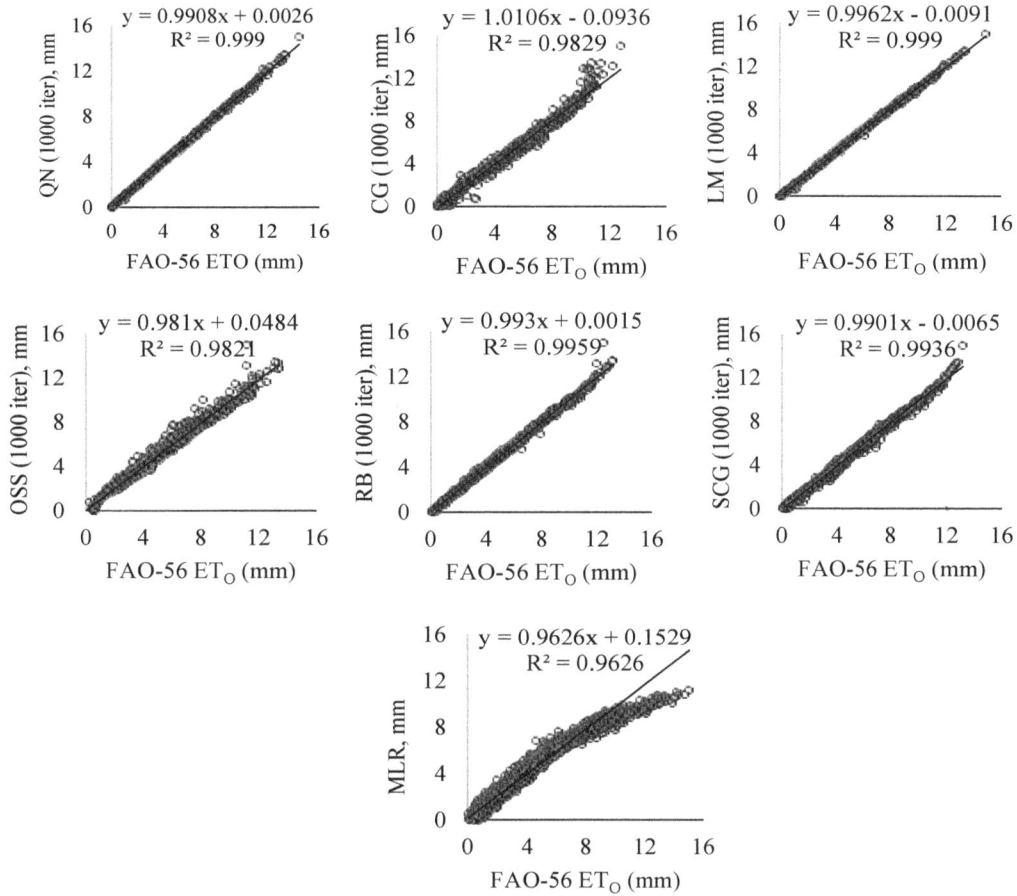

Figure 5: The FAO 56 PM ET_o and estimated values by different MLP algorithms for 1000 epochs and MLR in test period.

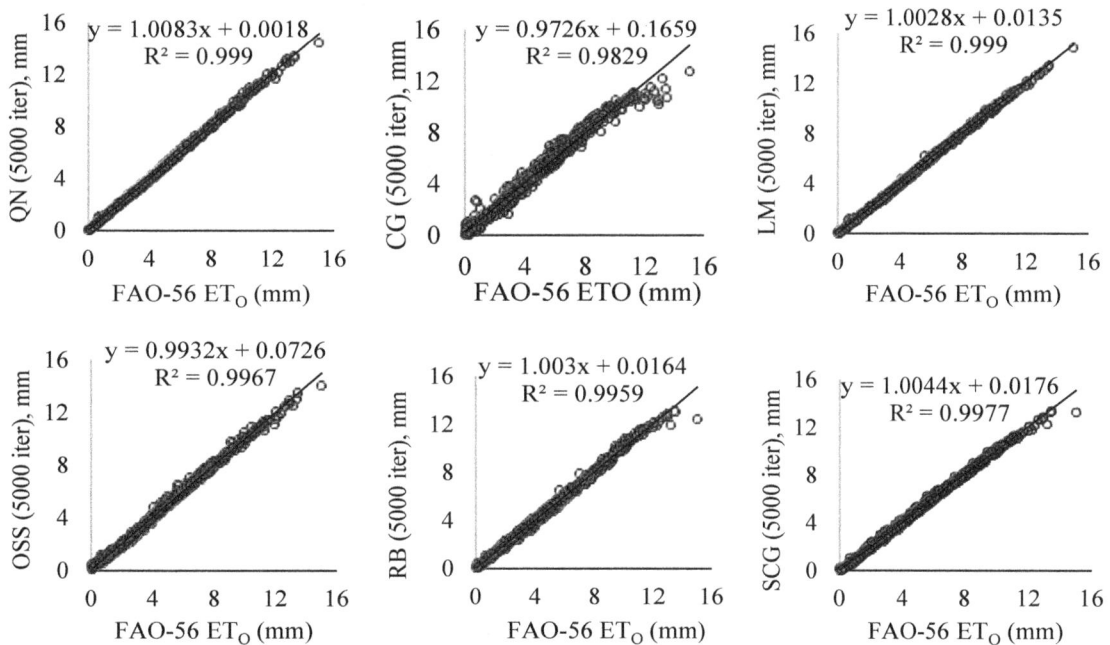

Figure 6: The FAO 56 PM ET0 and estimated values by different MLP algorithms for 5000 epochs and MLR in test period.

References

1. Allen R, Clemmens A (2005) Prediction accuracy for projectwide evapotranspiration using crop coefficients and reference evapotranspiration. J Irrig Drain Eng 131: 24-36.

2. Allen RG, Pruitt WO, Wright JL, Howell TA, Ventura F, et al. (2006) A recommendation on standardized surface resistance for hourly calculation of reference ETo by the FAO56 Penman-Monteith method. Agric Water Manag 81: 1-22.

3. Kumar M, Raghuwanshi NS, Singh R, Wallender WW, Pruitt WO (2002) Estimating evapotranspiration using artificial neural network. J Irrig Drain Eng 128: 224-233.

4. Trajkovic S (2005) Temperature-based approaches for estimating reference evapotranspiration. J Irrig Drain Eng 131: 316-323.

5. Kisi O (2007) Evapotranspiration modelling from climatic data using a neural computing technique. Hydrol Process 21: 1925-1934.

6. Rahimikhoob A (2010) Estimation of evapotranspiration based on only air temperature data using artificial neural networks for a subtropical climate in Iran. Theor Appl Climatol 101: 83-91.

7. Martí P, González-Altozano P, Gasque M (2011) Reference evapotranspiration estimation without local climatic data. Irrig Sci 29: 479-495.

8. Martí P, Manzano J, Royuela Á (2011) Assessment of a 4-input artificial neural network for ETo estimation through data set scanning procedures. Irrig Sci 29: 181-195.

9. Shrestha NK, Shukla S (2015) Support vector machine based modeling of evapotranspiration using hydro-climatic variables in a sub-tropical environment. Agric For Meteorol 200: 172-184.

10. Gocic M, Petković D, Shamshirband S, Kamsin A (2016) Comparative analysis of reference evapotranspiration equations modelling by extreme learning machine. Comput Electron Agric 127: 56-63.

11. Haykin S (1998) Neural Networks: A Comprehensive Foundation (2nd edn). Prentice Hall.

12. Ladlani I, Houichi L, Djemili L, Heddam S, Belouz K (2012) Modeling daily reference evapotranspiration (ET0) in the north of Algeria using generalized regression neural networks (GRNN) and radial basis function neural networks (RBFNN): a comparative study. Meteorol Atmos Phys 118: 163-178.

13. Adamala S, Raghuwanshi NS, Mishra A, Tiwari MK (2014) Evapotranspiration Modeling Using Second-Order Neural Networks. Journal of Hydrologic Engineering 19: 1131-1140.

14. Yassin MA, Alazba AA, Mattar MA (2016) Artificial neural networks versus gene expression programming for estimating reference evapotranspiration in arid climate. Agricultural Water Management 163: 110-124.

15. Trajkovic S, Todorovic B, Stankovic M (2003) Forecasting of reference evapotranspiration by artificial neural networks. J Irrig Drain Eng 129: 454-457.

16. Landeras G, Ortiz-Barredo A, López JJ (2008) Comparison of artificial neural network models and empirical and semi-empirical equations for daily reference evapotranspiration estimation in the Basque Country. Agric Water Manag 95: 553-565.

17. Kumar M, Raghuwanshi NS, Singh R (2011) Artificial neural networks approach in evapotranspiration modeling: a review. Irrig Sci 29: 11-25.

18. Kisi O (2016) Modeling reference evapotranspiration using three different heuristic regression approaches. Agric Water Manag 169: 162-172.

19. Kisi O, Sanikhani H, Zounemat-Kermani M, Niazi F (2015) Long-term monthly evapotranspiration modeling by several data-driven methods without climatic data. Comput Electron Agric 115: 66-77.

20. Jabloun M, Sahli A (2008) Evaluation of FAO-56 methodology for estimating reference evapotranspiration using limited climatic data. Application to Tunisia. Agric Water Manag 95: 707-715.

21. Perera KC, Western AW, Nawarathna B, George B (2015) Comparison of hourly and daily reference crop evapotranspiration equations across seasons and climate zones in Australia. Agric Water Manag 148: 84-96.

22. Kisi O (2013) Applicability of Mamdani and Sugeno fuzzy genetic approaches for modeling reference evapotranspiration. J Hydrol 504: 160-170.

23. Gharsallah O, Facchi A, Gandolfi C (2013) Comparison of six evapotranspiration models for a surface irrigated maize agro-ecosystem in Northern Italy. Agric Water Manag 130: 119-130.

24. Suleiman AA, Hoogenboom G (2009) A comparison of ASCE and FAO-56 reference evapotranspiration for a 15-min time step in humid climate conditions. J Hydrol 375: 326-333.

25. Adeloye AJ, Rustum R, Kariyama ID (2012) Neural computing modeling of the reference crop evapotranspiration. Environ Model Softw 29: 61-73.

26. Tabari H, Kisi O, Ezani A, Talaee PH (2012) SVM, ANFIS, regression and climate based models for reference evapotranspiration modeling using limited climatic data in a semi-arid highland environment. J Hydrol 444: 78-89.

27. Feng Y, Cui N, Zhao L, Hu X, Gong D (2016) Comparison of ELM, GANN, WNN and empirical models for estimating reference evapotranspiration in humid region of Southwest China. J Hydrol 536: 376-383.

28. Kisi O (2005) Suspended sediment estimation using neuro-fuzzy and neural network approaches. Hydrol Sci J 50: 683-696.

29. Burney SMA, Jilani TA, Ardil C (2007) A comparison of first and second order training algorithms for artificial neural networks. International Journal of Computer, Electrical, Automation, Control and Information Engineering 1: 145-151.

30. Yu H, Wilamowski B M (2012) Neural network training with second order algorithms. Berlin Heidelberg.

Permissions

All chapters in this book were first published in IDSE, by OMICS International; hereby published with permission under the Creative Commons Attribution License or equivalent. Every chapter published in this book has been scrutinized by our experts. Their significance has been extensively debated. The topics covered herein carry significant findings which will fuel the growth of the discipline. They may even be implemented as practical applications or may be referred to as a beginning point for another development.

The contributors of this book come from diverse backgrounds, making this book a truly international effort. This book will bring forth new frontiers with its revolutionizing research information and detailed analysis of the nascent developments around the world.

We would like to thank all the contributing authors for lending their expertise to make the book truly unique. They have played a crucial role in the development of this book. Without their invaluable contributions this book wouldn't have been possible. They have made vital efforts to compile up to date information on the varied aspects of this subject to make this book a valuable addition to the collection of many professionals and students.

This book was conceptualized with the vision of imparting up-to-date information and advanced data in this field. To ensure the same, a matchless editorial board was set up. Every individual on the board went through rigorous rounds of assessment to prove their worth. After which they invested a large part of their time researching and compiling the most relevant data for our readers.

The editorial board has been involved in producing this book since its inception. They have spent rigorous hours researching and exploring the diverse topics which have resulted in the successful publishing of this book. They have passed on their knowledge of decades through this book. To expedite this challenging task, the publisher supported the team at every step. A small team of assistant editors was also appointed to further simplify the editing procedure and attain best results for the readers.

Apart from the editorial board, the designing team has also invested a significant amount of their time in understanding the subject and creating the most relevant covers. They scrutinized every image to scout for the most suitable representation of the subject and create an appropriate cover for the book.

The publishing team has been an ardent support to the editorial, designing and production team. Their endless efforts to recruit the best for this project, has resulted in the accomplishment of this book. They are a veteran in the field of academics and their pool of knowledge is as vast as their experience in printing. Their expertise and guidance has proved useful at every step. Their uncompromising quality standards have made this book an exceptional effort. Their encouragement from time to time has been an inspiration for everyone.

The publisher and the editorial board hope that this book will prove to be a valuable piece of knowledge for researchers, students, practitioners and scholars across the globe.

List of Contributors

B.J. Pandian T. Sampathkumar and R. Chandrasekaran
Water Technology Centre, Tamil Nadu Agricultural University, Coimbatore, India

Zainudini MZ and Sardarzaei A
Faculty of Marine Science, Basic Science Department, Chabahar Maritime University, Iran

James C.Y. Guo
Professor and Director, Dept. of Civil Engineering, University of Colorado Denver, USA

Phogat V and Skewes MA
South Australian Research and Development Institute, Australia

Simunek J
Department of Environmental Sciences, University of California, USA

Cox JW
South Australian Research and Development Institute, Australia
The University of Adelaide, Australia

Tirzah Moreira de Melo, José Antônio Saldanha Louzada and Olavo Correa Pedrollo
Institute of Hydraulic Researches - Federal University of Rio Grande do Sul (IPH/UFRGS), Bento Gonçalves Av, 9500, P.O. Box 15029, Porto Alegre City, State of Rio Grande do Sul, Brazil

Alaa F. Abukila
Drainage Research Institute, National Water Research Center, El-Qanater El-Khairiya, Egypt

Joaquim Monserrat, Rubén García Ortiz, Lluis Cots and Javier Barragán
Department of Agroforestry Engineering, Universitat de Lleida, Lleida, Spain

Yadav RC
Ex Head of Research Centre, Soil and Water Conservation, Agra 282006, Uttar Pradesh, India

Luis Gurovich and Patricio Oyarce
Departamento de Fruticultura y Enología, Pontificia Universidad Católica de Chile. Santiago, Chile

Solmaz Javanbakht and Reza Dadmehr
Department of Water Engineering, Urmia University, Urmia, Iran

Giulio Lorenzini
University of Parma, Department of Industrial Engineering, viale G.P. Usberti no.181/A, Parma 43124, Italy

Alessandra Conti
Alma Mater Studiorum-University of Bologna, Department of Energetic Nuclear and Environmental Control Engineering, viale Risorgimento no. 2, Bologna 40136, Italy

Daniele De Wrachien
Department of Agricultural Hydraulics, University of Milan, via Celoria no.2, Milan 20133, Italy

Ravichandran VK and Prakash KC
Agronomy, Tamil Nadu Agricultural University, Chennai, India

Vibhu Nayar
IAMWARM, Chennai, India

Muya EM, Sijali IV and Radiro M
Food Crop Research Institute, KALRO-Kabete, Kenya

Okoth PFZ
International Institute for Rural Technologies, Nairobi, Kenya

Farai Malvern Simba, Alois Matorevhu, David Chikodzi and Talent Murwendo
Department of Physics, Geography and Environmental Science, Great Zimbabwe University, Zimbabwe

Wignyosukarto BS
Professor, Department of Civil and Environmental Engineering, Faculty of Engineering, Universitas Gadjah Mada, Yogyakarta, Indonesia

Mawandha HG
Graduate Student, Department of Civil and Environmental Engineering, Faculty of Engineering, Universitas Gadjah Mada, Yogyakarta, Indonesia

Jayadi R
Associate Professor, Department of Civil and Environmental Engineering, Faculty of Engineering, Universitas Gadjah Mada, Yogyakarta, Indonesia

Xinhua Jia, Thomas Scherer and Dongqing Lin
Department of Agricultural and Biosystems Engineering, North Dakota State University, Fargo, North Dakota-58108, USA

Xiaodong Zhang
Department of Earth System Science and Policy, University of North Dakota, Grand Forks, North Dakota-58202, USA

Ishara Rijal
Department of Geography, MichiganStateUniversity, East Lansing, Michigan-48824, USA

Luis Gurovich
Faculty of Agronomy and Forestry, Pontificia Universidad Católica de Chile, Chile

Patricio Oyarce
Agricultural Engineer, Departamento de Fruticultura y Enología, Pontificia Universidad Católica de Chile, Chile

Pande VC and Bagdi GL
Central Soil and Water Conservation Research and Training Institute, Research Centre, Vasad (Anand), Gujarat, India

Sena DR
Central Soil and Water Conservation Research and Training Institute, Dehradun, Uttarakhand, India

Upadhyaya A
Division of Land and Water Management, ICAR Research Complex for Eastern Region, Patna, India

Magalhaes ID and Lyra GB
Department of Plant production, Federal University of Alagoas, Brazil

Souza JL
Department of Agronomy, Federal University of Alagoas, Brazil

Teodora I, Cavalcante CA and Souza RC
Department of Agricultural engineering, Federal University of Alagoas, Brazil

Ferreira RA
Department of Phytotechnology, Federal University of Alagoas, Brazil

Upadhyaya A and Alok K Sikka
Division of Land and Water Management, Indian Council of Agricultural Research, India

Bassa Z, Abera A, Zeleke B, Alemu M, Bashe A and Areka MS
Southern Agricultural Research Institute, Areka Agricultural Research Center, Areka, Ethiopia

Vikas Kumar Singh, KN Tiwari and Santosh DT
Department of Agricultural and Food Engineering, Indian Institute of Technology, Kharagpur, India

Ozgur Kisi and Vahdettin Demir
Canik Basari University, Civil Engineering Department, Samsun, Turkey

Index

www.ingramcontent.com/pod-product-compliance
Lightning Source LLC
Chambersburg PA
CBHW080256230326
41458CB00097B/5027